The Proliferation of Nuclear Weapons

Edited by Paul F. Kisak

Contents

Chapter 1

Nuclear weapon

"A-bomb" redirects here. For other uses, see A-bomb (disambiguation).

The mushroom cloud of the atomic bombing of the Japanese city of Nagasaki on August 9, 1945 rose some 11 miles (18 km) above the bomb's hypocenter.

A **nuclear weapon** is an explosive device that derives its destructive force from nuclear reactions, either fission (fission bomb) or a combination of fission and fusion (thermonuclear weapon). Both reactions release vast quantities of energy from relatively small amounts of matter. The first test of a fission ("atomic") bomb released the same amount of energy as approximately 20,000 tons of TNT (84 TJ). The first thermonuclear ("hydrogen") bomb test released the same amount of energy as approximately 10 million tons of TNT (42 PJ).

A thermonuclear weapon weighing little more than 2,400 pounds (1,100 kg) can produce an explosive force comparable to the detonation of more than 1.2 million tons of TNT (5.0 PJ).[1] A nuclear device no larger than traditional bombs can devastate an entire city by blast, fire, and radiation. Nuclear weapons are considered weapons of mass destruction, and their use and control have been a major focus of international relations policy since their debut.

Nuclear weapons have been used twice in nuclear warfare, both times by the United States against Japan near the end of World War II. On August 6, 1945, the U.S. Army Air Forces detonated a uranium gun-type fission bomb nicknamed "Little Boy" over the Japanese city of Hiroshima; three days later, on August 9, the U.S. Army Air Forces detonated a plutonium implosion-type fission bomb codenamed "Fat Man" over the Japanese city of Nagasaki. The bombings resulted in the deaths of approximately 200,000 civilians and military personnel from acute injuries sustained from the explosions.[2] The ethics of the bombings and their role in Japan's surrender remain the subject of scholarly and popular debate.

Since the atomic bombings of Hiroshima and Nagasaki, nuclear weapons have been detonated on over two thousand occasions for the purposes of testing and demonstration. Only a few nations possess such weapons or are suspected of seeking them. The only countries known to have detonated nuclear weapons—and acknowledge possessing them—are (chronologically by date of first test) the United States, the Soviet Union (succeeded as a nuclear power by Russia), the United Kingdom, France, the People's Republic of China, India, Pakistan, and North Korea. Israel is also believed to possess nuclear weapons, though in a policy of deliberate ambiguity, it does not acknowledge having them. Germany, Italy, Turkey, Belgium and the Netherlands are nuclear weapons sharing states.[3][4][5]

The nuclear non-proliferation treaty aimed to reduce the spread of nuclear weapons, but its effectiveness has been questioned, and political tensions remained high in the 1970s and 1980s. As of 2016, 16,000 nuclear weapons are stored at sites in 14 countries and many are ready for immediate use. Modernisation of weapons continues to occur.[6]

1.1   Types

Main article: Nuclear weapon design

There are two basic types of nuclear weapons: those that

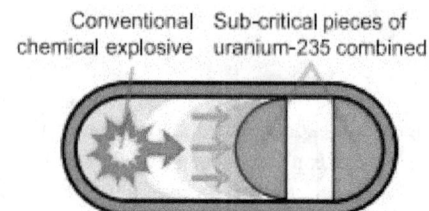

Conventional   Sub-critical pieces of
chemical explosive   uranium-235 combined

Gun-type assembly method

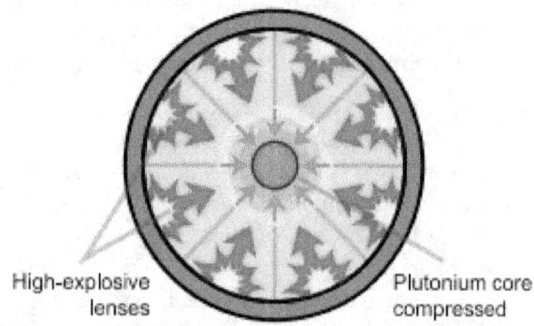

High-explosive   Plutonium core
lenses   compressed

Implosion assembly method

The two basic fission weapon designs

derive the majority of their energy from nuclear fission reactions alone, and those that use fission reactions to begin nuclear fusion reactions that produce a large amount of the total energy output.

1.1.1   Fission weapons

All existing nuclear weapons derive some of their explosive energy from nuclear fission reactions. Weapons whose explosive output is exclusively from fission reactions are commonly referred to as **atomic bombs** or **atom bombs** (abbreviated as **A-bombs**). This has long been noted as something of a misnomer, as their energy comes from the nucleus of the atom, just as it does with fusion weapons.

In fission weapons, a mass of fissile material (enriched uranium or plutonium) is assembled into a supercritical mass—the amount of material needed to start an exponentially growing nuclear chain reaction—either by shooting one piece of sub-critical material into another (the "gun"

method) or by compressing using explosive lenses a sub-critical sphere of material using chemical explosives to many times its original density (the "implosion" method). The latter approach is considered more sophisticated than the former and only the latter approach can be used if the fissile material is plutonium.

A major challenge in all nuclear weapon designs is to ensure that a significant fraction of the fuel is consumed before the weapon destroys itself. The amount of energy released by fission bombs can range from the equivalent of just under a ton to upwards of 500,000 tons (500 kilotons) of TNT (4.2 to 2.1×10^8 GJ).[7]

All fission reactions necessarily generate fission products, the radioactive remains of the atomic nuclei split by the fission reactions. Many fission products are either highly radioactive (but short-lived) or moderately radioactive (but long-lived), and as such are a serious form of radioactive contamination if not fully contained. Fission products are the principal radioactive component of nuclear fallout.

The most commonly used fissile materials for nuclear weapons applications have been uranium-235 and plutonium-239. Less commonly used has been uranium-233. Neptunium-237 and some isotopes of americium may be usable for nuclear explosives as well, but it is not clear that this has ever been implemented, and even their plausible use in nuclear weapons is a matter of scientific dispute.[8]

1.1.2   Fusion weapons

Main article: Thermonuclear weapon

The other basic type of nuclear weapon produces a large proportion of its energy in nuclear fusion reactions. Such fusion weapons are generally referred to as **thermonuclear weapons** or more colloquially as **hydrogen bombs** (abbreviated as **H-bombs**), as they rely on fusion reactions between isotopes of hydrogen (deuterium and tritium). All such weapons derive a significant portion, and sometimes a majority, of their energy from fission. This is because a fission reaction is required as a "trigger" for the fusion reactions, and the fusion reactions can themselves trigger additional fission reactions.[9]

Only six countries—United States, Russia, United Kingdom, People's Republic of China, France and India—have conducted thermonuclear weapon tests. (Whether India has detonated a "true", multi-staged thermonuclear weapon is controversial.)[10] North Korea claims to have tested a fusion weapon as of January 2016, though this claim is disputed.[11] Thermonuclear weapons are considered much more difficult to successfully design and execute than primitive fission weapons. Almost all of the nuclear weapons

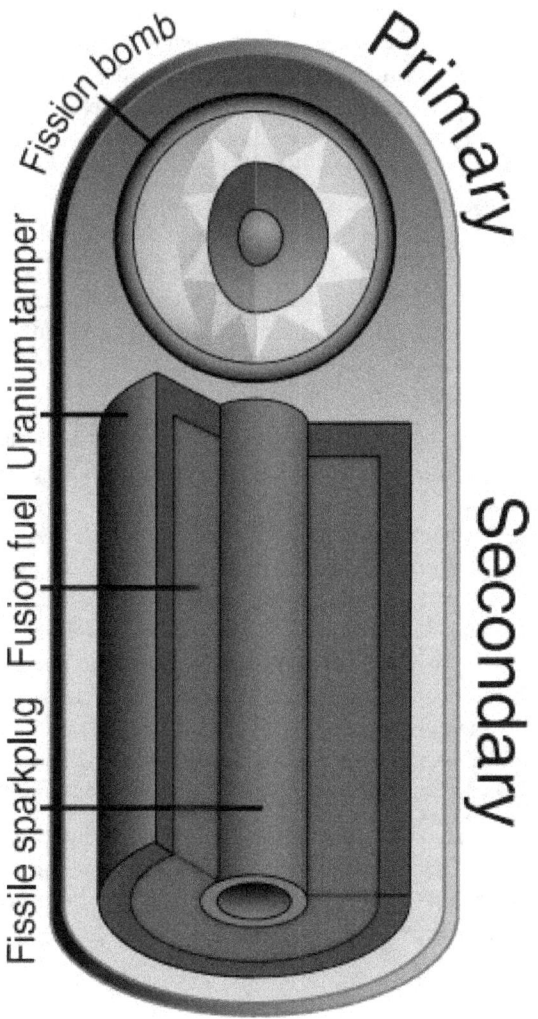

The basics of the Teller–Ulam design for a hydrogen bomb: a fission bomb uses radiation to compress and heat a separate section of fusion fuel.

deployed today use the thermonuclear design because it is more efficient.

Thermonuclear bombs work by using the energy of a fission bomb to compress and heat fusion fuel. In the Teller-Ulam design, which accounts for all multi-megaton yield hydrogen bombs, this is accomplished by placing a fission bomb and fusion fuel (tritium, deuterium, or lithium deuteride) in proximity within a special, radiation-reflecting container. When the fission bomb is detonated, gamma rays and X-rays emitted first compress the fusion fuel, then heat it to thermonuclear temperatures. The ensuing fusion reaction creates enormous numbers of high-speed neutrons, which can then induce fission in materials not normally prone to it, such as depleted uranium. Each of these components is known as a "stage", with the fission bomb as the "pri-

mary" and the fusion capsule as the "secondary". In large, megaton-range hydrogen bombs, about half of the yield comes from the final fissioning of depleted uranium.[7]

Virtually all thermonuclear weapons deployed today use the "two-stage" design described above, but it is possible to add additional fusion stages—each stage igniting a larger amount of fusion fuel in the next stage. This technique can be used to construct thermonuclear weapons of arbitrarily large yield, in contrast to fission bombs, which are limited in their explosive force. The largest nuclear weapon ever detonated, the Tsar Bomba of the USSR, which released an energy equivalent of over 50 megatons of TNT (210 PJ), was a three-stage weapon. Most thermonuclear weapons are considerably smaller than this, due to practical constraints from missile warhead space and weight requirements.[12]

Edward Teller, often referred to as the "father of the hydrogen bomb"

Fusion reactions do not create fission products, and thus contribute far less to the creation of nuclear fallout than fission reactions, but because all thermonuclear weapons contain at least one fission stage, and many high-yield thermonuclear devices have a final fission stage, thermonuclear weapons can generate at least as much nuclear fallout as fission-only weapons.

1.1.3   Other types

Main articles: boosted fission weapon, neutron bomb, and radiological bomb

There are other types of nuclear weapons as well. For example, a boosted fission weapon is a fission bomb that increases its explosive yield through a small amount of fusion reactions, but it is not a fusion bomb. In the boosted bomb, the neutrons produced by the fusion reactions serve primarily to increase the efficiency of the fission bomb. There are two types of boosted fission bomb: internally boosted, in which a deuterium-tritium mixture is injected into the bomb core, and externally boosted, in which concentric shells of lithium-deuteride and depleted uranium are layered on the outside of the fission bomb core.

Some weapons are designed for special purposes; a neutron bomb is a thermonuclear weapon that yields a relatively small explosion but a relatively large amount of neutron radiation; such a device could theoretically be used to cause massive casualties while leaving infrastructure mostly intact and creating a minimal amount of fallout. The detonation of any nuclear weapon is accompanied by a blast of neutron radiation. Surrounding a nuclear weapon with suitable materials (such as cobalt or gold) creates a weapon known as a salted bomb. This device can produce exceptionally large quantities of long-lived radioactive contamination. It has been conjectured that such a device could serve as a "doomsday weapon" because such a large quantity of radioactivies with half-lives of decades, lifted into the stratosphere where wind currents would distribute it around the globe, would make all life on the planet extinct.

In connection with the Strategic Defense Initiative, research into the Nuclear pumped laser was conducted under the Dod program Project Excalibur but this did not result in a working weapon. The concept involves the tapping of the energy of an exploding nuclear bomb to power a single-shot laser which is directed at a distant target.

During the Starfish Prime high-altitude nuclear test in 1962, an unexpected effect was produced which is called a Nuclear electromagnetic pulse. This is an intense flash of electromagnetic energy produced by a rain of high energy electrons which in turn are produced by a nuclear bomb's gamma rays. This flash of energy can permanently destroy or disrupt electronic equipment if insufficiently shielded. It has been proposed to use this effect to disable an enemy's military and civilian infrastructure as an adjunct to other nuclear or conventional military operations against that enemy. Because the effect is produced by very high altitude nuclear detonations, it can produce damage to electronics over a very wide, even continental, geographical area.

Research has been done into the possibility of pure fusion bombs: nuclear weapons that consist of fusion reactions without requiring a fission bomb to initiate them. Such a device might provide a simpler path to thermonuclear weapons than one that required development of fission weapons first, and pure fusion weapons would create significantly less nuclear fallout than other thermonuclear weapons, because they would not disperse fission products. In 1998, the United States Department of Energy divulged that the United States had, "...made a substantial investment" in the past to develop pure fusion weapons, but that, "The U.S. does not have and is not developing a pure fusion weapon", and that, "No credible design for a pure fusion weapon resulted from the DOE investment".[13]

Antimatter, which consists of particles resembling ordinary matter particles in most of their properties but having opposite electric charge, has been considered as a trigger mechanism for nuclear weapons.[14] A major obstacle is the difficulty of producing antimatter in large enough quantities, and there is no evidence that it is feasible beyond the military domain.[15] However, the U.S. Air Force funded studies of the physics of antimatter in the Cold War, and began considering its possible use in weapons, not just as a trigger, but as the explosive itself.[16] A fourth generation nuclear weapon design is related to, and relies upon, the same principle as Antimatter-catalyzed nuclear pulse propulsion.[17]

Most variation in nuclear weapon design is for the purpose of achieving different yields for different situations, and in manipulating design elements to attempt to minimize weapon size.[7]

1.2   Weapons delivery

See also: Nuclear weapons delivery, nuclear triad, Strategic bomber, Intercontinental ballistic missile, and Submarine-launched ballistic missile

Nuclear weapons delivery—the technology and systems used to bring a nuclear weapon to its target—is an important aspect of nuclear weapons relating both to nuclear weapon design and nuclear strategy. Additionally, development and maintenance of delivery options is among the most resource-intensive aspects of a nuclear weapons program: according to one estimate, deployment costs accounted for 57% of the total financial resources spent by the United States in relation to nuclear weapons since 1940.[18]

Historically the first method of delivery, and the method used in the two nuclear weapons used in warfare, was as a gravity bomb, dropped from bomber aircraft. This is usually the first method that countries developed, as it does not place many restrictions on the size of the weapon and *weapon miniaturization* requires considerable weapons design knowledge. It does, however, limit attack range, re-

The first nuclear weapons were gravity bombs, such as this "Fat Man" weapon dropped on Nagasaki, Japan. They were very large and could only be delivered by heavy bomber aircraft

A demilitarized and commercial launch of the Russian Strategic Rocket Forces R-36 ICBM; also known by the NATO reporting name: SS-18 Satan. Upon its first fielding in the late 1960s, the SS-18 remains the single highest throw weight missile delivery system ever built.

sponse time to an impending attack, and the number of weapons that a country can field at the same time.

With the advent of miniaturization, nuclear bombs can be delivered by both strategic bombers and tactical fighter-bombers, allowing an air force to use its current fleet with little or no modification. This method may still be considered the primary means of nuclear weapons delivery; the majority of U.S. nuclear warheads, for example, are free-fall gravity bombs, namely the B61.[7]

Montage of an inert test of a United States Trident SLBM (submarine launched ballistic missile), from submerged to the terminal, or re-entry phase, of the multiple independently targetable reentry vehicles

More preferable from a strategic point of view is a nuclear weapon mounted onto a missile, which can use a ballistic trajectory to deliver the warhead over the horizon. Although even short-range missiles allow for a faster and less vulnerable attack, the development of long-range intercontinental ballistic missiles (ICBMs) and submarine-launched ballistic missiles (SLBMs) has given some nations the ability to plausibly deliver missiles anywhere on the globe with a high likelihood of success.

More advanced systems, such as multiple independently targetable reentry vehicles (MIRVs), can launch multiple warheads at different targets from one missile, reducing the chance of a successful missile defense. Today, missiles are most common among systems designed for delivery of nuclear weapons. Making a warhead small enough to fit onto a missile, though, can be difficult.[7]

Tactical weapons have involved the most variety of delivery types, including not only gravity bombs and missiles but also artillery shells, land mines, and nuclear depth charges and torpedoes for anti-submarine warfare. An atomic mortar was also tested at one time by the United States. Small, two-man portable tactical weapons (somewhat misleadingly referred to as suitcase bombs), such as the Special Atomic Demolition Munition, have been developed, although the difficulty of combining sufficient yield with portability limits their military utility.[7]

1.3 Nuclear strategy

Main articles: Nuclear strategy and Deterrence theory
See also: Nuclear peace, Essentials of Post–Cold War
Deterrence, Single Integrated Operational Plan, nuclear
warfare, and On Thermonuclear War

Nuclear warfare strategy is a set of policies that deal with
preventing or fighting a nuclear war. The policy of trying to
prevent an attack by a nuclear weapon from another country by threatening nuclear retaliation is known as the strategy of nuclear deterrence. The goal in deterrence is to always maintain a second strike capability (the ability of a
country to respond to a nuclear attack with one of its own)
and potentially to strive for first strike status (the ability to
completely destroy an enemy's nuclear forces before they
could retaliate). During the Cold War, policy and military
theorists in nuclear-enabled countries worked out models
of what sorts of policies could prevent one from ever being attacked by a nuclear weapon, and developed weapon
game theory models that create the greatest and most stable
deterrence conditions.

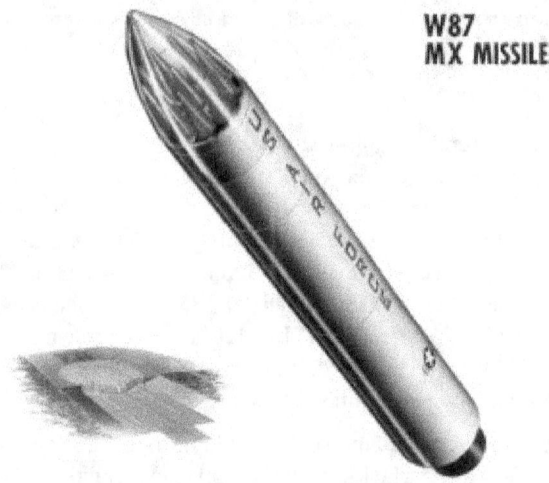

**W87
MX MISSILE**

*The now decommissioned United States' Peacekeeper missile was an
ICBM developed to entirely replace the minuteman missile in the late
1980s. Each missile, like the heavier lift Russian SS-18 Satan, could
contain up to ten nuclear warheads (shown in red), each of which
could be aimed at a different target. A factor in the development of
MIRVs was to make complete missile defense very difficult for an
enemy country.*

Different forms of nuclear weapons delivery (see above) allow for different types of nuclear strategies. The goals of
any strategy are generally to make it difficult for an enemy
to launch a pre-emptive strike against the weapon system
and difficult to defend against the delivery of the weapon
during a potential conflict. Sometimes this has meant keeping the weapon locations hidden, such as deploying them

on submarines or land mobile transporter erector launchers
whose locations are very hard for an enemy to track, and
other times, this means protecting them by burying them in
hardened missile silo bunkers.

Other components of nuclear strategies have included using
missile defense (to destroy the missiles before they land)
or implementation of civil defense measures (using early-
warning systems to evacuate citizens to safe areas before an
attack).

Note that weapons designed to threaten large populations,
or to generally deter attacks are known as *strategic weapons.*
Weapons designed for use on a battlefield in military situations are called *tactical weapons.*

There are critics of the very idea of nuclear strategy for
waging nuclear war who have suggested that a nuclear war
between two nuclear powers would result in mutual annihilation. From this point of view, the significance of nuclear weapons is purely to deter war because any nuclear
war would immediately escalate out of mutual distrust and
fear, resulting in mutually assured destruction. This threat
of national, if not global, destruction has been a strong motivation for anti-nuclear weapons activism.

Critics from the peace movement and within the military establishment have questioned the usefulness of such weapons
in the current military climate. According to an advisory
opinion issued by the International Court of Justice in 1996,
the use of (or threat of use of) such weapons would generally be contrary to the rules of international law applicable
in armed conflict, but the court did not reach an opinion as
to whether or not the threat or use would be lawful in specific extreme circumstances such as if the survival of the
state were at stake.

Another deterrence position in nuclear strategy is that
nuclear proliferation can be desirable. This view argues
that, unlike conventional weapons, nuclear weapons successfully deter all-out war between states, and they succeeded in doing this during the Cold War between the
U.S. and the Soviet Union.[19] In the late 1950s and early
1960s, Gen. Pierre Marie Gallois of France, an adviser to
Charles DeGaulle, argued in books like *The Balance of Terror: Strategy for the Nuclear Age* (1961) that mere possession of a nuclear arsenal, what the French called the *force
de frappe,* was enough to ensure deterrence, and thus concluded that the spread of nuclear weapons could increase
international stability. Some very prominent neo-realist
scholars, such as the late Kenneth Waltz, formerly a Political Science at UC Berkeley and Adjunct Senior Research
Scholar at Columbia University, and John Mearsheimer of
University of Chicago, have also argued along the lines of
Gallois. Specifically, these scholars have advocated some
forms of nuclear proliferation, arguing that it would decrease the likelihood of total war, especially in troubled

regions of the world where there exists a unipolar nuclear weapon state. Aside from the public opinion that opposes proliferation in any form, there are two schools of thought on the matter: those, like Mearsheimer, who favor selective proliferation,[20] and those of Kenneth Waltz, who was somewhat more non-interventionist.[21][22]

The threat of potentially suicidal terrorists possessing nuclear weapons (a form of nuclear terrorism) complicates the decision process. The prospect of mutually assured destruction may not deter an enemy who expects to die in the confrontation. Further, if the initial act is from a stateless terrorist instead of a sovereign nation, there is no fixed nation or fixed military targets to retaliate against. It has been argued by the New York Times, especially after the September 11, 2001 attacks, that this complication is the sign of the next age of nuclear strategy, distinct from the relative stability of the Cold War.[23] In 1996, the United States adopted a policy of allowing the targeting of its nuclear weapons at terrorists armed with weapons of mass destruction.[24]

Robert Gallucci, president of the John D. and Catherine T. MacArthur Foundation, argues that although traditional deterrence is not an effective approach toward terrorist groups bent on causing a nuclear catastrophe, Gallucci believes that "the United States should instead consider a policy of expanded deterrence, which focuses not solely on the would-be nuclear terrorists but on those states that may deliberately transfer or inadvertently lead nuclear weapons and materials to them. By threatening retaliation against those states, the United States may be able to deter that which it cannot physically prevent.".[25]

Graham Allison makes a similar case, arguing that the key to expanded deterrence is coming up with ways of tracing nuclear material to the country that forged the fissile material. "After a nuclear bomb detonates, nuclear forensics cops would collect debris samples and send them to a laboratory for radiological analysis. By identifying unique attributes of the fissile material, including its impurities and contaminants, one could trace the path back to its origin."[26] The process is analogous to identifying a criminal by fingerprints. "The goal would be twofold: first, to deter leaders of nuclear states from selling weapons to terrorists by holding them accountable for any use of their own weapons; second, to give leader every incentive to tightly secure their nuclear weapons and materials."[26]

1.4 Governance, control, and law

Main articles: Nuclear Non-Proliferation Treaty, Strategic Arms Limitation Talks, Intermediate-Range Nuclear Forces Treaty, START I, Strategic Offensive Reductions Treaty, Comprehensive Nuclear-Test-Ban Treaty, and New START

Because of the immense military power they can confer,

The International Atomic Energy Agency was created in 1957 to encourage peaceful development of nuclear technology while providing international safeguards against nuclear proliferation.

the political control of nuclear weapons has been a key issue for as long as they have existed; in most countries the use of nuclear force can only be authorized by the head of government or head of state.[27] Controls and regulations governing nuclear weapons are man-made, and so are imperfect. Therefore, there is an inherent danger of "accidents, mistakes, false alarms, blackmail, theft, and sabotage".[28]

In the late 1940s, lack of mutual trust was preventing the United States and the Soviet Union from making ground towards international arms control agreements. The Russell–Einstein Manifesto was issued in London on July 9, 1955 by Bertrand Russell in the midst of the Cold War. It highlighted the dangers posed by nuclear weapons and called for world leaders to seek peaceful resolutions to international conflict. The signatories included eleven pre-eminent intellectuals and scientists, including Albert Einstein, who signed it just days before his death on April 18, 1955. A few days after the release, philanthropist Cyrus S. Eaton offered to sponsor a conference—called for in the manifesto—in Pugwash, Nova Scotia, Eaton's birthplace. This conference was to be the first of the Pugwash Conferences on Science and World Affairs, held in July 1957.

By the 1960s steps were being taken to limit both the proliferation of nuclear weapons to other countries and the environmental effects of nuclear testing. The Partial Test Ban Treaty (1963) restricted all nuclear testing to underground nuclear testing, to prevent contamination from nuclear fallout, whereas the Nuclear Non-Proliferation Treaty (1968) attempted to place restrictions on the types of activities signatories could participate in, with the goal of allowing the transference of non-military nuclear technology to member countries without fear of proliferation.

In 1957, the International Atomic Energy Agency (IAEA) was established under the mandate of the United Nations to encourage development of peaceful applications for nuclear technology, provide international safeguards against its misuse, and facilitate the application of safety measures in its use. In 1996, many nations signed the Comprehensive Test Ban Treaty,[29] which prohibits all testing of nuclear weapons. A testing ban imposes a significant hindrance to nuclear arms development by any complying country.[30] The Treaty requires the ratification by 44 specific states before it can go into force; as of 2012, the ratification of eight of these states is still required.[29]

Additional treaties and agreements have governed nuclear weapons stockpiles between the countries with the two largest stockpiles, the United States and the Soviet Union, and later between the United States and Russia. These include treaties such as SALT II (never ratified), START I (expired), INF, START II (never ratified), SORT, and New START, as well as non-binding agreements such as SALT I and the Presidential Nuclear Initiatives[31] of 1991. Even when they did not enter into force, these agreements helped limit and later reduce the numbers and types of nuclear weapons between the United States and the Soviet Union/Russia.

Nuclear weapons have also been opposed by agreements between countries. Many nations have been declared Nuclear-Weapon-Free Zones, areas where nuclear weapons production and deployment are prohibited, through the use of treaties. The Treaty of Tlatelolco (1967) prohibited any production or deployment of nuclear weapons in Latin America and the Caribbean, and the Treaty of Pelindaba (1964) prohibits nuclear weapons in many African countries. As recently as 2006 a Central Asian Nuclear Weapon Free Zone was established amongst the former Soviet republics of Central Asia prohibiting nuclear weapons.

In the middle of 1996, the International Court of Justice, the highest court of the United Nations, issued an Advisory Opinion concerned with the "Legality of the Threat or Use of Nuclear Weapons". The court ruled that the use or threat of use of nuclear weapons would violate various articles of international law, including the Geneva Conventions, the Hague Conventions, the UN Charter, and the Universal Declaration of Human Rights. In view of the unique, destructive characteristics of nuclear weapons, the International Committee of the Red Cross calls on States to ensure that these weapons are never used, irrespective of whether they consider them lawful or not.[32]

Additionally, there have been other, specific actions meant to discourage countries from developing nuclear arms. In the wake of the tests by India and Pakistan in 1998, economic sanctions were (temporarily) levied against both countries, though neither were signatories with the Nuclear

Non-Proliferation Treaty. One of the stated *casus belli* for the initiation of the 2003 Iraq War was an accusation by the United States that Iraq was actively pursuing nuclear arms (though this was soon discovered not to be the case as the program had been discontinued). In 1981, Israel had bombed a nuclear reactor being constructed in Osirak, Iraq, in what it called an attempt to halt Iraq's previous nuclear arms ambitions; in 2007, Israel bombed another reactor being constructed in Syria.

In 2013, Mark Diesendorf says that governments of France, India, North Korea, Pakistan, UK, and South Africa have used nuclear power and/or research reactors to assist nuclear weapons development or to contribute to their supplies of nuclear explosives from military reactors.[33]

1.4.1 Disarmament

Main article: Nuclear disarmament
See also: Nuclear Tipping Point
For statistics on possession and deployment, see List of states with nuclear weapons.

Nuclear disarmament refers to both the act of reduc-

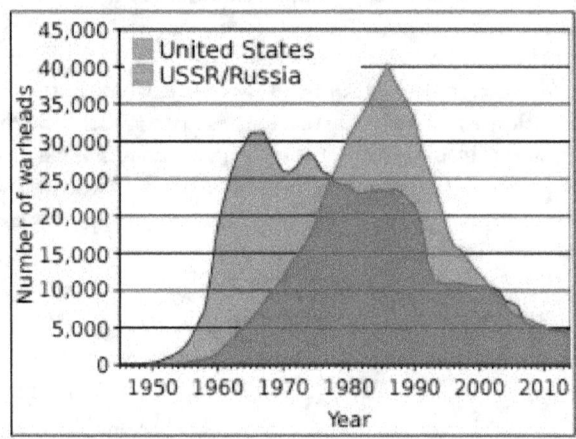

The USSR and United States nuclear weapon stockpiles throughout the Cold War until 2015, with a precipitous drop in total numbers following the end of the Cold War in 1991.

ing or eliminating nuclear weapons and to the end state of a nuclear-free world, in which nuclear weapons are completely eliminated.

Beginning with the 1963 Partial Test Ban Treaty and continuing through the 1996 Comprehensive Test Ban Treaty, there have been many treaties to limit or reduce nuclear weapons testing and stockpiles. The 1968 Nuclear Non-Proliferation Treaty has as one of its explicit conditions that all signatories must "pursue negotiations in good faith" towards the long-term goal of "complete disarmament". The nuclear weapon states have largely treated that aspect of the agreement as "decorative" and without force.[34]

Only one country—South Africa—has ever fully renounced nuclear weapons they had independently developed. The former Soviet republics of Belarus, Kazakhstan, and Ukraine returned Soviet nuclear arms stationed in their countries to Russia after the collapse of the USSR.

Proponents of nuclear disarmament say that it would lessen the probability of nuclear war occurring, especially accidentally. Critics of nuclear disarmament say that it would undermine the present nuclear peace and deterrence and would lead to increased global instability. Various American elder statesmen,[35] who were in office during the Cold War period, have been advocating the elimination of nuclear weapons. These officials include Henry Kissinger, George Shultz, Sam Nunn, and William Perry. In January 2010, Lawrence M. Krauss stated that "no issue carries more importance to the long-term health and security of humanity than the effort to reduce, and perhaps one day, rid the world of nuclear weapons".[36]

Ukrainian workers use equipment provided by the U.S. Defense Threat Reduction Agency to dismantle a Soviet-era missile silo. After the end of the Cold War, Ukraine and the other non-Russian, post-Soviet republics relinquished Soviet nuclear stockpiles to Russia.

In the years after the end of the Cold War, there have been numerous campaigns to urge the abolition of nuclear weapons, such as that organized by the Global Zero movement, and the goal of a "world without nuclear weapons" was advocated by United States President Barack Obama in an April 2009 speech in Prague.[37] A CNN poll from April 2010 indicated that the American public was nearly evenly split on the issue.[38]

Some analysts have argued that nuclear weapons have made the world relatively safer, with peace through deterrence and through the stability–instability paradox, including in south Asia.[39][40] Kenneth Waltz has argued that nuclear weapons have helped keep an uneasy peace, and further nu-

clear weapon proliferation might even help avoid the large scale conventional wars that were so common prior to their invention at the end of World War II.[22] But former Secretary Henry Kissinger says there is a new danger, which cannot be addressed by deterrence: "The classical notion of deterrence was that there was some consequences before which aggressors and evildoers would recoil. In a world of suicide bombers, that calculation doesn't operate in any comparable way".[41] George Shultz has said, "If you think of the people who are doing suicide attacks, and people like that get a nuclear weapon, they are almost by definition not deterrable".[42]

1.4.2 United Nations

Main article: United Nations Office for Disarmament Affairs

The UN Office for Disarmament Affairs (UNODA) is a department of the United Nations Secretariat established in January 1998 as part of the United Nations Secretary-General Kofi Annan's plan to reform the UN as presented in his report to the General Assembly in July 1997.[43]

Its goal is to promote nuclear disarmament and non-proliferation and the strengthening of the disarmament regimes in respect to other weapons of mass destruction, chemical and biological weapons. It also promotes disarmament efforts in the area of conventional weapons, especially land mines and small arms, which are often the weapons of choice in contemporary conflicts.

1.5 Controversy

See also: Nuclear weapons debate and History of the anti-nuclear movement

1.5.1 Ethics

Even before the first nuclear weapons had been developed, scientists involved with the Manhattan Project were divided over the use of the weapon. The role of the two atomic bombings of the country in Japan's surrender and the U.S.'s ethical justification for them has been the subject of scholarly and popular debate for decades. The question of whether nations should have nuclear weapons, or test them, has been continually and nearly universally controversial.[44]

1.5.2 Notable nuclear weapons accidents

Main article: Nuclear and radiation accidents

- February 13, 1950: a Convair B-36B crashed in northern British Columbia after jettisoning a Mark IV atomic bomb. This was the first such nuclear weapon loss in history.

- May 22, 1957: a 42,000-pound Mark-17 hydrogen bomb accidentally fell from a bomber near Albuquerque, New Mexico. The detonation of the device's conventional explosives destroyed it on impact and formed a crater 25-feet in diameter on land owned by the University of New Mexico. According to a researcher at the Natural Resources Defense Council, it was one of the most powerful bombs made to date.[45]

- June 7, 1960: the 1960 Fort Dix IM-99 accident destroyed a Boeing CIM-10 Bomarc nuclear missile and shelter and contaminated the BOMARC Missile Accident Site in New Jersey.

- January 24, 1961: the 1961 Goldsboro B-52 crash occurred near Goldsboro, North Carolina. A B-52 Stratofortress carrying two Mark 39 nuclear bombs broke up in mid-air, dropping its nuclear payload in the process.[46][47]

- 1965 Philippine Sea A-4 crash, where a Skyhawk attack aircraft with a nuclear weapon fell into the sea.[48] The pilot, the aircraft, and the B43 nuclear bomb were never recovered.[49] It was not until 1989 that the Pentagon revealed the loss of the one-megaton bomb.[50]

- January 17, 1966: the 1966 Palomares B-52 crash occurred when a B-52G bomber of the USAF collided with a KC-135 tanker during mid-air refuelling off the coast of Spain. The KC-135 was completely destroyed when its fuel load ignited, killing all four crew members. The B-52G broke apart, killing three of the seven crew members aboard.[51] Of the four Mk28 type hydrogen bombs the B-52G carried,[52] three were found on land near Almería, Spain. The non-nuclear explosives in two of the weapons detonated upon impact with the ground, resulting in the contamination of a 2-square-kilometer (490-acre) (0.78 square mile) area by radioactive plutonium. The fourth, which fell into the Mediterranean Sea, was recovered intact after a 2½-month-long search.[53]

- January 21, 1968: the 1968 Thule Air Base B-52 crash involved a United States Air Force (USAF) B-52 bomber. The aircraft was carrying four hydrogen bombs when a cabin fire forced the crew to abandon the aircraft. Six crew members ejected safely, but one who did not have an ejection seat was killed while trying to bail out. The bomber crashed onto sea ice in Greenland, causing the nuclear payload to rupture and disperse, which resulted in widespread radioactive contamination.

- September 18–19, 1980: the Damascus Accident, occurred in Damascus, Arkansas, where a Titan missile equipped with a nuclear warhead exploded. The accident was caused by a maintenance man who dropped a socket from a socket wrench down an 80-foot shaft, puncturing a fuel tank on the rocket. Leaking fuel resulted in a hypergolic fuel explosion, jettisoning the W-53 warhead beyond the launch site.[54][55][56]

1.5.3 Nuclear testing and fallout

Main article: Nuclear fallout
See also: Downwinders

Over 500 atmospheric nuclear weapons tests were con-

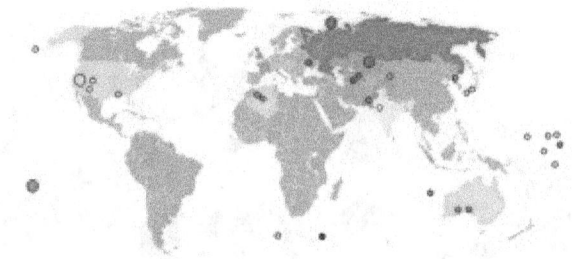

Over 2,000 nuclear tests have been conducted in over a dozen different sites around the world. Red Russia/Soviet Union, blue France, light blue United States, violet Britain, black Israel, orange China, yellow India, brown Pakistan, green North Korea and light green (territories exposed to nuclear bombs)

ducted at various sites around the world from 1945 to 1980. Radioactive fallout from nuclear weapons testing was first drawn to public attention in 1954 when the Castle Bravo hydrogen bomb test at the Pacific Proving Grounds contaminated the crew and catch of the Japanese fishing boat *Lucky Dragon*.[57] One of the fishermen died in Japan seven months later, and the fear of contaminated tuna led to a temporary boycotting of the popular staple in Japan. The incident caused widespread concern around the world, especially regarding the effects of nuclear fallout and atmospheric nuclear testing, and "provided a decisive impetus for the emergence of the anti-nuclear weapons movement in many countries".[57]

As public awareness and concern mounted over the possible health hazards associated with exposure to the nuclear fallout, various studies were done to assess the extent of the hazard. A Centers for Disease Control and Prevention/

This view of downtown Las Vegas shows a mushroom cloud in the background. Scenes such as this were typical during the 1950s. From 1951 to 1962 the government conducted 100 atmospheric tests at the nearby Nevada Test Site.

National Cancer Institute study claims that fallout from atmospheric nuclear tests would lead to perhaps 11,000 excess deaths amongst people alive during atmospheric testing in the United States from all forms of cancer, including leukemia, from 1951 to well into the 21st century.[58][59] As of March 2009, the U.S. is the only nation that compensates nuclear test victims. Since the Radiation Exposure Compensation Act of 1990, more than $1.38 billion in compensation has been approved. The money is going to people who took part in the tests, notably at the Nevada Test Site, and to others exposed to the radiation.[60][61]

In addition, leakage of byproducts of nuclear weapon production into groundwater has been an ongoing issue, particularly at the Hanford site.[62]

1.6 Effects of nuclear explosions on human health

Main article: Effects of nuclear explosions on human health

Some scientists estimate that if there were a nuclear war resulting in 100 Hiroshima-size nuclear explosions on cities, it could cause significant loss of life in the tens of millions from long term climatic effects alone. The climatology hypothesis is that *if* each city firestorms, a great deal of soot could be thrown up into the atmosphere which could blanket the earth, cutting out sunlight for years on end, causing the disruption of food chains, in what is termed a Nuclear Winter.[63][64]

The medical effects of the atomic bomb on Hiroshima upon humans can be put into the four categories below, with the effects of larger thermonuclear weapons producing blast and thermal effects so large that there would be a negligible number of survivors close enough to the center of the blast who would experience prompt/acute radiation effects, which were observed after the 16 kiloton yield Hiroshima bomb, due to its relatively low yield:[65][66]

- Initial stage—the first 1–9 weeks, in which are the greatest number of deaths, with 90% due to thermal injury and/or blast effects and 10% due to super-lethal radiation exposure.

- Intermediate stage—from 10–12 weeks. The deaths in this period are from ionizing radiation in the median lethal range - LD50

- Late period—lasting from 13–20 weeks. This period has some improvement in survivors' condition.

- Delayed period—from 20+ weeks. Characterized by numerous complications, mostly related to healing of thermal and mechanical injuries, and if the individual was exposed to a few hundred to a thousand Millisieverts of radiation, it is coupled with infertility, sub-fertility and blood disorders. Furthermore, ionizing radiation above a dose of around 50-100 Millisievert exposure has been shown to statistically begin increasing one's chance of dying of cancer sometime in their lifetime over the normal unexposed rate of ~25%, in the long term, a heightened rate of cancer, proportional to the dose received, would begin to be observed after ~5+ years, with lesser problems such as eye cataracts and other more minor effects in other organs and tissue also being observed over the long term.

Fallout exposure - Depending on if further afield individuals Shelter in place or evacuate perpendicular to the direction of the wind, and therefore avoid contact with the fallout plume, and stay there for the days and weeks after the nuclear explosion, their exposure to fallout, and therefore their total dose, will vary. With those who do shelter in place, and or evacuate, experiencing a total dose that would be negligible in comparison to someone who just went about their life as normal.[67][68]

Staying indoors until after the most hazardous fallout isotope, I-131 decays away to 0.1% of its initial quantity after ten half lifes - which is represented by 80 days in I-131s case, would make the difference between likely contracting Thyroid cancer or escaping completely from this substance depending on the actions of the individual.[69]

1.6.1 Public opposition

See also: History of the anti-nuclear movement and International Day against Nuclear Tests

Peace movements emerged in Japan and in 1954 they con-

Women Strike for Peace during the Cuban Missile Crisis

Demonstration against nuclear testing in Lyon, France, in the 1980s.

verged to form a unified "Japanese Council Against Atomic and Hydrogen Bombs". Japanese opposition to nuclear weapons tests in the Pacific Ocean was widespread, and "an estimated 35 million signatures were collected on petitions calling for bans on nuclear weapons".[70]

In the United Kingdom, the first Aldermaston March organised by the Campaign for Nuclear Disarmament(CND) took place at Easter 1958, when, according to the CND, several thousand people marched for four days from Trafalgar Square, London, to the Atomic Weapons Research Establishment close to Aldermaston in Berkshire, England, to demonstrate their opposition to nuclear weapons.[71][72] The Aldermaston marches continued into the late 1960s when tens of thousands of people took part in the four-day marches.[70]

In 1959, a letter in the *Bulletin of Atomic Scientists* was the start of a successful campaign to stop the Atomic Energy Commission dumping radioactive waste in the sea 19 kilometres from Boston.[73] In 1962, Linus Pauling won the Nobel Peace Prize for his work to stop the atmospheric testing of nuclear weapons, and the "Ban the Bomb" movement spread.[44]

In 1963, many countries ratified the Partial Test Ban Treaty prohibiting atmospheric nuclear testing. Radioactive fallout became less of an issue and the anti-nuclear weapons movement went into decline for some years.[57][74] A resurgence of interest occurred amid European and American fears of nuclear war in the 1980s.[75]

1.7 Costs and technology spin-offs

See also: Global Positioning System, Nuclear weapons delivery, History of computers, ENIAC, and Swords to ploughshares

According to an audit by the Brookings Institution, between 1940 and 1996, the U.S. spent $8.78 trillion in present-day terms[76] on nuclear weapons programs. 57 percent of which was spent on building nuclear weapons delivery systems. 6.3 percent of the total, $551 billion in present-day terms, was spent on environmental remediation and nuclear waste management, for example cleaning up the Hanford site, and 7 percent of the total, $617 billion was spent on making nuclear weapons themselves.[77]

1.8 Non-weapons uses

Main article: Peaceful nuclear explosion

Peaceful nuclear explosions are nuclear explosions conducted for non-military purposes, such as activities related to economic development including the creation of canals. During the 1960s and 70s, both the United States and the Soviet Union conducted a number of PNEs. Six of the explosions by the Soviet Union are considered to have been of an applied nature, not just tests.

Subsequently the United States and the Soviet Union halted their programs. Definitions and limits are covered in the

Peaceful Nuclear Explosions Treaty of 1976.[78][79] The Comprehensive Nuclear-Test-Ban Treaty of 1996 prohibits all nuclear explosions, regardless of whether they are for peaceful purposes or not.

1.9 See also

- *The Atomic Age* – Wikipedia book

1.9.1 History

- History of nuclear weapons
  - Atomic spies
  - German nuclear energy project
  - Japanese atomic program
  - Soviet atomic bomb project
  - Nuclear testing at Bikini Atoll
- Timeline of nuclear weapons development
- Los Alamos National Laboratory
- Lawrence Livermore National Laboratory
- Lists of nuclear disasters and radioactive incidents
- Nuclear and radiation accidents, including nuclear weapons accidents
  - Nevada Test Site
  - Project Gnome
- Military strategy
  - Civil Defense
  - Fractional Orbital Bombardment System
  - Mutual Assured Destruction
- Weapon of mass destruction
  - Nuclear strategy

1.9.2 More technical details

- Effects of nuclear explosions
- Intercontinental ballistic missile
- Nuclear blackout
- Neutron bomb
- Nuclear bombs and health
- Nuclear weapon yield

1.9.3 Popular culture

- Nuclear weapons in popular culture
- *The Butter Battle Book*

1.9.4 Proliferation and politics

- Agency for the Prohibition of Nuclear Weapons in Latin America and the Caribbean
- Comprehensive Test Ban Treaty
- International Court of Justice advisory opinion on legality of nuclear weapons
- List of states with nuclear weapons
- List of nuclear weapons
- Nth Country Experiment
- Nuclear close calls
- Nuclear Non-Proliferation Treaty
- Nuclear weapons and the United Kingdom
- The Letters of last resort (United Kingdom)
- Nuclear weapons and Russia
- Nuclear weapons and the United States
- Strategic Arms Limitation Talks
- Three Non-Nuclear Principles, of Japan

1.10 Notes and references

[1] Specifically the 1970 to 1980 designed and deployed US B83 nuclear bomb, with a yield of up to 1.2 megatons.

[2] "Frequently Asked Questions #1". Radiation Effects Research Foundation. Retrieved September 18, 2007. total number of deaths is not known precisely ... acute (within two to four months) deaths ... Hiroshima ... 90,000-166,000 ... Nagasaki ... 60,000-80,000

[3] "Federation of American Scientists: Status of World Nuclear Forces". Fas.org. Retrieved December 29, 2012.

[4] "Nuclear Weapons – Israel". Fas.org. January 8, 2007. Retrieved December 15, 2010.

[5] See also Mordechai Vanunu

[6] Ian Lowe, "Three minutes to midnight", *Australasian Science*, March 2016, p. 49.

[7] The best overall printed sources on nuclear weapons design are: Hansen, Chuck. *U.S. Nuclear Weapons: The Secret History*. San Antonio, TX: Aerofax, 1988; and the more-updated Hansen, Chuck, "Swords of Armageddon: U.S. Nuclear Weapons Development since 1945" (CD-ROM & download available). PDF. 2,600 pages, Sunnyvale, California, Chuklea Publications, 1995, 2007. ISBN 978-0-9791915-0-3 (2nd Ed.)

[8] David Albright and Kimberly Kramer (August 22, 2005). "Neptunium 237 and Americium: World Inventories and Proliferation Concerns" (PDF). Institute for Science and International Security. Retrieved October 13, 2011.

[9] Carey Sublette, Nuclear Weapons Frequently Asked Questions: 4.5.2 "Dirty" and "Clean" Weapons, accessed May 10, 2011.

[10] On India's alleged hydrogen bomb test, see Carey Sublette, What Are the Real Yields of India's Test?.

[11] McKirdy, Euan. "North Korea announces it conducted nuclear test". *CNN*. Retrieved 7 January 2016.

[12] Sublette, Carey. "The Nuclear Weapon Archive". Retrieved March 7, 2007.

[13] U.S. Department of Energy, Restricted Data Declassification Decisions, 1946 to the Present (RDD-8) (January 1, 2002), accessed November 20, 2011.

[14] "Page discussing the possibility of using antimatter as a trigger for a thermonuclear explosion". Cui.unige.ch. Retrieved May 30, 2013.

[15] Andre Gsponer; Jean-Pierre Hurni (1970). "Paper discussing the number of antiprotons required to ignite a thermonuclear weapon". *In G. Velarde and E. Minguez, eds., Proceedings of the International Conference on Emerging Nuclear Energy Systems, Madrid, June /July , (World Scientific, Singapore, 1987) 166–169*. Arxiv.org. **4** (30): arXiv:physics/0507114. arXiv:physics/0507114. Bibcode:2005physics...7114G.

[16] Keay Davidson; Chronicle Science Writer (October 4, 2004). "Air Force pursuing antimatter weapons: Program was touted publicly, then came official gag order". Sfgate.com. Retrieved May 30, 2013.

[17] "Fourth Generation Nuclear Weapons". Retrieved October 24, 2014.

[18] Stephen I. Schwartz, ed., *Atomic Audit: The Costs and Consequences of U.S. Nuclear Weapons Since 1940*. Washington, D.C.: Brookings Institution Press, 1998. See also Estimated Minimum Incurred Costs of U.S. Nuclear Weapons Programs, 1940–1996, an excerpt from the book. Archived November 21, 2008, at the Wayback Machine.

[19] Creveld, Martin Van (2000). "Technology and War II: Postmodern War?". In Charles Townshend. *The Oxford History of Modern War*. New York, USA: Oxford University Press. p. 349. ISBN 0-19-285373-2.

[20] Mearsheimer, John (2006). "Conversations in International Relations: Interview with John J. Mearsheimer (Part I)" (PDF). *International Relations*. **20** (1): 105–123. doi:10.1177/0047117806060939.See page 116

[21] Kenneth Waltz, "More May Be Better," in Scott Sagan and Kenneth Waltz, eds., *The Spread of Nuclear Weapons* (New York: Norton, 1995).

[22] Kenneth Waltz, "The Spread of Nuclear Weapons: More May Better," *Adelphi Papers*, no. 171 (London: International Institute for Strategic Studies, 1981).

[23] See, for example: Feldman, Noah. "Islam, Terror and the Second Nuclear Age," *New York Times Magazine* (October 29, 2006).

[24] Daniel Plesch & Stephen Young, "Senseless policy", *Bulletin of the Atomic Scientists*, November/December 1998, page 4. Fetched from URL on April 18, 2011.

[25] Gallucci, Robert (September 2006). "Averting Nuclear Catastrophe: Contemplating Extreme Responses to U.S. Vulnerability". *Annals of the American Academy of Political and Social Science*. **607**: 51–58. doi:10.1177/0002716206290457. Retrieved January 28, 2013.

[26] Allison, Graham (March 13, 2009). "How to Keep the Bomb From Terrorists". *Newsweek*. Retrieved January 28, 2013.

[27] In the United States, the President and the Secretary of Defense, acting as the National Command Authority, must *jointly* authorize the use of nuclear weapons.

[28] Eric Schlosser, Today's nuclear dillemma, *Bulletin of the Atomic Scientists*, November/December 2015, vol. 71 no. 6, 11-17.

[29] Preparatory Commission for the Comprehensive Nuclear-Test-Ban Treaty Organization (2010). "Status of Signature and Ratification". Accessed May 27, 2010. Of the "Annex 2" states whose ratification of the CTBT is required before it enters into force, China, Egypt, Iran, Israel, and the United States have signed but not ratified the Treaty. India, North Korea, and Pakistan have not signed the Treaty.

[30] Richelson, Jeffrey. *Spying on the bomb: American nuclear intelligence from Nazi Germany to Iran and North Korea*. New York: Norton, 2006.

[31] The Presidential Nuclear Initiatives (PNIs) on Tactical Nuclear Weapons At a Glance, Fact Sheet, Arms Control Association.

[32] Nuclear weapons and international humanitarian law International Committee of the Red Cross

[33] Mark Diesendorf (2013). "Book review: Contesting the future of nuclear power" (PDF). *Energy Policy*.

[34] Gusterson, Hugh, "Finding Article VI" *Bulletin of the Atomic Scientists* (January 8, 2007).

[35] Jim Hoagland (October 6, 2011). "Nuclear energy after Fukushima". *Washington Post*.

[36] Lawrence M. Krauss. The Doomsday Clock Still Ticks, *Scientific American*, January 2010, p. 26.

[37] Graham, Nick (April 5, 2009). "Obama Prague Speech On Nuclear Weapons". Huffingtonpost.com. Retrieved May 30, 2013.

[38] "CNN Poll: Public divided on eliminating all nuclear weapons". Politicalticker.blogs.cnn.com. April 12, 2010. Retrieved May 30, 2013.

[39] Krepon, Michael. "The Stability-Instability Paradox, Misperception, and Escalation Control in South Asia" (PDF). *Stimson*. Retrieved November 20, 2015.

[40] "Michael Krepon • The Stability-Instability Paradox". Retrieved October 24, 2014.

[41] Ben Goddard (January 27, 2010). "Cold Warriors say no nukes". *The Hill*.

[42] Hugh Gusterson (March 30, 2012). "The new abolitionists". *Bulletin of the Atomic Scientists*.

[43] ODS Team. "Renewing the United Nations: A Program for Reform (A/51/950)" (PDF). Daccess-dds-ny.un.org. Retrieved May 30, 2013.

[44] Jerry Brown and Rinaldo Brutoco (1997). *Profiles in Power: The Anti-nuclear Movement and the Dawn of the Solar Age*, Twayne Publishers, pp. 191–192.

[45] "Accident Revealed After 29 Years: H-Bomb Fell Near Albuquerque in 1957". Los Angeles Times. Associated Press. August 27, 1986. Retrieved 31 August 2014.

[46] Barry Schneider (May 1975). "Big Bangs from Little Bombs". *Bulletin of Atomic Scientists*. p. 28. Retrieved July 13, 2009.

[47] James C. Oskins; Michael H. Maggelet (2008). *Broken Arrow — The Declassified History of U.S. Nuclear Weapons Accidents*. lulu.com. ISBN 1-4357-0361-8. Retrieved December 29, 2008.

[48] "Ticonderoga Cruise Reports" (Navy.mil weblist of Aug 2003 compilation from cruise reports). Retrieved April 20, 2012. The National Archives hold[s] deck logs for aircraft carriers for the Vietnam Conflict.

[49] Broken Arrows at www.atomicarchive.com. Accessed August 24, 2007.

[50] "U.S. Confirms '65 Loss of H-Bomb Near Japanese Islands". *The Washington Post*. Reuters. May 9, 1989. p. A–27.

[51] Hayes, Ron (January 17, 2007). "H-bomb incident crippled pilot's career". Palm Beach Post. Archived from the original on June 16, 2011. Retrieved May 24, 2006.

[52] Maydew, Randall C. (1997). *America's Lost H-Bomb: Palomares, Spain, 1966*. Sunflower University Press. ISBN 978-0-89745-214-4.

[53] Long, Tony (January 17, 2008). "Jan. 17, 1966: H-Bombs Rain Down on a Spanish Fishing Village". WIRED. Retrieved February 16, 2008.

[54] Schlosser, Eric (2013). *Command and Control: Nuclear Weapons, the Damascus Accident, and the Illusion of Safety*. Penguin Press. ISBN 978-1-59420-227-8.

[55] Christ, Mark K. "Titan II Missile Explosion". *The Encyclopedia of Arkansas History & Culture*. Arkansas Historic Preservation Program. Retrieved 31 August 2014.

[56] Stumpf, David K. (2000). Christ, Mark K.; Slater, Cathryn H., eds. *"We Can Neither Confirm Nor Deny" Sentinels of History: Refelctions on Arkansas Properties on the National Register of Historic Places*. Fayetteville, Arkansas: University of Arkansas Press.

[57] Rudig, Wolfgang (1990). "Anti-nuclear Movements: A World Survey of Opposition to Nuclear Energy". Longman. pp. 54–55.

[58] "Report on the Health Consequences to the American Population from Nuclear Weapons Tests Conducted by the United States and Other Nations". CDC. Retrieved December 7, 2013.

[59] Committee to Review the CDC-NCI Feasibility Study of the Health Consequences Nuclear Weapons Tests, National Research Council. "Exposure of the American Population to Radioactive Fallout from Nuclear Weapons Tests". Retrieved October 24, 2014.

[60] ABC News. "What governments offer to victims of nuclear tests". *ABC News*. Retrieved October 24, 2014.

[61] Radiation Exposure Compensation System: Claims to Date Summary of Claims Received by 06/11/2009

[62] Coghlan, Andy. "US nuclear dump is leaking toxic waste". *New Scientist*. Retrieved 12 March 2016.

[63] Philip Yam. Nuclear Exchange, *Scientific American*, June 2010, p. 24.

[64] Alan Robock and Owen Brian Toon. Local Nuclear War, Global Suffering. *Scientific American*, January 2010, p. 74-81.

[65] "Remm.nlm.gov".

[66] "Nuclear Warfare" (PDF). *Nd.edu*. p. 3.

[67] 7 hour rule: At 7 hours after detonation the fission product activity will have decreased to about 1/10 (10%) of its amount at 1 hour. At about 2 days (49 hours-7X7) the activity will have decreased to 1% of the 1-hour value. Falloutradiation.com

[68] "Nuclear Warfare" (PDF). p. 22.

[69] Oak Ridge Reservation (USDOE). EPA Facility ID: TN1890090003; Site and Radiological Assessment Branch, Division of Health Assessment and Consultation, Agency for Toxic Substances and Disease Registry. "PUBLIC HEALTH ASSESSMENT Iodine-131 Releases" (PDF). *atsdr.cdc.gov*. U.S. Center for Disease Control. Retrieved 21 May 2016.

[70] Jim Falk (1982). *Global Fission: The Battle Over Nuclear Power*, Oxford University Press, pp. 96–97.

[71] "A brief history of CND". Cnduk.org. Retrieved May 30, 2013.

[72] "Early defections in march to Aldermaston". London: Guardian Unlimited. April 5, 1958.

[73] Jim Falk (1982). *Global Fission: The Battle Over Nuclear Power*, Oxford University Press, p. 93.

[74] Jim Falk (1982). *Global Fission: The Battle Over Nuclear Power*, Oxford University Press, p. 98.

[75] Spencer Weart, *Nuclear Fear: A History of Images* (Cambridge, Mass.: Harvard University Press, 1988), chapters 16 and 19.

[76] Federal Reserve Bank of Minneapolis Community Development Project. "Consumer Price Index (estimate) 1800–". Federal Reserve Bank of Minneapolis. Retrieved November 10, 2015.

[77] "Estimated Minimum Incurred Costs of U.S. Nuclear Weapons Programs, 1940–1996". *Brookings Institution*. Archived from the original on March 5, 2004. Retrieved November 20, 2015.

[78] "Announcement of Treaty on Underground Nuclear Explosions Peaceful Purposes (PNE Treaty)" (PDF). Gerald R. Ford Museum and Library. May 28, 1976.

[79] Peters, Gerhard; Woolley, John T. "Gerald R. Ford: "Message to the Senate Transmitting United States-Soviet Treaty and Protocol on the Limitation of Underground Nuclear Explosions," July 29, 1976". *The American Presidency Project*. University of California - Santa Barbara.

1.11   Bibliography

See also: List of books about nuclear issues

- Bethe, Hans Albrecht. *The Road from Los Alamos*. New York: Simon and Schuster, 1991. ISBN 0-671-74012-1

- DeVolpi, Alexander, Minkov, Vladimir E., Simonenko, Vadim A., and Stanford, George S. *Nuclear Shadowboxing: Contemporary Threats from Cold War Weaponry*. Fidlar Doubleday, 2004 (Two volumes, both accessible on Google Book Search) (Content of both volumes is now available in the 2009 trilogy by Alexander DeVolpi: *Nuclear Insights: The Cold War Legacy*)

- Glasstone, Samuel and Dolan, Philip J. *The Effects of Nuclear Weapons (third edition)*. Washington, D.C.: U.S. Government Printing Office, 1977. Available online (PDF).

- *NATO Handbook on the Medical Aspects of NBC Defensive Operations (Part 1 – Nuclear)*. Departments of the Army, Navy, and Air Force: Washington, D.C., 1996

- Hansen, Chuck. *U.S. Nuclear Weapons: The Secret History*. Arlington, TX: Aerofax, 1988

- Hansen, Chuck, "Swords of Armageddon: U.S. nuclear weapons development since 1945" (CD-ROM & download available). PDF. 2,600 pages, Sunnyvale, California, Chucklea Publications, 1995, 2007. ISBN 978-0-9791915-0-3 (2nd Ed.)

- Holloway, David. *Stalin and the Bomb*. New Haven: Yale University Press, 1994. ISBN 0-300-06056-4

- The Manhattan Engineer District, "The Atomic Bombings of Hiroshima and Nagasaki" (1946)

- (French) Jean-Hugues Oppel, *Réveillez le président*, Éditions Payot et rivages, 2007 (ISBN 978-2-7436-1630-4). The book is a fiction about the nuclear weapons of France; the book also contains about ten chapters on true historical incidents involving nuclear weapons and strategy.

- Smyth, Henry DeWolf. *Atomic Energy for Military Purposes*. Princeton, NJ: Princeton University Press, 1945. (Smyth Report – the first declassified report by the US government on nuclear weapons)

- *The Effects of Nuclear War*. Office of Technology Assessment, May 1979.

- Rhodes, Richard. *Dark Sun: The Making of the Hydrogen Bomb*. New York: Simon and Schuster, 1995. ISBN 0-684-82414-0

- Rhodes, Richard. *The Making of the Atomic Bomb*. New York: Simon and Schuster, 1986 ISBN 0-684-81378-5

- Schultz, George P. and Goodby, James E. *The War that Must Never be Fought*, Hoover Press, 2015, ISBN 978-0-8179-1845-3.

- Weart, Spencer R. *Nuclear Fear: A History of Images.* Cambridge, MA: Harvard University Press, 1988. ISBN 0-674-62836-5

- Weart, Spencer R. *The Rise of Nuclear Fear.* Cambridge, MA: Harvard University Press, 2012. ISBN 0-674-05233-1

1.12 External links

- Nuclear Weapon Archive from Carey Sublette is a reliable source of information and has links to other sources and an informative FAQ.

- The Federation of American Scientists provide solid information on weapons of mass destruction, including nuclear weapons and their effects

- Alsos Digital Library for Nuclear Issues—contains many resources related to nuclear weapons, including a historical and technical overview and searchable bibliography of web and print resources.

- Video archive of US, Soviet, UK, Chinese and French Nuclear Weapon Testing at sonicbomb.com

- The National Museum of Nuclear Science & History (United States)—located in Albuquerque, New Mexico; a Smithsonian Affiliate Museum

- Nuclear Emergency and Radiation Resources

- The Manhattan Project: Making the Atomic Bomb at AtomicArchive.com

- Los Alamos National Laboratory: History (U.S. nuclear history)

- *Race for the Superbomb,* PBS website on the history of the H-bomb

- Recordings of recollections of the victims of Hiroshima and Nagasaki

- The Woodrow Wilson Center's Nuclear Proliferation International History Project or NPIHP is a global network of individuals and institutions engaged in the study of international nuclear history through archival documents, oral history interviews and other empirical sources.

- NUKEMAP3D - a 3D nuclear weapons effects simulator powerd by Google Maps.

Chapter 2

History of nuclear weapons

A nuclear fireball lights up the night in the United States nuclear test Upshot-Knothole Badger on April 18, 1953.

Nuclear weapons possess enormous destructive power from nuclear fission or combined fission and fusion reactions. Starting with scientific breakthroughs made during the 1930s, the United States, the United Kingdom and Canada collaborated during World War II in what was called the Manhattan Project to counter the suspected Nazi German atomic bomb project. In August 1945 two fission bombs were dropped on Japan. The Soviet Union started development shortly thereafter with their own atomic bomb project, and not long after that both countries developed even more powerful fusion weapons known as "hydrogen bombs."

2.1 Physics and politics in the 1930s and 1940s

See also: History of physics § 20th century: birth of modern physics

In the first decades of the 20th century, physics was revolutionised with developments in the understanding of the

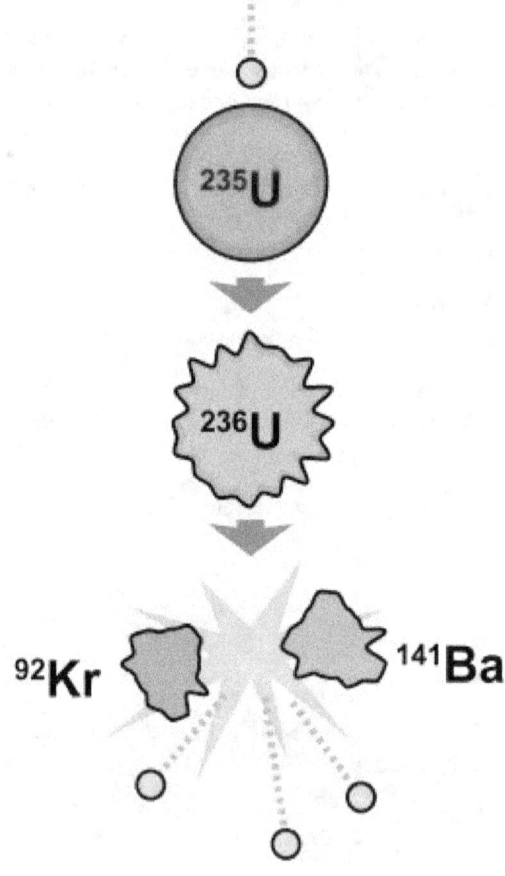

In nuclear fission, the nucleus of a fissile atom (in this case, enriched uranium) absorbs a thermal neutron, becomes unstable and splits into two new atoms, releasing some energy and between one and three new neutrons, which can perpetuate the process.

nature of atoms. In 1898, Pierre and Marie Curie discovered that pitchblende, an ore of uranium, contained a substance—which they named radium—that emitted large amounts of radioactivity. Ernest Rutherford and Frederick Soddy identified that atoms were breaking down and turning into different elements. Hopes were raised among scientists and laymen that the elements around us could contain tremendous amounts of unseen energy, waiting to be harnessed.

H. G. Wells was inspired to write about atomic weapons in a 1914 novel, *The World Set Free*, which appeared shortly before the First World War. In a 1924 article, Winston Churchill speculated about the possible military implications: "Might not a bomb no bigger than an orange be found to possess a secret power to destroy a whole block of buildings—nay to concentrate the force of a thousand tons of cordite and blast a township at a stroke?"[1]

In January 1933, Adolf Hitler was appointed Chancellor of Germany and it quickly became unsafe for Jewish scientists to remain in the country. Leó Szilárd fled to London where he proposed, and in 1934 patented, the idea of a nuclear chain reaction via neutrons. The patent also introduced the term critical mass to describe the minimum amount of material required to sustain the chain reaction and its potential to cause an explosion. (British patent 630,726.) He subsequently assigned the patent to the British Admiralty so that it could be covered by the Official Secrets Act.[2] In a very real sense, Szilárd was the father of the atomic bomb academically. Also in 1934, Irène and Frédéric Joliot-Curie discovered that artificial radioactivity could be induced in stable elements by bombarding them with alpha particles; Enrico Fermi reported similar results when bombarding uranium with neutrons.

In December 1938, Otto Hahn and Fritz Strassmann sent a manuscript to *Naturwissenschaften* reporting that they had detected the element barium after bombarding uranium with neutrons.[3] Lise Meitner and her nephew Otto Robert Frisch correctly interpreted these results as being due to the splitting of the uranium atom. (Frisch confirmed this experimentally on January 13, 1939.[4]) They gave the process the name "fission" because of its similarity to the splitting of a cell into two new cells.[5] Even before it was published, news of Meitner's and Frisch's interpretation crossed the Atlantic.[6] Scientists at Columbia University decided to replicate the experiment and on January 25, 1939, conducted the first nuclear fission experiment in the United States[7] in the basement of Pupin Hall.[8] The following year, they identified the active component of uranium as being the rare isotope uranium-235.[9]

Uranium appears in nature primarily in two isotopes: uranium-238 and uranium-235. When the nucleus of uranium-235 absorbs a neutron, it undergoes nuclear fission, releasing energy and, on average, 2.5 neutrons. Because uranium-235 releases more neutrons than it absorbs, it can support a chain reaction and so is described as fissile. Uranium-238, on the other hand, is not fissile as it does not normally undergo fission when it absorbs a neutron.

By the time Nazi Germany invaded Poland in 1939, beginning World War II, many of Europe's top scientists had already fled the imminent conflict. Physicists on both sides were well aware of the possibility of utilizing nuclear fission as a weapon, but no one was quite sure how it could be done. In August 1939, concerned that Germany might have its own project to develop fission-based weapons, Albert Einstein signed a letter to U.S. President Franklin D. Roosevelt warning him of the threat.[10] Roosevelt responded by setting up the Uranium Committee under Lyman James Briggs but, with little initial funding ($6,000), progress was slow. It was not until the Japanese attack on Pearl Harbor in December, 1941, that the U.S. decided to commit the necessary resources.[11]

Organized research first began in Britain as part of the Tube Alloys project. The Maud Committee was set up following the work of Frisch and Rudolf Peierls who calculated uranium-235's critical mass and found it to be much smaller than previously thought which meant that a deliverable bomb should be possible.[12] In the February 1940 Frisch–Peierls memorandum they stated that: "The energy liberated in the explosion of such a super-bomb...will, for an instant, produce a temperature comparable to that of the interior of the sun. The blast from such an explosion would destroy life in a wide area. The size of this area is difficult to estimate, but it will probably cover the centre of a big city."

Edgar Sengier, a director of Shinkolobwe Mine in Congo which produced by far the highest quality uranium ore in the world, had become aware of uranium's possible use in a bomb. In late 1940, fearful of its seizure by the Germans, he shipped the mine's entire stockpile of ore to a warehouse on Staten Island.[13]

For 18 months British research outpaced the American but by mid-1942, it became apparent that the industrial effort required was beyond Britain's already stretched wartime economy.[14]:204 In September 1942, General Leslie Groves was appointed to lead the U.S. project which became known as the Manhattan Project. Two of his first acts were to obtain authorization to assign the highest priority AAA rating on necessary procurements, and to put in train the purchase of all 1,250 tons of the Shinkolobwe ore.[13][15] The Tube Alloys project was quickly overtaken by the U.S. effort[14] and after Roosevelt and Churchill signed the Quebec Agreement in 1943, it was relocated and amalgamated into the Manhattan Project.

2.2 From Los Alamos to Hiroshima

Main article: Manhattan Project

With a scientific team led by J. Robert Oppenheimer, the Manhattan project brought together some of the top sci-

UC Berkeley physicist J. Robert Oppenheimer led the Allied scientific effort at Los Alamos.

entific minds of the day, including many exiles from Europe, with the production power of American industry for the goal of producing fission-based explosive devices before Germany. Britain and the U.S. agreed to pool their resources and information for the project, but the other Allied power, the Soviet Union (USSR), was not informed. The U.S. made an unprecedented investment in the project which at the time was the largest industrial enterprise ever seen,[14] spread across more than 30 sites in the U.S. and Canada. Scientific development was centralized in a secret laboratory at Los Alamos.

For a fission weapon to operate, there must be sufficient fissile material to support a chain reaction, a critical mass. To separate the fissile uranium-235 isotope from the non-fissile uranium-238, two methods were developed which took advantage of the fact that uranium-238 has a slightly greater atomic mass: electromagnetic separation and gaseous diffusion. Another secret site was erected at rural Oak Ridge, Tennessee, for the large-scale production and purification of the rare isotope, which required considerable investment. At the time, K-25, one of the Oak Ridge facilities, was the world's largest factory under one roof. The Oak Ridge site employed tens of thousands of people at its peak, most of whom had no idea what they were working on.

Although uranium-238 cannot be used for the initial stage of an atomic bomb, when it absorbs a neutron, it becomes uranium-239 which decays into neptunium-239, and finally the relatively stable plutonium-239, an element that does not exist naturally on Earth, but is fissile like uranium-235. After Fermi achieved the world's first sustained and controlled nuclear chain reaction with the creation of the first atomic pile, massive reactors were secretly constructed at what is now known as Hanford Site to transform uranium-238 into plutonium for a bomb.

The simplest form of nuclear weapon is a gun-type fission weapon, where a sub-critical mass would be shot at another sub-critical mass. The result would be a super-critical mass and an uncontrolled chain reaction that would create the desired explosion. The weapons envisaged in 1942 were the two gun-type weapons, Little Boy (uranium) and Thin Man (plutonium), and the Fat Man plutonium implosion bomb.

In early 1943 Oppenheimer determined that two projects should proceed forwards: the Thin Man project (plutonium gun) and the Fat Man project (plutonium implosion). The plutonium gun was to receive the bulk of the research effort, as it was the project with the most uncertainty involved. It was assumed that the uranium gun-type bomb could then be adapted from it.

In December 1943 the British mission of 19 scientists arrived in Los Alamos. Hans Bethe became head of the Theoretical Division.

In April 1944 it was found by Emilio Segrè that the plutonium-239 produced by the Hanford reactors had too high a level of background neutron radiation, and underwent spontaneous fission to a very small extent, due to the unexpected presence of plutonium-240 impurities. If such plutonium were used in a gun-type design, the chain reaction would start in the split second before the critical mass was fully assembled, blowing the weapon apart with a much lower yield than expected, in what is known as a fizzle.

As a result, development of Fat Man was given high priority. Chemical explosives were used to implode a sub-critical sphere of plutonium, thus increasing its density and making it into a critical mass. The difficulties with implosion centered on the problem of making the chemical explosives deliver a perfectly uniform shock wave upon the plutonium sphere— if it were even slightly asymmetric, the weapon would fizzle. This problem was solved by the use of explosive lenses which would focus the blast waves inside the imploding sphere, akin to the way in which an optical lens focuses light rays.[16]

After D-Day, General Groves ordered a team of scientists to follow eastward-moving victorious Allied troops into Europe to assess the status of the German nuclear program (and to prevent the westward-moving Soviets from gaining any materials or scientific manpower). They concluded that, while Germany had an atomic bomb program headed by Werner Heisenberg, the government had not made a significant investment in the project, and it had been nowhere near success.

Historians claim to have found a rough schematic showing a Nazi nuclear bomb.[17] In March 1945, a German scientific team was directed by the physicist Kurt Diebner to develop a primitive nuclear device in Ohrdruf, Thuringia.[17][18] Last ditch research was conducted in an experimental nuclear reactor at Haigerloch.

On April 12, after Roosevelt's death, Vice-President Harry S Truman assumed the presidency. At the time of the unconditional surrender of Germany on May 8, 1945, the Manhattan Project was still months away from producing a working weapon.

Because of the difficulties in making a working plutonium bomb, it was decided that there should be a test of the weapon. On July 16, 1945, in the desert north of Alamogordo, New Mexico, the first nuclear test took place, code-named "Trinity", using a device nicknamed "the gadget." The test, a plutonium implosion type device, released energy equivalent to 19 kilotons of TNT, far more powerful than any weapon ever used before. The news of the test's success was rushed to Truman at the Potsdam Conference, where Churchill was briefed and Soviet Premier Joseph Stalin was informed of the new weapon. On July 26, the Potsdam Declaration was issued containing an ultimatum for Japan: either surrender or suffer "complete and utter destruction", although nuclear weapons were not mentioned.[14]

After hearing arguments from scientists and military officers over the possible use of nuclear weapons against Japan (though some recommended using them as demonstrations in unpopulated areas, most recommended using them against built up targets, a euphemistic term for populated cities), Truman ordered the use of the weapons on Japanese cities, hoping it would send a strong message that would end in the capitulation of the Japanese leadership and avoid a lengthy invasion of the islands. On May 10–11, 1945, the Target Committee at Los Alamos, led by Oppenheimer, recommended Kyoto, Hiroshima, Yokohama, and Kokura as possible targets. Concerns about Kyoto's cultural heritage led to it being replaced by Nagasaki.

On August 6, 1945, a uranium-based weapon, Little Boy, was detonated above the Japanese city of Hiroshima. Three days later, a plutonium-based weapon, Fat Man, was detonated above the Japanese city of Nagasaki. The atomic bombing raids killed at least one hundred thousand Japanese civilians and military personnel outright, with the heat, radiation, and blast effects. Many tens of thousands would later die of radiation sickness and related cancers.[19][20] Truman promised a "rain of ruin" if Japan did not surrender immediately, threatening to systematically eliminate their ability to wage war.[21] On August 15, Emperor Hirohito announced Japan's surrender.[22]

2.3 Soviet atomic bomb project

Main article: Soviet atomic bomb project

The Soviet Union was not invited to share in the new weapons developed by the United States and the other Allies. During the war, information had been pouring in from a number of volunteer spies involved with the Manhattan Project (known in Soviet cables under the code-name of *Enormoz*), and the Soviet nuclear physicist Igor Kurchatov was carefully watching the Allied weapons development. It came as no surprise to Stalin when Truman had informed him at the Potsdam conference that he had a "powerful new weapon." Truman was shocked at Stalin's lack of interest.

The Soviet spies in the U.S. project were all volunteers and none were Soviet citizens. One of the most valuable, Klaus Fuchs, was a German émigré theoretical physicist who had been part of the early British nuclear efforts and the UK mission to Los Alamos. Fuchs had been intimately involved in the development of the implosion weapon, and passed on detailed cross-sections of the Trinity device to his Soviet contacts. Other Los Alamos spies—none of whom knew each other—included Theodore Hall and David Greenglass. The information was kept but not acted upon, as the Soviet Union was still too busy fighting the war in Europe to devote resources to this new project.

In the years immediately after World War II, the issue of who should control atomic weapons became a major international point of contention. Many of the Los Alamos scientists who had built the bomb began to call for "international control of atomic energy," often calling for either control by transnational organizations or the purposeful distribution of weapons information to all superpowers, but due to a deep distrust of the intentions of the Soviet Union, both in postwar Europe and in general, the policy-makers of the United States worked to attempt to secure an American nuclear monopoly.

A half-hearted plan for international control was proposed at the newly formed United Nations by Bernard Baruch (The Baruch Plan), but it was clear both to American commentators—and to the Soviets—that it was an attempt primarily to stymie Soviet nuclear efforts. The Soviets vetoed the plan, effectively ending any immediate postwar negotiations on atomic energy, and made overtures towards banning the use of atomic weapons in general.

The Soviets had put their full industrial might and manpower into the development of their own atomic weapons. The initial problem for the Soviets was primarily one of resources—they had not scouted out uranium resources in the Soviet Union and the U.S. had made deals to monopolise the largest known (and high purity) reserves in the Belgian Congo. The USSR used penal labour to mine the old deposits in Czechoslovakia—now an area under their control—and searched for other domestic deposits (which were eventually found).

Two days after the bombing of Nagasaki, the U.S. government released an official technical history of the Manhattan Project, authored by Princeton physicist Henry DeWolf Smyth, known colloquially as the Smyth Report. The sanitized summary of the wartime effort focused primarily on the production facilities and scale of investment, written in part to justify the wartime expenditure to the American public.

The Soviet program, under the suspicious watch of former NKVD chief Lavrenty Beria (a participant and victor in Stalin's Great Purge of the 1930s), would use the Report as a blueprint, seeking to duplicate as much as possible the American effort. The "secret cities" used for the Soviet equivalents of Hanford and Oak Ridge literally vanished from the maps for decades to come.

At the Soviet equivalent of Los Alamos, Arzamas-16, physicist Yuli Khariton led the scientific effort to develop the weapon. Beria distrusted his scientists, however, and he distrusted the carefully collected espionage information. As such, Beria assigned multiple teams of scientists to the same task without informing each team of the other's existence. If they arrived at different conclusions, Beria would bring them together for the first time and have them debate with their newfound counterparts. Beria used the espionage information as a way to double-check the progress of his scientists, and in his effort for duplication of the American project even rejected more efficient bomb designs in favor of ones that more closely mimicked the tried-and-true Fat Man bomb used by the U.S. against Nagasaki.

Working under a stubborn and scientifically ignorant administrator, the Soviet scientists struggled on. On August 29, 1949, the effort brought its results, when the USSR tested its first fission bomb, dubbed "Joe-1" by the U.S., years ahead of American predictions. The news of the first Soviet bomb was announced to the world first by the United States, which had detected the nuclear fallout it generated from its test site in Kazakhstan.

The loss of the American monopoly on nuclear weapons marked the first tit-for-tat of the nuclear arms race. The response in the U.S. was one of apprehension, fear, and scapegoating, which would lead eventually into the Red-baiting tactics of McCarthyism. Yet recent information from un-classified Venona intercepts and the opening of the KGB archives after the fall of the Soviet Union show that the USSR had useful spies that helped their program, although none were identified by McCarthy. Before this, though, President Truman announced a decision to begin a crash program that would develop a far more powerful weapon than those the U.S. used against Japan: the hydrogen bomb.

2.4 American developments after World War II

In 1946 Congress established the civilian Atomic Energy Commission (AEC) to take over the development of nuclear weapons from the military, and to develop nuclear power. The AEC made use of many private companies in processing uranium and thorium and in other urgent tasks related to the development of bombs. Many of these companies had very lax safety measures and employees were sometimes exposed to radiation levels far above what was allowed then or now.[23] In 1974, the Formerly Utilized Sites Remedial Action Program (FUSRAP) of the Army Corps of Engineers was set up to deal with contaminated sites left over from these operations.[24]

2.5 The first thermonuclear weapons

Main article: History of the Teller-Ulam design
The notion of using a fission weapon to ignite a process of nuclear fusion can be dated back to 1942. At the first major theoretical conference on the development of an atomic bomb hosted by J. Robert Oppenheimer at the University of California, Berkeley, participant Edward Teller directed the majority of the discussion towards Enrico Fermi's idea of a "Super" bomb that would use the same reactions that powered the Sun itself.

It was thought at the time that a fission weapon would be quite simple to develop and that perhaps work on a hydrogen bomb (thermonuclear weapon) would be possible to complete before the end of the Second World War. However, in reality the problem of a regular atomic bomb was large enough to preoccupy the scientists for the next few years, much less the more speculative "Super" bomb. Only Teller continued working on the project—against the will of project leaders Oppenheimer and Hans Bethe.

After the atomic bombings of Japan, many scientists at Los Alamos rebelled against the notion of creating a weapon thousands of times more powerful than the first atomic bombs. For the scientists the question was in part technical—the weapon design was still quite uncertain and unworkable—and in part moral: such a weapon, they argued, could only be used against large civilian populations, and could thus only be used as a weapon of genocide.

Many scientists, such as Bethe, urged that the United States should not develop such weapons and set an example towards the Soviet Union. Promoters of the weapon, including Teller, Ernest Lawrence, and Luis Alvarez, argued that such a development was inevitable, and to deny such protection to the people of the United States—especially when the Soviet Union was likely to create such a weapon themselves—was itself an immoral and unwise act.

Oppenheimer, who was now head of the General Advisory Committee of the successor to the Manhattan Project, the Atomic Energy Commission, presided over a recommendation against the development of the weapon. The reasons were in part because the success of the technology seemed limited at the time (and not worth the investment of resources to confirm whether this was so), and because Oppenheimer believed that the atomic forces of the United States would be more effective if they consisted of many large fission weapons (of which multiple bombs could be dropped on the same targets) rather than the large and unwieldy super bombs, for which there was a relatively limited number of targets of sufficient size to warrant such a development.

Furthermore, were such weapons developed by both the U.S. and the USSR, they would be more effectively used against the U.S. than by it, as the U.S. had far more regions of dense industrial and civilian activity as targets for large weapons than the Soviet Union.

In the end, President Truman made the final decision, looking for a proper response to the first Soviet atomic bomb test in 1949. On January 31, 1950, Truman announced a crash program to develop the hydrogen (fusion) bomb. At this point, however, the exact mechanism was still not known: the classical hydrogen bomb, whereby the *heat* of the fission bomb would be used to ignite the fusion material, seemed highly unworkable. However, an insight by Los Alamos mathematician Stanislaw Ulam showed that the fission bomb and the fusion fuel could be in separate parts of the bomb, and that *radiation* of the fission bomb could first work in a way to *compress* the fusion material before igniting it.

Teller pushed the notion further, and used the results of the boosted-fission "George" test (a boosted-fission device using a small amount of fusion fuel to boost the yield of a fission bomb) to confirm the fusion of heavy hydrogen elements before preparing for their first true multi-stage, Teller-Ulam hydrogen bomb test. Many scientists, initially against the weapon, such as Oppenheimer and Bethe, changed their previous opinions, seeing the development as being unstoppable.

The first fusion bomb was tested by the United States in *Operation Ivy* on November 1, 1952, on Elugelab Island in the Enewetak (or Eniwetok) Atoll of the Marshall Islands, code-named "Mike." Mike used liquid deuterium as its fusion fuel and a large fission weapon as its trigger. The device was a prototype design and not a deliverable weapon: standing over 20 ft (6 m) high and weighing at least 140,000 lb (64 t) (its refrigeration equipment added an additional 24,000 lb (11,000 kg) as well), it could not have been dropped from even the largest planes.

Its explosion yielded energy equivalent to 10.4 megatons of TNT—over 450 times the power of the bomb dropped onto Nagasaki— and obliterated Elugelab, leaving an underwater crater 6240 ft (1.9 km) wide and 164 ft (50 m) deep where the island had once been. Truman had initially tried to create a media blackout about the test—hoping it would not become an issue in the upcoming presidential election—but on January 7, 1953, Truman announced the development of the hydrogen bomb to the world as hints and speculations of it were already beginning to emerge in the press.

Not to be outdone, the Soviet Union exploded its first thermonuclear device, designed by the physicist Andrei Sakharov, on August 12, 1953, labeled "Joe-4" by the West. This created concern within the U.S. government and military, because, unlike Mike, the Soviet device was a deliverable weapon, which the U.S. did not yet have. This first device though was arguably not a true hydrogen bomb, and could only reach explosive yields in the hundreds of kilotons (never reaching the megaton range of a staged weapon). Still, it was a powerful propaganda tool for the Soviet Union, and the technical differences were fairly oblique to the American public and politicians.

Following the Mike blast by less than a year, Joe-4 seemed to validate claims that the bombs were inevitable and vindicate those who had supported the development of the fusion program. Coming during the height of McCarthyism, the effect was pronounced on the security hearings in early 1954, which revoked former Los Alamos director Robert Oppenheimer's security clearance on the grounds that he was unreliable, had not supported the American hydrogen bomb program, and had made long-standing left-wing ties in the 1930s. Edward Teller participated in the hearing as the only major scientist to testify against Oppenheimer, resulting in his virtual expulsion from the physics community.

On March 1, 1954, the U.S. detonated its first practical thermonuclear weapon (which used isotopes of lithium as its fusion fuel), known as the "Shrimp" device of the Castle Bravo test, at Bikini Atoll, Marshall Islands. The device yielded 15 megatons, more than twice its expected yield, and became the worst radiological disaster in U.S. history. The combination of the unexpectedly large blast and poor weather conditions caused a cloud of radioactive nuclear fallout to contaminate over 7,000 square miles (18,000 km²). 239 Marshall Island natives and 28 Americans were exposed to significant amounts of radiation, resulting in elevated levels of cancer and birth defects in the years to come.[25]

The crew of the Japanese tuna-fishing boat *Lucky Dragon 5*, who had been fishing just outside the exclusion zone, returned to port suffering from radiation sickness and skin burns; one crew member was terminally ill. Efforts were made to recover the cargo of contaminated fish but at least two large tuna were probably sold and eaten. A further 75 tons of tuna caught between March and December were found to be unfit for human consumption. When the crew member died and the full results of the contamination were made public by the U.S., Japanese concerns were reignited about the hazards of radiation.[26]

The hydrogen bomb age had a profound effect on the thoughts of nuclear war in the popular and military mind. With only fission bombs, nuclear war was something that possibly could be limited. Dropped by planes and only able to destroy the most built up areas of major cities, it was possible for many to look at fission bombs as a technological extension of large-scale conventional bombing—such as the extensive firebombing of German and Japanese cities during World War II. Proponents brushed aside as grave exaggeration claims that such weapons could lead to worldwide death or harm.

Even in the decades before fission weapons, there had been speculation about the possibility for human beings to end all life on the planet, either by accident or purposeful maliciousness—but technology had not provided the capacity for such action. The great power of hydrogen bombs made world-wide annihilation possible.

The Castle Bravo incident itself raised a number of questions about the survivability of a nuclear war. Government scientists in both the U.S. and the USSR had insisted that fusion weapons, unlike fission weapons, were cleaner, as fusion reactions did not produce the dangerously radioactive by-products of fission reactions. While technically true, this hid a more gruesome point: the last stage of a multi-staged hydrogen bomb often used the neutrons produced by the fusion reactions to induce fissioning in a jacket of natural uranium, and provided around half of the yield of the device itself.

This fission stage made fusion weapons considerably more dirty than they were made out to be. This was evident in the towering cloud of deadly fallout that followed the *Bravo* test. When the Soviet Union tested its first megaton device in 1955, the possibility of a *limited* nuclear war seemed even more remote in the public and political mind. Even cities and countries that were not direct targets would suffer

fallout contamination. Extremely harmful fission products would disperse via normal weather patterns and embed in soil and water around the planet.

Speculation began to run towards what fallout and dust from a full-scale nuclear exchange would do to the world as a whole, rather than just cities and countries directly involved. In this way, the fate of the world was now tied to the fate of the bomb-wielding superpowers.

2.6 Deterrence and brinkmanship

Main articles: Nuclear testing, Nuclear strategy, and Nuclear warfare
Throughout the 1950s and the early 1960s the U.S. and the USSR both endeavored, in a tit-for-tat approach, to prevent the other power from acquiring nuclear supremacy. This had massive political and cultural effects during the Cold War.

The first atomic bombs dropped on Hiroshima and Nagasaki were large, custom-made devices, requiring highly trained personnel for their arming and deployment. They could be dropped only from the largest bomber planes—at the time the B-29 Superfortress—and each plane could only carry a single bomb in its hold.

The first hydrogen bombs were similarly massive and complicated. This ratio of one plane to one bomb was still fairly impressive in comparison with conventional, non-nuclear weapons, but against other nuclear-armed countries it was considered a grave danger. In the immediate postwar years, the U.S. expended much effort on making the bombs "G.I.-proof"—capable of being used and deployed by members of the U.S. Army, rather than Nobel Prize–winning scientists. In the 1950s, the U.S. undertook a nuclear testing program to improve the nuclear arsenal.

Starting in 1951, the Nevada Test Site (in the Nevada desert) became the primary location for all U.S. nuclear testing (in the USSR, Semipalatinsk Test Site in Kazakhstan served a similar role). Tests were divided into two primary categories: "weapons related" (verifying that a new weapon worked or looking at exactly how it worked) and "weapons effects" (looking at how weapons behaved under various conditions or how structures behaved when subjected to weapons).

In the beginning, almost all nuclear tests were either atmospheric (conducted above ground, in the atmosphere) or underwater (such as some of the tests done in the Marshall Islands). Testing was used as a sign of both national and technological strength, but also raised questions about the safety of the tests, which released nuclear fallout into the atmosphere (most dramatically with the Castle Bravo test in 1954, but in more limited amounts with almost all atmospheric nuclear testing).

Because testing was seen as a sign of technological development (the ability to design usable weapons without some form of testing was considered dubious), halts on testing were often called for as stand-ins for halts in the nuclear arms race itself, and many prominent scientists and statesmen lobbied for a ban on nuclear testing. In 1958, the U.S., USSR, and the United Kingdom (a new nuclear power) declared a temporary testing moratorium for both political and health reasons, but by 1961 the Soviet Union had broken the moratorium and both the USSR and the U.S. began testing with great frequency.

As a show of political strength, the Soviet Union tested the largest-ever nuclear weapon in October 1961, the massive Tsar Bomba, which was tested in a reduced state with a yield of around 50 megatons—in its full state it was estimated to have been around 100 Mt. The weapon was largely impractical for actual military use, but was hot enough to induce third-degree burns at a distance of 62 mi (100 km) away. In its full, dirty, design it would have increased the amount of worldwide fallout since 1945 by 25%.

In 1963, all nuclear and many non-nuclear states signed the Limited Test Ban Treaty, pledging to refrain from testing nuclear weapons in the atmosphere, underwater, or in outer space. The treaty permitted underground tests.

Most tests were considerably more modest, and worked for direct technical purposes as well as their potential political overtones. Weapons improvements took on two primary forms. One was an increase in efficiency and power, and within only a few years fission bombs were developed that were many times more powerful than the ones created during World War II. The other was a program of miniaturization, reducing the size of the nuclear weapons.

Smaller bombs meant that bombers could carry more of them, and also that they could be carried on the new generation of rockets in development in the 1950s and 1960s. U.S. rocket science received a large boost in the postwar years, largely with the help of engineers acquired from the Nazi rocketry program. These included scientists such as Wernher von Braun, who had helped design the V-2 rockets the Nazis launched across the English Channel. An American program, Project Paperclip, had endeavored to move German scientists into American hands (and away from Soviet hands) and put them to work for the U.S.

2.7 Weapons improvement

Early nuclear-tipped rockets—such as the MGR-1 Honest John, first deployed by the U.S. in 1953—were surface-to-surface missiles with relatively short ranges (around 15 mi/25 km maximum) and yields around twice the size of the first fission weapons. The limited range meant they could only be used in certain types of military situations. U.S. rockets could not, for example, threaten Moscow with an immediate strike, and could only be used as tactical weapons (that is, for small-scale military situations).

Strategic weapons—weapons that could threaten an entire country—relied, for the time being, on long-range bombers that could penetrate deep into enemy territory. In the U.S., this requirement led, in 1946, to creation of the Strategic Air Command—a system of bombers headed by General Curtis LeMay (who previously presided over the firebombing of Japan during WWII). In operations like Chrome Dome, SAC kept nuclear-armed planes in the air 24 hours a day, ready for an order to attack Moscow.

These technological possibilities enabled nuclear strategy to develop a logic considerably different from previous military thinking. Because the threat of nuclear warfare was so awful, it was first thought that it might make any war of the future impossible. President Dwight D. Eisenhower's doctrine of "massive retaliation" in the early years of the Cold War was a message to the USSR, saying that if the Red Army attempted to invade the parts of Europe not given to the Eastern bloc during the Potsdam Conference (such as West Germany), nuclear weapons would be used against the Soviet troops and potentially the Soviet leaders.

With the development of more rapid-response technologies (such as rockets and long-range bombers), this policy began to shift. If the Soviet Union also had nuclear weapons and a policy of "massive retaliation" was carried out, it was reasoned, then any Soviet forces not killed in the initial attack, or launched while the attack was ongoing, would be able to serve their own form of nuclear retaliation against the U.S. Recognizing that this was an undesirable outcome, military officers and game theorists at the RAND think tank developed a nuclear warfare strategy that was eventually called Mutually Assured Destruction (MAD).

MAD divided potential nuclear war into two stages: *first strike* and *second strike*. First strike meant the first use of nuclear weapons by one nuclear-equipped nation against another nuclear-equipped nation. If the attacking nation did not prevent the attacked nation from a nuclear response, the attacked nation would respond with a second strike against the attacking nation. In this situation, whether the U.S. first attacked the USSR or the USSR first attacked the U.S., the end result would be that both nations would be damaged to the point of utter social collapse.

According to game theory, because starting a nuclear war was suicidal, no logical country would shoot first. However, if a country could launch a first strike that utterly destroyed the target country's ability to respond, that might give that country the confidence to initiate a nuclear war. The object of a country operating by the MAD doctrine is to deny the opposing country this first strike capability.

MAD played on two seemingly opposed modes of thought: cold logic and emotional fear. The English phrase MAD was often known by, "nuclear deterrence," was translated by the French as "dissuasion," and "terrorization" by the Soviets. This apparent paradox of nuclear war was summed up by British Prime Minister Winston Churchill as "the worse things get, the better they are"—the greater the threat of mutual destruction, the safer the world would be.

This philosophy made a number of technological and political demands on participating nations. For one thing, it said that it should always be assumed that an enemy nation may be trying to acquire first strike capability, which must always be avoided. In American politics this translated into demands to avoid "bomber gaps" and "missile gaps" where the Soviet Union could potentially outshoot the Americans. It also encouraged the production of thousands of nuclear weapons by both the U.S. and the USSR, far more than needed to simply destroy the major civilian and military infrastructures of the opposing country. These policies and strategies were satirized in the 1964 Stanley Kubrick film Dr. Strangelove, in which the Soviets, unable to keep up with the US's first strike capability, instead plan for MAD by building a Doomsday Machine, and thus, after a (literally) mad US General orders a nuclear attack on the USSR, the end of the world is brought about.

The policy also encouraged the development of the first early warning systems. Conventional war, even at its fastest, was fought over days and weeks. With long-range bombers, from the start of a nuclear attack to its conclusion was mere hours. Rockets could reduce a conflict to minutes. Planners reasoned that conventional command and control systems could not adequately react to a nuclear attack, so great lengths were taken to develop computer systems that could look for enemy attacks and direct rapid responses.

The U.S., poured massive funding into development of SAGE, a system that could track and intercept enemy bomber aircraft using information from remote radar stations. It was the first computer system to feature real-time processing, multiplexing, and display devices. It was the first general computing machine, and a direct predecessor of modern computers.

2.8 Emergence of the anti-nuclear movement

Main article: History of the anti-nuclear movement
The atomic bombings of Hiroshima and Nagasaki and the

end of World War II quickly followed the 1945 Trinity nuclear test, and the Little Boy device was detonated over the Japanese city of Hiroshima on 6 August 1945. Exploding with a yield equivalent to 12,500 tonnes of TNT, the blast and thermal wave of the bomb destroyed nearly 50,000 buildings and killed approximately 75,000 people.[27] Subsequently, the world's nuclear weapons stockpiles grew.[28]

Operation Crossroads was a series of nuclear weapon tests conducted by the United States at Bikini Atoll in the Pacific Ocean in the summer of 1946. Its purpose was to test the effect of nuclear weapons on naval ships. To prepare the Bikini atoll for the nuclear tests, Bikini's native residents were evicted from their homes and resettled on smaller, uninhabited islands where they were unable to sustain themselves.[29]

National leaders debated the impact of nuclear weapons on domestic and foreign policy. Also involved in the debate about nuclear weapons policy was the scientific community, through professional associations such as the Federation of Atomic Scientists and the Pugwash Conference on Science and World Affairs.[30] Radioactive fallout from nuclear weapons testing was first drawn to public attention in 1954 when a Hydrogen bomb test in the Pacific contaminated the crew of the Japanese fishing boat Lucky Dragon.[31] One of the fishermen died in Japan seven months later. The incident caused widespread concern around the world and "provided a decisive impetus for the emergence of the anti-nuclear weapons movement in many countries".[31] The anti-nuclear weapons movement grew rapidly because for many people the atomic bomb "encapsulated the very worst direction in which society was moving".[32]

Peace movements emerged in Japan and in 1954 they converged to form a unified "Japanese Council Against Atomic and Hydrogen Bombs". Japanese opposition to the Pacific nuclear weapons tests was widespread, and "an estimated 35 million signatures were collected on petitions calling for bans on nuclear weapons".[32] The Russell–Einstein Manifesto was issued in London on July 9, 1955 by Bertrand Russell in the midst of the Cold War. It highlighted the dangers posed by nuclear weapons and called for world leaders to seek peaceful resolutions to international conflict. The signatories included eleven pre-eminent intellectuals and scientists, including Albert Einstein, who signed it just days before his death on April 18, 1955. A few days after the release, philanthropist Cyrus S. Eaton offered to sponsor a conference—called for in the manifesto—in Pugwash, Nova Scotia, Eaton's birthplace. This conference was to be the first of the Pugwash Conferences on Science and World Affairs, held in July 1957.

In the United Kingdom, the first Aldermaston March organised by the Campaign for Nuclear Disarmament took place at Easter 1958, when several thousand people marched for four days from Trafalgar Square, London, to the Atomic Weapons Research Establishment close to Aldermaston in Berkshire, England, to demonstrate their opposition to nuclear weapons.[33][34] The Aldermaston marches continued into the late 1960s when tens of thousands of people took part in the four-day marches.[32]

In 1959, a letter in the *Bulletin of Atomic Scientists* was the start of a successful campaign to stop the Atomic Energy Commission dumping radioactive waste in the sea 19 kilometres from Boston.[35] On November 1, 1961, at the height of the Cold War, about 50,000 women brought together by Women Strike for Peace marched in 60 cities in the United States to demonstrate against nuclear weapons. It was the largest national women's peace protest of the 20th century.[36][37]

In 1958, Linus Pauling and his wife presented the United Nations with the petition signed by more than 11,000 scientists calling for an end to nuclear-weapon testing. The "Baby Tooth Survey," headed by Dr Louise Reiss, demonstrated conclusively in 1961 that above-ground nuclear testing posed significant public health risks in the form of radioactive fallout spread primarily via milk from cows that had ingested contaminated grass.[38][39][40] Public pressure and the research results subsequently led to a moratorium on above-ground nuclear weapons testing, followed by the Partial Test Ban Treaty, signed in 1963 by John F. Kennedy and Nikita Khrushchev.[30][41][42]

2.9 Cuban Missile Crisis

Main article: Cuban Missile Crisis
Bombers and short-range rockets were not reliable: planes could be shot down, and earlier nuclear missiles could cover only a limited range— for example, the first Soviet rockets' range limited them to targets in Europe. However, by the 1960s, both the United States and the Soviet Union had developed intercontinental ballistic missiles, which could be launched from extremely remote areas far away from their target. They had also developed submarine-launched ballistic missiles, which had less range but could be launched from submarines very close to the target without any radar warning. This made any national protection from nuclear missiles increasingly impractical.

The military realities made for a precarious diplomatic situation. The international politics of brinkmanship led leaders to exclaim their willingness to participate in a nuclear war rather than concede any advantage to their opponents, feeding public fears that their generation may be the last. Civil defense programs undertaken by both superpowers, exemplified by the construction of fallout shelters and urging civilians about the survivability of nuclear war, did little

to ease public concerns.

The climax of brinksmanship came in early 1962, when an American U-2 spy plane photographed a series of launch sites for medium-range ballistic missiles being constructed on the island of Cuba, just off the coast of the southern United States, beginning what became known as the Cuban Missile Crisis. The U.S. administration of John F. Kennedy concluded that the Soviet Union, then led by Nikita Khrushchev, was planning to station Soviet nuclear missiles on the island, which was under the control of communist Fidel Castro. On October 22, Kennedy announced the discoveries in a televised address. He announced a naval blockade around Cuba that would turn back Soviet nuclear shipments, and warned that the military was prepared "for any eventualities." The missiles had 2,400 mile (4,000 km) range, and would allow the Soviet Union to quickly destroy many major American cities on the Eastern Seaboard if a nuclear war began.

The leaders of the two superpowers stood nose to nose, seemingly poised over the beginnings of a third world war. Khrushchev's ambitions for putting the weapons on the island were motivated in part by the fact that the U.S. had stationed similar weapons in Britain, Italy, and nearby Turkey, and had previously attempted to sponsor an invasion of Cuba the year before in the failed Bay of Pigs Invasion. On October 26, Khrushchev sent a message to Kennedy offering to withdraw all missiles if Kennedy committed to a policy of no future invasions of Cuba. Khrushchev worded the threat of assured destruction eloquently:

> "You and I should not now pull on the ends of the rope in which you have tied a knot of war, because the harder you and I pull, the tighter the knot will become. And a time may come when this knot is tied so tight that the person who tied it is no longer capable of untying it, and then the knot will have to be cut. What that would mean I need not explain to you, because you yourself understand perfectly what dreaded forces our two countries possess."

A day later, however, the Soviets sent another message, this time demanding that the U.S. remove its missiles from Turkey before any missiles were withdrawn from Cuba. On the same day, a U-2 plane was shot down over Cuba and another almost intercepted over the Soviet Union, as Soviet merchant ships neared the quarantine zone. Kennedy responded by accepting the first deal publicly, and sending his brother Robert to the Soviet embassy to accept the second deal privately. On October 28, the Soviet ships stopped at the quarantine line and, after some hesitation, turned back towards the Soviet Union. Khrushchev announced that he had ordered the removal of all missiles in Cuba, and U.S.

Secretary of State Dean Rusk was moved to comment, "We went eyeball to eyeball, and the other fellow just blinked."

The Crisis was later seen as the closest the U.S. and the USSR ever came to nuclear war and had been narrowly averted by last-minute compromise by both superpowers. Fears of communication difficulties led to the installment of the first hotline, a direct link between the superpowers that allowed them to more easily discuss future military activities and political maneuverings. It had been made clear that missiles, bombers, submarines, and computerized firing systems made escalating any situation to Armageddon far more easy than anybody desired.

After stepping so close to the brink, both the U.S. and the USSR worked to reduce their nuclear tensions in the years immediately following. The most immediate culmination of this work was the signing of the Partial Test Ban Treaty in 1963, in which the U.S. and USSR agreed to no longer test nuclear weapons in the atmosphere, underwater, or in outer space. Testing underground continued, allowing for further weapons development, but the worldwide fallout risks were purposefully reduced, and the era of using massive nuclear tests as a form of saber-rattling ended.

In December 1979, NATO decided to deploy cruise and Pershing II missiles in Western Europe in response to Soviet deployment of intermediate range mobile missiles, and in the early 1980s, a "dangerous Soviet-US nuclear confrontation" arose.[43] In New York on June 12, 1982, one million people gathered to protest about nuclear weapons, and to support the second UN Special Session on Disarmament.[44][45] As the nuclear abolitionist movement grew, there were many protests at the Nevada Test Site. For example, on February 6, 1987, nearly 2,000 demonstrators, including six members of Congress, protested against nuclear weapons testing and more than 400 people were arrested.[46] Four of the significant groups organizing this renewal of anti-nuclear activism were Greenpeace, The American Peace Test, The Western Shoshone, and Nevada Desert Experience.

There have been at least four major false alarms, the most recent in 1995, that resulted in the activation of nuclear attack early warning protocols. They include the accidental loading of a training tape into the American early-warning computers; a computer chip failure that appeared to show a random number of attacking missiles; a rare alignment of the Sun, the U.S. missile fields and a Soviet early-warning satellite that caused it to confuse high-altitude clouds with missile launches; the launch of a Norwegian research rocket resulted in President Yeltsin activating his nuclear briefcase for the first time.[47]

2.10 Initial proliferation

In the fifties and sixties, three more countries joined the "nuclear club." The United Kingdom had been an integral part of the Manhattan Project following the Quebec Agreement in 1943. The passing of the McMahon Act by the United States in 1946 unilaterally broke this partnership and prevented the passage of any further information to the United Kingdom. The British Government, under Clement Attlee, determined that a British Bomb was essential. Because of British involvement in the Manhattan Project, Britain had extensive knowledge in some areas, but not in others.

An improved version of 'Fat Man' was developed, and on 26 February 1952, Prime Minister Winston Churchill announced that the United Kingdom also had an atomic bomb and a successful test took place on 3 October 1952. At first these were free-fall bombs, intended for use by the V Force of jet bombers. A Vickers Valiant dropped the first UK nuclear weapon on 11 October 1956 at Maralinga, South Australia. Later came a missile, Blue Steel, intended for carriage by the V Force bombers, and then the Blue Streak medium-range ballistic missile (later canceled). Anglo-American cooperation on nuclear weapons was restored by the 1958 US-UK Mutual Defence Agreement. As a result of this and the Polaris Sales Agreement, the United Kingdom has bought United States designs for submarine missiles and fitted its own warheads. It retains full independent control over the use of the missiles. It no longer possesses any free-fall bombs.

France had been heavily involved in nuclear research before World War II through the work of the Joliot-Curies. This was discontinued after the war because of the instability of the Fourth Republic and lack of finances.[48] However, in the 1950s, France launched a civil nuclear research program, which produced plutonium as a byproduct.

In 1956, France formed a secret Committee for the Military Applications of Atomic Energy and a development program for delivery vehicles. With the return of Charles de Gaulle to the French presidency in 1958, final decisions to build a bomb were made, which led to a successful test in 1960. Since then, France has developed and maintained its own nuclear deterrent independent of NATO.

In 1951, China and the Soviet Union signed an agreement whereby China supplied uranium ore in exchange for technical assistance in producing nuclear weapons. In 1953, China established a research program under the guise of civilian nuclear energy. Throughout the 1950s the Soviet Union provided large amounts of equipment. But as the relations between the two countries worsened the Soviets reduced the amount of assistance and, in 1959, refused to donate a bomb for copying purposes. Despite this, the Chinese made rapid progress and tested an atomic bomb on October 16, 1964, at Lop Nur. They tested a nuclear missile on October 25, 1966, and a hydrogen bomb on June 14, 1967.

Chinese nuclear warheads were produced from 1968 and thermonuclear warheads from 1974.[49] It is also thought that Chinese warheads have been successfully miniaturised from 2200 kg to 700 kg through the use of designs obtained by espionage from the United States. The current number of weapons is unknown owing to strict secrecy, but it is thought that up to 2000 warheads may have been produced, though far fewer may be available for use. China is the only nuclear weapons state to have guaranteed the non-first use of nuclear weapons.

2.11 Cold War

Main article: Cold War

After World War II, the balance of power between the Eastern and Western blocs and the fear of global destruction prevented the further military use of atomic bombs. This fear was even a central part of Cold War strategy, referred to as the doctrine of Mutually Assured Destruction. So important was this balance to international political stability that a treaty, the Anti-Ballistic Missile Treaty (or ABM treaty), was signed by the U.S. and the USSR in 1972 to curtail the development of defenses against nuclear weapons and the ballistic missiles that carry them. This doctrine resulted in a large increase in the number of nuclear weapons, as each side sought to ensure it possessed the firepower to destroy the opposition in all possible scenarios.

Early delivery systems for nuclear devices were primarily bombers like the United States B-29 Superfortress and Convair B-36, and later the B-52 Stratofortress. Ballistic missile systems, based on Wernher von Braun's World War II designs (specifically the V-2 rocket), were developed by both United States and Soviet Union teams (in the case of the U.S., effort was directed by the German scientists and engineers although the Soviet Union also made extensive use of captured German scientists, engineers, and technical data).

These systems were used to launch satellites, such as Sputnik, and to propel the Space Race, but they were primarily developed to create Intercontinental Ballistic Missiles (ICBMs) that could deliver nuclear weapons anywhere on the globe. Development of these systems continued throughout the Cold War—though plans and treaties, beginning with the Strategic Arms Limitation Treaty (SALT I), restricted deployment of these systems until, after the fall of the Soviet Union, system development essentially halted, and many weapons were disabled and destroyed. On Jan-

uary 27, 1967, more than 60 nations signed the Outer Space Treaty, banning nuclear weapons in space.

There have been a number of potential nuclear disasters. Following air accidents U.S. nuclear weapons have been lost near Atlantic City, New Jersey (1957); Savannah, Georgia (1958) (see Tybee Bomb); Goldsboro, North Carolina (1961); off the coast of Okinawa (1965); in the sea near Palomares, Spain (1966) (see 1966 Palomares B-52 crash); and near Thule, Greenland (1968) (see 1968 Thule Air Base B-52 crash). Most of the lost weapons were recovered, the Spanish device after three months' effort by the DSV Alvin and DSV Aluminaut.

The Soviet Union was less forthcoming about such incidents, but the environmental group Greenpeace believes that there are around forty non-U.S. nuclear devices that have been lost and not recovered, compared to eleven lost by America, mostly in submarine disasters. The U.S. has tried to recover Soviet devices, notably in the 1974 Project Azorian using the specialist salvage vessel *Hughes Glomar Explorer* to raise a Soviet submarine. After news leaked out about this boondoggle, the CIA would coin a favorite phrase for refusing to disclose sensitive information, called glomarization: *We can neither confirm nor deny the existence of the information requested but, hypothetically, if such data were to exist, the subject matter would be classified, and could not be disclosed.*[50]

The collapse of the Soviet Union in 1991 essentially ended the Cold War. However, the end of the Cold War failed to end the threat of nuclear weapon use, although global fears of nuclear war reduced substantially. In a major move of symbolic de-escalation, Boris Yeltsin, on January 26, 1992, announced that Russia planned to stop targeting United States cities with nuclear weapons.

2.12 Cost

The designing, testing, producing, deploying, and defending against nuclear weapons is one of the largest expenditures for the nations which possess nuclear weapons. In the United States during the Cold War years, between "one quarter to one third of all military spending since World War II [was] devoted to nuclear weapons and their infrastructure." [51] According to a retrospective Brookings Institution study published in 1998 by the Nuclear Weapons Cost Study Committee (formed in 1993 by the W. Alton Jones Foundation), the total expenditures for U.S. nuclear weapons from 1940 to 1998 was $5.5 trillion in 1996 Dollars.[52] The total public debt at the end of fiscal year 1998 was $5,478,189,000,000 in 1998 Dollars[53] or $5.3 trillion in 1996 Dollars. The *entire public debt* in 1998 was therefore equal to the cost of research, development, and deployment of U.S. nuclear weapons and nuclear weapons-related programs during the Cold War.[51][52][54]

2.13   Second nuclear age

See also: List of states with nuclear weapons

The *second nuclear age* can be regarded as proliferation of nuclear weapons among lesser powers and for reasons other than the American-Soviet-Chinese rivalry.

India embarked relatively early on a program aimed at nuclear weapons capability, but apparently accelerated this after border war with China in 1962. India's first atomic-test explosion was in 1974 with *Smiling Buddha*, which it described as a "peaceful nuclear explosion."

After the collapse of Eastern Military High Command and the disintegration of Pakistan as a result of the 1971 Winter war, Bhutto of Pakistan launched scientific research on nuclear weapons. The Indian test caused Pakistan to spur its programme, and the ISI conducted successful espionage operations in the Netherlands, while also developing the programme indigenously. India tested fission and perhaps fusion devices in 1998, and Pakistan successfully tested fission devices that same year, raising concerns that they would use nuclear weapons on each other.

All of the former Soviet bloc countries with nuclear weapons (Belarus, Ukraine, and Kazakhstan) returned their warheads to Russia by 1996.

South Africa also had an active program to develop uranium-based nuclear weapons, but dismantled its nuclear weapon program in the 1990s.[55] Experts do not believe it actually tested such a weapon, though it later claimed it constructed several crude devices that it eventually dismantled. In the late 1970s American spy satellites detected a "brief, intense, double flash of light near the southern tip of Africa."[56] Known as the Vela Incident, it was speculated to have been a South African or possibly Israeli nuclear weapons test, though some feel that it may have been caused by natural events or a detector malfunction.

Israel is widely believed to possess an arsenal of up to several hundred nuclear warheads, but this has never been officially confirmed or denied (though the existence of their Dimona nuclear facility was confirmed by Mordechai Vanunu in 1986).

In January 2004, Dr A. Q. Khan of Pakistan's programme confessed to having been a key mover in "proliferation activities",[57] seen as part of an international proliferation network of materials, knowledge, and machines from Pakistan to Libya, Iran, and North Korea.

North Korea announced in 2003 that it also had several nuclear explosives though it has not been confirmed and the validity of this has been a subject of scrutiny amongst weapons experts. The first claimed detonation of a nuclear weapon by the Democratic People's Republic of Korea was the 2006 North Korean nuclear test, conducted on October 9, 2006. On May 25, 2009, North Korea continued nuclear testing, violating United Nations Security Council Resolution 1718. A third test was conducted on 13 February 2013.

In Iran, Ayatollah Ali Khamenei issued a fatwa forbidding the production, stockpiling and use of nuclear weapons on August 9, 2005. The full text of the fatwa was released in an official statement at the meeting of the International Atomic Energy Agency (IAEA) in Vienna.[58] Despite this, however, there is mounting concern in many nations about Iran's refusal to halt its nuclear power program, which many (including some members of the US government) fear is a cover for weapons development.

2.14   See also

- ☞ *The Atomic Age* – Wikipedia book
- International Day against Nuclear Tests
- Japanese nuclear weapon program
- List of nuclear tests
- List of nuclear weapons
- Timeline of nuclear weapons development
- National Response Scenario Number One
- Nuclear weapon design
- Project-706
- Psychic numbing#Nuclear denial disorder
- The Bomb (film)
- Weapons of Mass Destruction

2.15   References

[1] Paul Kent Alkon (2006). *Winston Churchill's Imagination*. Associated University Presse. pp. 156–. ISBN 978-0-8387-5632-4. Retrieved 5 June 2013.

[2] Michael F. L'Annunziata (23 August 2007). *Radioactivity: Introduction and History*. Elsevier. pp. 240–. ISBN 978-0-08-054888-3. Retrieved 6 June 2013.

[3] O. Hahn and F. Strassmann *Über den Nachweis und das Verhalten der bei der Bestrahlung des Urans mittels Neutronen entstehenden Erdalkalimetalle (On the detection and characteristics of the alkaline earth metals formed by irradiation of uranium with neutrons)*, Naturwissenschaften Volume 27, Number 1, 11-15 (1939). The authors were identified as being at the Kaiser-Wilhelm-Institut für Chemie, Berlin-Dahlem. Received December 22, 1938.

[4] O. R. Frisch *Physical Evidence for the Division of Heavy Nuclei under Neutron Bombardment, Nature*, Volume 143, Number 3616, 276-276 (18 February 1939). The paper is dated 17 January 1939. The experiment for this letter to the editor was conducted on 13 January 1939; see Richard Rhodes *The Making of the Atomic Bomb* 263 and 268 (Simon and Schuster, 1986).

[5] Lise Meitner and O. R. Frisch *Disintegration of Uranium by Neutrons: a New Type of Nuclear Reaction, Nature*, Volume 143, Number 3615, 239-240 (11 February 1939). The paper is dated January 16, 1939. Meitner is identified as being at the Physical Institute, Academy of Sciences, Stockholm. Frisch is identified as being at the Institute of Theoretical Physics, University of Copenhagen.

[6] Richard Rhodes *The Making of the Atomic Bomb* 268 (Simon and Schuster, 1986).

[7] H. L. Anderson, E. T. Booth, J. R. Dunning, E. Fermi, G. N. Glasoe, and F. G. Slack *The Fission of Uranium, Phys. Rev.* Volume 55, Number 5, 511 - 512 (1839). Institutional citation: Pupin Physics Laboratories, Columbia University, New York, New York. Received February 16, 1939.

[8] Richard Rhodes *The Making of the Atomic Bomb* 267-270 (Simon and Schuster, 1886).

[9] "Early American Work on Fission". Retrieved 7 June 2013.

[10] Alice Calaprice; Trevor Lipscombe (2005). *Albert Einstein: a biography*. Greenwood Publishing Group. pp. 117–. ISBN 978-0-313-33080-3. Retrieved 16 May 2013.

[11] Geoffrey Lucas Herrera (2006). *Technology and International Transformation: The Railroad, the Atom Bomb, and the Politics of Technological Change*. SUNY Press. pp. 179–. ISBN 978-0-7914-6868-5. Retrieved 9 June 2013.

[12] Christoph Laucht (18 May 2012). *Elemental Germans: Klaus Fuchs, Rudolf Peierls and the Making of British Nuclear Culture 1939-59*. Palgrave Macmillan. pp. 31–. ISBN 978-1-137-22295-4. Retrieved 2 June 2013.

[13] Leslie R. Groves (1983). *Now It Can Be Told: The Story of the Manhattan Project*. Da Capo Press. pp. 33–. ISBN 978-0-7867-4822-8. Retrieved 9 June 2013.

[14] Geoffrey Best (15 November 2006). *Churchill and War*. Continuum International Publishing Group. pp. 206–. ISBN 978-1-85285-541-3. Retrieved 5 May 2013.

[15] Vincent C. Jones (1 December 1985). *Manhattan, the Army and the Atomic Bomb*. Government Printing Office. pp. 82–. ISBN 978-0-16-087288-4. Retrieved 13 June 2013.

[16] Lillian Hoddeson; Paul W. Henriksen; Roger A. Meade; Catherine L. Westfall (12 February 2004). *Critical Assembly: A Technical History of Los Alamos During the Oppenheimer Years, 1943-1945*. Cambridge University Press. pp. 168–. ISBN 978-0-521-54117-6. Retrieved 5 June 2013.

[17] Drawing uncovered of 'Nazi nuke'. BBC.com. Wednesday, 1 June 2005, 13:11 GMT 14:11 UK.

[18] Hitler 'tested small atom bomb'. BBC.com. Monday, 14 March 2005, 17:33 GMT.

[19] Rezelman, David; F.G. Gosling; Terrence R. Fehner (2000). "The atomic bombing of Hiroshima". *The Manhattan Project: An Interactive History*. U.S. Department of Energy. Archived from the original on 2007-09-11. Retrieved 2007-09-18. External link in |work= (help) page on Hiroshima casualties.

[20] *The Spirit of Hiroshima: An Introduction to the Atomic Bomb Tragedy*. Hiroshima Peace Memorial Museum. 1999.

[21] "First Atomic Bomb Dropped on Japan; Missile Is Equal to 20,000 Tons of TNT; Truman Warns Foe of a 'Rain of Ruin'". *The New York Times*. Retrieved 2 June 2013.

[22] "Japan surrenders". Retrieved 2 June 2013.

[23] "Poisoned Workers & Poisoned Places", USA Today, June 24, 2001.

[24] FUSRAP Chronology at Internet Archive.

[25] "The Marshall Islands: Tropical idylls scarred like Tohoku". Retrieved 25 October 2013.

[26] "Lucky Dragon's lethal catch". Retrieved 25 October 2013.

[27] Emsley, John (2001). "Uranium". *Nature's Building Blocks: An A to Z Guide to the Elements*. Oxford: Oxford University Press. p. 478. ISBN 0-19-850340-7.

[28] Mary Palevsky, Robert Futrell, and Andrew Kirk. Recollections of Nevada's Nuclear Past *UNLV FUSION*, 2005, p. 20.

[29] Niedenthal, Jack (2008), *A Short History of the People of Bikini Atoll*, retrieved 2009-12-05

[30] Jerry Brown and Rinaldo Brutoco (1997). *Profiles in Power: The Anti-nuclear Movement and the Dawn of the Solar Age*, Twayne Publishers, pp. 191-192.

[31] Wolfgang Rudig (1990). *Anti-nuclear Movements: A World Survey of Opposition to Nuclear Energy*, Longman, p. 54-55.

[32] Jim Falk (1982). *Global Fission: The Battle Over Nuclear Power*, Oxford University Press, pp. 96-97.

[33] A brief history of CND

[34] "Early defections in march to Aldermaston". London: Guardian Unlimited. 1958-04-05.

[35] Jim Falk (1982). *Global Fission: The Battle Over Nuclear Power*, Oxford University Press, p. 93.

[36] Woo, Elaine (January 30, 2011). "Dagmar Wilson dies at 94; organizer of women's disarmament protesters". *Los Angeles Times*.

[37] Hevesi, Dennis (January 23, 2011). "Dagmar Wilson, Anti-Nuclear Leader, Dies at 94". *The New York Times*.

[38] Louise Zibold Reiss (November 24, 1961). "Strontium-90 Absorption by Deciduous Teeth: Analysis of teeth provides a practicable method of monitoring strontium-90 uptake by human populations" (PDF). Science. Retrieved October 13, 2009.

[39] Thomas Hager (November 29, 2007). "Strontium-90". Oregon State University Libraries Special Collections. Retrieved December 13, 2007.

[40] Thomas Hager (November 29, 2007). "The Right to Petition". Oregon State University Libraries Special Collections. Retrieved December 13, 2007.

[41] Jim Falk (1982). *Global Fission: The Battle Over Nuclear Power*, Oxford University Press, p. 98.

[42] Linus Pauling (October 10, 1963). "Notes by Linus Pauling. October 10, 1963.". Oregon State University Libraries Special Collections. Retrieved December 13, 2007.

[43] The Forgotten Years of the World Nuclear Disarmament Movement, 1975-78.

[44] Jonathan Schell. The Spirit of June 12 *The Nation*, July 2, 2007.

[45] 1982 - A Million People March in New York City.

[46] 438 Protesters are Arrested at Nevada Nuclear Test Site.

[47] Forden, Geoffrey. "False Alarms on the Nuclear Front". Retrieved 1 October 2013.

[48] Nuclear Weapons - France Nuclear Forces.

[49] "Chinese Nuclear Tests Allegedly Cause 750,000 Deaths" *Epoch Times*. March 30, 2009. .

[50] "Neither Confirm Nor Deny". *Radiolab*. Radiolab, WNYC. 12 February 2014. Retrieved 18 February 2014.

[51] "...the total figure will likely be equal to the $5 trillion national debt. In short, one quarter to one third of all military spending since World War II has been devoted to nuclear weapons and their infrastructure..." page 33, Schwartz, Steven I.; Nuclear Weapons Cost Study Committee (November 1995). "Four Trillion Dollars and Counting". *Bulletin of Atomic Scientists*. Educational Foundation for Nuclear Science, Inc. **51** (6): 32–53. ISSN 0096-3402. Retrieved 2012-07-22. Cite uses deprecated parameter |coauthors= (help)

[52] Schwartz, Stephen I.; Bruce G. Blair, The Brookings Institution; Thomas S. Blanton and William Burr, the National Security Archive; Steven M. Kosiak, Center for Strategic and Budgetary Assessments; Arjun Makhijani, Institute for Energy and Environmental Research; Robert S. Norris, Natural Resources Defense Council; Kevin O'Neill, Institute for Science and International Security; John E. Pike, Federation of American Scientists; William J. Weida, Global Resource Action Center for the Environment. (1998). *Atomic audit: the costs and consequences of U.S. nuclear weapons since 1940*. Brookings Institution Press. ISBN 978-0815777748. Cite uses deprecated parameter |coauthors= (help)

[53] Historical Budget Tables

[54] "The peak U.S. inventory was around 35,000 nuclear weapons. The United States spent more than $5.5 trillion on the nuclear arms race, an amount equal to its national debt in 1998..." Graham, Jr., Thomas (2002). *Disarmament sketches: three decades of arms control and international law*. USA: University of Washington Press. p. 35. ISBN 978-0295982120.

[55] Von Wielligh, N. & von Wielligh-Steyn, L. (2015). The Bomb – South Africa's Nuclear Weapons Programme. Pretoria: Litera.

[56] CNS - South Africa's Nuclear Weapons Program: An Annotated Chronology, 1969-1994.

[57] I seek your pardon, *The Guardian* 5 Feb 2004

[58] Iran issues anti-nuke fatwa | World War 4 Report.

The first nuclear programs

- Gregg Herken, *Brotherhood of the Bomb: The Tangled Lives and Loyalties of Robert Oppenheimer, Ernest Lawrence, and Edward Teller* (New York: Henry Holt & Co., 2002).

- David Holloway, *Stalin and the Bomb: The Soviet Union and Atomic Energy 1939-1956* (New Haven: Yale University Press, 1995).

- Richard Rhodes, *Dark Sun: The Making of the Hydrogen Bomb* (New York: Simon and Schuster, 1995).

- Richard Rhodes, *The Making of the Atomic Bomb* (New York: Simon and Schuster, 1986).

- Henry DeWolf Smyth, *Atomic Energy for Military Purposes* (Princeton, NJ: Princeton University Press, 1945). (Smyth Report)

- Mark Walker, *German National Socialism and the Quest for Nuclear Power, 1939-1949* (London: Cambridge University Press, 1990).

Nuclear weapons and energy in culture

- Spencer Weart, *Nuclear Fear: A History of Images*, (Cambridge, MA: Harvard University Press, 1988); *The Rise of Nuclear Fear*, (Cambridge, MA: Harvard University Press, 2012).

Nuclear arsenals and capabilities

- Chuck Hansen, *U.S. Nuclear Weapons: The Secret History*, (Arlington, TX: Aerofax, 1988).

- Chuck Hansen, *The Swords of Armageddon: U.S. nuclear weapons development since 1945*, (Sunnyvale, CA: Chukelea Publications, 1995).

- Stephen Schwartz, ed., *Atomic Audit: The Costs and Consequences of U. S. Nuclear Weapons Since 1940* (Brookings Institution Press, 1998).

Second nuclear age

- Colin S. Gray, *The Second Nuclear Age*, (Lynne Rienner Publishers, 1999),

- Paul Bracken, *The Second Nuclear Age*, Foreign Affairs, January/February 2000,

2.16 Further reading

- "Presidency in the Nuclear Age", conference and forum at the JFK Library, Boston, October 12, 2009. Four panels: "The Race to Build the Bomb and the Decision to Use It", "Cuban Missile Crisis and the First Nuclear Test Ban Treaty", "The Cold War and the Nuclear Arms Race", and "Nuclear Weapons, Terrorism, and the Presidency".

2.17 External links

- Timeline of atomic age events

- Federation of American Scientists - Worldwide Nuclear Forces Guide

- The Genesis of the Atomic Bomb -

- Nuclear Weapons Archive - includes the nuclear weapon histories of many countries

- NDRC Nuclear Notebook: Nuclear pursuits - *"Bulletin of the Atomic Scientists"*. Comparative table of the histories and arsenals of the five NPT-designated nuclear powers as of 1993.

- NuclearFiles.org Timeline- from Atomic Discovery to the 2000s (decade)

- NuclearFiles.org A comprehensive history of nuclear weapons, including Pre, During, and Post Cold War

- Nevada Desert Experience Nevada Desert Experience

- Western States Legal Foundation Western States Legal Foundation

- Ariel E. Levite, "Heading for the Fourth Nuclear Age", *Proliferation Papers*, Paris, Ifri, Winter 2009

- The National Museum of Nuclear Science & History (United States) - located in Albuquerque, New Mexico; a Smithsonian Affiliate Museum

- Time-Lapse Map of All 2053 Nuclear Explosions on Planet Earth (7 Countries, 1945 - 1998) - Video (14:25).

- History of Nuclear Proliferation For more on the history of nuclear proliferation see the Woodrow Wilson Center's Nuclear Proliferation International History Project website.

Natural uranium
> 99.2% U-238
0.72% U-235

Electromagnetic U²³⁵ separation plant at Oak Ridge, Tenn. Massive new physics machines were assembled at secret installations around the United States for the production of enriched uranium and plutonium.

Low-enriched uranium
(reactor grade)
3-4% U-235

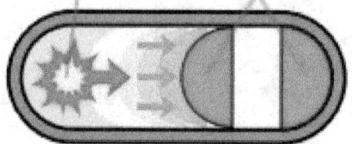

Conventional   Sub-critical pieces of
chemical explosive   uranium-235 combined

Gun-type assembly method

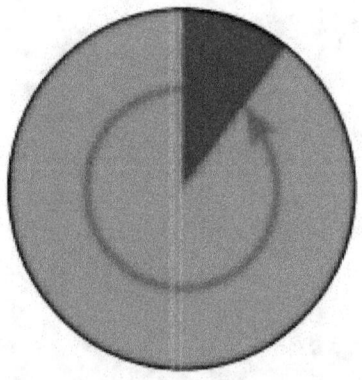

Highly enriched uranium
(weapons grade)
90% U-235

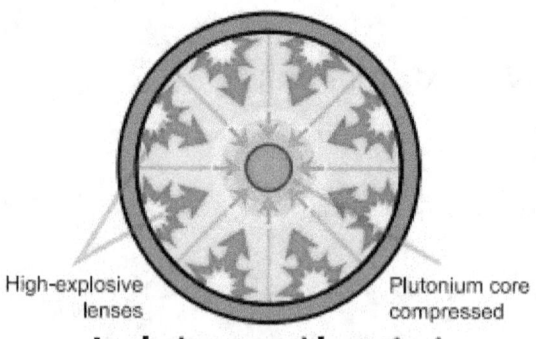

High-explosive   Plutonium core
lenses   compressed

Implosion assembly method

The two fission bomb assembly methods.

Proportions of uranium-238 (blue) and uranium-235 (red) found naturally versus grades that are enriched by separating the two isotopes atom-by-atom using various methods that all require a massive investment in time and money.

The atomic bombings of Hiroshima and Nagasaki killed tens of thousands of Japanese combatants and approximately one hundred thousand non-combatants and destroyed dozens of military bases and supply depots as well as hundreds (or thousands) of factories producing war materials.

Hungarian physicist Edward Teller toiled for years trying to discover a way to make a fusion bomb.

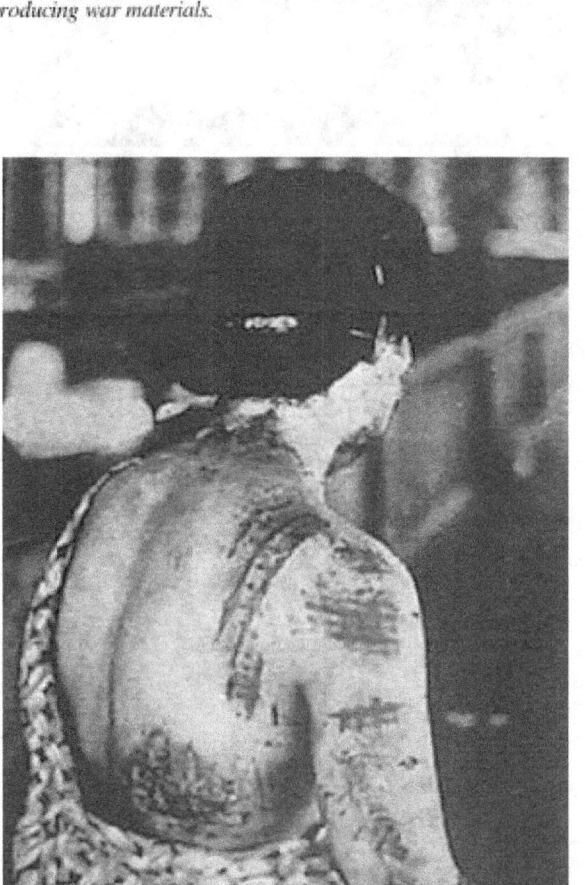

Hiroshima: burns from the intense thermal effect of the atomic bomb.

The "Mike" shot in 1952 inaugurated the age of fusion weapons.

November 1951 nuclear test at the Nevada Test Site, from Operation Buster, with a yield of 21 kilotons. It was the first U.S. nuclear field exercise conducted on land; troops shown are 6 mi (9.7 km) from the blast.

The introduction of nuclear-tipped rockets, like the MGR-1 Honest John, reflected a change in both nuclear technology and strategy.

Long-range bomber aircraft, such as the B-52 Stratofortress, allowed deployment of a wide range of strategic nuclear weapons.

With early warning systems, it was thought that the strikes of nuclear war would come from dark rooms filled with computers, not the battlefield of the wars of old.

Women Strike for Peace during the Cuban Missile Crisis

U-2 photographs revealed that the Soviet Union was stationing nuclear missiles on the island of Cuba in 1962, beginning the Cuban Missile Crisis.

ICBMs, like the American Minuteman missile, allowed nations to deliver nuclear weapons thousands of miles away with relative ease.

Submarine-launched ballistic missiles with multiple warheads made defending against nuclear attack impractical.

On 12 December 1982, 30,000 women held hands around the 6 miles (9.7 km) perimeter of the RAF Greenham Common base, in protest against the decision to site American cruise missiles there.

Chapter 3

Timeline of nuclear weapons development

This **timeline of nuclear weapons development** is a chronological catalog of the evolution of nuclear weapons rooting from the development of the science surrounding nuclear fission and nuclear fusion. In addition to the scientific advancements, this timeline also includes several political events relating to the development of nuclear weapons. The availability of intelligence on recent advancements in nuclear weapons of several major countries (such as United States and the Soviet Union) is limited because of the classification of technical knowledge of nuclear weapons development.

3.1   1930–1950

- **1933** - Leó Szilárd realizes the concept of the nuclear chain reaction, although no such reaction was known. He invented the idea of an atomic bomb in 1933 while crossing a London street. He patented it in 1934. (British patent 630,726)

- **1938** - December - German Chemists, Otto Hahn and Fritz Strassman detect barium after bombarding uranium with neutrons. This is correctly interpreted by Lise Meitner and her nephew Otto Robert Frisch as nuclear fission.

- **1939** - January - Otto Robert Frisch experimentally confirms Otto Hahn and Fritz Strassman's discovery of nuclear fission.

- **1939** - April - Nazi Germany begins the German nuclear energy project.

- **1939** - October - US President Franklin Delano Roosevelt receives the Einstein–Szilárd letter and authorizes the creation of the Advisory Committee on Uranium. The Uranium Committee has its first meeting on October 21, and $6,000 was budgeted for conducting neutron experiments.

- **1940** - April - MAUD Committee (Military Application of Uranium Detonation) established by Henry Tizard to investigate feasibility of an atomic bomb.

- **1941** - February - Plutonium discovered by Glenn Seaborg and Arthur Wahl.

- **1941** - October - President Roosevelt receives MAUD report on the design and costs to develop a nuclear weapon. Roosevelt approves project to confirm MAUD's finding

- **1942** - April - Joseph Stalin was first informed of the efforts to develop nuclear weapons based on a letter sent to him by Georgii Flerov pointing out that there was nothing being published on nuclear fission since its discovery, and the prominent physicists likely involved had not been publishing at all. This urged the Soviet Union to start a nuclear weapons program.

- **1942** - July - The *Heereswaffenamt* (HWA, Army Ordance Office) relinquishes control of the German nuclear energy project to the Reichsforschungsrat (RFR, Reich Research Council), essentially making it only a research project with objectives far short of making a weapon.

- **1942** - July through September - A summer conference at University of California, Berkeley is convened by physicist Robert Oppenheimer and discusses the design of a fission bomb. Edward Teller introduces the "Super" hydrogen bomb as a major discussion point.

- **1942** - August through November - The Manhattan Project is established under command of General Leslie Groves. "Site X" is chosen in Tennessee, for isotopic separation of uranium-235 from natural uranium, and will later become Oak Ridge National Laboratory. "Hanford Site is chosen in Washington, for making plutonium in nuclear reactors. "Site Y" is chosen in New Mexico, for bomb design and manufacture, and will later become Los Alamos National Laboratory.

- **1943** - March - The Japanese Committee on Research in the Application of Nuclear Physics, chaired by Yoshio Nishina concludes in a report that while an atomic bomb was feasible, it would be unlikely to produce one during the war. Japan then concentrated on research into radar.

- **1943** - April - Introductory lectures begin at Los Alamos, which later are compiled into The Los Alamos Primer.

- **1943** - August - Quebec Agreement signed by President Roosevelt and Winston Churchill. A team of British scientists join the Manhattan Project, including Klaus Fuchs.

- **1944** - April - Emilio Segrè discovers that the spontaneous fission rate of plutonium is too high to be used in a gun-type fission weapon. Leads to change in priority to the design of an implosion-type nuclear weapon.

- **1945** - May - Nazi Germany formally surrenders to Allied Powers, marking the end of World War II in Europe.

- **1945** - July 16 - The first nuclear explosion, the Trinity test of an implosion-type plutonium-based nuclear weapon known as "the gadget", near Alamogordo, New Mexico.

- **1945** - August 6 - "Little Boy", a gun-type uranium-235 weapon, is dropped on Hiroshima, Japan.

- **1945** - August 9 - "Fat Man", an implosion-type plutonium-239 weapon, is dropped on Nagasaki, Japan.

- **1945** - August - The Smyth Report is published detailing the efforts of the Manhattan Project.

- **1945** - August - Surrender of Japan to the Allied Powers.

- **1946** - January - The Atomic Energy Act of 1946 takes effect, officially turning over the Manhattan Project to the United States Atomic Energy Commission.

- **1946** - June - First meeting of the United Nations Atomic Energy Commission is held.

- **1946** - Soviet Union rejects the Baruch Plan.

- **1949** - August - The Soviet Union conducts its first atomic test, First Lightning (nicknamed Joe 1 by the Americans).

3.2   1950–1970

- **1951** - China and the Soviet Union sign an agreement whereby China would supply uranium ore in exchange for technical assistance in producing nuclear weapons.

- **1952** - October - The United Kingdom conducts Operation Hurricane, the first test of a British nuclear weapon. The plutonium implosion-type device was detonated in a lagoon between the Montebello Islands, Western Australia.

- **1952** - November - The United States test the first fusion bomb, "Ivy Mike".

- **1953** - The first nuclear-tipped rockets are deployed by the United States. The MGR-1 Honest John is such as example.

- **1953** - August - The Soviet Union conducts its first test of a hydrogen bomb, nicknamed Joe 4 by the Americans.

- **1954** - February - The United States detonates its first deliverable thermonuclear weapon at Bikini Atoll, Marshall Islands. The device had a yield almost three times as large as expected, leading to the worst radiological disaster in US history.

- **1956** - France establishes a secret committee for the Military Applications of Atomic Energy.

- **1956** - The Pakistan Atomic Energy Commission is established. This commission is responsible for the development of both the nuclear reactors and nuclear weapons in Pakistan.

- **1957** - July - The International Atomic Energy Agency is founded.

- **1958** - The United States and the United Kingdom sign the 1958 US-UK Mutual Defence Agreement. This is a bilateral treaty on nuclear weapons cooperation signed after the United Kingdom successfully tested a hydrogen bomb during Operation Grapple.

- **1960** - February - France successfully tests a nuclear weapon, called "Gerboise bleue", in the French Sahara.

- **1961** - The Israeli Prime Minister David Ben-Gurion informed the Canadian Prime Minister John Diefenbaker that a pilot plutonium-separation plant would be built at Dimona, the location of the nuclear reactor built with the aid of France and Great Britain. Intelligence would indicate from this and other information that Israel intended to produce nuclear weapons.

- **1961** - October - The Soviet Union detonates Tsar Bomba, the largest, most powerful nuclear weapon ever detonated.

- **1964** - October - China successfully tests an atomic bomb at Lop Nur.

- **1965** - January - The Soviet Union detonates Chagan as part of their Nuclear Explosions for the National Economy series to study the peaceful use of nuclear explosions.

- **1967** - June - China successfully tests a hydrogen bomb.

- **1967** - December - Japan, under Prime Minister Eisaku Satō, adopts the Three Non-Nuclear Principles.

- **1968** - July - The Nuclear Non-Proliferation Treaty opens for signatures. This treaty is intended to limit the spread of nuclear weapons. To date, 189 countries have signed the treaty, including the five permanent members of the UN Security Council. Only India, Israel, Pakistan, and North Korea have not signed the treaty (as sovereign states).

3.3   1970–1990

- **1972** - Zulfikar Ali Bhutto launched Pakistan's atomic program by making Munir Ahmad Khan as the program head.

- **1974** - May - India tests its first nuclear device, "Smiling Buddha", at Pokhran.

- **1974** - May - The Project-706 is established under command of General Zahid Ali Akbar.

- **1986** - September - Mordechai Vanunu divulges secrets about the Israeli nuclear weapons program to The Sunday Times in London.

3.4   1990–present

- **1991** - South Africa signs the Nuclear Non-Proliferation Treaty; they also announce that from 1979 to 1989, they had built and then dismantled a number of nuclear weapons.

- **1991** - June - The Brazilian-Argentine Agency for Accounting and Control of Nuclear Materials is established to play an active role in verifying the pacific use of nuclear materials that could be used for the manufacture of nuclear weapons in Argentina and Brazil.

- **1998** - May - India tests five more nuclear weapons as part of Operation Shakti at the Pokhran test site. This was India's second round of nuclear weapons testing.

- **1998** - May - Pakistan detonates five high-enriched uranium nuclear weapons in the Chagai Hills. A sixth nuclear test, at Kharan, was a plutonium device.

- **2003** - North Korea announces that it has several nuclear explosives.

- **2005** - August - In Iran, Ayatollah Ali Khamenei issued a fatwa forbidding the production, stockpiling and use of nuclear weapons.

- **2006** - October - North Korea tests a nuclear weapon for the first time.

- **2016** - January - North Korea Hydrogen bomb is 'tested' and confirmed by North Korea leader Kim Jong-Un

3.5   See also

- History of Nuclear Weapons

- Nuclear Weapons

- Nuclear Fission

- Nuclear Fusion

Chapter 4

Nuclear weapon design

The first nuclear weapons, cumbersome and inefficient, provided the basic design building blocks of all future weapons. Here the Gadget device is prepared for the first nuclear test: Trinity.

Nuclear weapon designs are physical, chemical, and engineering arrangements that cause the physics package[1] of a nuclear weapon to detonate. There are three existing basic design types. In most existing designs, the explosive energy of deployed devices is derived primarily from nuclear fission, not fusion.

- **Pure fission weapons** were the first nuclear weapons built and have so far been the only type ever used in warfare. The active material is fissile uranium (uranium with a high percentage of U-235) or plutonium (Pu-239), explosively assembled into a chain-reacting critical mass by one of two methods:

  - Gun assembly: one piece of fissile uranium is fired at a fissile uranium target at the end of the weapon, similar to firing a bullet down a gun barrel, achieving critical mass when combined.

  - Implosion: a fissile mass of either material (U-235, Pu-239, or a combination) is surrounded by high explosives that compress the mass, resulting in criticality.

The implosion method can use either uranium or plutonium as fuel. The gun method only uses uranium. Plutonium is considered impractical for the gun method because of early triggering due to Pu-240 contamination and due to its time constant for prompt critical fission being much shorter than that of U-235.

- **Boosted fission weapons** improve on the implosion design. The high pressure and temperature environment at the center of an exploding fission weapon compresses and heats a mixture of tritium and deuterium gas (heavy isotopes of hydrogen). The hydrogen fuses to form helium and free neutrons. The energy release from this fusion reaction is relatively negligible, but each neutron starts a new fission chain reaction, speeding up the fission and greatly reducing the amount of fissile material that would otherwise be wasted when expansion of the fissile material stops the chain reaction. Boosting can more than double the weapon's fission energy release.

- **Staged thermonuclear weapons** are essentially a chain of fusion-boosted fission weapons, usually with only two stages in the chain. The secondary stage is imploded by x-ray energy from the first stage, called the "primary." This radiation implosion is much more effective than the high-explosive implosion of the primary. Consequently, the secondary can be many times more powerful than the primary, without being bigger. The secondary can be designed to maximize fusion energy release, but in most designs fusion is employed only to drive or enhance fission, as it is in the primary. More stages could be added and conceptual designs incorporating up to seven have been produced, but the result would be a multi-megaton weapon too powerful to serve any plausible purpose.[2] (The United States briefly deployed a three-stage 25-megaton bomb, the B41, starting in 1961. Also in 1961, the Soviet Union tested, but did not deploy, a three-stage 50–100 megaton device, Tsar Bomba.)

- **Pure fusion weapons** have not been invented. Such weapons, though, would produce far less radioactive fallout than current designs, although they would release huge numbers of neutrons.

Pure fission weapons historically have been the first type to be built by a nation state. Large industrial states with well-developed nuclear arsenals have two-stage thermonuclear weapons, which are the most compact, scalable, and cost effective option once the necessary industrial infrastructure is built.

Most known innovations in nuclear weapon design originated in the United States, although some were later developed independently by other states;[3] the following descriptions feature U.S. designs.

In early news accounts, pure fission weapons were called atomic bombs or A-bombs, a misnomer since the energy comes only from the nucleus of the atom. Weapons involving fusion were called hydrogen bombs or H-bombs, also a misnomer since their energy comes mostly from fission. Practioners favored the terms nuclear and thermonuclear, respectively.

The term thermonuclear refers to the high temperatures required to initiate fusion. It omits the equally important factor of radiation pressure, which was considered secret at the time the term became widespread. Many nuclear weapon terms similarly obfuscate because of their origin in a classified environment.

4.1   Nuclear reactions

Nuclear fission separates or splits heavier atoms to form lighter atoms. Nuclear fusion combines together lighter atoms to form heavier atoms. Both reactions generate roughly a million times more energy than comparable chemical reactions, making nuclear bombs a million times more powerful than non-nuclear bombs, which a French patent claimed in May 1939.[4]

In some ways, fission and fusion are opposite and complementary reactions, but the particulars are unique for each. To understand how nuclear weapons are designed, it is useful to know the important similarities and differences between fission and fusion. The following explanation uses rounded numbers and approximations.[5]

4.1.1   Fission

Main article: nuclear fission

When a free neutron hits the nucleus of a fissile atom like uranium-235 (^{235}U), the uranium nucleus splits into two smaller nuclei called fission fragments, plus more neutrons. Fission can be self-sustaining because it produces more neutrons of the speed required to cause new fissions.

The U-235 nucleus can split in many ways, provided the atomic numbers add up to 92 and the atomic weights add to 236 (uranium plus the extra neutron). The following equation shows one possible split, namely into strontium-95 (^{95}Sr), xenon-139 (^{139}Xe), and two neutrons (n), plus energy:[6]

$$^{235}U + n \longrightarrow {}^{95}Sr + {}^{139}Xe + 2n + 180 \text{ MeV}$$

The immediate energy release per atom is about 180 million electron volts (MeV); i.e., 74 TJ/kg. Only 7% of this is gamma radiation and kinetic energy of fission neutrons. The remaining 93% is kinetic energy (or energy of motion) of the charged fission fragments, flying away from each other mutually repelled by the positive charge of their protons (38 for strontium, 54 for xenon). This initial kinetic energy is 67 TJ/kg, imparting an initial speed of about 12,000 kilometers per second. The charged fragments' high electric charge causes many inelastic collisions with nearby nuclei, and these fragments remain trapped inside the bomb's uranium pit and tamper until their motion is converted into heat. This takes about a millionth of a second (a microsecond), by which time the core and tamper of the bomb have expanded to plasma several meters in diameter with a temperature of tens of millions of degrees Celsius.

This is hot enough to emit black-body radiation in the X-ray spectrum. These X-rays are absorbed by the surrounding air, producing the fireball and blast of a nuclear explosion.

Most fission products have too many neutrons to be stable so they are radioactive by beta decay, converting neutrons into protons by throwing off beta particles (electrons) and gamma rays. Their half lives range from milliseconds to about 200,000 years. Many decay into isotopes that are themselves radioactive, so from 1 to 6 (average 3) decays may be required to reach stability.[7] In reactors, the radioactive products are the nuclear waste in spent fuel. In bombs, they become radioactive fallout, both local and global.

Meanwhile, inside the exploding bomb, the free neutrons released by fission carry away about 3% of the initial fission energy. Neutron kinetic energy adds to the blast energy of a bomb, but not as effectively as the energy from charged fragments, since neutrons are not slowed as quickly. The main contribution of fission neutrons to the bomb's power is the initiation of other fissions. Over half of the neutrons escape the bomb core, but the rest strike nearby U-235 nuclei causing them to fission in an exponentially growing chain reaction (1, 2, 4, 8, 16, etc.). Starting from one atom, the number of fissions can theoretically double a hundred times in a microsecond, which could consume all uranium or plutonium up to hundreds of tons by the hundredth link in the chain. In practice, bombs do not contain hundreds of tons of uranium or plutonium. Instead, typically (in a modern weapon) the core of a weapon contains only about 5 kilograms of plutonium, of which only 2 to 2.5 kilograms, representing 40 to 50 kilotons of energy, undergoes fission before the core blows itself apart.

Holding an exploding bomb together is the greatest challenge of fission weapon design. The heat of fission rapidly expands the fission core, spreading apart the target nuclei and making space for the neutrons to escape without being captured. The chain reaction stops.

Materials which can sustain a chain reaction are called fissile. The two fissile materials used in nuclear weapons are: U-235, also known as highly enriched uranium (HEU), oralloy (Oy) meaning Oak Ridge Alloy, or 25 (the last digits of the atomic number, which is 92 for uranium, and the atomic weight, here 235, respectively); and Pu-239, also known as plutonium, or 49 (from 94 and 239).

Uranium's most common isotope, U-238, is fissionable but not fissile (meaning that it cannot sustain a chain reaction by itself but can be made to fission with fast neutrons). Its aliases include natural or unenriched uranium, depleted uranium (DU), tuballoy (Tu), and 28. It cannot sustain a chain reaction, because its own fission neutrons are not powerful enough to cause more U-238 fission. The neutrons released by fusion will fission U-238. This U-238 fission reaction produces most of the energy in a typical two-stage thermonuclear weapon.

4.1.2 Fusion

Main article: nuclear fusion

Fusion produces neutrons which dissipate energy from the reaction.[8] In weapons, the most important fusion reaction is called the D-T reaction. Using the heat and pressure of fission, hydrogen-2, or deuterium (^{2}D), fuses with hydrogen-3, or tritium (^{3}T), to form helium-4 (^{4}He) plus one neutron (n) and energy:[9]

$$^2D + {}^3T \longrightarrow {}^4He + n + 17.6\ \text{MeV}$$

The total energy output, 17.6 MeV, is one tenth of that with fission, but the ingredients are only one-fiftieth as massive, so the energy output per unit mass is greater. In this fusion

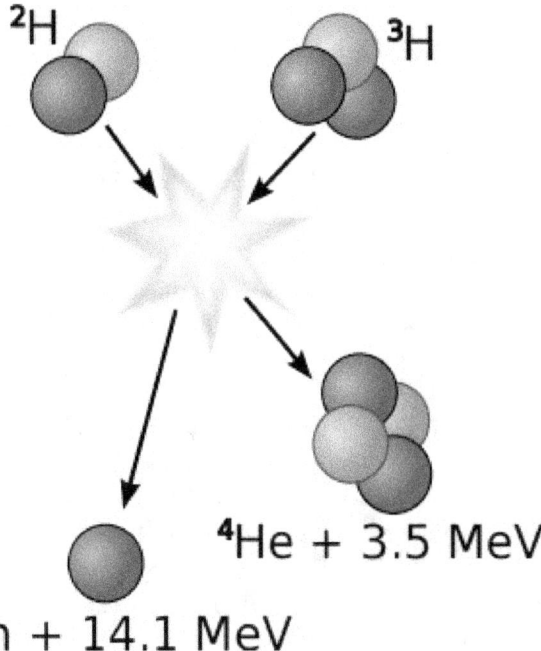

reaction 80% of the energy, or 14 MeV, is in the motion of the neutron which, having no electric charge and being almost as massive as the hydrogen nuclei that created it, can escape the scene without leaving its energy behind to help sustain the reaction – or to generate x-rays for blast and fire.

The only practical way to capture most of the fusion energy is to trap the neutrons inside a massive bottle of heavy material such as lead, uranium, or plutonium. If the 14 MeV neutron is captured by uranium (either type: 235 or 238) or plutonium, the result is fission and the release of 180 MeV of fission energy, multiplying the energy output tenfold.

Fission is thus necessary to start fusion, helps to sustain fusion, and captures and multiplies the energy released in fusion neutrons. In the case of a neutron bomb (see below) the last-mentioned does not apply since the escape of neutrons is the objective.

4.1.3 Tritium production

A third important nuclear reaction is the one that creates tritium, essential to the type of fusion used in weapons. Tritium, or hydrogen-3, is made by bombarding lithium-6 (^{6}Li) with a neutron (n). This neutron bombardment will cause the lithium-6 nucleus to fission, producing helium–4 (^{4}He) plus tritium (^{3}T) and energy:[9]

$$^6Li + n \longrightarrow {}^4He + {}^3T + 5\ \text{MeV}$$

A nuclear reactor is necessary to provide the neutrons if the tritium is to be provided before the weapon is used. The industrial-scale conversion of lithium-6 to tritium is very similar to the conversion of uranium-238 into plutonium-239. In both cases the feed material is placed inside a nuclear reactor and removed for processing after a period of time.

Alternatively, neutrons from earlier stage fusion reactions can be used to fission lithium-6 (in the form of lithium deuteride for example) and form tritium during detonation. This approach reduces the amount of tritium-based fuel in a weapon.[10]

The fission of one plutonium atom releases ten times more total energy than the fusion of one tritium atom. For this reason, tritium is included in nuclear weapon components only when it causes more fission than its production sacrifices, namely in the case of fusion-boosted fission.

Of the four basic types of nuclear weapon, the first, pure fission, uses the first of the three nuclear reactions above. The second, fusion-boosted fission, uses the first two. The third, two-stage thermonuclear, uses all three.

4.2   Pure fission weapons

The first task of a nuclear weapon design is to rapidly assemble a supercritical mass of fissile uranium or plutonium. A supercritical mass is one in which the percentage of fission-produced neutrons captured by another fissile nucleus is large enough that each fission event, on average, causes more than one additional fission event.

Once the critical mass is assembled, at maximum density, a burst of neutrons is supplied to start as many chain reactions as possible. Early weapons used an "urchin" inside the pit containing polonium−210 and beryllium separated by a thin barrier. Implosion of the pit crushed the urchin, mixing the two metals, thereby allowing alpha particles from the polonium to interact with beryllium to produce free neutrons. In modern weapons, the neutron generator is a high-voltage vacuum tube containing a particle accelerator which bombards a deuterium/tritium-metal hydride target with deuterium and tritium ions. The resulting small-scale fusion produces neutrons at a protected location outside the physics package, from which they penetrate the pit. This method allows better control of the timing of chain reaction initiation.

The critical mass of an uncompressed sphere of bare metal is 110 lb (50 kg) for uranium-235 and 35 lb (16 kg) for delta-phase plutonium-239. In practical applications, the amount of material required for criticallity is modified by shape, purity, density, and the proximity to neutron-reflecting material, all of which affect the escape or capture of neutrons.

To avoid a chain reaction during handling, the fissile material in the weapon must be sub-critical before detonation. It may consist of one or more components containing less than one uncompressed critical mass each. A thin hollow shell can have more than the bare-sphere critical mass, as can a cylinder, which can be arbitrarily long without ever reaching criticallity.

A *tamper* is an optional layer of dense material surrounding the fissile material. Due to its inertia it delays the expansion of the reacting material, increasing the efficiency of the weapon. Often the same layer serves both as tamper and as neutron reflector.

4.2.1   Gun-type assembly weapon

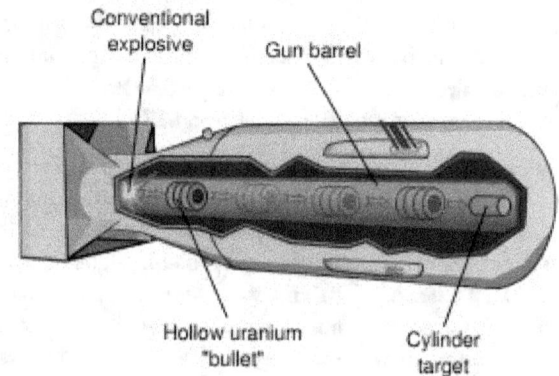

Diagram of a gun-type fission weapon

Main article: Gun-type fission weapon

Little Boy, the Hiroshima bomb, used 141 lb (64 kg) of uranium with an average enrichment of around 80%, or 112 lb (51 kg) of U-235, just about the bare-metal critical mass. (See Little Boy article for a detailed drawing.) When assembled inside its tamper/reflector of tungsten carbide, the 141 lb (64 kg) was more than twice critical mass. Before the detonation, the uranium-235 was formed into two sub-critical pieces, one of which was later fired down a gun barrel to join the other, starting the nuclear explosion. About 1% of the uranium underwent fission;[11] the remainder, representing most of the entire wartime output of the giant factories at Oak Ridge, scattered uselessly.[12]

The inefficiency was caused by the speed with which the uncompressed fissioning uranium expanded and became subcritical by virtue of decreased density. Despite its inefficiency, this design, because of its shape, was adapted for use in small-diameter, cylindrical artillery shells (a gun-type warhead fired from the barrel of a much larger gun). Such warheads were deployed by the United States until 1992, accounting for a significant fraction of the U-235 in the arsenal, and were some of the first weapons dismantled to comply with treaties limiting warhead numbers. The rationale for this decision was undoubtedly a combination of the lower yield and grave safety issues associated with the gun-type design.

4.2.2 Implosion-type weapon

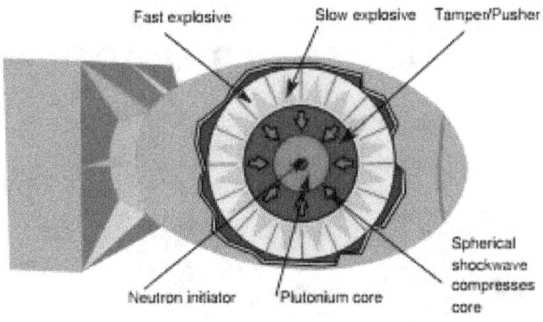

For both the Trinity device and the Fat Man, the Nagasaki bomb, nearly identical plutonium fission through implosion designs were used. The Fat Man device specifically used 13.6 lb (6.2 kg), about 12 US fl oz or 350 ml in volume) of Pu-239, which is only 41% of bare-sphere critical mass. (See Fat Man article for a detailed drawing.) Surrounded by a U-238 reflector/tamper, the Fat Man's pit was brought close to critical mass by the neutron-reflecting properties of the U-238. During detonation, criticality was achieved by implosion. The plutonium pit was squeezed to increase its density by simultaneous detonation, as with the "Trinity" test detonation three weeks earlier, of the conventional explosives placed uniformly around the pit. The explosives were detonated by multiple exploding-bridgewire detonators. It is estimated that only about 20% of the plutonium underwent fission; the rest, about 11 lb (5.0 kg), was scattered.

An implosion shock wave might be of such short duration that only part of the pit is compressed at any instant as the wave passes through it. To prevent this, a pusher shell may be needed. The pusher is located between the explosive lens and the tamper. It works by reflecting some of the shock wave backwards, thereby having the effect of lengthening its duration. It is made out of a low density metal—such as aluminium, beryllium, or an alloy of the two metals (aluminium being easier and safer to shape, and is two orders of magnitude cheaper; beryllium for its high-neutron-reflective capability). Fat Man used an aluminium pusher.

The series of RaLa Experiment tests of implosion-type fis-

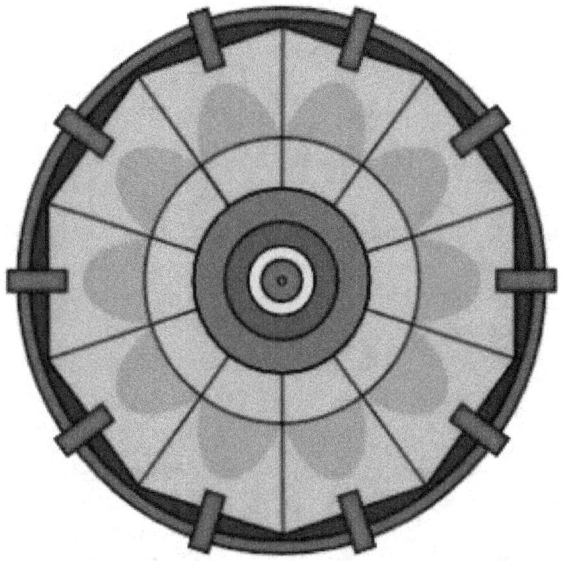

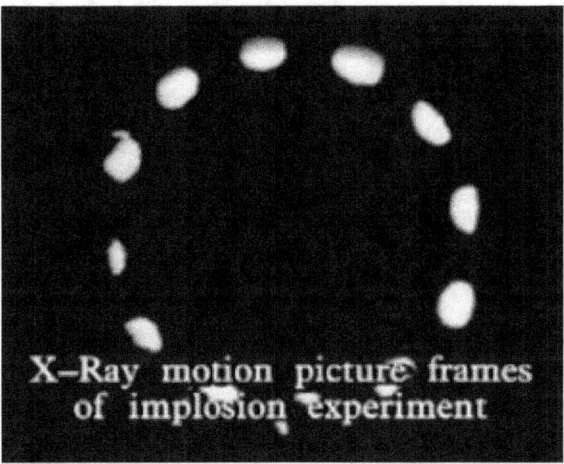

Flash X-Ray images of the converging shock waves formed during a test of the high explosive lens system.

sion weapon design concepts, carried out from July 1944 through February 1945 at the Los Alamos National Laboratory and a remote site 9 miles (14.3 km) east of it in Bayo Canyon, proved the practicality of the implosion design for a fission device, with the February 1945 tests positively determining its usability for the final Trinity/Fat Man plutonium implosion design.[13]

The key to Fat Man's greater efficiency was the inward momentum of the massive U-238 tamper. (The natural uranium tamper did not undergo fission from thermal neutrons, but did contribute perhaps 20% of the total yield from fission by fast neutrons). Once the chain reaction started in the plutonium, the momentum of the implosion had to be reversed before expansion could stop the fission. By holding everything together for a few hundred nanoseconds more, the efficiency was increased.

Plutonium pit

Main article: Pit (nuclear weapon)

The core of an implosion weapon – the fissile material and any reflector or tamper bonded to it – is known as the *pit*. Some weapons tested during the 1950s used pits made with U-235 alone, or in composite with plutonium,[14] but all-plutonium pits are the smallest in diameter and have been the standard since the early 1960s.

Casting and then machining plutonium is difficult not only because of its toxicity, but also because plutonium has many different metallic phases. As plutonium cools, changes in phase result in distortion and cracking. This distortion is normally overcome by alloying it with 3–3.5 molar% (0.9–1.0% by weight) gallium, forming a plutonium-gallium alloy, which causes it to take up its delta phase over a wide temperature range.[15] When cooling from molten it then has only a single phase change, from epsilon to delta, instead of the four changes it would otherwise pass through. Other trivalent metals would also work, but gallium has a small neutron absorption cross section and helps protect the plutonium against corrosion. A drawback is that gallium compounds are corrosive and so if the plutonium is recovered from dismantled weapons for conversion to plutonium dioxide for power reactors, there is the difficulty of removing the gallium.

Because plutonium is chemically reactive it is common to plate the completed pit with a thin layer of inert metal, which also reduces the toxic hazard.[16] The gadget used galvanic silver plating; afterwards, nickel deposited from nickel tetracarbonyl vapors was used,[16] but gold is now preferred.

4.2.3   Levitated-pit implosion

The first improvement on the Fat Man design was to put an air space between the tamper and the pit to create a hammer-on-nail impact. The pit, supported on a hollow cone inside the tamper cavity, was said to be levitated. The three tests of Operation Sandstone, in 1948, used Fat Man designs with levitated pits. The largest yield was 49 kilotons, more than twice the yield of the unlevitated Fat Man.[17]

It was immediately clear that implosion was the best design for a fission weapon. Its only drawback seemed to be its diameter. Fat Man was 5 feet (1.5 m) wide vs 2 feet (61 cm) for Little Boy.

Eleven years later, implosion designs had advanced suffi-

ciently that the 5-foot (1.5 m)-diameter sphere of Fat Man had been reduced to a 1-foot (0.30 m)-diameter cylinder 2 feet (0.61 m) long, the Swan device.

The Pu-239 pit of Fat Man was only 3.6 inches (9.1 cm) in diameter, the size of a softball. The bulk of Fat Man's girth was the implosion mechanism, namely concentric layers of U-238, aluminium, and high explosives. The key to reducing that girth was the two-point implosion design.

4.2.4   Two-point linear implosion

Linear Implosion

Fissile Pu-239 pit   High explosive

Inert wave shaper

Detonator

Detonation fronts emerging from around wave shapers

In the two-point linear implosion, the nuclear fuel is cast into a solid shape and placed within the center of a cylinder of high explosive. Detonators are placed at either end of the explosive cylinder, and a plate-like insert, or *shaper*, is placed in the explosive just inside the detonators. When the detonators are fired, the initial detonation is trapped between the shaper and the end of the cylinder, causing it to travel out to the edges of the shaper where it is refracted around the edges into the main mass of explosive. This causes the detonation to form into a ring that proceeds inwards from the shaper.[18]

Due to the lack of a tamper or lenses to shape the progression, the detonation does not reach the pit in a spherical shape. To produce the desired spherical implosion, the fissile material itself is shaped to produce the same effect. Due to the physics of the shock wave propagation within the explosive mass, this requires the pit to be an oblong shape, roughly egg shaped. The shock wave first reaches the pit at its tips, driving them inward and causing the mass to become spherical. The shock may also change plutonium from delta to alpha phase, increasing its density by 23%, but without the inward momentum of a true implosion.

The lack of compression makes such designs inefficient, but the simplicity and small diameter make it suitable for use in artillery shells and atomic demolition munitions – ADMs – also known as backpack or suitcase nukes; an example is the W48 artillery shell, the smallest nuclear weapon ever built or deployed. All such low-yield battlefield weapons, whether gun-type U-235 designs or linear implosion Pu-239 designs, pay a high price in fissile material in order to achieve diameters between six and ten inches (25 cm).

4.2.5 Two-point hollow-pit implosion

A more efficient two-point implosion system uses two high explosive lenses and a hollow pit.

A hollow plutonium pit was the original plan for the 1945 Fat Man bomb, but there was not enough time to develop and test the implosion system for it. A simpler solid-pit design was considered more reliable, given the time constraints, but it required a heavy U-238 tamper, a thick aluminium pusher, and three tons of high explosives.

After the war, interest in the hollow pit design was revived. Its obvious advantage is that a hollow shell of plutonium, shock-deformed and driven inward toward its empty center, would carry momentum into its violent assembly as a solid sphere. It would be self-tamping, requiring a smaller U-238 tamper, no aluminium pusher and less high explosive.

The Fat Man bomb had two concentric, spherical shells of high explosives, each about 10 in (25 cm) thick. The inner shell drove the implosion. The outer shell consisted of a soccer-ball pattern of 32 high explosive lenses, each of which converted the convex wave from its detonator into a concave wave matching the contour of the outer surface of the inner shell. If these 32 lenses could be replaced with only two, the high explosive sphere could become an ellipsoid (prolate spheroid) with a much smaller diameter.

A good illustration of these two features is a 1956 drawing from the Swedish nuclear weapon program (which was terminated before it produced a test explosion). The drawing shows the essential elements of the two-point hollow-pit design.

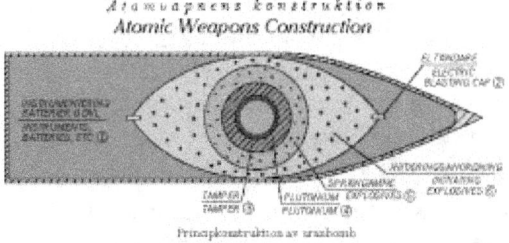

Atomvapens konstruktion
Atomic Weapons Construction

Principkonstruktion av uranbomb

There are similar drawings in the open literature from the French program, which produced an arsenal.

The mechanism of the high explosive lens (diagram item #6) is not shown in the Swedish drawing, but a standard lens made of fast and slow high explosives, as in Fat Man, would be much longer than the shape depicted. For a single high explosive lens to generate a concave wave that envelops an entire hemisphere, it must either be very long or the part of the wave on a direct line from the detonator to the pit must be slowed dramatically.

A slow high explosive is too fast, but the flying plate of an "air lens" is not. A metal plate, shock-deformed and pushed across an empty space, can be designed to move slowly enough.[19][20] A two-point implosion system using air lens technology can have a length no more than twice its diameter, as in the Swedish diagram above.

4.3 Fusion-boosted fission weapons

Main article: Boosted fission weapon

The next step in miniaturization was to speed up the fissioning of the pit to reduce the minimum inertial confinement time. This would allow the efficient fission of the fuel with less mass in the form of tamper or the fuel itself. The key to achieving faster fission would be to introduce more neutrons, and among the many ways to do this, adding a fusion reaction was relatively easy in the case of a hollow pit.

The easiest fusion reaction to achieve is found in a 50–50 mixture of tritium and deuterium. For fusion power experiments this mixture must be held at high temperatures for relatively lengthy times in order to have an efficient reaction. For explosive use, however, the goal is not to produce efficient fusion, but simply provide extra neutrons early in the process. Since a nuclear explosion is supercritical, any extra neutrons will be multiplied by the chain reaction, so even tiny quantities introduced early can have a large effect on the final outcome. For this reason, even the relatively low compression pressures and times (in fusion terms) found in the center of a hollow pit warhead are enough to create the desired effect.

In the boosted design, the fusion fuel in gas form is pumped into the pit during arming. This will fuse into helium and release free neutrons soon after fission begins. The neutrons will start a large number of new chain reactions while the pit is still critical or nearly critical. Once the hollow pit is perfected, there is little reason not to boost; deuterium and tritium is easily produced in the small quantities needed, and the technical aspects are trivial.

The concept of fusion-boosted fission was first tested on May 25, 1951, in the Item shot of Operation Greenhouse, Eniwetok, yield 45.5 kilotons.

Boosting reduces diameter in three ways, all the result of faster fission:

- Since the compressed pit does not need to be held together as long, the massive U-238 tamper can be replaced by a light-weight beryllium shell (to reflect escaping neutrons back into the pit). The diameter is reduced.

- The mass of the pit can be reduced by half, without reducing yield. Diameter is reduced again.

- Since the mass of the metal being imploded (tamper plus pit) is reduced, a smaller charge of high explosive is needed, reducing diameter even further.

Since boosting is required to attain full design yield, any reduction in boosting reduces yield. Boosted weapons are thus variable-yield weapons (also known as dial-a-yield); yield can be reduced any time before detonation simply by reducing the amount of tritium inserted into the pit during the arming procedure.

U.S. Swan Device - 1956

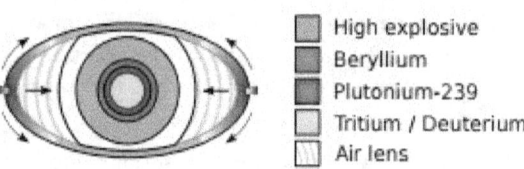

- ⬜ High explosive
- ⬜ Beryllium
- ⬜ Plutonium-239
- ⬜ Tritium / Deuterium
- ⬜ Air lens

The first device whose dimensions suggest employment of all these features (two-point, hollow-pit, fusion-boosted implosion) was the Swan device. It had a cylindrical shape with a diameter of 11.6 in (29 cm) and a length of 22.8 in (58 cm).

It was first tested standalone and then as the primary of a two-stage thermonuclear device during Operation Redwing. It was weaponized as the Robin primary and became the first off-the-shelf, multi-use primary, and the prototype for all that followed.

After the success of Swan, 11 or 12 inches (300 mm) seemed to become the standard diameter of boosted single-stage devices tested during the 1950s. Length was usually twice the diameter, but one such device, which became the W54 warhead, was closer to a sphere, only 15 inches (380 mm) long. It was tested two dozen times in the 1957–62 period before being deployed. No other design had such a long string of test failures.

One of the applications of the W54 was the Davy Crockett XM-388 recoilless rifle projectile. It had a dimension of just 11 inches, and is shown here in comparison to its Fat Man predecessor.

Another benefit of boosting, in addition to making weapons smaller, lighter, and with less fissile material for a given yield, is that it renders weapons immune to radiation interference (RI). It was discovered in the mid-1950s that plutonium pits would be particularly susceptible to partial predetonation if exposed to the intense radiation of a nearby nuclear explosion (electronics might also be damaged, but this was a separate problem). RI was a particular problem before effective early warning radar systems because a first strike attack might make retaliatory weapons useless. Boosting reduces the amount of plutonium needed in a weapon to below the quantity which would be vulnerable to this effect.

4.4 Two-stage thermonuclear weapons

Main article: Thermonuclear weapon

Pure fission or fusion-boosted fission weapons can be made to yield hundreds of kilotons, at great expense in fissile material and tritium, but by far the most efficient way to increase nuclear weapon yield beyond ten or so kilotons is to add a second independent stage, called a secondary.

In the 1940s, bomb designers at Los Alamos thought the secondary would be a canister of deuterium in liquefied or hydride form. The fusion reaction would be D-D, harder to achieve than D-T, but more affordable. A fission bomb at one end would shock-compress and heat the near end, and fusion would propagate through the canister to the far end. Mathematical simulations showed it would not work, even with large amounts of expensive tritium added.

The entire fusion fuel canister would need to be enveloped by fission energy, to both compress and heat it, as with the booster charge in a boosted primary. The design breakthrough came in January 1951, when Edward Teller and Stanislaw Ulam invented radiation implosion—for nearly three decades known publicly only as the Teller-Ulam H-

Ivy Mike, the first two-stage thermonuclear detonation, 10.4 megatons, November 1, 1952.

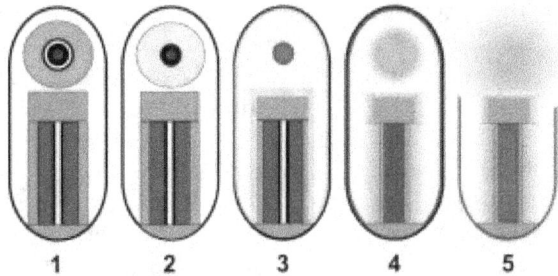

Ablation mechanism firing sequence.

1. *Warhead before firing. The nested spheres at the top are the fission primary; the cylinders below are the fusion secondary device.*

2. *Fission primary's explosives have detonated and collapsed the primary's fissile pit.*

3. *The primary's fission reaction has run to completion, and the primary is now at several million degrees and radiating gamma and hard X-rays, heating up the inside of the hohlraum and the shield and secondary's tamper.*

4. *The primary's reaction is over and it has expanded. The surface of the pusher for the secondary is now so hot that it is also ablating or expanding away, pushing the rest of the secondary (tamper, fusion fuel, and fissile spark plug) inwards. The spark plug starts to fission. Not depicted: the radiation case is also ablating and expanding outwards (omitted for clarity of diagram).*

5. *The secondary's fuel has started the fusion reaction and shortly will burn up. A fireball starts to form.*

bomb secret.

The concept of radiation implosion was first tested on May 9, 1951, in the George shot of Operation Greenhouse, Eniwetok, yield 225 kilotons. The first full test was on November 1, 1952, the Mike shot of Operation Ivy, Eniwetok, yield 10.4 megatons.

In radiation implosion, the burst of X-ray energy coming from an exploding primary is captured and contained within an opaque-walled radiation channel which surrounds the nuclear energy components of the secondary. The radiation quickly turns the plastic foam that had been filling the channel into a plasma which is mostly transparent to X-rays, and the radiation is absorbed in the outermost layers of the pusher/tamper surrounding the secondary, which ablates and applies a massive force[21] (much like an inside out rocket engine) causing the fusion fuel capsule to implode much like the pit of the primary. As the secondary implodes a fissile "spark plug" at its center ignites and provides neutrons and heat which enable the lithium deuteride fusion fuel to produce tritium and ignite as well. The fission and fusion chain reactions exchange neutrons with each other and boost the efficiency of both reactions. The greater implosive force, enhanced efficiency of the fissile "spark plug" due to boosting via fusion neutrons, and the fusion explosion itself provides significantly greater explosive yield from the secondary despite often not being much larger than the primary.

For example, for the Redwing Mohawk test on July 3, 1956, a secondary called the Flute was attached to the Swan primary. The Flute was 15 inches (38 cm) in diameter and 23.4 inches (59 cm) long, about the size of the Swan. But it weighed ten times as much and yielded 24 times as much energy (355 kilotons, vs 15 kilotons).

Equally important, the active ingredients in the Flute proba-

bly cost no more than those in the Swan. Most of the fission came from cheap U-238, and the tritium was manufactured in place during the explosion. Only the spark plug at the axis of the secondary needed to be fissile.

A spherical secondary can achieve higher implosion densities than a cylindrical secondary, because spherical implosion pushes in from all directions toward the same spot. However, in warheads yielding more than one megaton, the diameter of a spherical secondary would be too large for most applications. A cylindrical secondary is necessary in such cases. The small, cone-shaped re-entry vehicles in multiple-warhead ballistic missiles after 1970 tended to have warheads with spherical secondaries, and yields of a few hundred kilotons.

As with boosting, the advantages of the two-stage thermonuclear design are so great that there is little incentive not to use it, once a nation has mastered the technology.

In engineering terms, radiation implosion allows for the exploitation of several known features of nuclear bomb materials which heretofore had eluded practical application. For example:

- The optimal way to store deuterium in a reasonably dense state is to chemically bond it with lithium, as lithium deuteride. But the lithium-6 isotope is also the raw material for tritium production, and an exploding bomb is a nuclear reactor. Radiation implosion will hold everything together long enough to permit the complete conversion of lithium-6 into tritium, while the bomb explodes. So the bonding agent for deuterium permits use of the D-T fusion reaction without any pre-manufactured tritium being stored in the secondary. The tritium production constraint disappears.

- For the secondary to be imploded by the hot, radiation-induced plasma surrounding it, it must remain cool for the first microsecond, i.e., it must be encased in a massive radiation (heat) shield. The shield's massiveness allows it to double as a tamper, adding momentum and duration to the implosion. No material is better suited for both of these jobs than ordinary, cheap uranium-238, which also happens to undergo fission when struck by the neutrons produced by D-T fusion. This casing, called the pusher, thus has three jobs: to keep the secondary cool; to hold it, inertially, in a highly compressed state; and, finally, to serve as the chief energy source for the entire bomb. The consumable pusher makes the bomb more a uranium fission bomb than a hydrogen fusion bomb. Insiders never used the term "hydrogen bomb".[22]

- Finally, the heat for fusion ignition comes not from the primary but from a second fission bomb called the spark plug, embedded in the heart of the secondary. The implosion of the secondary implodes this spark plug, detonating it and igniting fusion in the material around it, but the spark plug then continues to fission in the neutron-rich environment until it is fully consumed, adding significantly to the yield.[23]

The initial impetus behind the two-stage weapon was President Truman's 1950 promise to build a 10-megaton hydrogen superbomb as the U.S. response to the 1949 test of the first Soviet fission bomb. But the resulting invention turned out to be the cheapest and most compact way to build small nuclear bombs as well as large ones, erasing any meaningful distinction between A-bombs and H-bombs, and between boosters and supers. All preferred techniques for fission and fusion explosions are incorporated into one all-encompassing, fully scalable design principle. Even 6 inches (150 mm) diameter nuclear artillery shells can be two-stage thermonuclears.

In the ensuing fifty years, nobody has come up with a more efficient way to build a nuclear bomb. It is the design of choice for the United States, Russia, the United Kingdom, China, and France, the five thermonuclear powers. According to Dr. Theodore Taylor after reviewing leaked photographs of disassembled weapons components taken before 1986, Israel possessed boosted weapons and would require supercomputers of that era to advance further toward full two-stage weapons in the megaton range without nuclear test detonations.[24] The other nuclear-armed nations, India, Pakistan, and North Korea, probably have single-stage weapons, possibly boosted.[23]

4.4.1   Interstage

In a two-stage thermonuclear weapon the energy from the primary impacts the secondary. An essential energy transfer modulator called the interstage, between the primary and the secondary, protects the secondary's fusion fuel from heating too quickly, which could cause it to explode in a conventional (and small) heat explosion before the fusion and fission reactions get a chance to start.

There is very little information in the open literature about the mechanism of the interstage. Its first mention in a U.S. government document formally released to the public appears to be a caption in a recent graphic promoting the Reliable Replacement Warhead Program. If built, this new design would replace "toxic, brittle material" and "expensive 'special' material" in the interstage.[25] This statement suggests the interstage may contain beryllium to moderate the flux of neutrons from the primary, and perhaps something to absorb and re-radiate the x-rays in a particular manner.[26] There is also some speculation that this interstage material, which may be code-named FOGBANK, might be an aerogel, possibly doped with beryllium and/or other substances.[27][28]

The interstage and the secondary are encased together inside a stainless steel membrane to form the canned subassembly (CSA), an arrangement which has never been depicted in any open-source drawing.[29] The most detailed illustration of an interstage shows a British thermonuclear weapon with a cluster of items between its primary and a cylindrical secondary. They are labeled "end-cap and neutron focus lens," "reflector/neutron gun carriage," and "reflector wrap." The origin of the drawing, posted on the internet by Greenpeace, is uncertain, and there is no accompanying explanation.[30]

4.5   Specific designs

While every nuclear weapon design falls into one of the above categories, specific designs have occasionally become the subject of news accounts and public discussion, often with incorrect descriptions about how they work and what they do. Examples:

4.5.1 Hydrogen bombs

Main article: Hydrogen bomb

While all modern nuclear weapons (fission and fusion alike) make some use of D-T fusion, in the public perception hydrogen bombs are multi-megaton devices a thousand times more powerful than Hiroshima's Little Boy. Such high-yield bombs are actually two-stage thermonuclears, scaled up to the desired yield, with uranium fission, as usual, providing most of their energy.

The idea of the hydrogen bomb first came to public attention in 1949, when prominent scientists openly recommended against building nuclear bombs more powerful than the standard pure-fission model, on both moral and practical grounds. Their assumption was that critical mass considerations would limit the potential size of fission explosions, but that a fusion explosion could be as large as its supply of fuel, which has no critical mass limit. In 1949, the Soviets exploded their first fission bomb, and in 1950 U.S. President Harry S. Truman ended the H-bomb debate by ordering the Los Alamos designers to build one.

In 1952, the 10.4-megaton Ivy Mike explosion was announced as the first hydrogen bomb test, reinforcing the idea that hydrogen bombs are a thousand times more powerful than fission bombs.

In 1954, J. Robert Oppenheimer was labeled a hydrogen bomb opponent. The public did not know there were two kinds of hydrogen bomb (neither of which is accurately described as a hydrogen bomb). On May 23, when his security clearance was revoked, item three of the four public findings against him was "his conduct in the hydrogen bomb program." In 1949, Oppenheimer had supported single-stage fusion-boosted fission bombs, to maximize the explosive power of the arsenal given the trade-off between plutonium and tritium production. He opposed two-stage thermonuclear bombs until 1951, when radiation implosion, which he called "technically sweet", first made them practical. The complexity of his position was not revealed to the public until 1976, nine years after his death.[31]

When ballistic missiles replaced bombers in the 1960s, most multi-megaton bombs were replaced by missile warheads (also two-stage thermonuclears) scaled down to one megaton or less.

4.5.2 Alarm Clock/Sloika

The first effort to exploit the symbiotic relationship between fission and fusion was a 1940s design that mixed fission and fusion fuel in alternating thin layers. As a single-stage device, it would have been a cumbersome application of boosted fission. It first became practical when incorporated into the secondary of a two-stage thermonuclear weapon.[32]

The U.S. name, Alarm Clock, came from Teller: he called it that because it might "wake up the world" to the possibility of the potential of the Super.[33] The Russian name for the same design was more descriptive: Sloika (Russian: Слойка), a layered pastry cake. A single-stage Soviet Sloika was tested on August 12, 1953. No single-stage U.S. version was tested, but the Union shot of Operation Castle, April 26, 1954, was a two-stage thermonuclear device code-named Alarm Clock. Its yield, at Bikini, was 6.9 megatons.

Because the Soviet Sloika test used dry lithium-6 deuteride eight months before the first U.S. test to use it (Castle Bravo, March 1, 1954), it was sometimes claimed that the USSR won the H-bomb race, even though the United States tested and developed the first hydrogen bomb: the Ivy Mike H-bomb test. The 1952 U.S. Ivy Mike test used cryogenically cooled liquid deuterium as the fusion fuel in the secondary, and employed the D-D fusion reaction. However, the first Soviet test to use a radiation-imploded secondary, the essential feature of a true H-bomb, was on November 23, 1955, three years after Ivy Mike. In fact, real work on the implosion scheme in the Soviet Union only commenced in the very early part of 1953, several months after the successful testing of Sloika.

4.5.3 Clean bombs

On March 1, 1954, the largest-ever U.S. nuclear test explosion, the 15-megaton Bravo shot of Operation Castle at Bikini Atoll, delivered a promptly lethal dose of fission-product fallout to more than 6,000 square miles (16,000 km²) of Pacific Ocean surface.[34] Radiation injuries to Marshall Islanders and Japanese fishermen made that fact public and revealed the role of fission in hydrogen bombs.

In response to the public alarm over fallout, an effort was made to design a clean multi-megaton weapon, relying almost entirely on fusion. The energy produced by the fissioning of unenriched natural uranium, when used as the tamper material in the secondary and subsequent stages in the Teller-Ulam design, can evidently dwarf the fusion yield output, as was the case in the Castle Bravo test; realizing that a non-fissionable tamper material is an essential requirement in a 'clean' bomb, it is clear that in such a bomb there will be a relatively massive amount of material that does not contribute energy by either fission or fusion. So for a given weight, 'dirty' weapons with fissionable tampers are much more powerful than a 'clean' weapon (or, for an equal yield, they are much lighter). The earliest known incidence of a three-stage device being tested, with the third stage, called the tertiary, being ignited by the secondary,

Bassoon, the prototype for a 9.3-megaton clean bomb or a 25-megaton dirty bomb. Dirty version shown here, before its 1956 test. The two attachments on the left are light pipes - see below for elaboration.

was May 27, 1956 in the Bassoon device. This device was tested in the Zuni shot of Operation Redwing. This shot used non fissionable tampers; an inert substitute material such as tungsten or lead was used. Its yield was 3.5 megatons, 85% fusion and only 15% fission.

The public records for devices that produced the highest proportion of their yield via fusion-only reactions are the Peaceful nuclear explosions of the 1970s, with the 3 detonations that excavated part of Pechora–Kama Canal, being cited as 98% fusion each in the *Taiga* test's 15 kiloton explosive yield devices, that is, a total fission fraction of 0.3 kilotons in a 15 kt device.[35] The 50 megaton Tsar Bomba at 97% fusion,[36] the 9.3 megaton Hardtack Poplar test at 95.2%,[37] and the 4.5 megaton Redwing Navajo test at 95% fusion.[38]

On July 19, 1956, AEC Chairman Lewis Strauss said that the Redwing Zuni shot clean bomb test "produced much of importance ... from a humanitarian aspect." However, less than two days after this announcement the dirty version of Bassoon, called Bassoon Prime, with a uranium-238 tamper in place, was tested on a barge off the coast of Bikini Atoll as the Redwing Tewa shot. The Bassoon Prime produced a 5-megaton yield, of which 87% came from fission. Data obtained from this test, and others, culminated in the eventual deployment of the highest yielding US nuclear weapon

known, and the highest yield-to-weight weapon ever made, a three-stage thermonuclear weapon with a maximum 'dirty' yield of 25-megatons designated as the B41 nuclear bomb, which was to be carried by U.S. Air Force bombers until it was decommissioned; this weapon was never fully tested.

As such, high-yield clean bombs appear to have been of little value from a military standpoint. The actual deployed weapons were the dirty versions, which maximized yield for the same size device. The need for low fission fraction nuclear devices was driven only by the likes of Project Orion and peaceful nuclear explosions - for earth excavation with little contamination of the resulting excavated area.

Newer 4th-generation nuclear weapons designs including pure fusion weapons and antimatter catalyzed nuclear pulse propulsion-like devices,[39] are being studied by the five largest nuclear weapon states.[40][41]

4.5.4   Cobalt bombs

Main article: Cobalt bomb

A fictional doomsday bomb, made popular by Nevil Shute's 1957 novel, and subsequent 1959 movie, *On the Beach*, the cobalt bomb was a hydrogen bomb with a jacket of cobalt. The neutron-activated cobalt would supposedly have maximized the environmental damage from radioactive fallout. These bombs were popularized in the 1964 film *Dr. Strangelove or: How I Learned to Stop Worrying and Love the Bomb*. The element added to the bombs is referred to in the film as 'cobalt-thorium G'

Such "salted" weapons were requested by the U.S. Air Force and seriously investigated, possibly built and tested, but not deployed. In the 1964 edition of the DOD/AEC book *The Effects of Nuclear Weapons*, a new section titled Radiological Warfare clarified the issue.[42] Fission products are as deadly as neutron-activated cobalt. The standard high-fission thermonuclear weapon is automatically a weapon of radiological warfare, as dirty as a cobalt bomb.

Initially, gamma radiation from the fission products of an equivalent size fission-fusion-fission bomb are much more intense than Co-60: 15,000 times more intense at 1 hour; 35 times more intense at 1 week; 5 times more intense at 1 month; and about equal at 6 months. Thereafter fission drops off rapidly so that Co-60 fallout is 8 times more intense than fission at 1 year and 150 times more intense at 5 years. The very long-lived isotopes produced by fission would overtake the ^{60}Co again after about 75 years.[43]

The triple "taiga" nuclear salvo test, as part of the preliminary March 1971 Pechora–Kama Canal project, produced substantial amounts of Co-60, with this fusion generated neutron activation product being responsible for about half of the gamma dose now(2011) at the test site, with green vegetation existing all around the lake that was formed.[44][45]

4.5.5 Fission-fusion-fission bombs vs. three-stage (tertiary) bombs

In 1954, to explain the surprising amount of fission-product fallout produced by hydrogen bombs, Ralph Lapp coined the term fission-fusion-fission to describe a process inside what he called a three-stage thermonuclear weapon. His process explanation was correct, but his choice of terms caused confusion in the open literature. The stages of a nuclear weapon are not fission, fusion, and fission. They are the primary, the secondary, and, in a very few exceptional and powerful weapons no longer in service, the tertiary. Tertiary (three-stage) designs, such as the U.S. B41 nuclear bomb and the Soviet Tsar Bomba (discussed above), were developed in the late 1950s and early 1960s; all have since been retired, as the typical multi-megaton yields of tertiary bombs do not destroy targets efficiently, since they waste energy in a sphere above and below an area of land. For this reason, all tertiary bombs have given way in modern nuclear arsenals to multiple smaller two-stage bomb tactics (see for example, MIRV). Such two-stage bombs, even though less efficient in yield, are nevertheless more destructive for their total bomb weight, because they can be distributed over a roughly two-dimensional area of land at the target.

All so-called "fission-fusion-fission" weapons (i.e., all conventional modern thermonuclear warheads) employ the additional step of "jacket fissioning," using fusion neutrons. This works as follows: the high-energy or "fast" neutrons generated by fusion are used to fission a fissionable jacket located around the fusion stage. In the past this jacket was often made of natural or depleted uranium; but today's weapons in which there is a premium on weight and size (i.e., virtually all modern strategic weapons) use moderately-to-highly enriched uranium as the jacketing material (see Oralloy thermonuclear warheads section below). The fast fission of the secondary jacket in fission-fusion-fission bombs is sometimes referred to as a "third stage" in the bomb, but it should not be confused with the obsolete true three-stage thermonuclear design, in which there existed another complete tertiary fusion stage.

In the era of open-air atomic testing, the fission jacket was sometimes omitted, in order to create so-called "clean bombs" (see above), or to reduce the amount of radioactive fallout from fission products in very large multi-megaton blasts. This was done most often in the testing of very large tertiary bomb designs, such as the Tsar Bomba and the Zuni test shot of Operation Redwing, as discussed above. In the testing of such weapons, it was assumed (and sometimes

shown operationally) that a jacket of natural uranium or enriched uranium could always be added to a given unjacketed bomb, if desired, to increase the yield from two to five times.

The fission jacket is not used in the enhanced radiation weapon, or neutron bomb, discussed later.

4.5.6 Arbitrarily large multi-staged devices

The idea of a device which has an arbitrarily large number of Teller-Ulam stages, with each driving a larger radiation-driven implosion than the preceding stage, is frequently suggested,[46][47] but technically disputed.[48] There are "well-known sketches and some reasonable-looking calculations in the open literature about two-stage weapons, but no similarly accurate descriptions of true three stage concepts."[48]

According to George Lemmer's 1967 *Air Force and Strategic Deterrence 1951-1960* paper, in 1957, LANL stated that a 1000 megaton warhead could be built.[49] Apparently there were three of these US designs analyzed in the Gigaton(1000 megaton) range; LLNL's GNOMON and SUNDIAL - objects that cast shadows - and LANL's "TAV". SUNDIAL attempting to have a 10 Gt yield, while the Gnomon and TAV designs attempted to produce a yield of 1 Gt.[50] A Freedom of information request has been filed(FOIA 13-00049-K) for information on the three above US designs and as of 2014, the request remains open.[51]

In a 1995 meeting at Lawrence Livermore National Laboratory (LLNL), Edward Teller proposed to a collective of U.S. and Russian ex-Cold War weapons designers, that they collaborate on designing a 1000 megaton nuclear explosive device for diverting extinction class asteroids(10+ km in dia) which would be employed in the event that one were on an impact trajectory with earth.[52][53][54]

There have also been some calculations made in 1979 by Lowell Wood, Teller's protégé, that Teller's "classical Super" design could potentially be ignited by a large enough Teller-Ulam device.[55]

4.5.7 Neutron bombs

Main article: Neutron bomb

A neutron bomb, technically referred to as an enhanced radiation weapon (ERW), is a type of tactical nuclear weapon designed specifically to release a large portion of its energy as energetic neutron radiation. This contrasts with standard thermonuclear weapons, which are designed to capture this intense neutron radiation to increase its overall explosive yield. In terms of yield, ERWs typically produce about one-tenth that of a fission-type atomic weapon. Even with their significantly lower explosive power, ERWs are still capable of much greater destruction than any conventional bomb. Meanwhile, relative to other nuclear weapons, damage is more focused on biological material than on material infrastructure (though extreme blast and heat effects are not eliminated).

ERWs are more accurately described as suppressed yield weapons. When the yield of a nuclear weapon is less than one kiloton, its lethal radius from blast, 700 m (2,300 ft), is less than that from its neutron radiation. However, the blast is more than potent enough to destroy most structures, which are less resistant to blast effects than even unprotected human beings. Blast pressures of upwards of 20 PSI are survivable, whereas most buildings will collapse with a pressure of only 5 PSI.

Commonly misconceived as a weapon designed to kill populations and leave infrastructure intact, these bombs (as mentioned above) are still very capable of leveling buildings over a large radius. The intent of their design was to kill tank crews – tanks giving excellent protection against blast and heat, surviving (relatively) very close to a detonation. And with the Soviets' vast tank battalions during the Cold War, this was the perfect weapon to counter them. The neutron radiation could instantly incapacitate a tank crew out to roughly the same distance that the heat and blast would incapacitate an unprotected human (depending on design). The tank chassis would also be rendered highly radioactive (temporarily) preventing its re-use by a fresh crew.

Neutron weapons were also intended for use in other applications, however. For example, they are effective in anti-nuclear defenses – the neutron flux being capable of neutralising an incoming warhead at a greater range than heat or blast. Nuclear warheads are very resistant to physical damage, but are very difficult to harden against extreme neutron flux.

ERWs were two-stage thermonuclears with all non-essential uranium removed to minimize fission yield. Fusion provided the neutrons. Developed in the 1950s, they were first deployed in the 1970s, by U.S. forces in Europe. The last ones were retired in the 1990s.

A neutron bomb is only feasible if the yield is sufficiently high that efficient fusion stage ignition is possible, and if the yield is low enough that the case thickness will not absorb too many neutrons. This means that neutron bombs have a yield range of 1–10 kilotons, with fission proportion varying from 50% at 1-kiloton to 25% at 10-kilotons (all of which comes from the primary stage). The neutron output per kiloton is then 10–15 times greater than for a pure fission implosion weapon or for a strategic warhead like a W87 or W88.[56]

4.5.8   Oralloy thermonuclear warheads

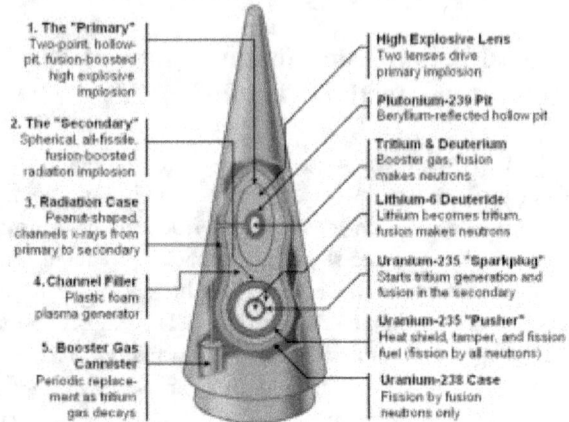

W88 Warhead for Trident D-5 Ballistic Missile

1. The "Primary"
Two-point, hollow-pit, fusion-boosted high explosive implosion

2. The "Secondary"
Spherical, all-fissile, fusion-boosted radiation implosion

3. Radiation Case
Peanut-shaped, channels x-rays from primary to secondary

4. Channel Filler
Plastic foam plasma generator

5. Booster Gas Cannister
Periodic replacement as tritium gas decays

High Explosive Lens
Two lenses drive primary implosion

Plutonium-239 Pit
Beryllium-reflected hollow pit

Tritium & Deuterium
Booster gas, fusion makes neutrons

Lithium-6 Deuteride
Lithium becomes tritium, fusion makes neutrons

Uranium-235 "Sparkplug"
Starts tritium generation and fusion in the secondary

Uranium-235 "Pusher"
Heat shield, tamper, and fission fuel (fission by all neutrons)

Uranium-238 Case
Fission by fusion neutrons only

drawing of W-88

In 1999, nuclear weapon design was in the news again, for the first time in decades. In January, the U.S. House of Representatives released the Cox Report (Christopher Cox R-CA) which alleged that China had somehow acquired classified information about the U.S. W88 warhead. Nine months later, Wen Ho Lee, a Taiwanese immigrant working at Los Alamos, was publicly accused of spying, arrested, and served nine months in pre-trial detention, before the case against him was dismissed. It is not clear that there was, in fact, any espionage.

In the course of eighteen months of news coverage, the W88 warhead was described in unusual detail. *The New York Times* printed a schematic diagram on its front page.[57] The most detailed drawing appeared in *A Convenient Spy*, the 2001 book on the Wen Ho Lee case by Dan Stober and Ian Hoffman, adapted and shown here with permission.

Designed for use on Trident II (D-5) submarine-launched ballistic missiles, the W88 entered service in 1990 and was the last warhead designed for the U.S. arsenal. It has been described as the most advanced, although open literature accounts do not indicate any major design features that were not available to U.S. designers in 1958.

The above diagram shows all the standard features of ballistic missile warheads since the 1960s, with two exceptions that give it a higher yield for its size.

- The outer layer of the secondary, called the "pusher", which serves three functions: heat shield, tamper, and fission fuel, is made of U-235 instead of U-238, hence the name Oralloy (U-235) Thermonuclear. Being fissile, rather than merely fissionable, allows the pusher to fission faster and more completely, increasing yield.

This feature is available only to nations with a great wealth of fissile uranium. The United States is estimated to have 500 tons.

- The secondary is located in the wide end of the re-entry cone, where it can be larger, and thus more powerful. The usual arrangement is to put the heavier, denser secondary in the narrow end for greater aerodynamic stability during re-entry from outer space, and to allow more room for a bulky primary in the wider part of the cone. (The W87 warhead drawing in the W87 article shows the usual arrangement.) Because of this new geometry, the W88 primary uses compact conventional high explosives (CHE) to save space,[58] rather than the more usual, and bulky but safer, insensitive high explosives (IHE). The re-entry cone probably has ballast in the nose for aerodynamic stability.[59]

The alternating layers of fission and fusion material in the secondary are an application of the Alarm Clock/Sloika principle.

4.5.9 Reliable replacement warhead

Main article: Reliable Replacement Warhead

The United States has not produced any nuclear warheads since 1989, when the Rocky Flats pit production plant, near Boulder, Colorado, was shut down for environmental reasons. With the end of the Cold War two years later, the production line was idled except for inspection and maintenance functions.

The National Nuclear Security Administration, the latest successor for nuclear weapons to the Atomic Energy Commission and the Department of Energy, has proposed building a new pit facility and starting the production line for a new warhead called the Reliable Replacement Warhead (RRW).[60] Two advertised safety improvements of the RRW would be a return to the use of "insensitive high explosives which are far less susceptible to accidental detonation", and the elimination of "certain hazardous materials, such as beryllium, that are harmful to people and the environment."[61] Because of the U.S. moratorium on nuclear explosive testing, any new design would rely on previously tested concepts.

4.6 Weapon design laboratories

All the nuclear weapon design innovations discussed in this article originated from the following three labs in the manner described. Other nuclear weapon design labs in other countries duplicated those design innovations independently, reverse-engineered them from fallout analysis, or acquired them by espionage.[62]

4.6.1 Berkeley

Main article: Lawrence Berkeley National Laboratory

The first systematic exploration of nuclear weapon design concepts took place in mid-1942 at the University of California, Berkeley. Important early discoveries had been made at the adjacent Lawrence Berkeley Laboratory, such as the 1940 cyclotron-made production and isolation of plutonium. A Berkeley professor, J. Robert Oppenheimer, had just been hired to run the nation's secret bomb design effort. His first act was to convene the 1942 summer conference.

By the time he moved his operation to the new secret town of Los Alamos, New Mexico, in the spring of 1943, the accumulated wisdom on nuclear weapon design consisted of five lectures by Berkeley professor Robert Serber, transcribed and distributed as the Los Alamos Primer. The Primer addressed fission energy, neutron production and capture, nuclear chain reactions, critical mass, tampers, predetonation, and three methods of assembling a bomb: gun assembly, implosion, and "autocatalytic methods," the one approach that turned out to be a dead end.

4.6.2 Los Alamos

Main article: Los Alamos National Laboratory

At Los Alamos, it was found in April 1944 by Emilio G. Segrè that the proposed Thin Man Gun assembly type bomb would not work for plutonium because of predetonation problems caused by Pu-240 impurities. So Fat Man, the implosion-type bomb, was given high priority as the only option for plutonium. The Berkeley discussions had generated theoretical estimates of critical mass, but nothing precise. The main wartime job at Los Alamos was the experimental determination of critical mass, which had to wait until sufficient amounts of fissile material arrived from the production plants: uranium from Oak Ridge, Tennessee, and plutonium from the Hanford site in Washington.

In 1945, using the results of critical mass experiments, Los Alamos technicians fabricated and assembled components for four bombs: the *Trinity* Gadget, Little Boy, Fat Man, and an unused spare Fat Man. After the war, those who could, including Oppenheimer, returned to university teaching positions. Those who remained worked on levitated and hollow pits and conducted weapon effects tests such as Crossroads Able and Baker at Bikini Atoll in 1946.

All of the essential ideas for incorporating fusion into nuclear weapons originated at Los Alamos between 1946 and 1952. After the Teller-Ulam radiation implosion breakthrough of 1951, the technical implications and possibilities were fully explored, but ideas not directly relevant to making the largest possible bombs for long-range Air Force bombers were shelved.

Because of Oppenheimer's initial position in the H-bomb debate, in opposition to large thermonuclear weapons, and the assumption that he still had influence over Los Alamos despite his departure, political allies of Edward Teller decided he needed his own laboratory in order to pursue H-bombs. By the time it was opened in 1952, in Livermore, California, Los Alamos had finished the job Livermore was designed to do.

4.6.3   Livermore

Main article: Lawrence Livermore National Laboratory

With its original mission no longer available, the Livermore lab tried radical new designs that failed. Its first three nuclear tests were fizzles: in 1953, two single-stage fission devices with uranium hydride pits, and in 1954, a two-stage thermonuclear device in which the secondary heated up prematurely, too fast for radiation implosion to work properly.

Shifting gears, Livermore settled for taking ideas Los Alamos had shelved and developing them for the Army and Navy. This led Livermore to specialize in small-diameter tactical weapons, particularly ones using two-point implosion systems, such as the Swan. Small-diameter tactical weapons became primaries for small-diameter secondaries. Around 1960, when the superpower arms race became a ballistic missile race, Livermore warheads were more useful than the large, heavy Los Alamos warheads. Los Alamos warheads were used on the first intermediate-range ballistic missiles, IRBMs, but smaller Livermore warheads were used on the first intercontinental ballistic missiles, ICBMs, and submarine-launched ballistic missiles, SLBMs, as well as on the first multiple warhead systems on such missiles.[63]

In 1957 and 1958, both labs built and tested as many designs as possible, in anticipation that a planned 1958 test ban might become permanent. By the time testing resumed in 1961 the two labs had become duplicates of each other, and design jobs were assigned more on workload considerations than lab specialty. Some designs were horse-traded. For example, the W38 warhead for the Titan I missile started out as a Livermore project, was given to Los Alamos when it became the Atlas missile warhead, and in 1959 was given back to Livermore, in trade for the W54 Davy Crockett warhead, which went from Livermore to Los Alamos.

The period of real innovation was ending by then, anyway. Warhead designs after 1960 took on the character of model changes, with every new missile getting a new warhead for marketing reasons. The chief substantive change involved packing more fissile uranium into the secondary, as it became available with continued uranium enrichment and the dismantlement of the large high-yield bombs.

4.7   Explosive testing

Nuclear weapons are in large part designed by trial and error. The trial often involves test explosion of a prototype.

In a nuclear explosion, a large number of discrete events, with various probabilities, aggregate into short-lived, chaotic energy flows inside the device casing. Complex mathematical models are required to approximate the processes, and in the 1950s there were no computers powerful enough to run them properly. Even today's computers and simulation software are not adequate.[64]

It was easy enough to design reliable weapons for the stockpile. If the prototype worked, it could be weaponized and mass-produced.

It was much more difficult to understand how it worked or why it failed. Designers gathered as much data as possible during the explosion, before the device destroyed itself, and used the data to calibrate their models, often by inserting fudge factors into equations to make the simulations match experimental results. They also analyzed the weapon debris in fallout to see how much of a potential nuclear reaction had taken place.

4.7.1   Light pipes

An important tool for test analysis was the diagnostic light pipe. A probe inside a test device could transmit information by heating a plate of metal to incandescence, an event that could be recorded at the far end of a long, very straight pipe.

The picture below shows the Shrimp device, detonated on March 1, 1954 at Bikini, as the Castle Bravo test. Its 15-megaton explosion was the largest ever by the United States. The silhouette of a man is shown for scale. The device is supported from below, at the ends. The pipes going into the shot cab ceiling, which appear to be supports, are diagnostic light pipes. The eight pipes at the right end (1) sent information about the detonation of the primary. Two in the middle (2) marked the time when x-radiation from the primary reached the radiation channel around the secondary. The last two pipes (3) noted the time radiation reached the far end of the radiation channel, the difference between (2) and (3) being the radiation transit time for the channel.[65]

From the shot cab, the pipes turned horizontal and traveled 7500 ft (2.3 km), along a causeway built on the Bikini reef, to a remote-controlled data collection bunker on Namu Island.

While x-rays would normally travel at the speed of light through a low density material like the plastic foam channel filler between (2) and (3), the intensity of radiation from the exploding primary created a relatively opaque radiation front in the channel filler which acted like a slow-moving logjam to retard the passage of radiant energy. While the secondary is being compressed via radiation induced ablation, neutrons from the primary catch up with the x-rays, penetrate into the secondary and start breeding tritium with the third reaction noted in the first section above. This Li-6 + n reaction is exothermic, producing 5 MeV per event. The spark plug is not yet compressed and thus is not critical, so there won't be significant fission or fusion. But if enough neutrons arrive before implosion of the secondary is complete, the crucial temperature difference will be degraded. This is the reported cause of failure for Livermore's first thermonuclear design, the Morgenstern device, tested as Castle Koon, April 7, 1954.

These timing problems are measured by light-pipe data. The mathematical simulations which they calibrate are called radiation flow hydrodynamics codes, or channel codes. They are used to predict the effect of future design modifications.

It is not clear from the public record how successful the Shrimp light pipes were. The data bunker was far enough back to remain outside the mile-wide crater, but the 15-megaton blast, two and a half times greater than expected, breached the bunker by blowing its 20-ton door off the hinges and across the inside of the bunker. (The nearest people were twenty miles (32 km) farther away, in a bunker that survived intact.)[66]

4.7.2 Fallout analysis

See also: Nuclear forensics

The most interesting data from Castle Bravo came from radio-chemical analysis of weapon debris in fallout. Because of a shortage of enriched lithium-6, 60% of the lithium in the Shrimp secondary was ordinary lithium-7, which doesn't breed tritium as easily as lithium-6 does. But it does breed lithium-6 as the product of an (n, 2n) reaction (one neutron in, two neutrons out), a known fact, but with unknown probability. The probability turned out to be high.

Fallout analysis revealed to designers that, with the (n, 2n) reaction, the Shrimp secondary effectively had two and half times as much lithium-6 as expected. The tritium, the fusion yield, the neutrons, and the fission yield were all increased accordingly.[67]

As noted above, Bravo's fallout analysis also told the outside world, for the first time, that thermonuclear bombs are more fission devices than fusion devices. A Japanese fishing boat, *Daigo Fukuryū Maru*, sailed home with enough fallout on her decks to allow scientists in Japan and elsewhere to determine, and announce, that most of the fallout had come from the fission of U-238 by fusion-produced 14 MeV neutrons.

4.7.3 Underground testing

Subsidence Craters at Yucca Flat, Nevada Test Site.

The global alarm over radioactive fallout, which began with the Castle Bravo event, eventually drove nuclear testing literally underground. The last U.S. above-ground test took place at Johnston Island on November 4, 1962. During the next three decades, until September 23, 1992, the United States conducted an average of 2.4 underground nuclear explosions per month, all but a few at the Nevada Test Site (NTS) northwest of Las Vegas.

The Yucca Flat section of the NTS is covered with subsidence craters resulting from the collapse of terrain over radioactive underground caverns created by nuclear explosions (see photo).

After the 1974 Threshold Test Ban Treaty (TTBT), which limited underground explosions to 150 kilotons or less, warheads like the half-megaton W88 had to be tested at less than full yield. Since the primary must be detonated at full yield in order to generate data about the implosion of the secondary, the reduction in yield had to come from the secondary. Replacing much of the lithium-6 deuteride fusion fuel with lithium-7 hydride limited the tritium available for fusion, and thus the overall yield, without changing the dynamics of the implosion. The functioning of the device could be evaluated using light pipes, other sensing devices, and analysis of trapped weapon debris. The full yield of the stockpiled weapon could be calculated by extrapolation.

4.8   Production facilities

When two-stage weapons became standard in the early 1950s, weapon design determined the layout of the new, widely dispersed U.S. production facilities, and vice versa.

Because primaries tend to be bulky, especially in diameter, plutonium is the fissile material of choice for pits, with beryllium reflectors. It has a smaller critical mass than uranium. The Rocky Flats plant near Boulder, Colorado, was built in 1952 for pit production and consequently became the plutonium and beryllium fabrication facility.

The Y-12 plant in Oak Ridge, Tennessee, where mass spectrometers called Calutrons had enriched uranium for the Manhattan Project, was redesigned to make secondaries. Fissile U-235 makes the best spark plugs because its critical mass is larger, especially in the cylindrical shape of early thermonuclear secondaries. Early experiments used the two fissile materials in combination, as composite Pu-Oy pits and spark plugs, but for mass production, it was easier to let the factories specialize: plutonium pits in primaries, uranium spark plugs and pushers in secondaries.

Y-12 made lithium-6 deuteride fusion fuel and U-238 parts, the other two ingredients of secondaries.

The Hanford Site near Richland WA operated Plutonium production nuclear reactors and separations facilities during World War 2 and the Cold War. Nine Plutonium production reactors were built and operated there. The first being the B-Reactor which began operations in September 1944 and the last being the N-Reactor which ceased operations in January 1987.

The Savannah River Site in Aiken, South Carolina, also built in 1952, operated nuclear reactors which converted U-238 into Pu-239 for pits, and converted lithium-6 (produced at Y-12) into tritium for booster gas. Since its reactors were moderated with heavy water, deuterium oxide, it also made deuterium for booster gas and for Y-12 to use in making lithium-6 deuteride.

4.9   Warhead design safety

Because even low-yield nuclear warheads have astounding destructive power, weapon designers have always recognised the need to incorporate mechanisms and associated procedures intended to prevent accidental detonation.

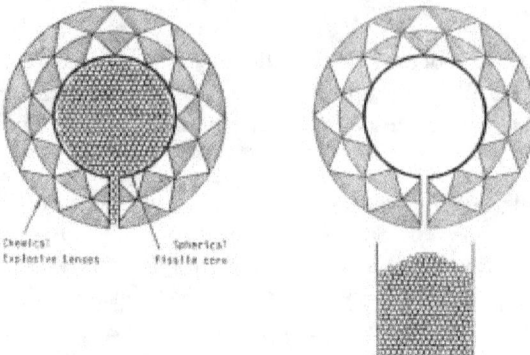

A diagram of the Green Grass *warhead's steel ball safety device, shown left, filled (safe) and right, empty (live). The steel balls were emptied into a hopper underneath the aircraft before flight, and could be re-inserted using a funnel by rotating the bomb on its trolley and raising the hopper.*

Gun-type weapons

It is inherently dangerous to have a weapon containing a quantity and shape of fissile material which can form a critical mass through a relatively simple accident. Because of this danger, the propellant in Little Boy (four bags of cordite) was inserted into the bomb in flight, shortly after takeoff on August 6, 1945. This was the first time a gun-type nuclear weapon had ever been fully assembled.

If the weapon falls into water, the moderating effect of the water can also cause a criticality accident, even without the weapon being physically damaged. Similarly, a fire caused by an aircraft crashing could easily ignite the propellant, with catastrophic results. Gun-type weapons have always been inherently unsafe.

In-flight pit insertion

Neither of these effects is likely with implosion weapons since there is normally insufficient fissile material to form a critical mass without the correct detonation of the lenses. However, the earliest implosion weapons had pits so close to criticality that accidental detonation with some nuclear yield was a concern.

On August 9, 1945, Fat Man was loaded onto its airplane fully assembled, but later, when levitated pits made a space between the pit and the tamper, it was feasible to use in-flight pit insertion. The bomber would take off with no fissile material in the bomb. Some older implosion-type weapons, such as the US Mark 4 and Mark 5, used this system.

In-flight pit insertion will not work with a hollow pit in contact with its tamper.

Steel ball safety method

As shown in the diagram above, one method used to decrease the likelihood of accidental detonation employed metal balls. The balls were emptied into the pit: this prevented detonation by increasing the density of the hollow pit, thereby preventing symmetrical implosion in the event of an accident. This design was used in the Green Grass weapon, also known as the Interim Megaton Weapon, which was used in the Violet Club and Yellow Sun Mk.1 bombs.

One-Point Safety Test

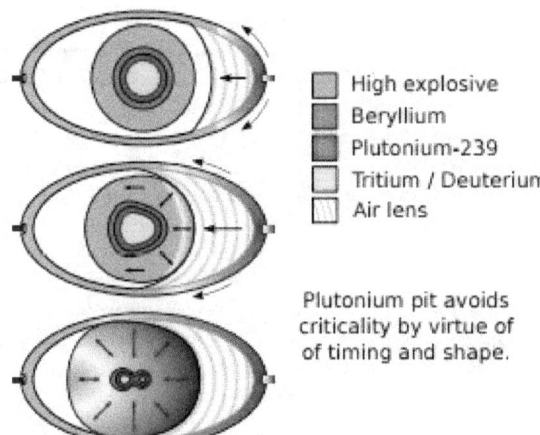

High explosive
Beryllium
Plutonium-239
Tritium / Deuterium
Air lens

Plutonium pit avoids criticality by virtue of of timing and shape.

Chain safety method

Alternatively, the pit can be "safed" by having its normally hollow core filled with an inert material such as a fine metal chain, possibly made of cadmium to absorb neutrons.

While the chain is in the center of the pit, the pit can not be compressed into an appropriate shape to fission; when the weapon is to be armed, the chain is removed. Similarly, although a serious fire could detonate the explosives, destroying the pit and spreading plutonium to contaminate the surroundings as has happened in several weapons accidents, it could not cause a nuclear explosion.

One-point safety

While the firing of one detonator out of many will not cause a hollow pit to go critical, especially a low-mass hollow pit that requires boosting, the introduction of two-point implosion systems made that possibility a real concern.

In a two-point system, if one detonator fires, one entire hemisphere of the pit will implode as designed. The high-explosive charge surrounding the other hemisphere will explode progressively, from the equator toward the opposite pole. Ideally, this will pinch the equator and squeeze the second hemisphere away from the first, like toothpaste in a tube. By the time the explosion envelops it, its implosion will be separated both in time and space from the implosion of the first hemisphere. The resulting dumbbell shape, with each end reaching maximum density at a different time, may not become critical.

Unfortunately, it is not possible to tell on the drawing board how this will play out. Nor is it possible using a dummy pit of U-238 and high-speed x-ray cameras, although such tests are helpful. For final determination, a test needs to be made with real fissile material. Consequently, starting in 1957, a year after Swan, both labs began one-point safety tests.

Out of 25 one-point safety tests conducted in 1957 and 1958, seven had zero or slight nuclear yield (success), three had high yields of 300 t to 500 t (severe failure), and the rest had unacceptable yields between those extremes.

Of particular concern was Livermore's W47, which generated unacceptably high yields in one-point testing. To prevent an accidental detonation, Livermore decided to use mechanical safing on the W47. The wire safety scheme described below was the result.

When testing resumed in 1961, and continued for three decades, there was sufficient time to make all warhead designs inherently one-point safe, without need for mechanical safing.

Wire safety method

One particularly dangerous warhead was Livermore's W47, designed for the Polaris submarine missile. The last test before the 1958 moratorium was a one-point test of the W47 primary, which had an unacceptably high nuclear yield of

400 lb (180 kg) of TNT equivalent (Hardtack II Titania). With the test moratorium in force, there was no way to refine the design and make it inherently one-point safe. Los Alamos had a suitable primary that was one-point safe, but rather than share with Los Alamos the credit for designing the first SLBM warhead, Livermore chose to use mechanical safing on its own inherently unsafe primary. The result was a safety scheme consisting of a boron-coated wire inserted into the hollow pit at manufacture. The warhead was armed by withdrawing the wire onto a spool driven by an electric motor. Once withdrawn, the wire could not be reinserted.[68] The wire had a tendency to become brittle during storage, and break or get stuck during arming, preventing complete removal and rendering the warhead a dud.[69] It was estimated that 50-75% of warheads would fail. This required a complete rebuild of the W47 primaries.[70] The oil used for lubricating the wire also promoted corrosion of the pit.[71]

Strong link weak link

A strong link/weak link and exclusion zone nuclear detonation mechanism is a form of automatic safety interlock.

Permissive Action Links

In addition to the above steps to reduce the probability of a nuclear detonation arising from a single fault, locking mechanisms referred to by NATO states as Permissive Action Links are sometimes attached to the control mechanisms for nuclear warheads. Permissive Action Links act solely to prevent the unauthorised use of a nuclear weapon.

4.10   References

4.10.1   Bibliography

- Cohen, Sam, *The Truth About the Neutron Bomb: The Inventor of the Bomb Speaks Out*, William Morrow & Co., 1983

- Coster-Mullen, John, "Atom Bombs: The Top Secret Inside Story of Little Boy and Fat Man", Self-Published, 2011

- Glasstone, Samuel and Dolan, Philip J., editors, *The Effects of Nuclear Weapons (third edition)* (PDF), U.S. Government Printing Office, 1977.

- Grace, S. Charles, *Nuclear Weapons: Principles, Effects and Survivability (Land Warfare: Brassey's New Battlefield Weapons Systems and Technology, vol 10)*

- Hansen, Chuck, "Swords of Armageddon: U.S. Nuclear Weapons Development since 1945" (CD-ROM & download available). PDF. 2,600 pages, Sunnyvale, California, Chucklea Publications, 1995, 2007. ISBN 978-0-9791915-0-3 (2nd Ed.)

- *The Effects of Nuclear War*. Office of Technology Assessment (May 1979).

- Rhodes, Richard. *The Making of the Atomic Bomb*. Simon and Schuster, New York, (1986 ISBN 978-0-684-81378-3)

- Rhodes, Richard. *Dark Sun: The Making of the Hydrogen Bomb*. Simon and Schuster, New York, (1995 ISBN 978-0-684-82414-7)

- Smyth, Henry DeWolf, *Atomic Energy for Military Purposes*, Princeton University Press, 1945. (see: Smyth Report)

4.10.2   Notes

[1] The physics package is the nuclear explosive module inside the bomb casing, missile warhead, or artillery shell, etc., which delivers the weapon to its target. While photographs of weapon casings are common, photographs of the physics package are quite rare, even for the oldest and crudest nuclear weapons. For a photograph of a modern physics package see W80.

[2] Life Editors (1961), "To the Outside World, a Superbomb more Bluff than Bang", *Life*, New York (Vol. 51, No. 19, November 10, 1961), pp. 34–37, retrieved 2010-06-28. Article on the Soviet Tsar Bomba test. Because explosions are spherical in shape and targets are spread out on the relatively flat surface of the earth, numerous smaller weapons cause more destruction. From page 35: ". . .five five-megaton weapons would demolish a greater area than a single 50-megatonner."

[3] The United States and the Soviet Union were the only nations to build large nuclear arsenals with every possible type of nuclear weapon. The U.S. had a four-year head start and was the first to produce fissile material and fission weapons, all in 1945. The only Soviet claim for a design first was the Joe 4 detonation on August 12, 1953, said to be the first deliverable hydrogen bomb. However, as Herbert York first revealed in *The Advisors: Oppenheimer, Teller and the Superbomb* (W.H. Freeman, 1976), it was not a true hydrogen bomb (it was a boosted fission weapon of the Sloika/Alarm Clock type, not a two-stage thermonuclear). Soviet dates for the essential elements of warhead miniaturization – boosted, hollow-pit, two-point, air lens primaries – are not available in the open literature, but the larger size of Soviet ballistic missiles is often explained as evidence of an initial Soviet difficulty in miniaturizing warheads.

[4] fr 971324, Caisse Nationale de la Recherche Scientifique (National Fund for Scientific Research), "Perfectionnements aux charges explosives (Improvements to explosive charges)", published 16 January 1951, issued 12 July 1950.

[5] The main source for this section is Samuel Glasstone and Philip Dolan, *The Effects of Nuclear Weapons*, Third Edition, 1977, U.S. Dept of Defense and U.S. Dept of Energy (see links in General References, below), with the same information in more detail in Samuel Glasstone, *Sourcebook on Atomic Energy*, Third Edition, 1979, U.S. Atomic Energy Commission, Krieger Publishing.

[6] Glasstone and Dolan, *Effects*, p. 12.

[7] Glasstone, *Sourcebook*, p. 503.

[8] "neutrons carry off most of the reaction energy," Glasstone and Dolan, *Effects*, p. 21.

[9] Glasstone and Dolan, *Effects*, p. 21.

[10] Martin, James E. *Physics for Radiation Protection*. WILEY-VCH Verlag GmbH & Co. KGaA, Weinheim, 2006, p. 195.

[11] Glasstone and Dolan, *Effects*, p. 12–13. When one pound (454 g) of U-235 undergoes complete fission, the yield is 8 kilotons. The 13 to 16-kiloton yield of the Little Boy bomb was therefore produced by the fission no more than 2 pounds (910 g) of U-235, out of the 141 pounds (64,000 g) in the pit. The remaining 139 pounds (63 kg), 98.5% of the total, contributed nothing to the energy yield.

[12] Compere, A.L., and Griffith, W.L. 1991. "The U.S. Calutron Program for Uranium Enrichment: History,. Technology, Operations, and Production. Report," ORNL-5928, as cited in John Coster-Mullen, "Atom Bombs: The Top Secret Inside Story of Little Boy and Fat Man," 2003, footnote 28, p. 18. The total wartime output of Oralloy produced at Oak Ridge by July 28, 1945 was 165 pounds (74.68 kg). Of this amount, 84% was scattered over Hiroshima (see previous footnote).

[13] Hoddeson, Lillian; et al. (2004). *Critical Assembly: A Technical History of Los Alamos During the Oppenheimer Years, 1943-1945*. Cambridge University Press. p. 271. ISBN 0-521-54117-4.

[14] "Restricted Data Declassification Decisions from 1945 until Present" – "Fact that plutonium and uranium may be bonded to each other in unspecified pits or weapons."

[15] "Restricted Data Declassification Decisions from 1946 until Present". Retrieved 7 October 2014.

[16] Fissionable Materials section of the Nuclear Weapons FAQ, Carey Sublette, accessed Sept 23, 2006

[17] All information on nuclear weapon tests comes from Chuck Hansen, *The Swords of Armageddon: U.S. Nuclear Weapons Development since 1945*, October 1995, Chucklea Productions, Volume VIII, p. 154, Table A-1, "U.S. Nuclear Detonations and Tests, 1945–1962."

[18] Nuclear Weapons FAQ: 4.1.6.3 Hybrid Assembly Techniques, accessed December 1, 2007. Drawing adapted from the same source.

[19] Nuclear Weapons FAQ: 4.1.6.2.2.4 Cylindrical and Planar Shock Techniques, accessed December 1, 2007.

[20] "Restricted Data Declassification Decisions from 1946 until Present", Section V.B.2.k "The fact of use in high explosive assembled (HEA) weapons of spherical shells of fissile materials, sealed pits; air and ring HE lenses," declassified November 1972.

[21] 4.4 Elements of Thermonuclear Weapon Design. Nuclearweaponarchive.org. Retrieved on 2011-05-01.

[22] Until a reliable design was worked out in the early 1950s, the hydrogen bomb (public name) was called the superbomb by insiders. After that, insiders used a more descriptive name: two-stage thermonuclear. Two examples. From Herb York, *The Advisors*, 1976, "This book is about ... the development of the H-bomb, or the superbomb as it was then called." p. ix, and "The rapid and successful development of the superbomb (or super as it came to be called) . . ." p. 5. From National Public Radio Talk of the Nation, November 8, 2005, Siegfried Hecker of Los Alamos, "the hydrogen bomb – that is, a two-stage thermonuclear device, as we referred to it – is indeed the principal part of the US arsenal, as it is of the Russian arsenal."

[23] Howard Morland, "Born Secret," **Cardozo Law Review**, March 2005, pp. 1401–1408.

[24] http://www.wisconsinproject.org/countries/israel/nuke.html

[25] "Improved Security, Safety & Manufacturability of the Reliable Replacement Warhead," NNSA March 2007.

[26] A 1976 drawing which depicts an interstage that absorbs and re-radiates x-rays. From Howard Morland, "The Article," **Cardozo Law Review**, March 2005, p 1374.

[27] Ian Sample (6 March 2008). "Technical hitch delays renewal of nuclear warheads for Trident". *The Guardian*.

[28] "ArmsControlWonk: FOGBANK", March 7, 2008. (Accessed 2010-04-06)

[29] "SAND8.8 – 1151 Nuclear Weapon Data – Sigma I," Sandia Laboratories, September 1988.

[30] The Greenpeace drawing. From Morland, **Cardozo Law Review**, March 2005, p 1378.

[31] Herbert York, *The Advisors: Oppenheimer, Teller and the Superbomb* (1976).

[32] "The 'Alarm Clock' ... became practical only by the inclusion of Li6 (in 1950) and its combination with the radiation implosion." Hans A. Bethe, Memorandum on the History of Thermonuclear Program, May 28, 1952.

[33] Rhodes 1995, p. 256.

[34] See map.

[35] The Soviet Program for Peaceful Uses of Nuclear Explosions by Milo D. Nordyke. Science & Global Security, 1998, Volume 7, pp. 1-117

[36] 4.5 Thermonuclear Weapon Designs and Later Subsections. Nuclearweaponarchive.org. Retrieved on 2011-05-01.

[37] Operation Hardtack I. Nuclearweaponarchive.org. Retrieved on 2011-05-01.

[38] Operation Redwing. Nuclearweaponarchive.org. Retrieved on 2011-05-01.

[39] Weapon and Technology: 4th Generation Nuclear Nanotech Weapons. Weapons.technology.youngester.com (2010-04-19). Retrieved on 2011-05-01.

[40] Fourth Generation Nuclear Weapons. Nuclearweaponarchive.org. Retrieved on 2011-05-01.

[41] Never say "never". Whyfiles.org. Retrieved on 2011-05-01.

[42] Samuel Glasstone, *The Effects of Nuclear Weapons*, 1962, Revised 1964, U.S. Dept of Defense and U.S. Dept of Energy, pp.464–5. This section was removed from later editions, but, according to Glasstone in 1978, not because it was inaccurate or because the weapons had changed.

[43] "Nuclear Weapons FAQ: 1.6".

[44] "Radiological investigations at the "Taiga" nuclear explosion site: Site description and in situ measurements V Ramzaev, V Repin, A Medvedev, E Khramtsov… - Journal of environmental …, 2011 - Elsevier". Retrieved 7 October 2014.

[45] "Radiological investigations at the "Taiga" nuclear explosion site, part II: man-made γ-ray emitting radionuclides in the ground and the resultant kerma rate in air V Ramzaev, V Repin, A Medvedev, E Khramtsov… - Journal of environmental …, 2012 - Elsevier". Retrieved 7 October 2014.

[46] Winterberg, Friedwardt (2010). *The Release of Thermonuclear Energy by Inertial Confinement: Ways Towards Ignition*. World Scientific. pp. 192–193. ISBN 9814295914.

[47] Croddy, Eric A.; Wirtz, James J.; Larsen, Jeffrey, Eds. (2005). *Weapons of Mass Destruction: An Encyclopedia of Worldwide Policy, Technology, and History*. ABC-CLIO, Inc. p. 376. ISBN 1851094903.

[48] *How much large can bombs be made through staging? One often finds claims on the public Internet that multiple stages could be combined one after the other, in an arbitrary large number, and that therefore the in-principle yield of a thermonuclear could be increased without limit. Such authors usually conclude this argument with the wise statement that nuclear weapons were made already so destructive, that no one could possibly think of increasing their yield even further, or that their military use would be pointless…The idea of adding four, ten, a hundred stages, in a disciplined and well orderly way, driving a larger radiation-driven implosion after the other sounds much more like a sheer nonsense than an in-principle design for an Armageddon-class weapon. It should be added that, to the best knowledge of this author, statements about the actual yield of the most powerful weapons in the U.S. nuclear arsenal, either deployed or envisaged at some stage, were declassified, but no detailed hints at triple staging were released in the open from official sources. Also, there are (convincing) well-known sketches and some reasonable-looking calculations in the open literature about two-stage weapons, but no similarly accurate descriptions of true three stages concepts.* http://www.ieri.be/fr/publications/ierinews/2011/juillet/fission-fusion-and-staging.

[49] The Air Force and Strategic Detterence 1951-1960. USAF historical division LIAISON OFFICE by George F. Lemmer 1967. Formerly restricted data

[50] Bowen and Little, "AF Atomic Energy Program" Vol I-V RS. Authors: Lee Bowen and Stuart Little.

[51] http://documents.theblackvault.com/documents/foia/FOIA%2014-00108-H.pdf

[52] "A new use for nuclear weapons: hunting rogue asteroids A persistent campaign by weapons designers to develop a nuclear defense against extraterrestrial rocks slowly wins government support 2013". *Center for Public Integrity*. Retrieved 7 October 2014.

[53] Jason Mick (October 17, 2013). "The mother of all bombs would sit in wait in an orbitary platform".

[54] planetary defense workshop LLNL 1995

[55] "Necessary conditions for the initiation and propagation of nuclear-detonation waves in plane atmospheres Phys. Rev. A 20, 316 – Published 1 July 1979 Thomas A. Weaver and Lowell Wood.". *Physical Review A*. Retrieved 7 October 2014.

[56] "Neutron bomb: Why 'clean' is deadly". *BBC News*. July 15, 1999. Retrieved January 6, 2010.

[57] Broad, William J. (7 September 1999), "Spies versus sweat, the debate over China's nuclear advance," *The New York Times*, p 1. The front page drawing was similar to one that appeared four months earlier in the *San Jose Mercury News*.

[58] Jonathan Medalia, "The Reliable Replacement Warhead Program: Background and Current Developments," CRS Report RL32929, Dec 18, 2007, p CRS-11.

[59] Richard Garwin, "Why China Won't Build U.S. Warheads", *Arms Control Today*, April–May 1999. Archived November 5, 2005, at the Wayback Machine.

[60] Home – NNSA Archived April 1, 2007, at the Wayback Machine.

[61] DoE Fact Sheet: Reliable Replacement Warhead Program Archived August 20, 2008, at the Wayback Machine.

[62] William J. Broad, "The Hidden Travels of The Bomb: Atomic insiders say the weapon was invented only once, and its secrets were spread around the globe by spies, scientists and the covert acts of nuclear states," *New York Times*, December 9, 2008, p D1.

[63] Sybil Francis, *Warhead Politics: Livermore and the Competitive System of Nuclear Warhead Design*, UCRL-LR-124754, June 1995, Ph.D. Dissertation, Massachusetts Institute of Technology, available from National Technical Information Service. This 233-page thesis was written by a weapons-lab outsider for public distribution. The author had access to all the classified information at Livermore that was relevant to her research on warhead design; consequently, she was required to use non-descriptive code words for certain innovations.

[64] Walter Goad, Declaration for the Wen Ho Lee case, May 17, 2000. Goad began thermonuclear weapon design work at Los Alamos in 1950. In his Declaration, he mentions "basic scientific problems of computability which cannot be solved by more computing power alone. These are typified by the problem of long range predictions of weather and climate, and extend to predictions of nuclear weapons behavior. This accounts for the fact that, after the enormous investment of effort over many years, weapons codes can still not be relied on for significantly new designs."

[65] Chuck Hansen, *The Swords of Armageddon*, Volume IV, pp. 211–212, 284.

[66] Dr. John C. Clark, as told to Robert Cahn, "We Were Trapped by Radioactive Fallout," *The Saturday Evening Post*, July 20, 1957, pp. 17–19, 69–71.

[67] Rhodes, Richard (1995). *Dark Sun; the Making of the Hydrogen Bomb*. Simon and Schuster. p. 541.

[68] Chuck Hansen, *The Swords of Armageddon*, Volume VII, pp. 396–397.

[69] Sybil Francis, *Warhead Politics*, pp. 141, 160.

[70] http://www.princeton.edu/sgs/publications/sgs/pdf/4_3harvey.pdf

[71] https://books.google.com/books?id=95eoQSNDp6gC&pg=PA214&dq=warhead+corrosion&lr=&num=50&as_brr=3&ei=C65gS9CtDYLmzAS4i_CLCQ&cd=2#v=onepage&q=&f=false

4.11 External links

- Carey Sublette's Nuclear Weapon Archive is a reliable source of information and has links to other sources.

    - Nuclear Weapons Frequently Asked Questions: Section 4.0 Engineering and Design of Nuclear Weapons

- The Federation of American Scientists provides solid information on weapons of mass destruction, including nuclear weapons and their effects

- Globalsecurity.org provides a well-written primer in nuclear weapons design concepts (site navigation on righthand side).

- More information on the design of two-stage fusion bombs

- Militarily Critical Technologies List (MCTL), Part II (1998) (PDF) from the US Department of Defense at the Federation of American Scientists website.

- "Restricted Data Declassification Decisions from 1946 until Present", Department of Energy report series published from 1994 until January 2001 which lists all known declassification actions and their dates. Hosted by Federation of American Scientists.

- The Holocaust Bomb: A Question of Time is an update of the 1979 court case *USA v. The Progressive*, with links to supporting documents on nuclear weapon design.

- Annotated bibliography on nuclear weapons design from the Alsos Digital Library for Nuclear Issues

- The Woodrow Wilson Center's Nuclear Proliferation International History Project or NPIHP is a global network of individuals and institutions engaged in the study of international nuclear history through archival documents, oral history interviews and other empirical sources.

Chapter 5

List of nuclear weapons

This is a **list of nuclear weapons** listed according to country of origin, and then by type within the states.

This list is incomplete; you can help by expanding it.

5.1 United States

Main article: United States and weapons of mass destruction

US nuclear weapons of **all types** – bombs, warheads, shells, and others – are numbered in the same sequence starting with the Mark 1 and (as of March 2006) ending with the W-91 (which was canceled prior to introduction into service). All designs which were formally intended to be weapons at some point received a number designation. Pure test units which were experiments (and not intended to be weapons) are not numbered in this sequence.

Early weapons were very large and could only be used as free fall bombs. These were known by "Mark" designators, like the Mark 4 which was a development of the Fat Man weapon. As weapons became more sophisticated they also became much smaller and lighter, allowing them to be used in many roles. At this time the weapons began to receive designations based on their role; bombs were given the prefix "B", while the same warhead used in other roles, like missiles, would normally be prefixed "W". For instance, the W-53 warhead was also used as the basis for the B53 nuclear bomb. Such examples share the same sequence number.

In other cases, when the modifications are more significant, variants are assigned their own number. An example is the B61 nuclear bomb, which was the parent design for the W80, W81, and W84. There are also examples of out-of-sequence numbering and other prefixes used in special occasions.

This list includes weapons which were developed to the point of being assigned a model number (and in many cases, prototypes were test fired), but which were then canceled prior to introduction into military service. Those models are listed as canceled, along with the year or date of cancellation of their program.

- Bombs – designated with Mark ("Mk") numbers until 1968, and with "B" numbers after that. "Test Experimental" bombs designated with "TX".

  - Mark 1 – "Little Boy" gun-type uranium weapon (used against Hiroshima). (13–18 kilotons, 1945–1950)

  - Mark 2 – "Thin Man" plutonium gun design—cancelled in 1944

    - Implosion Mark 2 – Another Manhattan Project plutonium implosion weapon, a hollow pit implosion design, was also sometimes referred to as Mark 2. Also cancelled 1944.

  - Mark 3 – "Fat Man" plutonium implosion weapon (used against Nagasaki). (21 kilotons, 1945–1950)

  - Mark 4 – Post-war "Fat Man" redesign. Bomb designed with weapon characteristics as the foremost criteria. (1949–1953)

  - Mark 5 – Significantly smaller high efficiency nuclear bomb. (1–120 kilotons, 1952–1963)

  - Mark 6 – Improved version of Mk-4. (8–160 kilotons, 1951–1962)

  - Mark 7 – Multi-purpose tactical bomb. (8–61 kilotons, 1952–1967)

  - Mark 8 – Gun-assembly, HEU weapon designed for penetrating hardened targets. (25–30 kilotons, 1951–1957)

  - Mark 10 – Improved version of Mk-8 (12–15 kilotons, cancelled May 1952).

  - Mark 11 – Re-designed Mk-8. Gun-type (8–30 kilotons).

- Mark 12 – Light-weight bomb to be carried by fighter planes (12–14 kilotons).

- Mark 13 – Improved version of Mk-6 (cancelled August 1954).

- TX/Mark 14 – First deployable solid-fuel thermonuclear bomb (Castle Union device). Only five produced. (5 Megatons)

- Mark 15 – First "lightweight" thermonuclear weapon. (1.7–3.8 Megatons, 1955–1965)

- TX/Mark 16 – First weaponized thermonuclear weapon (Ivy Mike device). Only cryogenic weapon ever deployed. Only five produced. (6–8 Megatons)

- Mark 17 – High-yield thermonuclear. Heaviest U.S. weapon, second highest yield of any U.S. weapon. Very similar to Mk-24. (10–15 Megatons)

- Mark 18 – Very high yield fission weapon (Ivy King device).

- Mark 20 – Improved Mark 13 (cancelled 1954)

- Mark 21 – Re-designed variant of Castle Bravo test

- Mark 22 – Failed thermonuclear design (Castle Koon device, cancelled April 1954).

- Mark 24 – High-yield thermonuclear, very similar to Mk-17 but had a different secondary.

- Mark 26 – Similar design to Mk 21 (cancelled 1956).

- Mark 27 – Navy nuclear bomb (1958–1965)

- B28 nuclear bomb (Mark 28) (1958–1991)

- Mark 36 – Strategic nuclear bomb (1956–1961) 9–10 Megatons

- B39 nuclear bomb (Mark 39) (1957–1966)

- B41 nuclear bomb (Mark 41) (1960–1976); highest yield US nuclear weapon (25 Megatons).

- B43 nuclear bomb (Mark 43) (1961–1991)

- B46 nuclear bomb or (Mark 46); experimental, design evolved into B53 nuclear bomb and W-53 warhead (cancelled 1958)

- Mk 101 Lulu

- B53 nuclear bomb (1962–1997; dismantled 2010–2011)

- B57 nuclear bomb (1963–1993)

- B61 nuclear bomb (1966–present)

- B77 nuclear bomb (cancelled 1977)

- B83 nuclear bomb (1983–present)

- B90 nuclear bomb (cancelled 1991)

- Nuclear artillery shells

  - 16-inch (406 mm)
    - W23 (1956–1962) Gun-type

  - 280 mm:
    - W9 (1952–1957) Gun-type
    - W19 (1953–1956) Gun-type, W9 derivative

  - 8-inch (203 mm)
    - W33 (1956–1980s) Gun-type
    - W75 (cancelled 1973)
    - W79 (1981–1992)

  - 155mm
    - W48 (1963–1992)
    - W74 (cancelled 1973)
    - W82 (cancelled 1983 (W-82-0 Enhanced Radiation) and 1990 (W-82-1 fission only))

- Atomic Demolition Munitions

  - W-7/ADM-B (c. 1954–1967)
  - T4 ADM (1957–1963) Gun-type
  - W30/Tactical Atomic Demolition Munition (1961–1966)
  - W31/ADM (1960–1965)
  - W45/Medium Atomic Demolition Munition (1964–1984)
  - W54/Special Atomic Demolition Munition (1965–1989)

- Missile warheads

  - W4 for SM-62 Snark missile (cancelled 1951)
  - W5 for MGM-1 Matador (1954–1963)
  - W7 for MGR-1 Honest John (1954–1960), Corporal SRBM (1955–1964), Nike Hercules SAM (1958–1960s)
  - W8 for SSM-N-8 Regulus, Gun-type (cancelled 1955)
  - W12 for RIM-8 Talos missile (cancelled 1955)
  - W13 for SM-62 Snark missile and Redstone MRBM (cancelled 1954)
  - W15 for missiles (cancelled 1957)
  - W21 for B-58 bomber, SM-64 Navaho missile (cancelled 1957)
  - W25 for MB-1 "Ding Dong", later AIR-2 Genie (1957–1984)

- W27 for SSM-N-8 Regulus missile (1958–1965)

- W28 for AGM-28 Hound Dog missile, MGM-13 Mace missile (1958–1976)

- W29 for (cancelled 1955)

- W30 for RIM-8 Talos missile (1959–1979)

- W31 for Honest John (1961–1985), Nike Hercules (1960s–1988)

- W34 for Mk101 Lulu nuclear depth charge, Mk45 ASTOR torpedo, Mk105 bomb (1958–1976)

- W35 for Atlas ICBM, Titan I ICBM, Thor IRBM, PGM-19 Jupiter (cancelled 1958)

- W37 (cancelled 1956)

- W38 for Atlas ICBM and Titan I ICBM (1961–1965)

- W39 for Redstone MRBM (1958–1964)

- W40 for MGM-18 Lacrosse SRBM (1959–1964)

- W41 for (cancelled 1957)

- W42 for Air to Air and Surface to Air missiles (cancelled 1961)

- W44 for ASROC (1961–1989)

1962 test of an ASROC antisubmarine rocket armed with the W44

- W45 for Little John rocket, RIM-2 Terrier and AGM-12 Bullpup missiles, MADM (1961–1969 (some 1988))

- W46 for Redstone, Snark, B-58 (cancelled 1958)

- W47 for Polaris SLBM (1960–1974)

- W49 for PGM-19 Jupiter (1959–1963) and Thor IRBM (1959–1963)

- W50 for MGM-31 Pershing (1960–1990)

- W51 for various (program converted to W54 in 1959)

- W52 for MGM-29 Sergeant (1962–1977)

- W53 for LGM-25C Titan II (1962–1987)

- W54 for Davy Crockett recoilless rifle and AIM-26 Falcon AAM (1961–1972)

- W55 for Subroc (1965–1989)

- W56 for Minuteman I and II ICBM (1963–1993)

- W58 for Polaris A-3 SLBM (1964–1982)

- W59 for Minuteman I ICBM and Skybolt missile (1962–1969)

- W60 for Typhon SAM (cancelled 1963)

- W62 for Minuteman III ICBM, (1970–2010)

- W63 for Lance SRBM (cancelled 1966)

- W64 for Lance SRBM (cancelled 1964)

- W65 for Sprint ABM (cancelled 1968)

- W66 for Sprint ABM (1970–1975)

- W67 for Poseidon SLBM and Minuteman III ICBM (cancelled 1967)

- W68 for Poseidon SLBM (1970–1991)

- W69 for AGM-69 SRAM (1972–1990)

- W70 for Lance SRBM (1973–1992)

- W71 for LIM-49A Spartan ABM (1974–1975; dismantled 1992)

- W72 for AGM-62 Walleye (1970–1979)

- W73 for Condor missile (cancelled 1970)

- W76 for Trident I SLBM (1978–present)

- W78 for LGM-30 Minuteman III (1979–present)

- W80 for AGM-86 ALCM, AGM-129 ACM and BGM-109 Tomahawk (1981–present)

- W81 for RIM-67 Standard ER, based on B61 (cancelled 1986)

- W84 for BGM-109G Gryphon GLCM (1983–1991)

- W85 for Pershing II Weapon System (1983–1991)

- W86 for Pershing II Weapon System Earth penetrating warhead option (cancelled 1980)

- W87 for Peacekeeper ICBM (1986–2005) and Minuteman III ICBM (2007–present)

  - W87-1 for MGM-134 Midgetman ICBM (cancelled 1992)

- W88 for Trident II SLBM (1988–present)
- W89 for AGM-131 SRAM II (cancelled 1991)
- W91 for SRAM-T (cancelled 1991)
- RNEP (Robust Nuclear Earth Penetrator) design program (2001–2005)

- Cancelled design projects

  - Reliable Replacement Warhead (RRW1) design program (2004–2008)

See also Enduring Stockpile.

5.1.1 Common nuclear primaries

A number of American weapons designs shared common components between several designs. These include publicly identified models listed below.

5.2 Soviet Union/Russian Federation

Main article: Russia and weapons of mass destruction

At the peak of its arsenal in 1988, Russia possessed around 45,000 nuclear weapons in its stockpile, roughly 13,000 more than the United States arsenal, the second largest in the world, which peaked in 1966.[1]

- Tests

  - Joe-1

- Torpedoes

  - 53-58 torpedo with 10 kilotons RDS-9 warhead
  - 65-73 torpedo with 20 kilotons
  - VA-111 Shkval with 150 kilotons

- Bombs

  - RDS-1, 22 kiloton bomb. Tested 29 August 1949 as "First Light" (Joe 1). Total of 5 stockpiled
  - RDS-2, 38 kiloton bomb. Tested 24 September 1951 as "Second Light." The RDS-2 was an entirely Russian design, delayed by development of the RDS-1
  - RDS-3, 42 kiloton bomb. First Soviet bomb tested in an airdrop on 18 October 1951. First 'mass-produced" Soviet bomb

  - RDS-3I, 62 kiloton bomb. Tested 24 October 1954. The RDS-31 was an improved RDS-3 with external neutron generator
  - RDS-4, "Tatyana" 42 kiloton bomb. The RDS-4 was smaller and lighter than previous Soviet Bombs.
  - RDS-6, also known as RDS-6S, or "sloika" or 'layer cake" gaining about 20% of its yield from fusion. RDS-6 was tested on 12 August 1953. Yield 400 kilotons
  - RDS-7, a backup for the RDS-6, the RDS-7 was a 500 kiloton all fission bomb comparable to the US Mk-18, development dropped after success of the RDS-6S
  - RDS-27, 250 kiloton bomb, a 'boosted' fission bomb tested 6 November 1955.
  - RDS-37, 1.6 megaton bomb, the first Soviet two-stage hydrogen bomb, tested 22 November 1955
  - RDS-220 Tsar Bomba an extremely large three stage bomb, initially designed as a 100-megaton-bomb, but was scaled down to 50 megatons for testing.

- Intercontinental Ballistic Missiles

  - RDS-9, 40 kiloton warhead[2] for R-5M MRBM (SS-3)
  - RDS-37 3 megaton warhead[3] for R7 Semyorka / SS-6 ICBM
  - RDS-46 5 megaton warhead[3] for R-7A Semyorka / SS-6 ICBM
  - 8F17 3 megaton[4] warhead for R-16 / SS-7 ICBM
  - 8F115 and 8F116 5-6 megaton[4] warhead for R-16 / SS-7 ICBM
  - Unknown model warheads for R-9 / SS-8 Sasin ICBM
  - 15F42 1.2 megaton warhead for UR 100U / SS-11 Mod 3 Sego ICBM
  - Unknown model 750 kiloton to 1.0 megaton warhead for RT-2 / SS-13 Mod 1 Savage ICBM
  - 15F1r 750 kiloton to 1.65 megaton warhead for RT-2 / SS-13 Mod 2 Savage ICBM
  - Unknown model 466 kiloton warhead for RT-2 / SS-13 Mod 3 Savage ICBM
  - Unknown model 500 kiloton warhead for RT-20 / SS-15 Scrooge ICBM
  - Unknown model 1.5 megaton warhead for RT-20 / SS-15 Scrooge ICBM

- Unknown model 650 kiloton to 1.5 megaton warheads for RT-21 Temp 2S SS-16 Sinner ICBM

- Unknown model 300–750 kiloton warheads for MR-UR-100 Sotka / SS-17 Spanker Mod 1 ICBM

- Unknown model 4–6 megaton warhead for MR-UR-100 Sotka / SS-17 Spanker Mod 2 ICBM

- 8F675 (Mod2) 20 megaton warhead for R-36M2 / SS-18 Satan ICBM

- 8F021 2 or 5 megaton warheads for R-36MP / SS-18 Satan ICBM (3 MIRV warheads)

- unknown 550 kiloton warheads for R-36M2 / SS-18 Satan ICBM (10 MIRV warheads)

- Unknown model 750 kiloton warheads for R-36M2 / SS-18 Satan ICBM (10 MIRV warheads)

- Unknown model 550 kiloton warheads for UR-100N / SS-19 Mod 1 Stilleto ICBM (6 MIRV warheads)

- Unknown model 2.5–5 megaton warhead for UR-100N / SS-19 Mod 2 Stilleto ICBM

- Unknown model 550 kiloton warheads for RT-23 Molodets / SS-24 Scalpel ICBM (10 MIRV warheads)

- Unknown model 550 kiloton warhead for RT-2PM Topol / SS-25 Sickle ICBM

- Unknown model 550 kiloton warhead for RT-2UTTH Topol M / SS-27 ICBM

- Various tactical nuclear weapons including "suitcase bombs" (RA-115 or RA-115-01 as examples)

5.3 United Kingdom

Main article: United Kingdom and Weapons of Mass Destruction

Blue Steel Yellow Sun productionised air-delivered Thermonuclear bomb casing.

- Warheads
  - Blue Danube Tallboy casing with Fission warhead.
  - Red Snow for Yellow Sun Mk.2.
  - Green Grass For Yellow Sun Mk.1.
  - Red Beard, tactical nuclear weapon.

- WE.177 (also used as a nuclear depth charge).

- Blue Cat - nuclear warhead a.k.a. Tony - UK version of US W44, a.k.a. *Tsetse*.

- Blue Fox - kiloton range nuclear weapon, later renamed Indigo Hammer - not to be confused with the later Blue Fox radar.

- Blue Peacock ten-kiloton nuclear land mine, a.k.a. the "chicken-powered nuclear bomb", originally 'Blue Bunny' It used the Blue Danube physics package.

- Blue Rosette - short-case nuclear weapon bomb casing for reconnaissance bomber to spec R156T, including the Avro 730, Handley Page HP.100, English Electric P10, Vickers SP4 and various others.

- Blue Slug - nuclear ship-to-ship missile using Sea Slug launcher.

- Blue Water - nuclear tipped surface to surface missile.

- Green Bamboo - nuclear weapon.

- Green Cheese - nuclear anti-ship missile.

- Green Flash - Green Cheese's replacement.

- Green Granite - nuclear weapons - Green Granite (small) & Green Granite (large).

- Green Grass - nuclear weapon

- Indigo Hammer - nuclear weapon

- Violet Club - nuclear weapon

5.4 France

Main articles: France and weapons of mass destruction and Force de Frappe

France is said to have an arsenal of 350 nuclear weapons stockpiled as of 2002.

- Bombs
  - AN 11
  - AN 22
  - AN 52 (MR 50 CTC)

- Warheads (and missiles)
  - MR 31 (S2)
  - MR 41 (M1 and M2)
  - MR 50 CTC (AN 51 CTC and AN 52 CTC)

- AN 51 CTC (Pluton)
- AN 52 CTC (AN 52)
- TN 60 (M20)
- TN 61 (M20 and S3)
- TN 70 MIRV (M4)
- TN 71 MIRV (M4)
- TN 75 MIRV (M45 and M51)
- TN 76 MIRV (M5)
- TN 80 (ASMP)
- TN 81 (ASMP)
- TN 90 (Hàdes)
- TNA (ASMP-A)
- TNO MIRV (M51)

5.5 China

Main article: China and weapons of mass destruction

China is believed to possess around 250 nuclear weapons, but has released very little information about the contents of its arsenal.

- Tests:
  - 596 (nuclear test)
  - Test No. 6
- Ballistic Missiles:
  - DF-1
  - DF-2
  - DF-3A
  - DF-4
  - DF-5
  - DF-11
  - DF-15
  - DF-21
  - DF-31
  - DF-31B
  - DF-41
  - JL-1
  - JL-2
  - B-611
  - P-12

- Cruise Missiles
  - DH-10
  - CJ-10
  - HN1
  - HN2
  - HN3
  - CF-2
  - CF-1
  - SS-N-2

5.6 India

Main article: India and weapons of mass destruction

India is believed to possess between 90-110 nuclear weapons (March 2010 estimate). The specifications of its weapon production are not disclosed to the public.

- Tests:
  - Smiling Buddha
  - Operation Shakti
- Missiles
  - Agni-I
  - Agni-II
  - Agni-III
  - Agni-IV
  - Agni-V
  - Prithvi I
  - Prithvi II
  - Prithvi III
  - Shaurya
  - K-4
  - K-5
  - K-15 Sagarika (missile)
- Cruise Missiles
  - Brahmos
  - Nirbhay
  - P-70 Ametist
  - P-270 Moskit
  - Popeye

5.7   Israel

Main article: Israel and weapons of mass destruction

Israel is widely believed to possess a substantial arsenal of nuclear weapons and missiles, estimated at 75-130 and 100-200[5] warheads, but refuses officially to confirm or deny whether it has a nuclear weapon program, leaving the details of any such weapons unclear. Mordechai Vanunu, a former nuclear technician for Israel, confirmed the existence of a nuclear weapons program in 1986.

Unconfirmed rumors have hinted at tactical nuclear artillery shells, light fission bombs and missile warheads, and perhaps thermonuclear missile warheads.[6]

The BBC News Online website published an article[7] on 28 May 2008, which quotes former U.S. President Jimmy Carter as stating that Israel has at least 150 nuclear weapons. The article continues to state that this is the second confirmation of Israel's nuclear capability by a U.S. spokesman following comments from U.S. Defense Secretary Robert Gates at a Senate hearing and had apparently been confirmed a short time later by Israeli Prime Minister Ehud Olmert.[8]

5.8   Pakistan

Main article: Pakistan and weapons of mass destruction

As of March 2010, Pakistan is believed to possess between 90-120 nuclear weapons. The specifications of its weapon production are not disclosed to the public. The main series for nuclear transportation is Hatf.

- Tests:
    - Chagai-I
    - Chagai-II
- Missiles:
    - Abdali-I (BRBM)
    - Ghaznavi (SRBM)
    - Ghauri (missile) (MRBM)
    - Ghauri-II (MRBM)
    - Ghauri-III (Close IRBM)
    - Hatf-I/IA (BRBM)
    - Shaheen missile (MRBM)
    - Shaheen-II (IRBM)
    - Shaheen-III (IRBM)

- Cruise Missiles:

    - Babur missile (Cruise Missile)
    - Ra'ad (Air Launched Cruise Missile)
    - Nasr

The first two in the above-mentioned series are not confirmed to be capable for nuclear standoff

5.9   North Korea

Main article: North Korea and weapons of mass destruction

North Korea claims to possess nuclear weapons, however, the specifications of its systems are not public. It is estimated to have 6-18 low yield nuclear weapons (August 2012 estimate).[9] On 9 October 2006, North Korea carried out an alleged nuclear test. (See 2006 North Korean nuclear test) Nuclear weapons produced by North Korea are known to have failed.

On 25 May 2009, North Korea conducted a second test of nuclear weapons at the same location as the original test. The test weapon was of the same magnitude as the atomic bombs dropped on Japan in the 2nd World War. At the same time of the test, North Korea tested two short range ballistic missiles. North Korea is continuously carrying on nuclear tests, such as on 2 February 2013, when it tested a 7 kt nuclear weapon.

5.10   South Africa

Main article: South Africa and weapons of mass destruction

South Africa built six or seven gun-type weapons. All constructed weapons were verified by International Atomic Energy Agency and other international observers to have been dismantled, along with the complete weapons program, and their highly enriched uranium was reprocessed back into low enriched form unsuitable for weapons.

5.11   See also

- Lists of nuclear disasters and radioactive incidents

- Nuclear weapon yield

5.12 References

[1] Robert S. Norris and Hans M. Kristensen, "Global nuclear stockpiles, 1945-2006," Bulletin of the Atomic Scientists 62, no. 4 (July/August 2006), 64-66.

[2] "de beste bron van informatie over Nuclear weapons. Deze website is te koop!". atomicforum.org. Retrieved 2012-08-14.

[3] "R-7 - SS-6 SAPWOOD Russian / Soviet Nuclear Forces". Fas.org. Retrieved 2012-08-14.

[4] "R-16 / SS-7 SADDLER - Russian / Soviet Nuclear Forces". Fas.org. Retrieved 2012-08-14.

[5] Normark, Magnus, Anders Lindblad, Anders Norqvist, Björn Sandström and Louise Waldenström. "Israel and WMD: Incentives and Capabilities." Swedish Defence Research Agency FOI-R-—1734--SE December 2005 <http://www.foi.se/FOI/templates/Page____4657.aspx>

[6] *The Samson option: Israel's nuclear arsenal and American foreign policy.* Hersh, Seymour M., New York, Random House, 1991, ISBN 0-394-57006-5

[7] "Middle East | Israel 'has 150 nuclear weapons'". BBC News. 2008-05-26. Retrieved 2012-08-14.

[8] "Israel 'has 150 nuclear weapons'", *BBC News Online* May 28, 2008

[9] "North Korea could have fuel for 48 nuclear weapons by 2015". The Daily Telegraph. 20 August 2012. Retrieved 8 November 2012.

5.12.1 Bibliography

• Holloway, David, *Stalin and the Bomb*, New Haven & London, Yale University Press, 1994, ISBN 0-300-06056-4.

• Zaloga, Steven J., *The Kremlin's Nuclear Sword* Washington, D.C., Smithsonian Institution Press, 2002, ISBN 1-58834-007-4.

• Hansen, Chuck. *U.S. Nuclear Weapons.* Arlington, Texas, Areofax, Inc., 1988. ISBN 0-517-56740-7.

• Gibson, James N. *Nuclear Weapons of the United States*, Altglen, PA, Schiffer Publishing, 1996, ISBN 978-0-7643-0063-9.

• Cochran, Thomas, Arkin, William, Hoenig, Milton "Nuclear Weapons Databook, Volume I, U.S. Nuclear Forces and Capabilities," Cambridge, Massachusetts, Ballinger Pub. Co., 1984, ISBN 0-88410-173-8.

• Hansen, Chuck, "Swords of Armageddon" (CD-ROM & download available). PDF. 2,600 pages, Sunnyvale, California, Chucklea Publications, 1995, 2007. ISBN 978-0-9791915-0-3 (2nd Ed.)

5.13 External links

• CNS Resources on South Africa's Nuclear Weapons Program indicates that "most international experts conclude that South Africa has completed its nuclear disarmament. South Africa is the first and to date only country to build nuclear weapons and then entirely dismantle its nuclear weapons program."

Chapter 6

Thermonuclear weapon

"H Bomb" redirects here. For the American chess player so nicknamed, see Hikaru Nakamura.

A **thermonuclear weapon** is a nuclear weapon that uses the energy from a primary nuclear fission reaction to compress and ignite a secondary nuclear fusion reaction. The result is greatly increased explosive power when compared to single-stage fission weapons. It is colloquially referred to as a **hydrogen bomb** or **H-bomb** because it employs fusion of isotopes of hydrogen. The fission stage in such weapons is required to cause the fusion that occurs in thermonuclear weapons.[1]

The first full scale thermonuclear test was done by the United States in 1952; the concept has since been employed by most of the world's nuclear powers in the design of their weapons.[2] The modern design of all thermonuclear weapons in the United States is known as the *Teller–Ulam configuration* for its two chief contributors, Edward Teller and Stanislaw Ulam, who developed it in 1951[3] for the United States, with certain concepts developed with the contribution of John von Neumann. Similar devices were developed by the Soviet Union, United Kingdom, China, and France.

As thermonuclear weapons represent the most efficient design for weapon energy yield in weapons with yields above 50 kilotons of TNT (210 TJ), virtually all the nuclear weapons deployed by the five nuclear-weapon states under the NPT today are thermonuclear weapons using the Teller–Ulam design.[4]

The radiation implosion mechanism is a heat engine that exploits the temperature difference between the secondary stage's hot, surrounding radiation channel and its relatively cool interior. This temperature difference is briefly maintained by a massive heat barrier called the "pusher", which also serves as an implosion tamper, increasing and prolonging the compression of the secondary. If made of uranium, as is almost always the case, it can capture neutrons produced by the fusion reaction and undergo fission itself, increasing the overall explosive yield. In many Teller–Ulam weapons, fission of the pusher dominates the explosion and

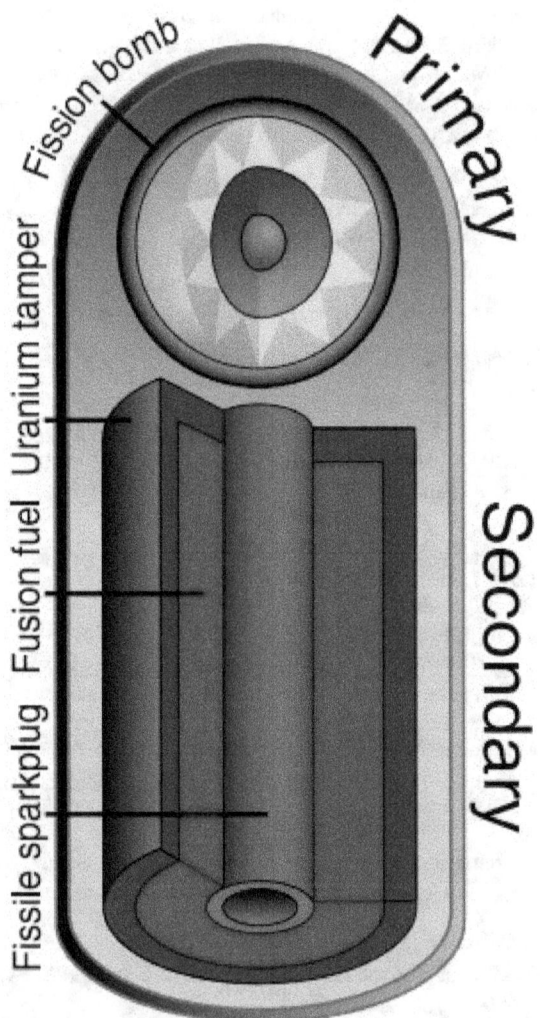

The basics of the Teller–Ulam design for a thermonuclear weapon. Radiation from a primary fission bomb compresses a secondary section containing both fission and fusion fuel. The compressed secondary is heated from within by a second fission explosion.

produces radioactive fission product fallout.

6.1 Public knowledge concerning nuclear weapon design

Edward Teller in 1958

Detailed knowledge of fission and fusion weapons is classified to some degree in virtually every industrialized nation. In the United States, such knowledge can by default be classified as "Restricted Data", even if it is created by persons who are not government employees or associated with weapons programs, in a legal doctrine known as "born secret" (though the constitutional standing of the doctrine has been at times called into question; see *United States v. Progressive, Inc.*). Born secret is rarely invoked for cases of private speculation. The official policy of the United States Department of Energy has been to not acknowledge the leaking of design information, as such acknowledgment would potentially validate the information as accurate. In a small number of prior cases, the U.S. government has attempted to censor weapons information in the public press, with limited success.[5] According to the *New York Times*, physicist Kenneth Ford defied government orders to remove classified information from his new book, *Building the H Bomb: A Personal History*. Ford claims he only used pre-existing information and even submitted a manuscript to the government who wanted to remove entire sections of the book for concern that foreign nations could use the information.[6]

Though large quantities of vague data have been officially released, and larger quantities of vague data have been unofficially leaked by former bomb designers, most public descriptions of nuclear weapon design details rely to some degree on speculation, reverse engineering from known information, or comparison with similar fields of physics (inertial confinement fusion is the primary example). Such processes have resulted in a body of unclassified knowledge about nuclear bombs that is generally consistent with official unclassified information releases, related physics, and is thought to be internally consistent, though there are some points of interpretation that are still considered open. The state of public knowledge about the Teller–Ulam design has been mostly shaped from a few specific incidents outlined in a section below.

6.2 Basic principle

The basic principle of the Teller–Ulam configuration is the idea that different parts of a thermonuclear weapon can be chained together in "stages", with the detonation of each stage providing the energy to ignite the next stage. At a bare minimum, this implies a *primary* section that consists of an implosion-type fission bomb (a "trigger"), and a *secondary* section that consists of fusion fuel. The energy released by the *primary* compresses the *secondary* through a process called "radiation implosion", at which point it is heated and undergoes nuclear fusion. Because of the staged design, it is thought that a *tertiary* section, again of fusion fuel, could be added as well, based on the same principle as the *secondary*; the AN602 "Tsar Bomba" is thought to have been a three-stage device.

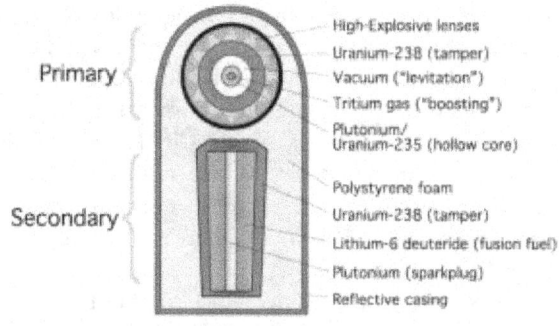

One possible version of the Teller–Ulam configuration

Surrounding the other components is a hohlraum or *radiation case*, a container that traps the first stage or primary's energy inside temporarily. The outside of this radiation case, which is also normally the outside casing of the bomb, is the only direct visual evidence publicly available of any thermonuclear bomb component's configuration. Numerous photographs of various thermonuclear bomb exteriors have been declassified.[7]

The primary is thought to be a standard implosion method fission bomb, though likely with a core boosted by small amounts of fusion fuel (usually 50/50% deuterium/tritium gas) for extra efficiency; the fusion fuel releases excess neutrons when heated and compressed, inducing additional fission. Generally, a research program with the capacity to create a thermonuclear bomb has already mastered the ability to engineer boosted fission. When fired, the plutonium-239 (Pu-239) and/or uranium-235 (U-235) core would be compressed to a smaller sphere by special layers of conventional high explosives arranged around it in an explosive lens pattern, initiating the nuclear chain reaction that powers the conventional "atomic bomb".

The secondary is usually shown as a column of fusion fuel and other components wrapped in many layers. Around the column is first a "pusher-tamper", a heavy layer of uranium-238 (U-238) or lead that serves to help compress the fusion fuel (and, in the case of uranium, may eventually undergo fission itself). Inside this is the fusion fuel itself, usually a form of lithium deuteride, which is used because it is easier to weaponize than liquified tritium/deuterium gas. This dry fuel, when bombarded by neutrons, produces tritium, a heavy isotope of hydrogen which can undergo nuclear fusion, along with the deuterium present in the mixture. (See the article on nuclear fusion for a more detailed technical discussion of fusion reactions.) Inside the layer of fuel is the "spark plug", a hollow column of fissile material (plutonium-239 or uranium-235) that, when compressed, can itself undergo nuclear fission (because of the shape, it is not a critical mass without compression). The tertiary, if one is present, would be set below the secondary and probably be made up of the same materials.[8][9]

Separating the secondary from the primary is the interstage. The fissioning primary produces four types of energy: 1) expanding hot gases from high explosive charges that implode the primary; 2) superheated plasma that was originally the bomb's fissile material and its tamper; 3) the electromagnetic radiation; and 4) the neutrons from the primary's nuclear detonation. The interstage is responsible for accurately modulating the transfer of energy from the primary to the secondary. It must direct the hot gases, plasma, electromagnetic radiation and neutrons toward the right place at the right time. Less than optimal interstage designs have resulted in the secondary failing to work entirely on multiple shots, known as a "fissile fizzle". The Koon shot of Operation Castle is a good example; a small flaw allowed the neutron flux from the primary to prematurely begin heating the secondary, weakening the compression enough to prevent any fusion.

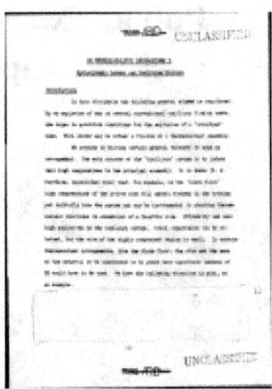

Declassified paper by Teller and Ulam on March 9, 1951: *On Heterocatalytic Detonations I: Hydrodynamic Lenses and Radiation Mirrors*, in which they proposed their revolutionary staged implosion idea. This declassified version is extensively redacted.

There is very little detailed information in the open literature about the mechanism of the interstage. One of the best sources is a simplified diagram of a British thermonuclear weapon similar to the American W80 warhead. It was released by Greenpeace in a report titled *"Dual Use Nuclear Technology"*.[10] The major components and their arrangement are in the diagram, though details are almost absent; what scattered details it does include, likely have intentional omissions and/or inaccuracies. They are labeled "End-cap and Neutron Focus Lens" and "Reflector Wrap"; the former channels neutrons to the U-235/Pu-239 Spark Plug while the latter refers to an X-ray reflector; typically a cylinder made out of an X-ray opaque material such as uranium with the primary and secondary at either end. It does not reflect like a mirror; instead, it gets heated to a high temperature by the X-ray flux from the primary, then it emits more evenly spread X-rays that travel to the secondary, causing what is known as radiation implosion. In Ivy Mike, gold was used as a coating over the uranium to enhance the blackbody effect.[11] Next comes the "Reflector/Neutron Gun Carriage". The reflector seals the gap between the Neutron Focus Lens (in the center) and the outer casing near the primary. It separates the primary from the secondary and performs the same function as the previous reflector. There are about six neutron guns (seen here from Sandia National Laboratories[12]) each poking through the outer edge of the reflector with one end in each section; all are clamped to the carriage and arranged more or less evenly around the casing's circumference. The neutron guns are tilted so the neutron emitting end of each gun end is pointed towards the central axis of the bomb. Neutrons from each neutron gun pass through and are focused by the neutron focus lens towards the centre of primary in order to boost the initial fissioning of the plutonium. A "Polystyrene Polarizer/Plasma Source" is also shown (see below).

The first U.S. government document to mention the interstage was only recently released to the public promoting the 2004 initiation of the Reliable Replacement Warhead Program. A graphic includes blurbs describing the potential advantage of a RRW on a part by part level, with the interstage blurb saying a new design would replace "toxic, brittle material" and "expensive 'special' material... [which require] unique facilities".[13] The "toxic, brittle material" is widely assumed to be beryllium, which fits that description and would also moderate the neutron flux from the primary. Some material to absorb and re-radiate the X-rays in a particular manner may also be used.[14]

The "special material" is thought to be a substance called "FOGBANK", an unclassified codename, though it is often referred to as "*THE* fogbank" (or "*A* Fogbank") as if it were a subassembly instead of a material. Its composition is classified, though aerogel has been suggested as a possibility. Manufacture stopped for many years; however, the *Life Extension Program* required it to start up again – Y-12 currently being the sole producer (the "unique facility" referenced). The manufacturing process used acetonitrile as a solvent, which led to at least three evacuations in 2006. Acetonitrile is widely used in the petroleum and pharmaceutical industries. Like most solvents, it is flammable and can be toxic.[15]

6.2.1 Summary

A simplified summary of the above explanation is:

1. An implosion assembly type of fission bomb is exploded. This is the primary stage. If a small amount of deuterium/tritium gas is placed inside the primary's core, it will be compressed during the explosion and a nuclear fusion reaction will occur; the released neutrons from this fusion reaction will induce further fission in the plutonium-239 or uranium-235 used in the primary stage. The use of fusion fuel to enhance the efficiency of a fission reaction is called boosting. Without boosting, a large portion of the fissile material will remain unreacted; the Little Boy and Fat Man bombs had an efficiency of only 1.4% and 17%, respectively, because they were unboosted.

2. Energy released in the primary stage is transferred to the secondary (or fusion) stage. The exact mechanism whereby this happens is secret. This energy compresses the fusion fuel and sparkplug; the compressed sparkplug becomes critical and undergoes a fission chain reaction, further heating the compressed fusion fuel to a high enough temperature to induce fusion, and also supplying neutrons that react with lithium to create tritium for fusion.

3. The fusion fuel of the secondary stage may be surrounded by depleted uranium or natural uranium, whose U-238 is not fissile and cannot sustain a chain reaction, but which is fissionable when bombarded by the high-energy neutrons released by fusion in the secondary stage. This process provides considerable energy yield (as much as half of the total yield in large devices), but is not considered a tertiary "stage". Tertiary stages are further fusion stages (see below), which have been only rarely used, and then only in the most powerful bombs ever made.

Thermonuclear weapons may or may not use a boosted primary stage, use different types of fusion fuel, and may surround the fusion fuel with beryllium (or another neutron reflecting material) instead of depleted uranium to prevent early premature fission from occurring before the secondary is optimally compressed.

6.3 Compression of the secondary

The basic idea of the Teller–Ulam configuration is that each "stage" would undergo fission or fusion (or both) and release energy, much of which would be transferred to another stage to trigger it. How exactly the energy is "transported" from the *primary* to the *secondary* has been the subject of some disagreement in the open press, but is thought to be transmitted through the X-rays that are emitted from the fissioning *primary*. This energy is then used to compress the *secondary*. The crucial detail of *how* the X-rays create the pressure is the main remaining disputed point in the unclassified press. There are three proposed theories:

- Radiation pressure exerted by the X-rays. This was the first idea put forth by Howard Morland in the article in *The Progressive*.

- X-rays creating a plasma in the radiation case's filler (a polystyrene or "FOGBANK" plastic foam). This was a second idea put forward by Chuck Hansen and later by Howard Morland.

- Tamper/Pusher ablation. This is the concept best supported by physical analysis.

6.3.1 Radiation pressure

The radiation pressure exerted by the large quantity of X-ray photons inside the closed casing might be enough to compress the secondary. Electromagnetic radiation such as X-rays or light carries momentum and exerts a force on any surface it strikes. The pressure of radiation at the intensities seen in everyday life, such as sunlight striking a surface, is usually imperceptible, but at the extreme intensities found in a thermonuclear bomb the pressure is enormous.

For two thermonuclear bombs for which the general size and primary characteristics are well understood, the Ivy Mike test bomb and the modern W-80 cruise missile warhead variant of the W-61 design, the radiation pressure was calculated to be 73 million bar (atmospheres) (7.3 T Pa) for the Ivy Mike design and 1,400 million bar (140 TPa) for the W-80.[16]

6.3.2   Foam plasma pressure

Foam plasma pressure is the concept that Chuck Hansen introduced during the Progressive case, based on research that located declassified documents listing special foams as liner components within the radiation case of thermonuclear weapons.

The sequence of firing the weapon (with the foam) would be as follows:

1. The high explosives surrounding the core of the primary fire, compressing the fissile material into a supercritical state and beginning the fission chain reaction.

2. The fissioning primary emits X-rays, which "reflect" along the inside of the casing, irradiating the polystyrene foam.

3. The irradiated foam becomes a hot plasma, pushing against the tamper of the secondary, compressing it tightly, and beginning the fission reaction in the spark plug.

4. Pushed from both sides (from the primary and the spark plug), the lithium deuteride fuel is highly compressed and heated to thermonuclear temperatures. Also, by being bombarded with neutrons, each lithium−6 atom splits into one tritium atom and one alpha particle. Then begins a fusion reaction between the tritium and the deuterium, releasing even more neutrons, and a huge amount of energy.

5. The fuel undergoing the fusion reaction emits a large flux of neutrons, which irradiates the U-238 tamper (or the U-238 bomb casing), causing it to undergo a fission reaction, providing about half of the total energy.

This would complete the fission-fusion-fission sequence. Fusion, unlike fission, is relatively "clean"—it releases energy but no harmful radioactive products or large amounts of nuclear fallout. The fission reactions though, especially the last fission reaction, release a tremendous amount of fission products and fallout. If the last fission stage is omitted, by replacing the uranium tamper with one made of lead, for example, the overall explosive force is reduced by approximately half but the amount of fallout is relatively low. The neutron bomb is a hydrogen bomb with an intentionally thin tamper, allowing as much radiation as possible to escape.

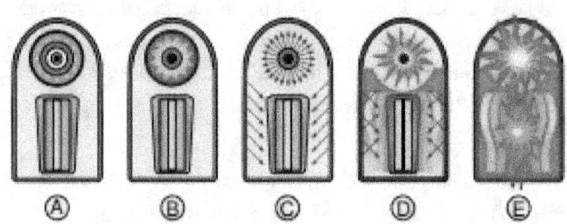

Foam plasma mechanism firing sequence.

1. Warhead before firing; primary (fission bomb) at top, secondary (fusion fuel) at bottom, all suspended in polystyrene foam.

2. High-explosive fires in primary, compressing plutonium core into supercriticality and beginning a fission reaction.

3. Fission primary emits X-rays that are scattered along the inside of the casing, irradiating the polystyrene foam.

4. Polystyrene foam becomes plasma, compressing secondary, and plutonium sparkplug begins to fission.

5. Compressed and heated, lithium-6 deuteride fuel produces tritium and begins the fusion reaction. The neutron flux produced causes the U-238 tamper to fission. A fireball starts to form.

Current technical criticisms of the idea of "foam plasma pressure" focus on unclassified analysis from similar high energy physics fields that indicate that the pressure produced by such a plasma would only be a *small multiplier* of the basic photon pressure within the radiation case, and also that the known foam materials intrinsically have a very low absorption efficiency of the gamma ray and X-ray radiation from the primary. Most of the energy produced would be absorbed by either the walls of the radiation case and/or the tamper around the secondary. Analyzing the effects of that absorbed energy led to the third mechanism: ablation.

6.3.3   Tamper-pusher ablation

The proposed tamper-pusher ablation mechanism is that the primary compression mechanism for the thermonuclear secondary is that the outer layers of the tamper-pusher, or heavy metal casing around the thermonuclear fuel, are heated so much by the X-ray flux from the primary that they ablate away, exploding outwards at such high speed that the rest of the tamper recoils inwards at a tremendous velocity, crushing the fusion fuel and the spark plug.

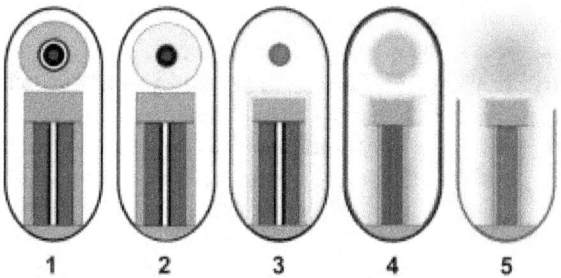

Ablation mechanism firing sequence.

1. *Warhead before firing. The nested spheres at the top are the fission primary; the cylinders below are the fusion secondary device.*

2. *Fission primary's explosives have detonated and collapsed the primary's fissile pit.*

3. *The primary's fission reaction has run to completion, and the primary is now at several million degrees and radiating gamma and hard X-rays, heating up the inside of the hohlraum and the shield and secondary's tamper.*

4. *The primary's reaction is over and it has expanded. The surface of the pusher for the secondary is now so hot that it is also ablating or expanding away, pushing the rest of the secondary (tamper, fusion fuel, and fissile spark plug) inwards. The spark plug starts to fission. Not depicted: the radiation case is also ablating and expanding outwards (omitted for clarity of diagram).*

5. *The secondary's fuel has started the fusion reaction and shortly will burn up. A fireball starts to form.*

Rough calculations for the basic ablation effect are relatively simple: the energy from the primary is distributed evenly onto all of the surfaces within the outer radiation case, with the components coming to a thermal equilibrium, and the effects of that thermal energy are then analyzed. The energy is mostly deposited within about one X-ray optical thickness of the tamper/pusher outer surface, and the temperature of that layer can then be calculated. The velocity at which the surface then expands outwards is calculated and, from a basic Newtonian momentum balance, the velocity at which the rest of the tamper implodes inwards.

Applying the more detailed form of those calculations to the Ivy Mike device yields vaporized pusher gas expansion velocity of 290 kilometers per second and an implosion velocity of perhaps 400 kilometers per second if 3/4 of the total tamper/pusher mass is ablated off, the most energy efficient proportion. For the W-80 the gas expansion velocity is roughly 410 kilometers per second and the implosion velocity 570 kilometers per second. The pressure due to the ablating material is calculated to be 5.3 billion bar (530 TPa) in the Ivy Mike device and 64 billion bar (6.4 PPa) in the W-80 device.[16]

6.3.4 Comparing implosion mechanisms

Comparing the three mechanisms proposed, it can be seen that:

The calculated ablation pressure is one order of magnitude greater than the higher proposed plasma pressures and nearly two orders of magnitude greater than calculated radiation pressure. No mechanism to avoid the absorption of energy into the radiation case wall and the secondary tamper has been suggested, making ablation apparently unavoidable. The other mechanisms appear to be unneeded.

United States Department of Defense official declassification reports indicate that foamed plastic materials are or may be used in radiation case liners, and despite the low direct plasma pressure they may be of use in delaying the ablation until energy has distributed evenly and a sufficient fraction has reached the secondary's tamper/pusher.[17]

Richard Rhodes' book *Dark Sun* stated that a 1-inch-thick (25 mm) layer of plastic foam was fixed to the lead liner of the inside of the Ivy Mike steel casing using copper nails. Rhodes quotes several designers of that bomb explaining that the plastic foam layer inside the outer case is to delay ablation and thus recoil of the outer case: if the foam were not there, metal would ablate from the inside of the outer case with a large impulse, causing the casing to recoil outwards rapidly. The purpose of the casing is to contain the explosion for as long as possible, allowing as much X-ray ablation of the metallic surface of the secondary stage as possible, so it compresses the secondary efficiently, maximizing the fusion yield. Plastic foam has a low density, so causes a smaller impulse when it ablates than metal does.[17]

6.4 Design variations

A number of possible variations to the weapon design have been proposed:

- Either the tamper or the casing have been proposed to be made of uranium-235 (highly enriched uranium) in the final fission jacket. The far more expensive U-235 is also fissionable with fast neutrons like the standard U-238, but its fission-efficiency is higher than natural uranium, which is almost entirely U-238. Using a final fissionable jacket of U-235 would thus be expected to increase the yield of any Teller–Ulam bomb above a U-238 (depleted uranium) or natural uranium jacket design.

- In some descriptions, additional internal structures exist to protect the secondary from receiving excessive neutrons from the primary.

- The inside of the casing may or may not be specially machined to "reflect" the X-rays. X-ray "reflection" is not like light reflecting off of a mirror, but rather the reflector material is heated by the X-rays, causing the material itself to emit X-rays, which then travel to the secondary.

Two special variations exist that will be discussed in a further section: the cryogenically cooled liquid deuterium device used for the Ivy Mike test, and the putative design of the W88 nuclear warhead—a small, MIRVed version of the Teller–Ulam configuration with a prolate (egg or watermelon shaped) primary and an elliptical secondary.

Most bombs do not apparently have tertiary "stages"—that is, third compression stage(s), which are additional fusion stages compressed by a previous fusion stage (the fissioning of the last blanket of uranium, which provides about half the yield in large bombs, does not count as a "stage" in this terminology).

The U.S. tested three-stage bombs in several explosions (see Operation Redwing) but is only thought to have fielded one such tertiary model, i.e., a bomb in which a fission stage, followed by a fusion stage, finally compresses yet another fusion stage. This U.S. design was the heavy but highly efficient (i.e., nuclear weapon yield per unit bomb weight) 25 Mt B41 nuclear bomb.[18] The Soviet Union is thought to have used multiple stages (including more than one tertiary fusion stages) in their 50 megaton (100 Mt in intended use) Tsar Bomba (however, as with other bombs, the fissionable jacket could be replaced with lead in such a bomb, and in this one, for demonstration, it was). If any hydrogen bombs have been made from configurations other than those based on the Teller–Ulam design, the fact of it is not publicly known. (A possible exception to this is the Soviet early *Sloika* design).

In essence, the Teller–Ulam configuration relies on at least two instances of implosion occurring: first, the conventional (chemical) explosives in the primary would compress the fissile core, resulting in a fission explosion many times more powerful than that which chemical explosives could achieve alone (first stage). Second, the radiation from the fissioning of the primary would be used to compress and ignite the secondary fusion stage, resulting in a fusion explosion many times more powerful than the fission explosion alone. This chain of compression could then be continued with an arbitrary number of tertiary fusion stages.[19][20] although this is debated (see more: Arbitrarily large yield debate). Finally, efficient bombs (but not so-called neutron bombs) end with the fissioning of the final natural uranium tam-

per, something that could not normally be achieved without the neutron flux provided by the fusion reactions in secondary or tertiary stages. Such designs are suggested to be capable of being scaled up to an arbitrary large yield (with apparently as many fusion stages as desired),[19][20] potentially to the level of a "doomsday device." However, usually such weapons were not more than a dozen megatons, which was generally considered enough to destroy even most hardened practical targets (for example, a control facility such as the Cheyenne Mountain Complex). Even such large bombs have been replaced by smaller-yield bunker buster type nuclear bombs (see more: nuclear bunker buster).

As discussed above, for destruction of cities and non-hardened targets, breaking the mass of a single missile payload down into smaller MIRV bombs, in order to spread the energy of the explosions into a "pancake" area, is far more efficient in terms of area-destruction per unit of bomb energy. This also applies to single bombs deliverable by cruise missile or other system, such as a bomber, resulting in most operational warheads in the U.S. program having yields of less than 500 kilotons.

6.5   History

Main article: History of the Teller–Ulam design

6.5.1   United States

Main articles: Ivy Mike and Operation Castle

The idea of a thermonuclear fusion bomb ignited by a smaller fission bomb was first proposed by Enrico Fermi to his colleague Edward Teller in 1941 at the start of what would become the Manhattan Project.[3] Teller spent most of the Manhattan Project attempting to figure out how to make the design work, to some degree neglecting his assigned work on the fission bomb program. His difficult and devil's advocate attitude in discussions led Robert Oppenheimer to sidetrack him and other "problem" physicists into the super program to smooth his way.

Stanislaw Ulam, a co-worker of Teller, made the first key conceptual leaps towards a workable fusion design. Ulam's two innovations that rendered the fusion bomb practical were that compression of the thermonuclear fuel before extreme heating was a practical path towards the conditions needed for fusion, and the idea of staging or placing a separate thermonuclear component outside a fission primary component, and somehow using the primary to compress the secondary. Teller then realized that the gamma and X-

Operation Castle *thermonuclear test*, Castle Romeo *shot*

ray radiation produced in the primary could transfer enough energy into the secondary to create a successful implosion and fusion burn, if the whole assembly was wrapped in a *hohlraum* or radiation case.[3] Teller and his various proponents and detractors later disputed the degree to which Ulam had contributed to the theories underlying this mechanism. Indeed, shortly before his death, and in a last-ditch effort to discredit Ulam's contributions, Teller claimed that one of his own "graduate students" had proposed the mechanism.

The "George" shot of Operation Greenhouse of 9 May 1951 tested the basic concept for the first time on a very small scale. As the first successful (uncontrolled) release of nuclear fusion energy, which made up a small fraction of the 225 kt total yield,[21] it raised expectations to a near certainty that the concept would work.

On November 1, 1952, the Teller–Ulam configuration was tested at full scale in the "Ivy Mike" shot at an island in the Enewetak Atoll, with a yield of 10.4 megatons (over 450 times more powerful than the bomb dropped on Nagasaki during World War II). The device, dubbed the *Sausage*, used an extra-large fission bomb as a "trigger" and liquid deuterium—kept in its liquid state by 20 short tons (18 metric tons) of cryogenic equipment—as its fusion fuel, and weighed around 80 short tons (70 metric tons) altogether.

The liquid deuterium fuel of Ivy Mike was impractical for a deployable weapon, and the next advance was to use a solid lithium deuteride fusion fuel instead. In 1954 this

was tested in the "Castle Bravo" shot (the device was codenamed the *Shrimp*), which had a yield of 15 megatons (2.5 times expected) and is the largest U.S. bomb ever tested.

Efforts in the United States soon shifted towards developing miniaturized Teller–Ulam weapons that could easily outfit intercontinental ballistic missiles and submarine-launched ballistic missiles. By 1960, with the W47 warhead[22] deployed on Polaris ballistic missile submarines, megatonclass warheads were as small as 18 inches (0.5 m) in diameter and 720 pounds (320 kg) in weight. It was later found in live testing that the Polaris warhead did not work reliably and had to be redesigned. Further innovation in miniaturizing warheads was accomplished by the mid-1970s, when versions of the Teller–Ulam design were created that could fit ten or more warheads on the end of a small MIRVed missile (see the section on the W88 below).[7]

6.5.2 Soviet Union

Main articles: Joe 4 and RDS-37
See also: Soviet atomic bomb project

The first Soviet fusion design, developed by Andrei Sakharov and Vitaly Ginzburg in 1949 (before the Soviets had a working fission bomb), was dubbed the *Sloika*, after a Russian layer cake, and was not of the Teller–Ulam configuration. It used alternating layers of fissile material and lithium deuteride fusion fuel spiked with tritium (this was later dubbed Sakharov's "First Idea"). Though nuclear fusion might have been technically achievable, it did not have the scaling property of a "staged" weapon. Thus, such a design could not produce thermonuclear weapons whose explosive yields could be made arbitrarily large (unlike U.S. designs at that time). The fusion layer wrapped around the fission core could only moderately multiply the fission energy (modern Teller–Ulam designs can multiply it 30-fold). Additionally, the whole fusion stage had to be imploded by conventional explosives, along with the fission core, multiplying the bulk of chemical explosives needed substantially.

Their first Sloika design test, RDS-6s, was detonated in 1953 with a yield equivalent to 400 kilotons of TNT (15–20% from fusion). Attempts to use a *Sloika* design to achieve megaton-range results proved unfeasible. After the United States tested the "Ivy Mike" bomb in November 1952, proving that a multimegaton bomb could be created, the Soviets searched for an additional design. The "Second Idea", as Sakharov referred to it in his memoirs, was a previous proposal by Ginzburg in November 1948 to use lithium deuteride in the bomb, which would, in the course of being bombarded by neutrons, produce tritium and free deuterium.[23] In late 1953 physicist Viktor Davidenko achieved the first breakthrough, that of keeping the *primary* and *secondary* parts of the bombs in separate pieces ("staging"). The next breakthrough was discovered and developed by Sakharov and Yakov Zel'dovich, that of using the X-rays from the fission bomb to compress the *secondary* before fusion ("radiation implosion"), in early 1954. Sakharov's "Third Idea", as the Teller–Ulam design was known in the USSR, was tested in the shot "RDS-37" in November 1955 with a yield of 1.6 megatons.

The Soviets demonstrated the power of the "staging" concept in October 1961, when they detonated the massive and unwieldy *Tsar Bomba*, a 50 megaton hydrogen bomb that derived almost 97% of its energy from fusion. It was the largest nuclear weapon developed and tested by any country.

6.5.3   United Kingdom

Operation Grapple on Christmas Island was the first British hydrogen bomb test.

In 1954 work began at Aldermaston to develop the British fusion bomb, with Sir William Penney in charge of the project. British knowledge on how to make a thermonuclear fusion bomb was rudimentary, and at the time the United States was not exchanging any nuclear knowledge because of the Atomic Energy Act of 1946. However, the British were allowed to observe the American Castle tests and used sampling aircraft in the mushroom clouds, providing them with clear, direct evidence of the compression produced in the secondary stages by radiation implosion.[24]

Because of these difficulties, in 1955 British prime minister Anthony Eden agreed to a secret plan, whereby if the Aldermaston scientists failed or were greatly delayed in developing the fusion bomb, it would be replaced by an extremely large fission bomb.[24]

In 1957 the Operation Grapple tests were carried out. The first test, Green Granite was a prototype fusion bomb, but failed to produce equivalent yields compared to the Americans and Soviets, only achieving approximately 300 kilotons. The second test Orange Herald was the modified fission bomb and produced 720 kilotons—making it the largest fission explosion ever. At the time almost everyone (including the pilots of the plane that dropped it) thought that this was a fusion bomb. This bomb was put into service in 1958. A second prototype fusion bomb Purple Granite was used in the third test, but only produced approximately 150 kilotons.[24]

A second set of tests was scheduled, with testing recommencing in September 1957. The first test was based on a "… new simpler design. A two stage thermonuclear bomb that had a much more powerful trigger". This test Grapple X Round C was exploded on November 8 and yielded approximately 1.8 megatons. On April 28, 1958 a bomb was dropped that yielded 3 megatons—Britain's most powerful test. Two final air burst tests on September 2 and September 11, 1958, dropped smaller bombs that yielded around 1 megaton each.[24]

American observers had been invited to these kinds of tests. After their successful detonation of a megaton-range device (and thus demonstrating their practical understanding of the Teller–Ulam design "secret"), the United States agreed to exchange some of their nuclear designs with the United Kingdom, leading to the 1958 US–UK Mutual Defence Agreement. Instead of continuing with their own design, the British were given access to the design of the smaller American Mk 28 warhead and were able to manufacture copies.[24]

The United Kingdom had worked closely with the Americans on the Manhattan Project. British access to nuclear weapons information was cut-off by the United States at one point due to concerns about Soviet espionage. Full cooperation was not reestablished until an agreement governing the handling of secret information and other issues was signed.[24]

6.5.4   China

Main article: Test No. 6

The People's Republic of China detonated its first hydrogen bomb on June 17, 1967, 32 months after detonating its first fission weapon, with a yield of 3.31 Mt. It took place in the Lop Nor Test Site, in northwest China.[25] China received help from the Soviet Union to develop nuclear weapons. Some scholars claim this assistance nearly rose to the level of handing over plans for a bomb. The Chinese viewed the fact that the Soviet Union held back any nuclear technology at all as a betrayal of socialist solidarity and were enraged when the USSR withdrew its nuclear advisors. It is said that the date of the Soviet pull-out is stamped on security badges at Chinese nuclear sites in order to remind personnel of the need for an independent nuclear program. Even as late as the 2000s, Chinese nuclear weapons were considered relatively unsophisticated when held up to the standards of other nuclear powers.[24]

A story in *The New York Times* by William Broad[26] reported that in 1995, a supposed double agent from the People's Republic of China delivered information indicating that China knew secret details of the W88 warhead, supposedly through espionage.[27] (This line of investigation eventually resulted in the abortive trial of Wen Ho Lee.)

6.5.5 France

Very little is known about France's development of the Teller–Ulam design beyond the fact that France detonated a 2.6 Mt device in the "Canopus" test in August 1968. France had great difficulty with its initial development of the Teller-Ulam design but later overcame these difficulties. France is believed to have nuclear weapons of equal sophistication to the other major nuclear powers.[24]

As of 2009, France had four nuclear submarines armed with 16 M45 nuclear-tipped missiles each. These missiles were scheduled to be replaced by the M51 model in 2010. These missiles both carry warheads of the TN-75 design. France also has about 60 air-launched missiles tipped with TN 80/81 warheads with a yield of about 300 kilotons each. France's nuclear program has been carefully designed to ensure that these weapons remain usable decades into the future.[24]

6.5.6 Other countries

India

Main article: India and weapons of mass destruction

On May 11, 1998, India reportedly detonated a thermonuclear bomb in its Operation Shakti tests ("Shakti-1", specifically).[29] Dr. Samar Mubarakmand, a Pakistani, as-

The detonation of Shakti-1 produced a nuclear yield of 45 kt.[28]

serted that Shakti-1 was a successful thermonuclear test.[29] The yield of India's hydrogen bomb remains highly debatable among the Indian science community and the international scholars.[30] The question of politicisation and disputes between Indian scientists further complicated the matter.[31]

Director for the 1998 test site preparations, Dr. K. Santhanam, reported the yield of the thermonuclear explosion was lower than expected, although his statement has been disputed by other Indian scientists involved in the test.[32] International sources, using local data and citing a United States Geological Survey report compiling seismic data from 125 IRIS stations across the world, argue that the magnitudes suggested a combined yield of up to 60 kilotonnes, consistent with the Indian announced total yield of 56 kilotonnes.[33][34]

Israel

Main articles: Nuclear weapons and Israel and Vela Incident

Israel is alleged to possess thermonuclear weapons of the

Teller–Ulam design,[35] but it is not known to have tested any nuclear devices, although it is widely speculated that the Vela Incident of 1979 may have been a joint Israeli-South African nuclear test.[36][37]

It is well established that American scientist, Edward Teller (father of the hydrogen bomb), advised and guided the Israeli establishment on general nuclear matters for some twenty years.[38] Between 1964 and 1967, Teller made six visits to Israel where he lectured at the Tel Aviv University on general topics in theoretical physics.[39] It took him a year to convince the CIA about Israel's capability and finally in 1976, Carl Duckett of the CIA testified in the U.S. Congress, after receiving credible information from an "American scientist" (Edward Teller), on Israel's nuclear capability.[37] Sometime in 1990, Teller came to confirm the speculations in media that it was during his visits, three decades ago, that he concluded to the CIA that Israel was in possession of nuclear weapons.[37] After he conveyed the matter to the higher level of the U.S. government, Teller reportedly said: "They [Israel] have it, and they were clever enough to trust their research and not to test, they know that to test would get them into trouble."[37]

Pakistan

Main article: Pakistan and weapons of mass destruction

According to the scientific data received and published by PAEC, the Corps of Engineers, and Kahuta Research Laboratories (KRL), in May 1998, Pakistan carried out six underground nuclear tests in Chagai Hills and Kharan Desert in Balochistan Province (see the code-names of the tests, *Chagai-I* and *Chagai-II*).[29] None of these boosted fission devices was the thermonuclear weapon design, according to KRL and PAEC.[29]

North Korea

Main article: North Korea and weapons of mass destruction

North Korea claimed to have tested its miniaturised thermonuclear bomb on 6 January 2016. North Korea's first three nuclear tests (2006, 2009 and 2013) were relatively low yield and do not appear to have been of a thermonuclear weapon design. In 2013, the South Korean Defense Ministry has speculated that North Korea may be trying to develop a "hydrogen bomb" and such a device may be North Korea's next weapons test.[40][41] In January 2016, North Korea claimed to have successfully tested a hydrogen bomb,[42] though only a magnitude 5.1 seismic event was detected at the time of the test,[43] a similar magnitude to the 2013 test of a 6-9 kt atomic bomb. These seismic recordings have scientists worldwide doubting North Korea's claim that a hydrogen bomb was tested and suggest it was a non-fusion nuclear test.[44]

6.6 Public knowledge

The Teller–Ulam design was for many years considered one of the top nuclear secrets, and even today it is not discussed in any detail by official publications with origins "behind the fence" of classification. United States Department of Energy (DOE) policy has been, and continues to be, that they do not acknowledge when "leaks" occur, because doing so would acknowledge the accuracy of the supposed leaked information.

Photographs of warhead casings, such as this one of the W80 nuclear warhead, allow for some speculation as to the relative size and shapes of the primaries and secondaries in U.S. thermonuclear weapons.

Aside from images of the warhead casing, most information in the public domain about this design is relegated to a few terse statements by the DOE and the work of a few individual investigators.

6.6.1 DOE statements

In 1972 the United States government declassified a document stating "[I]n thermonuclear (TN) weapons, a fission 'primary' is used to trigger a TN reaction in thermonuclear fuel referred to as a 'secondary'", and in 1979 added, "[I]n thermonuclear weapons, radiation from a fission explosive can be contained and used to transfer energy to compress and ignite a physically separate component containing thermonuclear fuel." To this latter sentence they specified that "*Any elaboration of this statement will be classified.*"[45] The only information that may pertain to the *spark plug* was declassified in 1991: "Fact that fissile and/or fissionable materials are present in some secondaries, material unidentified, location unspecified, use unspecified, and weapons undesignated." In 1998 the DOE declassified the statement that "The fact that materials may be present in channels and the term 'channel filler,' with no elaboration", which may refer to the polystyrene foam (or an analogous substance).[46]

Whether these statements vindicate some or all of the models presented above is up for interpretation, and official U.S. government releases about the technical details of nuclear weapons have been purposely equivocating in the past (see, e.g., Smyth Report). Other information, such as the types of fuel used in some of the early weapons, has been declassified, though precise technical information has not been.

6.6.2 *The Progressive* case

Main article: United States v. The Progressive

Most of the current ideas on the workings of the Teller–Ulam design came into public awareness after the Department of Energy (DOE) attempted to censor a magazine article by U.S. antiweapons activist Howard Morland in 1979 on the "secret of the hydrogen bomb". In 1978, Morland had decided that discovering and exposing this "last remaining secret" would focus attention onto the arms race and allow citizens to feel empowered to question official statements on the importance of nuclear weapons and nuclear secrecy. Most of Morland's ideas about how the weapon worked were compiled from highly accessible sources—the drawings that most inspired his approach came from none other than the *Encyclopedia Americana*. Morland also interviewed (often informally) many former Los Alamos scientists (including Teller and Ulam, though neither gave him any useful information), and used a variety of interpersonal strategies to encourage informative responses from them (i.e., asking questions such as "Do they still use spark plugs?" even if he was not aware what the latter term specifically referred to).[47]

Morland eventually concluded that the "secret" was that the *primary* and *secondary* were kept separate and that radiation pressure from the *primary* compressed the *secondary* before igniting it. When an early draft of the article, to be published in *The Progressive* magazine, was sent to the DOE after falling into the hands of a professor who was opposed to Morland's goal, the DOE requested that the article not be published, and pressed for a temporary injunction. The DOE argued that Morland's information was (1) likely derived from classified sources, (2) if not derived from classified sources, itself counted as "secret" information under the "born secret" clause of the 1954 Atomic Energy Act,

and (3) was dangerous and would encourage nuclear proliferation.

Morland and his lawyers disagreed on all points, but the injunction was granted, as the judge in the case felt that it was safer to grant the injunction and allow Morland, et al., to appeal, which they did in *United States v. The Progressive* (1979).

Through a variety of more complicated circumstances, the DOE case began to wane as it became clear that some of the data they were attempting to claim as "secret" had been published in a students' encyclopedia a few years earlier. After another H-bomb speculator, Chuck Hansen, had his own ideas about the "secret" (quite different from Morland's) published in a Wisconsin newspaper, the DOE claimed that *The Progressive* case was moot, dropped its suit, and allowed the magazine to publish its article, which it did in November 1979. Morland had by then, however, changed his opinion of how the bomb worked, suggesting that a foam medium (the polystyrene) rather than radiation pressure was used to compress the *secondary*, and that in the *secondary* there was a *spark plug* of fissile material as well. He published these changes, based in part on the proceedings of the appeals trial, as a short erratum in *The Progressive* a month later.[48] In 1981, Morland published a book about his experience, describing in detail the train of thought that led him to his conclusions about the "secret".[47][49]

Morland's work is interpreted as being at least partially correct because the DOE had sought to censor it, one of the few times they violated their usual approach of not acknowledging "secret" material that had been released; however, to what degree it lacks information, or has incorrect information, is not known with any confidence. The difficulty that a number of nations had in developing the Teller–Ulam design (even when they apparently understood the design, such as with the United Kingdom), makes it somewhat unlikely that this simple information alone is what provides the ability to manufacture thermonuclear weapons. Nevertheless, the ideas put forward by Morland in 1979 have been the basis for all the current speculation on the Teller–Ulam design.

6.7 Variations

6.7.1 Ivy Mike

In his 1995 book *Dark Sun: The Making of the Hydrogen Bomb*, author Richard Rhodes describes in detail the internal components of the "Ivy Mike" *Sausage* device, based on information obtained from extensive interviews with the scientists and engineers who assembled it. According to Rhodes, the actual mechanism for the compression of the secondary was a combination of the radiation pressure, foam plasma pressure, and tamper-pusher ablation theories described above; the radiation from the primary heated the polyethylene foam lining the casing to a plasma, which then re-radiated radiation into the secondary's pusher, causing its surface to ablate and driving it inwards, compressing the secondary and causing the fusion reaction; the general applicability of this principle is unclear.[11]

6.7.2   W88

In 1999 a reporter for the *San Jose Mercury News* reported that the U.S. W88 nuclear warhead, a small MIRVed warhead used on the Trident II SLBM, had a prolate (egg or watermelon shaped) *primary* (code-named *Komodo*) and a spherical *secondary* (code-named *Cursa*) inside a specially shaped radiation case (known as the "peanut" for its shape).[50]

The reentry cones for the two warheads are the same size, 1.75 meters (69 in) long, with a maximum diameter of 55 cm. (22 in).[51] The higher yield of the W88 implies a larger secondary, which produces most of the yield. Putting the secondary, which is heavier than the primary, in the wider part of the cone allows it to be larger, but it also moves the center of mass aft, potentially causing aerodynamic stability problems during reentry. Dead-weight ballast must be added to the nose to move the center of mass forward.

To make the primary small enough to fit into the narrow part of the cone, its bulky insensitive high explosive charges must be replaced with more compact "non-insensitive" high explosives that are more hazardous to handle. The higher yield of the W88, which is the last new warhead produced by the United States, thus comes at a price of higher warhead weight and higher workplace hazard. The W88 also contains tritium, which has a half life of only 12.32 years and must be repeatedly replaced.[52] If these stories are true, it would explain the reported higher yield of the W88, 475 kilotons, compared with only 300 kilotons for the earlier W87 warhead.

6.8   See also

- Pure fusion weapon

6.9   References

[1] The misleading term "hydrogen bomb" was already in wide public use before fission product fallout from the Castle Bravo test in 1954 revealed the extent to which the design relies on fission.

[2] From National Public Radio Talk of the Nation, November 8, 2005, Siegfried Hecker of Los Alamos, "the hydrogen bomb – that is, a two-stage thermonuclear device, as we referred to it – is indeed the principal part of the U.S. arsenal, as it is of the Russian arsenal."

[3] Teller, Edward; Ulam, Stanislaw (March 9, 1951). "On Heterocatalytic Detonations I. Hydrodynamic Lenses and Radiation Mirrors" (PDF). LAMS-1225. Los Alamos Scientific Laboratory. Retrieved September 26, 2014, on the Nuclear Non-Proliferation Institute website. This is the original classified paper by Teller and Ulam proposing staged implosion. This declassified version is heavily redacted, leaving only a few paragraphs.

[4] Carey Sublette (July 3, 2007). "Nuclear Weapons FAQ Section 4.4.1.4 The Teller–Ulam Design". *Nuclear Weapons FAQ*. Retrieved 17 July 2011. "So far as is known all high yield nuclear weapons today (>50 kt or so) use this design."

[5] Broad, William J. (23 March 2015). "Hydrogen Bomb Physicist's Book Runs Afoul of Energy Department". *New York Times*. Retrieved 20 November 2015.

[6] Greene, Jes (25 March 2015). "A physicist might be in trouble for what he revealed in his new book about the H bomb". *Business Insider*. Retrieved 20 November 2015.

[7] "Complete List of All U.S. Nuclear Weapons". 1 October 1997. Retrieved 2006-03-13.

[8] Hansen, Chuck (1988). *U.S. nuclear weapons: The secret history*. Arlington, TX: Aerofax. ISBN 0-517-56740-7.

[9] Hansen, Chuck (2007). *Swords of Armageddon: U.S. Nuclear Weapons Development Since 1945* (PDF) (CD-ROM & download available) (2 ed.). Sunnyvale, California: Chukelea Publications. ISBN 978-0-9791915-0-3. 2,600 pages.

[10] "Figure 5 – Thermonuclear Warhead Components". Archived from the original on July 12, 2010. Retrieved 27 August 2010. A cleaned up version: "British H-bomb posted on the Internet by Greenpeace". Federation of American Scientists. Retrieved 27 August 2010.

[11] Rhodes, Richard (1995). *Dark Sun: The Making of the Hydrogen Bomb*. New York: Simon & Schuster. ISBN 0-684-80400-X.

[12] http://nuclearweaponarchive.org/Usa/Weapons/ W76NeutronTube1200c20.jpg

[13] "Improved Security, Safety & Manufacturability of the Reliable Replacement Warhead", NNSA March 2007.

[14] A 1976 drawing that depicts an interstage that absorbs and re-radiates X-rays. From Howard Morland, "The Article", *Cardozo Law Review*, March 2005, p 1374.

[15] [Fogbank] Speculation on Fogbank, *Arms Control Wonk*

[16] "Nuclear Weapons Frequently Asked Questions 4.4.3.3 The Ablation Process". 2.04. 20 February 1999. Retrieved 2006-03-13.

[17] "Nuclear Weapons Frequently Asked Questions 4.4.4 Implosion Systems". 2.04. 20 February 1999. Retrieved 2006-03-13.

[18] "The B-41 (Mk-41) Bomb – High yield strategic thermonuclear bomb". 21 October 1997. Retrieved 2006-03-13.

[19] Winterberg, Friedwardt (2010). *The Release of Thermonuclear Energy by Inertial Confinement: Ways Towards Ignition*. World Scientific. pp. 192–193. ISBN 9814295914.

[20] Croddy, Eric A.; Wirtz, James J.; Larsen, Jeffrey, Eds. (2005). *Weapons of Mass Destruction: An Encyclopedia of Worldwide Policy, Technology, and History*. ABC-CLIO, Inc. p. 376. ISBN 1851094903.

[21] The "George" shot, Comprehensive Test Ban Treaty Organisation website

[22] "Photograph of a W47 warhead" (JPG). Retrieved 2006-03-13.

[23] Holloway, David (1994). *Stalin and the bomb: The Soviet Union and atomic energy, 1939–1956*. New Haven, CT: Yale University Press. p. 299. ISBN 0-300-06056-4.

[24] Younger, Stephen (2009). *The Bomb: A New History*. New York: Harper Collins. ISBN 978-0-06-173614-8.

[25] https://www.ctbto.org/specials/testing-times/ 17-june-1967-chinas-first-thermonuclear-test

[26] "Spies versus sweat, the debate over China's nuclear advance". *The New York Times*. 7 September 1999. Retrieved 2011-04-18.

[27] Christopher Cox, chairman (1999). *Report of the United States House of Representatives Select Committee on U.S. National Security and Military/Commercial Concerns with the People's Republic of China.*, esp. Ch. 2, "PRC Theft of U.S. Thermonuclear Warhead Design Information".

[28] "Forces gung-ho on N-arsenal". Times of India. Retrieved 21 July 2012.

[29] Khan, Kamran (30 May 1998). "Tit-for-Tat: Pakistan tested 6 nuclear devices in response to Indian's tests.". *The News International*. Retrieved 10 August 2011. "None of these explosions were thermonuclear, we are doing research and can do a fusion test if asked, said by Abdul Qadeer Khan. "These boosted devices are like a half way stage towards a thermonuclear bomb. They use elements of the thermonuclear process, and are effectively stronger Atom bombs", quoted by Munir Ahmad Khan.

[30] PTI, Press Trust of India (September 25, 2009). "AEC ex-chief backs Santhanam on Pokhran-II". *The Hindu, 2009*. Retrieved 18 January 2013.

[31] Carey Sublette, et. al. "What are the real yield of India's Test?". What Are the Real Yields of India's Test?. Retrieved 18 January 2013.

[32] "Former NSA disagrees with scientist, says Pokhran II successful". *The Times of India*. 27 August 2009. Archived from the original on 30 August 2009. Retrieved 20 November 2015.

[33] India tested H-bomb, says New Scientist

[34] "?". Rediff.com. Retrieved 27 August 2010.

[35] Samdani, Zafar (25 March 2000). "India, Pakistan can build hydrogen bomb: Scientist". *Dawn News Interviews*. Retrieved 23 December 2012.

[36] Hersh 1991, p. 271.

[37] Cohen, Avner (October 15, 1999). "The Battle over the NPT: America Learns the Truth". *Israel and the bomb*. (google Book). New York: Columbia University Press. pp. 297–300. ISBN 978-0231104838.

[38] Karpin, Michael (2005). *The Bomb in the Basement*. New York: Simon & Schuster Paperbacks. pp. 289–293. ISBN 0-7432-6595-5.

[39] Gábor Palló (2000). "The Hungarian Phenomenon in Israeli Science" (PDF). *Hungarian Academy of Science*. **25** (1). Retrieved 11 December 2012.

[40] Kim Kyu-won (February 7, 2013). "North Korea could be developing a hydrogen bomb". The Hankyoreh. Retrieved February 8, 2013.

[41] Kang Seung-woo; Chung Min-uck (February 4, 2013). "North Korea may detonate H-bomb". Korea Times. Retrieved February 8, 2013.

[42] "North Korea claims fully successful hydrogen bomb test". Russia Today. January 6, 2016. Retrieved January 6, 2016.

[43] M5.1 - 21km ENE of Sungjibaegam, North Korea (Report). USGS. January 6, 2016. Retrieved January 6, 2016.

[44] "North Korea nuclear H-bomb claims met by scepticism".

[45] emphasis in original

[46] *Restricted Data Declassification Decisions, 1946 to the present, Volume 7*. United States Department of Energy. January 2001.

[47] Morland, Howard (1981). *The secret that exploded*. New York: Random House. ISBN 0-394-51297-9.

[48] "The H-Bomb Secret: How we got it and why we're telling it". *The Progressive*. **43** (11). November 1979.

[49] Alexander De Volpi; Jerry Marsh; Ted Postol & George Stanford (1981). *Born secret: the H-bomb, the Progressive case and national security*. New York: Pergamon Press. ISBN 0-08-025995-2.

[50] Dan Stober & Ian Hoffman (2001). *A convenient spy: Wen Ho Lee and the politics of nuclear espionage*. New York: Simon & Schuster. ISBN 0-7432-2378-0.

[51] "The W88 Warhead – Intermediate yield strategic SLBM MIRV warhead". 1 October 1997. Retrieved 2006-03-13.

[52] Morland, Howard (February 2003). *The holocaust bomb: A question of time*.

6.9.1   Bibliography

Basic principles

- "Engineering and Design of Nuclear Weapons" from Carey Sublette's Nuclear Weapons FAQ.

- Hansen, Chuck, *U.S. nuclear weapons: The secret history* (Arlington, TX: Aerofax, 1988). ISBN 0-517-56740-7

- Hansen, Chuck (2007). *Swords of Armageddon: U.S. Nuclear Weapons Development Since 1945* (PDF) (CD-ROM & download available) (2 ed.). Sunnyvale, California: Chukelea Publications. ISBN 978-0-9791915-0-3. 2,600 pages.

- Dalton E. G. Barroso, *The physics of nuclear explosives*, in Portuguese. (São Paulo, Brazil: Editora Livraria da Física, 2009). ISBN 978-85-7861-016-6

History

- DeGroot, Gerard, "The Bomb: A History of Hell on Earth", London: Pimlico, 2005. ISBN 0-7126-7748-8

- Peter Galison and Barton Bernstein, "In any light: Scientists and the decision to build the Superbomb, 1942–1954" *Historical Studies in the Physical and Biological Sciences* Vol. 19, No. 2 (1989): 267–347.

- German A. Goncharov, "American and Soviet H-bomb development programmes: historical background" (trans. A.V. Malyavkin), *Physics—Uspekhi* Vol. 39, No. 10 (1996): 1033–1044. *Available online (PDF)*

- David Holloway, *Stalin and the bomb: The Soviet Union and atomic energy, 1939–1956* (New Haven, CT: Yale University Press, 1994). ISBN 0-300-06056-4

- Richard Rhodes, *Dark sun: The making of the hydrogen bomb* (New York: Simon and Schuster, 1995). ISBN 0-684-80400-X

- S.S. Schweber, *In the shadow of the bomb: Bethe, Oppenheimer, and the moral responsibility of the scientist* (Princeton, N.J.: Princeton University Press, 2000). ISBN 0-691-04989-0

- Gary Stix, "Infamy and honor at the Atomic Café: Edward Teller has no regrets about his contentious career", *Scientific American* (October 1999): 42–43.

Analyzing fallout

- De Geer, Lars-Erik (1991). "The radioactive signature of the hydrogen bomb" (PDF). *Science and Global Security*. **2**: 351–363. doi:10.1080/08929889108426372.

- Yulii Borisovich Khariton and Yuri Smirnov, The Khariton version *Bulletin of the Atomic Scientists* Vol. 49, No. 4 (May 1993): 20–31.

6.10   External links

Principles

- "Hydrogen bomb / Fusion weapons" at GlobalSecurity.org (see also links on right)

- "Basic Principles of Staged Radiation Implosion (Teller–Ulam)" from Carey Sublette's NuclearWeaponArchive.org.

- "Matter, Energy, and Radiation Hydrodynamics" from Carey Sublette's Nuclear Weapons FAQ.

- "Engineering and Design of Nuclear Weapons" from Carey Sublette's Nuclear Weapons FAQ.

- "Elements of Thermonuclear Weapon Design" from Carey Sublette's Nuclear Weapons FAQ.

- Annotated bibliography for nuclear weapons design from the Alsos Digital Library for Nuclear Issues

History

- PBS: Race for the Superbomb: Interviews and Transcripts (with U.S. and USSR bomb designers as well as historians).

- Howard Morland on how he discovered the "H-bomb secret" (includes many slides).

- *The Progressive* November 1979 issue – "The H-Bomb Secret: How we got it, why we're telling" (entire issue online).

- Annotated bibliography on the hydrogen bomb from the Alsos Digital Library

- University of Southampton, Mountbatten Centre for International Studies, Nuclear History Working Paper No5.

- Peter Kuran's "Trinity and Beyond" – documentary film on the history of nuclear weapon testing.

Chapter 7

Boosted fission weapon

"Fission-fusion-fission" redirects here. For the term as applied to multistage H-bombs, see Thermonuclear weapon.

A **boosted fission weapon** usually refers to a type of nuclear bomb that uses a small amount of fusion fuel to increase the rate, and thus yield, of a fission reaction. The neutrons released by the fusion reactions add to the neutrons released due to fission, allowing for more neutron-induced fission reactions to take place. The rate of fission is thereby greatly increased such that much more of the fissile material is able to undergo fission before the core explosively disassembles. The fusion process itself adds only a small amount of energy to the process, perhaps 1%.[1]

The alternative meaning is an obsolete type of single-stage nuclear bomb that uses thermonuclear fusion on a large scale to create fast neutrons that can cause fission in depleted uranium, but which is not a two-stage hydrogen bomb. This type of bomb was referred to by Edward Teller as "Alarm Clock", and by Andrei Sakharov as "Sloika" or "Layer Cake" (Teller and Sakharov developed the idea independently, as far as is known).[2]

The idea of boosting was originally developed between late 1947 and late 1949 at Los Alamos.[3] The primary benefit of boosting is further miniaturization of nuclear weapons as it reduces the minimum inertial confinement time required for a supercritical nuclear explosion by providing a sudden influx of fast neutrons before the critical mass would blow itself apart. This would eliminate the need for an aluminum pusher and uranium tamper and the explosives needed to push them and the fissile material into a supercritical state. While the bulky Fat Man had a diameter of 5 feet (1.5 m) and required 3 tons of high explosives for implosion, a boosted fission primary can be fitted on a small nuclear warhead (such as the W88) to ignite the thermonuclear secondary.

7.1 Gas boosting in modern nuclear weapons

In a fission bomb, the fissile fuel is "assembled" quickly by a uniform spherical implosion created with conventional explosives, producing a supercritical mass. In this state, many of the neutrons released by the fissioning of a nucleus will induce fission of other nuclei in the fuel mass, also releasing additional neutrons, leading to a chain reaction. This reaction consumes at most 20% of the fuel before the bomb blows itself apart, or possibly much less if conditions are not ideal: the Little Boy (gun type mechanism) and Fat Man (implosion type mechanism) bombs had efficiencies of 1.38% and 13%, respectively.

Fusion boosting is achieved by introducing tritium and deuterium gas (solid lithium deuteride-tritide has also been used in some cases, but gas allows more flexibility and can be stored externally) into a hollow cavity at the center of the sphere of fission fuel, or into a gap between an outer layer and a "levitated" inner core, sometime before implosion. By the time about 1% of the fission fuel has fissioned, the temperature rises high enough to cause thermonuclear fusion, which produces relatively large numbers of neutrons speeding up the late stages of the chain reaction and approximately doubling its efficiency.

Deuterium-tritium fusion neutrons are extremely energetic, seven times more energetic than an average fission neutron, which makes them much more likely to be captured in the fissile material and lead to fission. This is due to several reasons:

1. Their high velocity creates the opposite of time absorption: time magnification.

2. When these energetic neutrons strike a fissile nucleus, a much larger number of secondary neutrons are released by the fission (e.g. 4.6 vs 2.9 for Pu-239).

3. The fission cross section is larger both in absolute terms, and in proportion to the scattering and capture

cross sections.

Taking these factors into account, the maximum alpha value for D-T fusion neutrons in plutonium (density 19.8 g/cm^3) is some 8 times higher than for an average fission neutron (2.5×10^9 vs 3×10^8).

A sense of the potential contribution of fusion boosting can be gained by observing that the complete fusion of one mole of tritium (3 grams) and one mole of deuterium (2 grams) would produce one mole of neutrons (1 gram), which, neglecting escape losses and scattering for the moment, could fission one mole (239 grams) of plutonium directly, producing 4.6 moles of secondary neutrons, which can in turn fission another 4.6 moles of plutonium (1,099 g). The fission of this 1,338 g of plutonium in the first two generations would release 23[4] kilotons of TNT equivalent (97 TJ) of energy, and would by itself result in a 29.7% efficiency for a bomb containing 4.5 kg of plutonium (a typical small fission trigger). The energy released by the fusion of the 5 g of fusion fuel itself is only 1.73% of the energy released by the fission of 1,338 g of plutonium. Larger total yields and higher efficiency are possible, since the chain reaction can continue beyond the second generation after fusion boosting.[5]

Fusion-boosted fission bombs can also be made immune to neutron radiation from nearby nuclear explosions, which can cause other designs to predetonate, blowing themselves apart without achieving a high yield. The combination of reduced weight in relation to yield and immunity to radiation has ensured that most modern nuclear weapons are fusion-boosted.

The fusion reaction rate typically becomes significant at 20 to 30 megakelvins. This temperature is reached at very low efficiencies, when less than 1% of the fissile material has fissioned (corresponding to a yield in the range of hundreds of tons of TNT). Since implosion weapons can be designed that will achieve yields in this range even if neutrons are present at the moment of criticality, fusion boosting allows the manufacture of efficient weapons that are immune to predetonation. Elimination of this hazard is a very important advantage in using boosting. It appears that every weapon now in the U.S. arsenal is a boosted design.[5]

According to one weapons designer, boosting is mainly responsible for the remarkable 100-fold increase in the efficiency of fission weapons since 1945.[6]

7.2 Some early non-staged thermonuclear weapon designs

Early thermonuclear weapon designs such as the Joe-4, the Soviet "Layer Cake" ("Sloika", Russian: Слойка), used large amounts of fusion to induce fission in the uranium-238 atoms that make up depleted uranium. These weapons had a fissile core surrounded by a layer of lithium-6 deuteride, in turn surrounded by a layer of depleted uranium. Some designs (including the layer cake) had several alternate layers of these materials. The Soviet *Layer Cake* was similar to the American *Alarm Clock* design, which was never built, and the British *Green Bamboo* design, which was built but never tested.

When this type of bomb explodes, the fission of the highly enriched uranium or plutonium core creates neutrons, some of which escape and strike atoms of lithium-6, creating tritium. At the temperature created by fission in the core, tritium and deuterium can undergo thermonuclear fusion without a high level of compression. The fusion of tritium and deuterium produces a neutron with an energy of 14 MeV—a much higher energy than the 1 MeV of the neutron that began the reaction. This creation of high-energy neutrons, rather than energy yield, is the main purpose of fusion in this kind of weapon. This 14 MeV neutron then strikes an atom of uranium-238, causing fission: without this fusion stage, the original 1 MeV neutron hitting an atom of uranium-238 would probably have just been absorbed. This fission then releases energy and also neutrons, which then create more tritium from the remaining lithium-6, and so on, in a continuous cycle. Energy from fission of uranium-238 is useful in weapons: both because depleted uranium is much cheaper than highly enriched uranium and because it cannot go critical and is therefore less likely to be involved in a catastrophic accident.

This kind of thermonuclear weapon can produce up to 20% of its yield from fusion, with the rest coming from fission, and is limited in yield to less than one megaton of TNT (4 PJ) equivalent. Joe-4 yielded 400 kilotons of TNT (1.7 PJ). In comparison, a "true" hydrogen bomb can produce up to 97% of its yield from fusion, and its explosive yield is limited only by device size.

7.3 Maintenance of gas boosted nuclear weapons

Tritium is a radioactive isotope with a half-life of 12.355 years. Its main decay product is Helium-3, which has the largest cross-section for neutron capture of any nuclide. Therefore, periodically the weapon must have its helium waste flushed out and its tritium supply recharged. This is because any helium-3 in the weapon's tritium supply would act as a poison during the weapon's detonation, absorbing neutrons meant to collide with the nuclei of its fission fuel.[7]

Tritium is relatively expensive to produce because each triton produced requires production of at least one free neutron which is used to bombard a feedstock material (lithium-6, deuterium, or helium-3). Actually, because of losses and inefficiencies, the number of free neutrons needed is closer to two for each triton produced (and tritium begins decaying immediately, so there are losses during collection, storage, and transport from the production facility to the weapons in the field.) The production of free neutrons demands the operation of either a breeder reactor or a particle accelerator (with a spallation target) dedicated to the tritium production facility.[8] [9]

7.4   See also

- Nuclear weapon design

7.5   References

[1] "Facts about Nuclear Weapons: Boosted Fission Weapons", Indian Scientists Against Nuclear Weapons Archived July 8, 2008, at the Wayback Machine.

[2] Rhodes, Richard, *Dark Sun*: The Making of the Hydrogen Bomb, *New York, Simon & Schuster (1996)*

[3] Bethe, Hans A. (28 May 1952).   Chuck Hansen, ed. "Memorandum on the History Of Thermonuclear Program". Federation of American Scientists. Retrieved 19 May 2010.

[4] "Nuclear Weapon Archive: 12.0 Useful Tables".

[5] "Nuclear Weapon Archive: 4.3 Fission-Fusion Hybrid Weapons".

[6] Olivier Coutard (2002). *The Governance of Large Technical Systems*. Taylor & Francis. p. 177.

[7] "Section 6.3.1.2 Nuclear Materials Tritium". *High Energy Weapons Archive FAQ*. Carey Sublette. Retrieved June 7, 2016.

[8] "Section 6.3.1.2 Nuclear Materials Tritium". *High Energy Weapons Archive FAQ*. Carey Sublette. Retrieved June 7, 2016.

[9] "Section 4.3.1 Fusion Boosted Fission Weapons". *High Energy Weapons Archive FAQ*. Carey Sublette. Retrieved June 7, 2016.

Chapter 8

Neutron bomb

A **neutron bomb**, officially termed as a type of **Enhanced Radiation Weapon** (ERW), is a low yield thermonuclear weapon in which a burst of neutrons generated by a nuclear fusion reaction is intentionally allowed to escape the weapon, rather than being absorbed by its other components.[3] The neutron bomb was to be used as a tactical nuclear weapon intended for use against armored forces. Originally conceived by the U.S. military, their design goals were to stop massed Soviet armored divisions from overrunning allied nations with less civilian and structural collateral damage to these allied nations.[4][5]

The weapon's radiation case, usually made from relatively thick uranium, lead or steel in a standard bomb, is, instead, made of as thin a material as possible, to facilitate the greatest escape of fusion-produced neutrons. The *usual* nuclear weapon yield—expressed as kilotons of TNT equivalent—is not a measure of a neutron weapon's destructive power. It refers only to the energy released (mostly heat and blast), and does not express the lethal effect of neutron radiation on living organisms.

Compared to a pure fission bomb with an identical explosive yield, a neutron bomb would emit about ten times[6] the amount of neutron radiation. In a fission bomb, at sea level, the total radiation pulse energy which is composed of both gamma rays and neutrons is approximately 5% of the entire energy released; in the neutron bomb it would be closer to 40%. Furthermore, the neutrons emitted by a neutron bomb have a much higher average energy level (close to 14 MeV) than those released during a fission reaction (1–2 MeV).[7] Technically speaking, all low yield nuclear weapons are radiation weapons, including non-enhanced variants. Up to about 10 kilotons in yield, all nuclear weapons have prompt neutron radiation[2] as their furthest-reaching lethal component, after which point the lethal blast and thermal effects radius begins to out-range the lethal ionizing radiation radius.[8][9][10] Enhanced radiation weapons also fall into this same yield range and simply enhance the intensity and range of the neutron dose for a given yield.

8.1 History and deployment to present

Conception of the neutron bomb is generally credited to Samuel T. Cohen of the Lawrence Livermore National Laboratory, who developed the concept in 1958.[11] Initial development was carried out as part of projects DOVE and STARLING, and an early device was tested underground in early 1962. Designs of a "weaponized" version were carried out in 1963.[12][13]

Development of two production designs for the Army's MGM-52 Lance short-range missile began in July 1964, the W63 at Livermore and the W64 at Los Alamos. Both entered Phase 3 testing in July 1964, and the W64 was cancelled in favor of the W63 in September 1964. The W63 was in turn cancelled in November 1965 in favor of the W70 (Mod 0), a conventional design.[12] By this time, the same concepts were being used to develop warheads for the Sprint missile, an anti-ballistic missile (ABM), with Livermore designing the W65 and Los Alamos the W66. Both entered Phase 3 testing in October 1965, but the W65 was cancelled in favor of the W66 in November 1968. Testing of the W66 was carried out in the late 1960s, and entered production in June 1974,[12] the first neutron bomb to do so. Approximately 120 were built, with about 70 of these being on active duty during 1975 and 1976 as part of the Safeguard Program. When that program was shut down they were placed in storage, and eventually decommissioned in the early 1980s.[12]

Development of ER warheads for Lance continued, but in the early 1970s attention had turned to using modified versions of the W70, the W70 Mod 3.[12] Development was subsequently postponed by President Jimmy Carter in 1978 following protests against his administration's plans to deploy neutron warheads to ground forces in Europe.[14] On November 17, 1978, in a test the USSR detonated its first similar-type bomb.[15] President Ronald Reagan restarted production in 1981.[14] The Soviet Union began a propaganda campaign against the US's neutron bomb in 1981 following Reagan's announcement. In 1983 Reagan then announced the Strategic Defense Initiative, which surpassed neutron bomb production in ambition and vision and with that the neutron bomb quickly faded from the center of the public's attention.[15]

Three types of enhanced radiation weapons (ERW) were deployed by the United States.[16] The W66 warhead, for the anti-ICBM Sprint missile system, was deployed in 1975 and retired the next year, along with the missile system. The W70 Mod 3 warhead was developed for the short-range, tactical MGM-52 Lance missile, and the W79 Mod 0 was developed for nuclear artillery shells. The latter two types were retired by President George H. W. Bush in 1992, following the end of the Cold War.[17][18] The last W70 Mod 3 warhead was dismantled in 1996,[19] and the last W79 Mod 0 was dismantled by 2003, when the dismantling of all W79 variants was completed.[20]

According to the Cox Report, as of 1999 the United States had never deployed a neutron weapon. The nature of this statement is not clear; it reads "The stolen information also includes classified design information for an enhanced radiation weapon (commonly known as the "neutron bomb"), which neither the United States, nor any other nation, has ever deployed."[21] However, the fact that neutron bombs had been produced by the US was well known at this time and part of the public record. Sam Cohen suggests the report is playing with the definitions; the US bombs were never deployed to Europe, they remained stockpiled in the US.[22]

In addition to the two superpowers, France and China are known to have tested neutron or enhanced radiation bombs. France conducted an early test of the technology in 1967[23] and tested an "actual" neutron bomb in 1980.[24] China conducted a successful test of neutron bomb principles in 1984 and a successful test of a neutron bomb in 1988. However, neither of those countries chose to deploy the neutron bomb. Chinese nuclear scientists stated before the 1988 test that China had no need for the neutron bomb, but it was developed to serve as a "technology reserve", in case the need arose in the future.[25]

Although no country is currently known to deploy them in an offensive manner, all thermonuclear dial-a-yield warheads that have about 10 kiloton and lower as one dial option, with a considerable fraction of that yield derived from fusion reactions, can be considered able to be neutron bombs in use, if not in name. The only country definitely known to deploy dedicated (that is, not dial-a-yield) neutron warheads for any length of time is Russia, which inherited the USSR's neutron warhead equipped ABM-3 Gazelle missile program. This anti-ballistic missile (ABM) system contains at least 68 neutron warheads with a 10 kiloton yield each and it has been in service since 1995, with

inert missile testing approximately every other year since then (2014). The system is designed to destroy incoming *endoatmospheric* level nuclear warheads aimed at Moscow and other targets and is the lower-tier/last umbrella of the A-135 anti-ballistic missile system (NATO reporting name: ABM-3).[26]

By 1984, according to Mordechai Vanunu, Israel was mass-producing neutron bombs.[27] A number of analysts believe that the Vela incident was an Israeli neutron bomb experiment.[28]

Considerable controversy arose in the U.S. and Western Europe following a June 1977 *Washington Post* exposé describing U.S. government plans to purchase the bomb. The article focused on the fact that it was the first weapon specifically intended to kill humans with radiation.[29][30] Lawrence Livermore National Laboratory director Harold Brown and Soviet General Secretary Leonid Brezhnev both described the neutron bomb as a "capitalist bomb", because it was designed to destroy people while preserving property.[31][32] Science fiction author and commentator Isaac Asimov also stated that "Such a neutron bomb or N bomb seems desirable to those who worry about property and hold life cheap."[33][34]

8.2   Use

Neutron bombs are purposely designed with explosive yields lower than other nuclear weapons. Since neutrons are scattered and absorbed by air,[2] neutron radiation effects drop off rapidly with distance in air. As such, there is a sharper distinction, relative to thermal effects, between areas of high lethality and areas with minimal radiation doses.[3] All high yield (more than ~10 kiloton) neutron bombs, such as the extreme example of a device that derived 97% of its energy from fusion, the 50 megaton Tsar Bomba, are not able to radiate sufficient neutrons beyond their lethal blast range when detonated as a surface burst or low altitude air burst and so are no longer classified as neutron bombs, thus limiting the yield of neutron bombs to a maximum of about 10 kilotons. The intense pulse of high-energy neutrons generated by a neutron bomb is the principal killing mechanism, not the fallout, heat or blast.

The inventor of the neutron bomb, Sam Cohen, criticized the description of the W70 as a neutron bomb since it could be configured to yield 100 kilotons:

> the W-70 ... is not even remotely a "neutron bomb." Instead of being the type of weapon that, in the popular mind, "kills people and spares buildings" it is one that both kills and physically destroys on a massive scale. The W-70

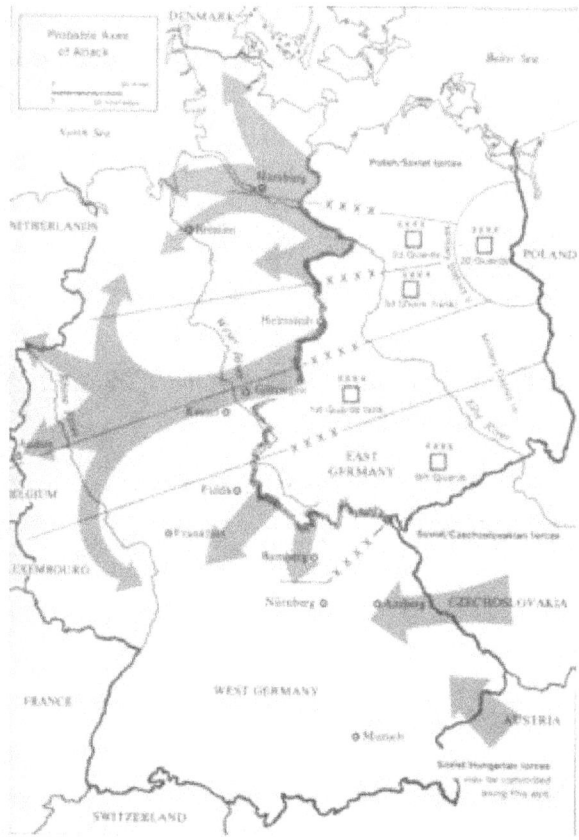

U.S. Army M110 howitzers in a 1984 REFORGER staging area before transport. Variants of this "dual capable"[41] nuclear artillery howitzer would launch the W79 neutron bomb.[42]

The Soviet/Warsaw pact invasion plan, "Seven Days to the River Rhine" to seize West Germany. According to proponents, neutron bombs would blunt an invasion by Soviet tanks and armored vehicles without causing as much damage as other nuclear weapons would.[35] Neutron bombs would have been used if the REFORGER conventional response of NATO to the invasion was too slow or ineffective.[36][37]

is not a discriminate weapon, like the neutron bomb—which, incidentally, should be considered a weapon that "kills enemy personnel while sparing the physical fabric of the attacked populace, and even the populace too."[38]

Although neutron bombs are commonly believed to "leave the infrastructure intact", with current designs that have explosive yields in the low kiloton range,[39] detonation in a built up area would still cause considerable, although not total, destruction through blast and heat effects out to a considerable radius.[40]

As the Warsaw Pact tank strength was over twice that of NATO, and Soviet Deep Battle doctrine was likely to be to use this numerical advantage to rapidly sweep across continental Europe if the Cold War ever turned hot, any weapon that could break up their intended mass tank formation deployments and force them to deploy their tanks in a thinner, more easily dividable manner,[4] would aid ground

forces in the task of hunting down solitary tanks and firing anti-tank missiles upon them,[43] such as the contemporary M47 Dragon and BGM-71 TOW missiles, which NATO had hundreds of thousands of.[44]

Rather than making extensive preparations for battlefield nuclear combat in Central Europe, "The Soviet military leadership believed that conventional superiority provided the Warsaw Pact with the means to approximate the effects of nuclear weapons and achieve victory in Europe without resort to those weapons."[45]

Neutron bombs, or more precisely, enhanced [neutron] radiation weapons were also to find use as strategic anti-ballistic missile weapons,[40] and in this role they are believed to remain in active service within Russia's Gazelle (missile).[46]

8.2.1   Effects

Upon detonation, a near-ground airburst of a 1 kiloton neutron bomb would produce a large blast wave and a powerful pulse of both thermal radiation and ionizing radiation, the latter mostly in the form of fast (14.1 MeV) neutrons. The thermal pulse would cause third degree burns to unprotected skin out to approximately 500 meters. The blast would create at least 4.6 psi out to a radius of 600 meters, which would severely damage all non-reinforced concrete structures. At the conventional effective combat range against modern main battle tanks and armored personnel carriers (<690–900 m), the blast from a 1 kt neutron bomb would destroy or damage to the point of non-usability almost all un-reinforced civilian buildings.

Using neutron bombs to stop an enemy armored attack by rapidly incapacitating crews with a dose of 8000+ rads of radiation[48] would require exploding large numbers of them to blanket the enemy forces, destroying all normal

Wood frame house in 1953 nuclear test, 5 pounds per square inch (psi) overpressure, full collapse. Although neutron bombs, such as that fitted on the MGM-52 Lance missile would cause similar levels of destruction as depicted here within the zone were ~1970s tank crews would also be incapacitated by neutron radiation. When compared to the range of destruction that would be caused by the comparatively higher yield conventional nuclear weapons that it supplanted (e.g., MGR-1 Honest John), which had been needed to deliver the same range and intensity of neutron dose to neutralize tank crews, the range of civilian destruction and amount of fission product fallout generated by a neutron bomb is far more constrained.[47] Sparing the destruction of West Germany more than would otherwise be the case.

civilian buildings within ~600 meters of the immediate area.[48][49] Neutron activation from the explosions could make many building materials in the city radioactive, such as zinc coated steel/galvanized steel (see area denial use below).

Because liquid-filled objects like the human body are resistant to gross overpressure, the 4-5 psi blast overpressure would cause very few direct casualties at the ~600 m range. The powerful winds produced by this overpressure, however, could throw bodies into objects or throw debris at high velocity, including window glass, both with potentially lethal results. Casualties would be highly variable depending on surroundings, including potential building collapses.[50]

The pulse of neutron radiation would cause immediate and permanent incapacitation to unprotected outdoor humans in the open out to 900 meters,[6] with death occurring in one or two days. The median lethal dose (LD_{50}) of 600 rads would extend to between 1350 and 1400 meters for those unprotected and outdoors,[48] where approximately half of those exposed would die of radiation sickness after several weeks.

A human residing within, or simply shielded by, at least one concrete building with walls and ceilings 30 cm (12 in) thick, or alternatively of damp soil 24 inches thick, would receive a neutron radiation exposure reduced by a factor of 10.[51][52] Even near ground zero, basement sheltering or buildings with similar radiation shielding characteristics would drastically reduce the radiation dose.[53]

Furthermore, the neutron absorption spectrum of air is disputed by some authorities, and depends in part on absorption by hydrogen from water vapor. Thus, absorption might vary exponentially with humidity, making neutron bombs far more deadly in desert climates than in humid ones.[48]

8.2.2 Questionable effectiveness in modern anti-tank role

See also: Centurion Tank § Nuclear tests, Object 279, and Neutron transport

The questionable effectiveness of ER weapons against

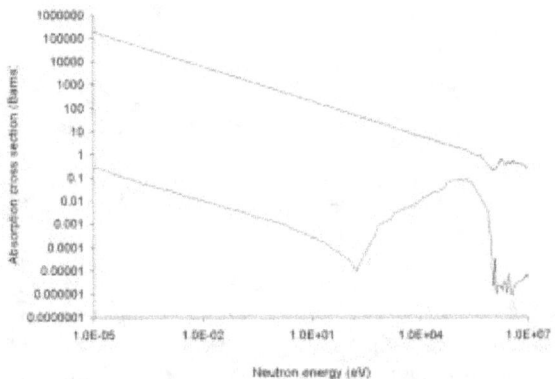

The neutron cross section and absorption probability in barns of the two natural boron isotopes found in nature (top curve is for 10B and bottom curve for 11B. As neutron energy increases to 14 MeV, the absorption effectiveness, in general, decreases. Thus, for boron-containing armor to be effective, fast neutrons must first be slowed by another element by neutron scattering.

modern tanks is cited as one of the main reasons that these weapons are no longer fielded or stockpiled. With the increase in average tank armor thickness since the first ER weapons were fielded, tank armor protection approaches the level where tank crews are now almost fully protected from radiation effects. Thus, for an ER weapon to incapacitate a modern tank crew through irradiation, the weapon must now be detonated at such a close proximity to the tank that the nuclear explosion's blast would now be equally effective at incapacitating it and its crew.[54] However this assertion was regarded as dubious in a reply in 1986[55] by a member of the Royal Military College of Science as neutron radiation from a 1 kiloton neutron bomb would incapacitate the crew of a tank with a protection factor of 35 out to a range of 280 meters, but the incapacitating blast range, depending on the exact weight of the tank, is much less, from

70 to 130 meters. However although the author did note that effective neutron absorbers and neutron poisons such as boron carbide can be incorporated into conventional armor and strap on neutron moderating hydrogenous material (hydrogen atom containing substances), such as explosive reactive armor, can both increase the protection factor, the author holds that in practice combined with neutron scattering, the actual average total tank area protection factor is rarely higher than 15.5 to 35.[56] According to the Federation of American Scientists, the neutron protection factor of a "tank" can be as low as 2,[2] without qualifying whether the statement implies a light tank, medium tank, or main battle tank.

A composite high density concrete, or alternatively, a laminated Graded Z shield, 24 units thick of which 16 units are iron and 8 units are polyethylene containing boron (BPE), and additional mass behind it to attenuate neutron capture gamma rays is more effective than just 24 units of pure iron or BPE alone, due to the advantages of both iron and BPE in combination. Iron is effective in slowing down/scattering high-energy neutrons in the 14-MeV energy range and attenuating gamma rays, while the hydrogen in polyethylene is effective in slowing down these now slower fast neutrons in the few MeV range, and boron 10 has a high absorption cross section for thermal neutrons and a low production yield of gamma rays when it absorbs a neutron.[57][58][59][60] The Soviet T72 tank, in response to the neutron bomb threat, is cited as having fitted a boronated,[61] polyethylene liner, which has had its neutron shielding properties simulated.[52][62]

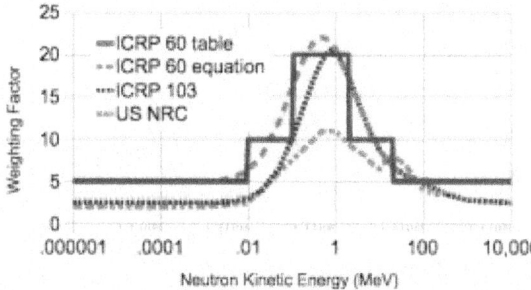

Radiation Weighting Factors for Neutrons

The radiation weighting factor for neutrons of various energy has been revised over time and certain agencies have different weighting factors, however despite the variation amongst the agencies, from the graph, for a given energy, A fusion neutron (14.1 MeV) although more energetic, is less biologically harmful as rated in Sieverts, than a fission generated thermal neutron or a fusion neutron slowed to that energy, ~0.8 MeV.

However, some tank armor material contains depleted uranium (DU), common in the US's M1A1 Abrams tank, which "incorporates steel-encased depleted uranium armour",[63] a substance that will fast fission when it captures a fast, fusion-generated neutron, and thus on fissioning will produce fission neutrons and fission products embedded within the armor, products which emit among other things, penetrating gamma rays. Although the neutrons emitted by the neutron bomb may not penetrate to the tank crew in lethal quantities, the fast fission of DU within the armor could still ensure a lethal environment for the crew and maintenance personnel by fission neutron and gamma ray exposure,[64] largely depending on the exact thickness and elemental composition of the armor—information usually hard to attain. Despite this, Ducrete—which has an elemental composition similar to, but not identical to the ceramic 2nd generation heavy metal Chobham armor of the Abrams tank—is an effective radiation shield, to both *fission* neutrons and gamma rays due to it being a graded Z material.[65][66] Uranium being about twice as dense as lead is thus nearly twice as effective at shielding gamma ray radiation per unit thickness.[67]

8.2.3 Use against ballistic missiles

As an anti-ballistic missile weapon, the first fielded ER warhead, the W66, was developed for the Sprint missile system as part of the Safeguard Program to protect United States cities and missile silos from incoming Soviet warheads by damaging their electronic components with the intense neutron flux.[40] Ionization greater than 5,000 rads in silicon chips delivered over seconds to minutes will degrade the function of semiconductors for long periods.[68] Due to the rarefied atmosphere encountered high above the earth at the most likely intercept point of an incoming warhead by a neutron bomb/warhead, whether it be the retired Sprint missile's W66 neutron warhead or the still in service Russian counterpart, the ABM-3 Gazelle, at the Terminal phase point (10–30 km) of the incoming warheads flight, the neutrons generated by a mid- to high-altitude nuclear explosion (HANE) have an even greater range than that encountered after a low altitude air burst, as in the high altitude case, there is a lower density of air molecules that produces, by comparison, an appreciable reduction in the air shielding effect/half-value thickness.

However, although this neutron transparency advantage attained only increases at increased altitudes, neutron effects lose importance in the exoatmospheric environment, being overtaken by the range of another effect of a nuclear detonation, at approximately the same altitude as the end of the incoming missile's boost phase (~150 km), ablation producing soft X-rays are the chief nuclear effects threat to the survival of incoming missiles and warheads rather than neutrons.[69] A factor exploited by the other warhead of the Safeguard Program, the enhanced (X-ray) radiation W71 and its USSR/Russian counterpart, the warhead on the A-

135 Gorgon missile.

Another method by which neutron radiation can be used to destroy incoming nuclear warheads is by serving as an intense neutron generator and to thus initiate fission in the incoming warhead's fissionable components by fast fission, potentially causing the incoming warhead to prematurely detonate in a fizzle if within sufficient proximity, but in most likely interception ranges, requiring only that enough fissionable material in the warhead fissions to interfere with the functioning of the incoming warhead when it is later fuzed to explode (see related physics: Subcritical reactor).

Lithium-6 hydride (Li6H) is cited as being used as a countermeasure to reduce the vulnerability and "harden" nuclear warheads from the effects of externally generated neutrons.[70][71] Radiation hardening of the warhead's electronic components as a countermeasure to high altitude neutron warheads somewhat reduces the range that a neutron warhead could successfully cause an unrecoverable glitch by the *transient radiation effects on electronics* (TREE) effects.[72][73]

8.2.4   Use as an area denial weapon

In November 2012, during the planning stages of Operation Hammer of God, British Labour Peer Lord Gilbert suggested that multiple enhanced radiation reduced blast (ERRB) warheads could be detonated in the mountain region of the Afghanistan/Pakistan border to prevent infiltration.[74] He proposed to warn the inhabitants to evacuate, then irradiate the area, making it unusable and impassable.[75] Used in this manner, the neutron bomb(s), regardless of burst height, would release neutron activated casing materials used in the bomb, and depending on burst height, create radioactive soil activation products.

In much the same fashion as the area denial effect resulting from fission product (the substances that make up most fallout) contamination in an area following a conventional surface burst nuclear explosion, as considered in the Korean War by Douglas MacArthur, it would thus be a form of radiological warfare - with the difference that neutron bombs produce half, or less, of the quantity of fission products relative to the same-yield pure fission bomb. Radiological warfare with neutron bombs that rely on fission primaries would thus still produce fission fallout, albeit a comparatively *cleaner* and shorter lasting version of it in the area if air bursts were used, as little to no fission products would be deposited on the direct immediate area, instead becoming diluted global fallout.

However the most effective use of a neutron bomb with respect to area denial would be to encase it in a thick shell of material that could be neutron activated, and use a surface burst. In this manner the neutron bomb would be turned into a *salted bomb*; a case of zinc-64, produced as a byproduct of depleted zinc oxide enrichment, would for example probably be the most attractive for military use, as when activated, the zinc-65 so formed is a gamma emitter, with a half life of 244 days.[76]

8.2.5   Maintenance

Neutron bombs-warheads require considerable maintenance for their abilities, requiring some tritium for fusion boosting and tritium in the secondary stage (yielding more neutrons), in amounts on the order of a few tens of grams[77] (10–30 grams[78] estimated). Because tritium has a relatively short half-life of 12.32 years (after that time, half the tritium has decayed), it is necessary to replenish it periodically to keep the bomb effective. (For instance: to maintain a constant level of 24 grams of tritium in a warhead, about 1.3 grams per bomb per year[79] must be supplied.) Moreover, tritium decays into helium-3, which absorbs neutrons[80] and will thus further reduce the bomb's neutron yield.

8.3   See also

- Atomic demolition munitions - similar strategic use, low yield nuclear weapons.
- Neutron transport
- Nuclear fallout
- Nuclear strategy
- Nuclear warfare
- Nuclear weapon design
- W54

8.4   References

[1] "Sci/Tech Neutron bomb: Why 'clean' is deadly".

[2] "Chapter 2 Conventional and Nuclear Weapons - Energy Production and Atomic Physics Section I - General. Figure 2-IX, Table 2-III".

[3] "The Neutron Bomb".

[4] "Neutron bomb an explosive issue, 1981".

[5] Muller, Richard A. (2009). *Physics for Future Presidents: The Science Behind the Headlines.* W.W. Norton & Company. p. 148. ISBN 978-0-393-33711-2.

[6] Kistiakovsky, George (Sep 1978). "The folly of the neutron bomb". *Bulletin of the Atomic Scientists.* **34**: 27. Retrieved 11 February 2011.

[7] Hafemeister, David W. (2007). *Physics of societal issues: calculations on national security, environment, and energy.* Springer. p. 18. ISBN 978-0-387-95560-5.

[8] "Mock up". Remm.nlm.gov. Retrieved 2013-11-30.

[9] "Range of weapons effects". Johnstonsarchive.net. Retrieved 2013-11-30.

[10] "Weapon designer Robert Christy discussing scaling laws, that is, how injuries from ionizing radiation do not linearly scale in lock step with the range of thermal flash injuries, especially as higher and higher yield nuclear weapons are used". Webofstories.com. Retrieved 2013-11-30.

[11] Robert D. McFadden (December 1, 2010). "Samuel T. Cohen, Neutron Bomb Inventor, Dies at 89". *The New York Times.* Retrieved 2010-12-02. After the war, he joined the RAND Corporation and in 1958 designed the neutron bomb as a way to strike a cluster of enemy forces while sparing infrastructure and distant civilian populations.

[12] Cochran, Thomas; Arkin, William; Hoenig, Milton (1987). *Nuclear Weapons Databook: U.S. nuclear warhead production. Volume 2.* Ballinger Publishing. p. 23.

[13] "About: Chemistry article", by Anne Marie Helmenstine, Ph. D

[14] "On this Day: 7 April". *BBC.* 1978-04-07. Retrieved 2010-07-02. Jimmy Carter's successor, Ronald Reagan, changed US policy and gave the order for the production of neutron warheads to start in 1981. ...

[15] "The Soviet neutron bomb at 30. March 07 2010. RT".

[16] "Nuclear Weapon News and Background". Archived from the original on 2007-09-29. Retrieved 2012-10-11.

[17] Christopher Ruddy (June 15, 1997). "Bomb inventor says U.S. defenses suffer because of politics". *Tribune-Review.* Retrieved 2010-07-03. With the fall of the Berlin Wall and the end of communism as we knew it, the Bush administration moved to dismantle all of our tactical nuclear weapons, including the Reagan stockpile of neutron bombs. In Cohen's mind, America was brought back to Square One. Without tactical weapons like the neutron bomb, America would be left with two choices if an enemy was winning a conventional war: surrender, or unleash the holocaust of strategic nuclear weapons.

[18] "Types of Nuclear Weapons". Nuclearweaponarchive.org. Retrieved 2012-10-12.

[19] John Pike. "March 13, 1996". Globalsecurity.org. Retrieved 2012-10-12.

[20] "NNSA Dismantles Last Nuclear Artillery Shell" (PDF). National Nuclear Security Administration. Retrieved 2012-10-12.

[21] "Report Of The Select Committee On U.S. National Security And Military/Commercial Concerns With The People's Republic Of China: Chapter 2 - PRC Theft Of U.S. Thermonuclear Warhead Design Information".

[22] Cohen, Samuel (9 August 1999). "Check Your Facts: Cox Report Bombs". *Insight on the News.*

[23] "Neutron bomb: Why 'clean' is deadly". BBC News. 1999-07-15. Retrieved 2012-10-12.

[24] UK parliamentary question on whether condemnation was considered by Thatcher government

[25] Ray, Jonathan (January 2015). "Red China's "Capitalist Bomb": Inside the Chinese Neutron Bomb Program" (PDF). *China Strategic Perspectives.* Washington, DC: National Defense University Press. **8**.

[26] http://www.globalsecurity.org/wmd/world/russia/gazelle.htm

[27] *The Nuclear Express: A Political History of the Bomb and Its Proliferation*, By Thomas C. Reed, Danny B. Stillman (2010), page 181

[28] *The Nuclear Express: A Political History of the Bomb and Its Proliferation*, By Thomas C. Reed, Danny B. Stillman (2010), page 177

[29] Wittner, Lawrence S. (2009). *Confronting the bomb: a short history of the world nuclear disarmament movement.* Stanford University Press. pp. 132–133. ISBN 978-0-8047-5632-7.

[30] Auten, Brian J. (2008). *Carter's conversion: the hardening of American defense policy.* University of Missouri Press. p. 134. ISBN 978-0-8262-1816-2.

[31] National security for a new era: globalization and geopolitics after Iraq, Donald Snow

[32] Herken, Greff (2003). *Brotherhood of the Bomb: The Tangled Lives and Loyalties of Robert Oppenheimer, Ernest Lawrence, and Edward Teller.* Macmillan. p. 332. ISBN 978-0-8050-6589-3.

[33] Asimov, Isaac. The New Intelligent Man's Guide to Science. Basic Books, New York, 1965. Page 410.

[34] Dewar, Dale; Oelck, Florian (2014). *From Hiroshima to Fukushima to You: A Primer on Radiation and Health.* Between the Lines. p. 29. ISBN 9781771131285.

[35] "Neutron bomb an explosive issue, 1981".

[36] "Neutron bomb an explosive issue, 1981".

[37] Healy, Melissa (October 3, 1987). "Senate Permits Study for New Tactical Nuclear Missile". *Los Angeles Times.* Retrieved 2012-08-08.

[38] "Check Your Facts: Cox Report Bombs". Insight on the News. 9 August 1999. Retrieved 5 June 2015. – via Questia (subscription required)

[39] "List of All U.S. Nuclear Weapons". Nuclearweaponarchive.org. 2006-10-14. Retrieved 2012-10-12.

[40] "What Is a Neutron Bomb? By Anne Marie Helmenstine, Ph.D.".

[41] Netherlands dual capable artillery, 1985

[42] LLNL achievements in the 1970s

[43] "what is a neutron bomb "In strategic terms, the neutron bomb has a theoretical deterrent effect: discouraging an armoured ground assault by arousing the fear of neutron bomb counterattack"".

[44] "Neutron bomb an explosive issue, 1981".

[45] http://www.gwu.edu/~{}nsarchiv//nukevault/ebb285/

[46] Soviet Ballistic Missile Defense and the Western Alliance, By David Scott Yost, pg 67-68

[47] "Neutron bomb an explosive issue, 1981".

[48] "Fact-index, neutron bomb".

[49] Calculated from http://nuclearweaponarchive.org/Nwfaq/ Nfaq5.html assuming 0.5 kt combined blast and thermal

[50] "1) Effects of blast pressure on the human body" (PDF). Retrieved 2012-10-12.

[51] "Field manual 3-4 chapter 4".

[52] "Applications of the Monte Carlo Adjoint Shielding Methodology - MIT".

[53] "Neutron bomb an explosive issue, 1981".

[54] New Scientist March 13, 1986 pg 45. 1986-03-13. Retrieved 2012-10-12.

[55]

[56] New Scientist June 12, 1986 pg 62.

[57] "Monte Carlo Calculations Using MCNP4B for an Optimal Shielding Design of a 14-MeV Neutron Source, Submitted to the Journal of Radiation Protection Dosimetry 1998" (PDF).

[58] "Neutron Interactions – Part 2 George Starkschall, Ph.D. Department of Radiation Physics." (PDF).

[59] "22.55 "Principles of Radiation Interactions"" (PDF).

[60] "The Preparation of Polyethylene and Mineral Material Composites, and Experimental and Theoretical (Using MCNP Code) Verification of Their Characteristics for Neutron Beam Attenuation" (PDF). Journal of Science and Engineering Vol. 1 (2), 2013, 95-101.

[61] "What is a neutron bomb".

[62] Terror Reigns Again By Ronan Strobing. pg 418.

[63] "M1A1/2 Abrams Main Battle Tank, United States of America".

[64] ""For example, M-1 tank armor includes depleted uranium, which can undergo fast fission and can be made to be radioactive when bombarded with neutrons".".

[65] http://web.ead.anl.gov/uranium/pdf/ducretecosteffec.pdf Paper Summary Submitted to Spectrum 2000, Sept 24-28, 2000, Chattanooga, TN. Ducrete: A Cost Effective Radiation Shielding Material. Quote- "The Ducrete/DUAGG replaces the conventional aggregate in concrete producing concrete with a density of 5.6 to 6.4 g/cm3 (compared to 2.3 g/cm3 for conventional concrete). This shielding material has the unique feature of having both high Z and low Z elements in a single matrix. Consequently, it is very effective for the attenuation of gamma and neutron radiation..."

[66] M. J. Haire and S. Y. Lobach, "Cask size and weight reduction through the use of depleted uranium dioxide (DUO$_2$)-concrete material", Waste Management 2006 Conference,Tucson, Arizona, February 26–March 2, 2006.

[67] "Half-Value Layer (Shielding)".

[68] "FAS Nuclear Weapon Radiation Effects".

[69] "Nuclear Matters Handbook". Nuclear weapon-generated X-rays are the chief threat to the survival of strategic missiles in-flight above the atmosphere and to satellites...The Neutron and gamma ray effects dominate at lower altitudes where the air absorbs most of the X-rays.

[70] "Section 12.0 Useful Tables Nuclear Weapons Frequently Asked Questions". Due to moderating ability and light weight, used to harden weapons against outside neutron fluxes (especially in combination with Li-6)...The very high cross section of this reaction for thermalized neutrons, combined with the light weight of the Li-6 atom, make it useful in the form of lithium hydride for hardening of nuclear weapons against external neutron fluxes.

[71] "Restricted Data Declassification Policy, 1946 to the Present RDD-1". The fact that Li6H is used in unspecified weapons for hardening

[72] "The Nuclear Matters Handbook, F.13".

[73] "Transient Radiation Effects on Electronics (TREE) Handbook Formerly Design Handbook for TREE, Chapters 1-6".

[74] "Huffington Post". Retrieved 2012-11-27.

[75] "Lord Gilbert obituary, by Andrew Roth, 3 June 2013. "Nobody lives up in the mountains on the border between Afghanistan and Pakistan except for a few goats and a handful of people herding them," he observed. "If you told them that some ... warheads were going to be dropped there and that it would be a very unpleasant place to go, they would not go there."".

[76] "1.6 Cobalt Bombs and other Salted Bombs. Nuclear Weapons Archive, Carey Sublette.".

[77] Kalinowski, Martin (2004). *International control of tritium for nuclear nonproliferation and disarmament*. CRC Press. p. 10. ISBN 978-0-415-31615-6.

[78] Zerriffi, Hisham (January 1996). "Tritium: The environmental, health, budgetary, and strategic effects of the Department of Energy's decision to produce tritium". Institute for Energy and Environmental Research.

[79] After a year the initial amount of 24 grams of Tritium decays to 2^(−1/12.32)x 24=22.68 grams.

[80] When absorbing neutrons, helium-3 produces back some tritium, but it comes too late in the reaction for fusion boosting and doesn't compensate for the decayed tritium missing at the reaction start.

8.5 Further reading

- Cohen, Sam, *The Truth About the Neutron Bomb: The Inventor of the Bomb Speaks Out*, William Morrow & Co., 1983, ISBN 0-688-01646-4

- Cohen, Sam, *F*** You! Mr. President: Confessions of the Father of the Neutron Bomb*, Xlibris Corporation, 2000

8.6 External links

- Strategic Implications of Enhanced Radiation Weapons

- Nuclear Files.org Definition and history of the neutron bomb

- Creator of Neutron Bomb Leaves an Explosive Legacy

- The Woodrow Wilson Center's Nuclear Proliferation International History Project or NPIHP is a global network of individuals and institutions engaged in the study of international nuclear history through archival documents, oral history interviews and other empirical sources.

Chapter 9

Radiological weapon

A **radiological weapon** or **radiological dispersion device** (**RDD**) is any weapon that is designed to spread radioactive material with the intent to kill and cause disruption. According to the U.S. Department of Defense, an RDD is "any device, including any weapon or equipment, other than a nuclear explosive device, specifically designed to employ radioactive material by disseminating it to cause destruction, damage, or injury by means of the radiation produced by the decay of such material".[1][2]

One type of RDD is a "conventional explosive combined with some type of radiological material", also known as a dirty bomb. It is not a true nuclear weapon and does not yield the same explosive power. It uses conventional explosives to spread radioactive material, most commonly the spent fuels from nuclear power plants or radioactive medical waste. "It is not a Weapon of Mass Destruction (WMD), but rather, as researcher Peter Probst calls it, a "weapon of mass disruption" (Hughes, 2002). In fact, effective dispersal ranges are rather limited. Most deaths (if any) would come from the initial explosion (non-nuclear), but it does depend on the type of radiological material used. (Department of Homeland Security [DHS], 2003)."[1][3][4]

Another version is the salted bomb, a true nuclear weapon designed to produce larger amounts of nuclear fallout than a regular nuclear weapon.

9.1 Explanation

Radiological weapons of mass destruction have been suggested as a possible weapon of terrorism used to create panic and casualties in densely populated areas. They could also render a great deal of property uninhabitable for an extended period, unless costly remediation were undertaken. The radiological source and quality greatly impacts the effectiveness of a radiological weapon.

Factors such as: energy and type of radiation, half-life, longevity, availability, shielding, portability, and the role of the environment will determine the effect of the radiolog-ical weapon. Radioisotopes that pose the greatest security risk include: 137Cs, used in radiological medical equipment, 60Co, 241Am, 252Cf, 192Ir, 238Pu, 90Sr, 226Ra, and 238U.

All of these isotopes, except for the final one, are created in nuclear power plants. While the amount of radiation dispersed from the event will likely be minimal, the fact of any radiation may be enough to cause panic and disruption.

9.2 History

The professional history of radioactive weaponry may be traced to a 1940 science fiction story, "Solution Unsatisfactory"[5] by Robert A. Heinlein and a 1943 memo from James Bryant Conant, Arthur Holly Compton and Harold Urey to Brigadier General Leslie Groves, head of the Manhattan Project.

Transmitting a report entitled, "Use of Radioactive Materials as a Military Weapon," the Groves memo states:

> As a gas warfare instrument the material would ... be inhaled by personnel. The amount necessary to cause death to a person inhaling the material is extremely small. It has been estimated that one millionth of a gram accumulating in a person's body would be fatal. There are no known methods of treatment for such a casualty.... It cannot be detected by the senses; It can be distributed in a dust or smoke form so finely powdered that it will permeate a standard gas mask filter in quantities large enough to be extremely damaging....
>
> Radioactive warfare can be used [...] To make evacuated areas uninhabitable; To contaminate small critical areas such as rail-road yards and airports; As a radioactive poison gas to create casualties among troops; Against large cities, to promote panic, and create casualties among civil-

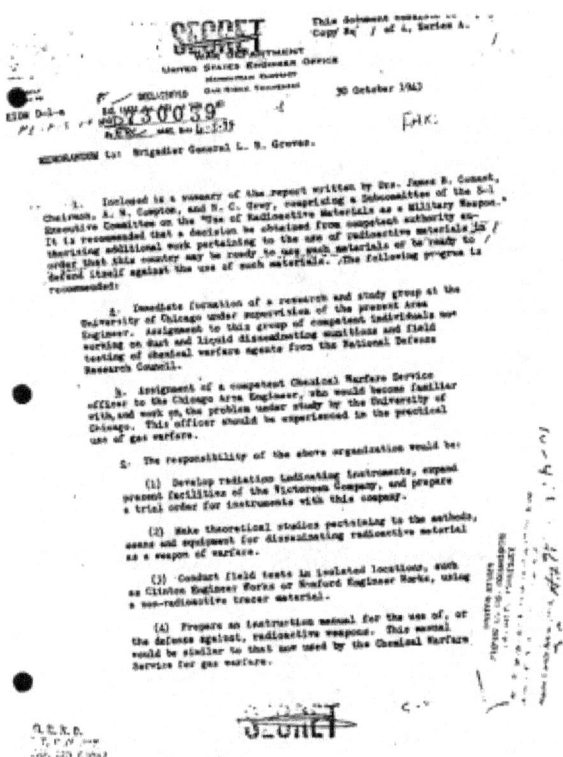

October 30, 1943 memo from Drs. Conant, Compton, and Urey to Brigadier General L. R. Groves, Manhattan District, Oak Ridge, Tennessee; declassified June 5, 1974.

ian populations.

Areas so contaminated by radioactive dusts and smokes, would be dangerous as long as a high enough concentration of material could be maintained.... they can be stirred up as a fine dust from the terrain by winds, movement of vehicles or troops, etc., and would remain a potential hazard for a long time.

These materials may also be so disposed as to be taken into the body by ingestion instead of inhalation. Reservoirs or wells would be contaminated or food poisoned with an effect similar to that resulting from inhalation of dust or smoke. For days production could contaminate a million gallons of water to an extent that a quart drunk in one day would probably result in complete incapacitation or death in about a month's time.

The United States, however, chose not to pursue radiological weapons during World War II, though early on in the project considered it as a backup plan in case nuclear fission proved impossible to tame. Some US policymakers and scientists involved in the project felt that radiological weapons would qualify as chemical weapons and thus violate international law.

9.3 Deployment

One possible way of dispersing the material is by using a dirty bomb, a conventional explosive which disperses radioactive material. Dirty bombs are not a type of nuclear weapon, which requires a nuclear chain reaction and the creation of a critical mass. Whereas a nuclear weapon will usually create mass casualties immediately following the blast, a dirty bomb scenario would initially cause only minimal casualties from the conventional explosion.

Means of radiological warfare that do not rely on any specific weapon, but rather on spreading radioactive contamination via a food chain or water table, seem to be more effective in some ways, but share many of the same problems as chemical warfare.

9.4 Military uses

Radiological weapons are widely considered to be militarily useless for a state-sponsored army and are initially not hoped to be used by any military forces. Firstly, the use of such a weapon is of no use to an occupying force, as the target area becomes uninhabitable (due to the fallout caused by radioactive poisoning of the involved environment).

Furthermore, area-denial weapons are generally of limited use to an attacking army, as it slows the rate of advance.

9.5 Dirty bombs

A dirty bomb is a radiological weapon dispersed with conventional explosives.

There is currently (as of 2007) an ongoing debate about the damage that terrorists using such a weapon might inflict. Many experts believe that a dirty bomb such that terrorists might reasonably be able to construct would be unlikely to harm more than a few people and hence it would be no more deadly than a conventional bomb. Furthermore, the casualties would be a result of the initial explosion, because alpha and beta emitting material needs to be inhaled to do damage to the human body. Gamma radiation emitting material is so radioactive that it can't be deployed without wrapping an amount of shielding material around the bomb that would make transport by car or plane impossible without risking detection. Because of this a dirty bomb with radioactive material around an explosive device would be almost useless, unless said shielding was removed shortly before detonation. This is not only because of the effectiveness but also because this material would be easy to clean up. Furthermore, the possibility of terrorists making a gas or aerosol that is radioactive is very unlikely because of the complex chemical work to achieve this goal.[6]

Hence, this line of argument goes, the objectively dominant effect would be the moral and economic damage due to the massive fear and panic such an incident would spur. On the other hand, some believe that the fatalities and injuries might be in fact much more severe. This point was made by physicist Peter D. Zimmerman (King's College London) who reexamined the Goiânia accident which is arguably comparable.[7] and popularized in a subsequent fictionalized account produced by the BBC and broadcast in the United States by PBS.[8] The latter program showed how shielding might be used to minimize the detection risk.

9.6 Salted bomb

Main article: Salted bomb

A salted bomb is a theoretical nuclear weapon designed to produce enhanced quantities of radioactive fallout, rendering a large area uninhabitable. As far as is publicly known none have ever been built.

9.7 See also

- Biological warfare
- Chemical warfare
- Cobalt bomb
- Lists of nuclear disasters and radioactive incidents
- Nuclear fallout
- Nuclear weapon
- Radioactive contamination
- Weapon of mass destruction
- Nuclear terrorism

9.8 References

[1] Rickert, Paul (2005-12-31). "The Likely Effect of a Radiological Dispersion Device". Liberty University. pp. 2, 3. Retrieved 21 October 2014.

[2] Ford, J. (March 1998). "Radiological Dispersion Devices: Assessing the transnational threat". National Defense University - Institute for National Strategic Studies - Strategic Forum. Archived from the original on December 12, 2005. Retrieved December 31, 2005.

[3] Hughes, D. (4 March 2002). "When terrorists go nuclear". Popular Mechanics. Archived from the original on September 19, 2005. Retrieved December 31, 2005.

[4] "Radiological Dispersion Devices Fact Sheet". Department of Homeland Security. 10 February 2003. Archived from the original on December 29, 2005. Retrieved December 31, 2005.

[5] Full story at publisher's web site

[6]

[7] Dirty Bombs: The Threat Revisited in Defense Horizons, Feb. 2004, a publication of the National Defense University

[8] Dirty Bomb

9.9 External links

- Annotated bibliography for radiological dispersal devices (RDD) from the Alsos Digital Library for Nuclear Issues. This page has no results.

Chapter 10

Nuclear electromagnetic pulse

This article is about nuclear-generated EMP. For other types, see Electromagnetic pulse

An **electromagnetic pulse** (commonly abbreviated as EMP, pronounced /iː.ɛm.piː/) is a burst of electromagnetic radiation. Nuclear explosions create a characteristic pulse of electromagnetic radiation called a nuclear EMP or NEMP.

The resulting rapidly changing electric and magnetic fields may couple with electrical and electronic systems to produce damaging current and voltage surges. The specific characteristics of any particular nuclear EMP event vary according to a number of factors, the greatest of which is the altitude of the detonation.

The term "electromagnetic pulse" generally excludes optical (infrared, visible, ultraviolet) and ionizing (such as X-ray and gamma radiation) ranges.

In military terminology, a nuclear warhead detonated hundreds of kilometers above the Earth's surface is known as a high-altitude electromagnetic pulse (HEMP) device. Effects of a HEMP device depend on factors including the altitude of the detonation, energy yield, gamma ray output, interactions with the Earth's magnetic field and electromagnetic shielding of targets.

10.1 History

The fact that an electromagnetic pulse is produced by a nuclear explosion was known in the earliest days of nuclear weapons testing. The magnitude of the EMP and the significance of its effects, however, were not immediately realized.[1]

During the first United States nuclear test on 16 July 1945, electronic equipment was shielded due to Enrico Fermi's expectation of the electromagnetic pulse. The official technical history for that first nuclear test states, "All signal lines were completely shielded, in many cases doubly shielded.

In spite of this many records were lost because of spurious pickup at the time of the explosion that paralyzed the recording equipment."[2] During British nuclear testing in 1952–1953 instrumentation failures were attributed to "radioflash", which was their term for EMP.[3][4]

The first openly reported observation of the unique aspects of high-altitude nuclear EMP occurred during the helium balloon lofted Yucca nuclear test of the Hardtack I series on 28 April 1958. In that test, the electric field measurements from the 1.7 kiloton weapon went off the scale of the test instruments and was estimated to be about 5 times the oscilloscope limits. The Yucca EMP was initially positive-going whereas low-altitude bursts were negative pulses. Also, the polarization of the Yucca EMP signal was horizontal, whereas low-altitude nuclear EMP was vertically polarized. In spite of these many differences, the unique EMP results were dismissed as a possible wave propagation anomaly.[5]

The high-altitude nuclear tests of 1962, as discussed below, confirmed the unique results of the Yucca high-altitude test and increased the awareness of high-altitude nuclear EMP beyond the original group of defense scientists.

The larger scientific community became aware of the significance of the EMP problem after a three-article series on nuclear EMP was published in 1981 by William J. Broad in *Science*.[1][6][7]

10.1.1 Starfish Prime

Main article: Starfish Prime

In July 1962, the US carried out the Starfish Prime test, exploding a 1.44 megaton bomb 400 kilometres (250 mi) above the mid-Pacific Ocean. This demonstrated that the effects of a high-altitude nuclear explosion were much larger than had been previously calculated. Starfish Prime made those effects known to the public by causing electrical damage in Hawaii, about 1,445 kilometres (898 mi) away from the detonation point, knocking out about 300 streetlights, setting off numerous burglar alarms and damaging a microwave link.[8]

Starfish Prime was the first success in the series of United States high-altitude nuclear tests in 1962 known as Operation Fishbowl. Subsequent tests gathered more data on the high-altitude EMP phenomenon.

The Bluegill Triple Prime and Kingfish high-altitude nuclear tests of October and November 1962 in Operation Fishbowl provided data that was clear enough to enable physicists to accurately identify the physical mechanisms behind the electromagnetic pulses.[9]

The EMP damage of the Starfish Prime test was quickly repaired because of the ruggedness (compared to today)[10] of Hawaii's electrical and electronic infrastructure.

The relatively small magnitude of the Starfish Prime EMP in Hawaii (about 5.6 kilovolts/metre) and the relatively small amount of damage (for example, only 1 to 3 percent of streetlights extinguished)[11] led some scientists to believe, in the early days of EMP research, that the problem might not be significant. Newer calculations[10] showed that if the Starfish Prime warhead had been detonated over the northern continental United States, the magnitude of the EMP would have been much larger (22 to 30 kv/m) because of the greater strength of the Earth's magnetic field over the United States, as well as its different orientation at high latitudes. These calculations, combined with the accelerating reliance on EMP-sensitive microelectronics, heightened awareness that EMP could be a significant problem.

10.1.2 Soviet Test 184

Main article: Soviet Project K nuclear tests

In 1962, the Soviet Union also performed three EMP-producing nuclear tests in space over Kazakhstan, the last in the "Soviet Project K nuclear tests".[12] Although these weapons were much smaller (300 kiloton) than the Starfish Prime test, they were over a populated, large land mass and at a location where the Earth's magnetic field was greater, the damage caused by the resulting EMP was reportedly much greater than in Starfish Prime. The geomagnetic storm–like E3 pulse from Test 184 induced a current surge in a long underground power line that caused a fire in the power plant in the city of Karaganda.

After the collapse of the Soviet Union, the level of this damage was communicated informally to U.S. scientists.[13] After the 1991 collapse of the Soviet Union, there was a period of a few years of cooperation between United States and Russian scientists on the HEMP phenomenon. In addition, funding was secured to enable Russian scientists to formally report on some of the Soviet EMP results in international

scientific journals.[14] As a result, formal documentation of some of the EMP damage in Kazakhstan exists[15][16] but is still sparse in the open scientific literature, especially in relation to the level of damage that was indicated in the open reports.

For one of the K Project tests, Soviet scientists instrumented a 570-kilometer (350 mi) section of telephone line in the area that they expected to be affected by the pulse. The monitored telephone line was divided into sub-lines of 40 to 80 kilometres (25 to 50 mi) in length, separated by repeaters. Each sub-line was protected by fuses and by gas-filled overvoltage protectors. The EMP from the 22 October (K-3) nuclear test (also known as Test 184) blew all of the fuses and fired all of the overvoltage protectors in all of the sub-lines.[15]

Published reports, including a 1998 IEEE article,[15] have stated that there were significant problems with ceramic insulators on overhead electrical power lines during the tests. A 2010 technical report written for Oak Ridge National Laboratory stated that "Power line insulators were damaged, resulting in a short circuit on the line and some lines detaching from the poles and falling to the ground."[17]

10.2   Characteristics of nuclear EMP

Nuclear EMP is a complex multi-pulse, usually described in terms of three components, as defined by the International Electrotechnical Commission (IEC).[18]

The three components of nuclear EMP, as defined by the IEC, are called "E1", "E2" and "E3".

10.2.1   E1

The E1 pulse is the very fast component of nuclear EMP. E1 is a very brief but intense electromagnetic field that induces very high voltages in electrical conductors. E1 causes most of its damage by causing electrical breakdown voltages to be exceeded. E1 can destroy computers and communications equipment and it changes too quickly for ordinary surge protectors to provide effective protection against it, although there are special fast-acting surge protectors that will block the E1 pulse.

E1 is produced when gamma radiation from the nuclear detonation ionizes (strips electrons from) atoms in the upper atmosphere. This is known as the Compton effect and the resulting current is called the "Compton current". The electrons travel in a generally downward direction at relativistic speeds (more than 90 percent of the speed of light). In the absence of a magnetic field, this would produce a large, radial pulse of electric current propagating outward from

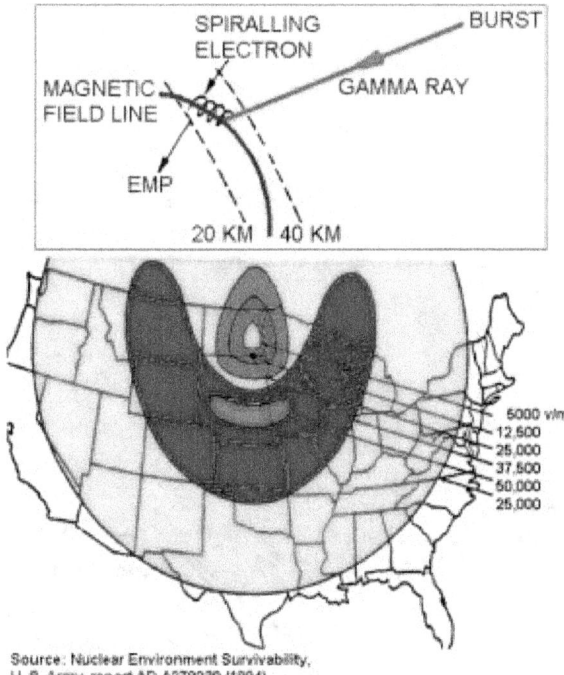

Source: Nuclear Environment Survivability,
U. S. Army, report AD-A278230 (1994)

The mechanism for a 400 km high-altitude burst EMP: gamma rays hit the atmosphere between 20–40 km altitude, ejecting electrons which are then deflected sideways by the Earth's magnetic field. This makes the electrons radiate EMP over a massive area. Because of the curvature and downward tilt of Earth's magnetic field over the USA, the maximum EMP occurs south of the detonation and the minimum occurs to the north.[19]

the burst location confined to the source region (the region over which the gamma photons are attenuated). The Earth's magnetic field deflects the electron flow at a right angle to the field, leading to synchrotron radiation emitted by the electrons. Because the outward traveling gamma pulse is propagating at the speed of light, the synchrotron radiation of the Compton electrons adds coherently, leading to a radiated electromagnetic signal. This interaction produces a very large, but very brief, electromagnetic pulse over the affected area.[20]

Several physicists worked on the problem of identifying the mechanism of the uniquely large E1 pulse produced by a nuclear weapon detonated at high altitude (HEMP). The correct mechanism was finally identified by Conrad Longmire of Los Alamos National Laboratory in 1963.[9]

Conrad Longmire gives numerical values for a typical case of E1 pulse produced by a second-generation nuclear weapon such as those of Operation Fishbowl in 1962. The typical gamma rays given off by the weapon have an energy of about 2 MeV (mega-electron volts). The gamma rays transfer about half of their energy to the ejected free electrons, giving an energy of about 1 MeV.[20]

In a vacuum and absent a magnetic field, the electrons would travel with a current density of tens of amperes per square metre.[20] Because of the downward tilt of the Earth's magnetic field at high latitudes, the area of peak field strength is a U-shaped region to the equatorial side of the nuclear detonation. As shown in the diagram at the right, for nuclear detonations over the continental United States, this U-shaped region is south of the detonation point. Near the equator, where the Earth's magnetic field is more nearly horizontal, the E1 field strength is more nearly symmetrical around the burst location.

At geomagnetic field strengths typical of the central United States, central Europe or Australia, these initial electrons spiral around the magnetic field lines with a typical radius of about 85 metres (about 280 feet). These initial electrons are stopped by collisions with other air molecules at an average distance of about 170 metres (a little less than 580 feet). This means that most of the electrons are stopped by collisions with air molecules before completing a full spiral around the field lines.[20]

This interaction of the very rapidly moving negatively charged electrons with the magnetic field radiates a pulse of electromagnetic energy. The pulse typically rises to its peak value in some 5 nanoseconds. Its magnitude typically decays to half of its peak value within 200 nanoseconds. (By the IEC definition, this E1 pulse ends 1000 nanoseconds after it begins.) This process occurs simultaneously on about 10^{25} electrons.[20] The simultaneous action of the very large number of electrons causes the resulting electromagnetic pulses from each electron to radiate coherently, thus adding to produce a single very large amplitude, but very narrow, radiated electromagnetic pulse.

Secondary collisions cause subsequent electrons to lose energy before they reach ground level. The electrons generated by these subsequent collisions have such reduced energy that they do not contribute significantly to the E1 pulse.[20]

These 2 MeV gamma rays typically produce an E1 pulse near ground level at moderately high latitudes that peaks at about 50,000 volts per metre. This is a peak power density of 6.6 megawatts per square metre.

The ionization process in the mid-stratosphere causes this region to become an electrical conductor, a process that blocks the production of further electromagnetic signals and causes the field strength to saturate at about 50,000 volts per metre. The strength of the E1 pulse depends upon the number and intensity of the gamma rays and upon the rapidity of the gamma ray burst. Strength is also somewhat dependent upon altitude.

There are reports of "super-EMP" nuclear weapons that are able to exceed the 50,000 volt per metre limit by the nearly instantaneous release of a burst of much higher gamma radiation levels than are known to be produced by second-generation nuclear weapons. The reality and possible construction details of these weapons are classified and unconfirmed in the open scientific literature.[21]

10.2.2   E2

The E2 component is generated by scattered gamma rays and inelastic gammas produced by neutrons. This E2 component is an "intermediate time" pulse that, by the IEC definition, lasts from about 1 microsecond to 1 second after the explosion. E2 has many similarities to lightning, although lightning-induced E2 may be considerably larger than a nuclear E2. Because of the similarities and the widespread use of lightning protection technology, E2 is generally considered to be the easiest to protect against.

According to the United States EMP Commission, the main problem with E2 is the fact that it immediately follows E1, which may have damaged the devices that would normally protect against E2.

The EMP Commission Executive Report of 2004 states, "In general, it would not be an issue for critical infrastructure systems since they have existing protective measures for defense against occasional lightning strikes. The most significant risk is synergistic, because the E2 component follows a small fraction of a second after the first component's insult, which has the ability to impair or destroy many protective and control features. The energy associated with the second component thus may be allowed to pass into and damage systems."[22]

10.2.3   E3

Main article: Geomagnetically induced current

The E3 component is very different from E1 and E2. E3 is a very slow pulse, lasting tens to hundreds of seconds. It is caused by the nuclear detonation's temporary distortion of the Earth's magnetic field. The E3 component has similarities to a geomagnetic storm caused by a solar flare.[23][24] Like a geomagnetic storm, E3 can produce geomagnetically induced currents in long electrical conductors, damaging components such as power line transformers.[25]

Because of the similarity between solar-induced geomagnetic storms and nuclear E3, it has become common to refer to solar-induced geomagnetic storms as "solar EMP."[26] "Solar EMP", however, does not include an E1 or E2 component.

See also: Coronal mass ejection and Solar flare

10.3   Generation

Factors that control weapon effectiveness include altitude, yield, construction details, target distance, intervening geographical features, and local strength of the Earth's magnetic field.

10.3.1   Weapon altitude

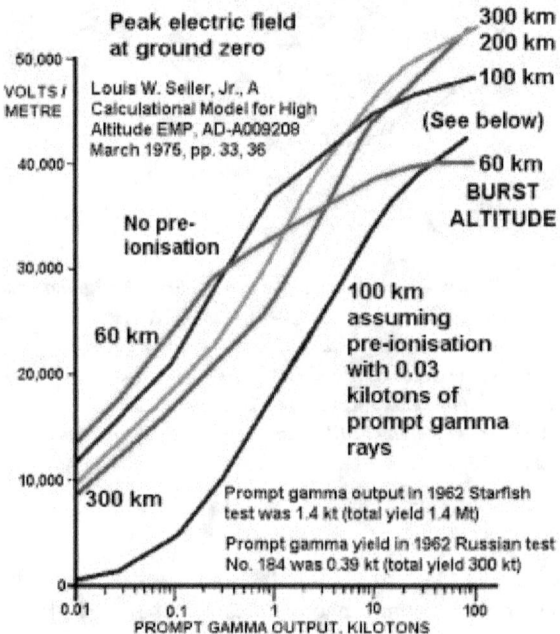

*How the peak EMP on the ground varies with the weapon yield and burst altitude. The yield here is the prompt gamma ray output measured in kilotons. This varies from 0.115–0.5% of the total weapon yield, depending on weapon design. The 1.4 Mt total yield 1962 Starfish Prime test had a gamma output of 0.1%, hence 1.4 kt of prompt gamma rays. (The **blue** 'pre-ionisation' curve applies to certain types of thermonuclear weapon, where gamma and x-rays from the primary fission stage ionise the atmosphere and make it electrically conductive before the main pulse from the thermonuclear stage. The pre-ionisation in some situations can literally short out part of the final EMP, by allowing a conduction current to immediately oppose the Compton current of electrons.)[27][28]*

According to an internet primer published by the Federation of American Scientists[29]

> A high-altitude nuclear detonation produces an immediate flux of gamma rays from the nuclear reactions within the device. These photons in turn produce high energy free electrons by Compton

scattering at altitudes between (roughly) 20 and 40 km. These electrons are then trapped in the Earth's magnetic field, giving rise to an oscillating electric current. This current is asymmetric in general and gives rise to a rapidly rising radiated electromagnetic field called an electromagnetic pulse (EMP). Because the electrons are trapped essentially simultaneously, a very large electromagnetic source radiates coherently.

The pulse can easily span continent-sized areas, and this radiation can affect systems on land, sea, and air. ... A large device detonated at 400–500 km (250 to 312 miles) over Kansas would affect all of the continental U.S. The signal from such an event extends to the visual horizon as seen from the burst point.

Thus, for equipment to be affected, the weapon needs to be above the visual horizon.

The altitude indicated above is greater than that of the International Space Station and many low Earth orbit satellites. Large weapons could have a dramatic impact on satellite operations and communications such as occurred during Operation Fishbowl. The damaging effects on orbiting satellites are usually due to factors other than EMP. In the Starfish Prime nuclear test, most damage was to the satellites' solar panels while passing through radiation belts created by the explosion.[30]

For detonations within the atmosphere, the situation is more complex. Within the range of gamma ray deposition, simple laws no longer hold as the air is ionised and there are other EMP effects, such as a radial electric field due to the separation of Compton electrons from air molecules, together with other complex phenomena. For a surface burst, absorption of gamma rays by air would limit the range of gamma ray deposition to approximately 10 miles, while for a burst in the lower-density air at high altitudes, the range of deposition would be far greater.

10.3.2 Weapon yield

Typical nuclear weapon yields used during Cold War planning for EMP attacks were in the range of 1 to 10 megatons[31] This is roughly 50 to 500 times the size of the Hiroshima and Nagasaki bombs. Physicists have testified at United States Congressional hearings that weapons with yields of 10 kilotons or less can produce a large EMP.[32]

The EMP at a fixed distance from an explosion increases at most as the square root of the yield (see the illustration to the right). This means that although a 10 kiloton weapon has only 0.7% of the energy release of the 1.44-megaton

Starfish Prime test, the EMP will be at least 8% as powerful. Since the E1 component of nuclear EMP depends on the prompt gamma ray output, which was only 0.1% of yield in Starfish Prime but can be 0.5% of yield in low yield pure nuclear fission weapons, a 10 kiloton bomb can easily be 5 x 8% = 40% as powerful as the 1.44 megaton Starfish Prime at producing EMP.[33]

The total prompt gamma ray energy in a fission explosion is 3.5% of the yield, but in a 10 kiloton detonation the triggering explosive around the bomb core absorbs about 85% of the prompt gamma rays, so the output is only about 0.5% of the yield. In the thermonuclear Starfish Prime the fission yield was less than 100% and the thicker outer casing absorbed about 95% of the prompt gamma rays from the pusher around the fusion stage. Thermonuclear weapons are also less efficient at producing EMP because the first stage can pre-ionize the air[33] which becomes conductive and hence rapidly shorts out the Compton currents generated by the fusion stage. Hence, small pure fission weapons with thin cases are far more efficient at causing EMP than most megaton bombs.

This analysis, however, only applies to the fast E1 and E2 components of nuclear EMP. The geomagnetic storm-like E3 component of nuclear EMP is more closely proportional to the total energy yield of the weapon.[34]

10.3.3 Target distance

In nuclear EMP all of the components of the electromagnetic pulse are generated outside of the weapon.[29]

For high-altitude nuclear explosions, much of the EMP is generated far from the detonation (where the gamma radiation from the explosion hits the upper atmosphere). This electric field from the EMP is remarkably uniform over the large area affected.

According to the standard reference text on nuclear weapons effects published by the U.S. Department of Defense, "The peak electric field (and its amplitude) at the Earth's surface from a high-altitude burst will depend upon the explosion yield, the height of the burst, the location of the observer, and the orientation with respect to the geomagnetic field. As a general rule, however, the field strength may be expected to be tens of kilovolts per metre over most of the area receiving the EMP radiation."[35]

The text also states that, "... over most of the area affected by the EMP the electric field strength on the ground would exceed $0.5E_{max}$. For yields of less than a few hundred kilotons, this would not necessarily be true because the field strength at the Earth's tangent could be substantially less than $0.5E_{max}$."[35]

(E_{max} refers to the maximum electric field strength in the affected area.)

In other words, the electric field strength in the entire area that is affected by the EMP will be fairly uniform for weapons with a large gamma ray output. For smaller weapons, the electric field may fall at a faster rate as distance increases.

10.4   Effects

10.4.1   On aircraft

Many nuclear detonations have taken place using aerial bombs. The B-29 aircraft that delivered the nuclear weapons at Hiroshima and Nagasaki did not lose power due to electrical damage, because electrons (ejected from the air by gamma rays) are stopped quickly in normal air for bursts below roughly 10 kilometres (6.2 mi), so they are not significantly deflected by the Earth's magnetic field.[36]

If the aircraft carrying the Hiroshima and Nagasaki bombs had been within the intense nuclear radiation zone when the bombs exploded over those cities, then they would have suffered effects from the charge separation (radial) EMP. But this only occurs within the severe blast radius for detonations below about 10 km altitude.

During Operation Fishbowl, EMP disruptions were suffered aboard a KC-135 photographic aircraft flying 300 km (190 mi) from the 410 kt (1,700 TJ) detonations at 48 and 95 km (30 and 59 mi) burst altitudes.[33] The vital electronics were less sophisticated than today's and the aircraft was able to land safely.

10.4.2   Vacuum tube versus solid state electronics

Older, vacuum tube (valve) based equipment is generally much less vulnerable to nuclear EMP than newer solid state equipment. Soviet Cold War–era military aircraft often had avionics based on vacuum tubes due to limited solid-state capabilities and a belief that the vacuum-tube gear would be more likely to survive.[1]

Other components in vacuum tube circuitry can be damaged by EMP. Vacuum tube equipment was damaged in the 1962 testing.[16] The solid state PRC-77 VHF manpackable 2-way radio survived extensive EMP testing.[37] The earlier PRC-25, nearly identical except for a vacuum tube final amplification stage, was tested in EMP simulators, but was not certified to remain fully functional.

10.5   Post–Cold War attack scenarios

The United States military services developed, and in some cases published, hypothetical EMP attack scenarios.[38]

The United States EMP Commission was created by the United States Congress in 2001. The commission is formally known as the Commission to Assess the Threat to the United States from Electromagnetic Pulse (EMP) Attack.[39]

The Commission brought together notable scientists and technologists to compile several reports. In 2008, the EMP Commission released the "Critical National Infrastructures Report".[34] This report describes the likely consequences of a nuclear EMP on civilian infrastructure. Although this report covered the United States, most of the information can be generalized to other industrialized countries. The 2008 report was a followup to a more generalized report issued by the commission in 2004.[24][40]

In written testimony delivered to the United States Senate in 2005, an EMP Commission staff member reported:

> The EMP Commission sponsored a worldwide survey of foreign scientific and military literature to evaluate the knowledge, and possibly the intentions, of foreign states with respect to electromagnetic pulse (EMP) attack. The survey found that the physics of EMP phenomenon and the military potential of EMP attack are widely understood in the international community, as reflected in official and unofficial writings and statements. The survey of open sources over the past decade finds that knowledge about EMP and EMP attack is evidenced in at least Britain, France, Germany, Israel, Egypt, Taiwan, Sweden, Cuba, India, Pakistan, Iraq under Saddam Hussein, Iran, North Korea, China and Russia.
>
> Many foreign analysts–particularly in Iran, North Korea, China, and Russia–view the United States as a potential aggressor that would be willing to use its entire panoply of weapons, including nuclear weapons, in a first strike. They perceive the United States as having contingency plans to make a nuclear EMP attack, and as being willing to execute those plans under a broad range of circumstances.
>
> Russian and Chinese military scientists in open source writings describe the basic principles of nuclear weapons designed specifically to generate an enhanced-EMP effect, that they term "Super-EMP" weapons. "Super-EMP" weapons, according to these foreign open source writings,

can destroy even the best protected U.S. military and civilian electronic systems.[21]

The United States EMP Commission determined that long-known protections are almost completely absent in the civilian infrastructure of the United States and that large parts of US military services were less-protected against EMP than during the Cold War. In public statements, the EMP experts on the EMP Commission recommended making electronic equipment and electrical components resistant to EMP — and maintaining spare parts inventories that would enable prompt repairs.[24][34][41] The United States EMP Commission did not look at the civilian infrastructures of other nations.

In 2011 the Defense Science Board published a report about the ongoing efforts to defend critical military and civilian systems against EMP and other nuclear weapons effects.[42]

10.6 Common misconceptions

A 2010 technical report written for the US government's Oak Ridge National Laboratory included a brief section addressing common EMP myths.[43] The remainder of this section is a direct quotation from that Oak Ridge report regarding common HEMP Myths:

> Much of the literature on HEMP is either classified or not easily accessible. Probably because of this, some of what is openly available tends to vary in accuracy – some, especially from the Internet, has major inaccuracies. Some discussions of HEMP have the right words and concepts, but do not quite have them put together right, or have inaccurate interpretations. Here we will discuss some common misunderstandings. HEMP has also appeared in some movies, and there are on-line discussions about possible errors in their depiction of HEMP. Here we will be concerned with E1 HEMP, and ignore misunderstandings about other types of EMP.
>
> **Extremists:** Some general emphasis of comments fall into either "the world as we know it will come to an end" if there is a high altitude nuclear burst, or the other extreme: "it's not a big deal, nothing much will happen".

Since we really have never had a nuclear burst over anything like our current modern infrastructure, no one really knows for sure what would happen, but both extremes are not very believable.

Yield: There appears to be an assumption that yield is important – it is not for E1. The assumption that E1 is an issue only for cold war type situations, but not for terrorists or rogue nations, is false. Very big bombs might have better area coverage of high fields by going to higher burst heights, but for peak fields the burst yield is only a very minor consideration.

1962 experience: Some point to the Starfish event, and the rather minor HEMP effects produced at Hawaii by it. However, there are many problems with extrapolating that experience:

1. That was about half a century ago. Since then, the use of electronics has increased greatly, and the type of sensitive electronics we currently use did not really exist back then.

2. The burst was fairly far away from Hawaii, and the incident E1 HEMP was much less than worst case.

3. The island is small – if over the continental U.S., long transmission lines would be exposed (especially an issue for late-time HEMP). In addition, widely separated substations would have been exposed, although with electromechanical relays (not solid state). Also the yield argument has been used – Starfish was a very big weapon, yet it did very little – see the previous item, yield is not really very significant.

Cars dying: Some say that all vehicles traveling will come to a halt, with all modern vehicles damaged because of their use of modern electronics (and one movie even had a bulk, non-electronic part dying). Most likely there will be some vehicles affected, but probably just a small fraction of them (although this could create traffic jams in large cities). A car does not have very long cabling to act as antennas, and there is some protection from metallic construction. As nonmetallic materials are used more and more in the future to decrease weight and increase fuel efficiency, this advantage may disappear.

Wristwatch dying: One movie critic pointed out that electronics in a helicopter were affected, but not the star's electronic watch. A watch is much too small for HEMP to affect it.

Electrons present: One critic, with some awareness of the generation process, said that HEMP could not be present unless there were also energetic electrons present. This is true when one is within the source region, which exists for all types of EMP — there are energetic electrons present. However for the HEMP, the radiation and energetic electrons are present at altitudes of 20 to 40 km, not at the ground.

Turn equipment off: There is truth to this recommendation (if there were a way to know that a burst was about to happen). Equipment is more vulnerable if it is operating, because some failure modes involving E1 HEMP trigger the system's energy to damage itself. However, damage can also happen, but not as easily, to systems that are turned off.

Maximum conductor length: There is a suggestion that equipment will be OK if all connected conductors are less than a specific length. Certainly shorter lengths are generally better, but there is no magic length value, with shorter always being better and longer not. Coupling is much too complex for such a blanket statement — instead it should be "the shorter the better, in general". (There can be exceptions,

such as resonance effects, which depend on line lengths.)

Stay away from metal: There is a recommendation to be some distance away from any metal when a HEMP event occurs (assuming there was warning), because very high voltages could be generated. Metal can collect E1 HEMP energy, and easily generate high voltages. However, the "skin effect" (a term not really derived from the skin of humans or any other animal) means that if a human were touching a large "antenna" during an E1 HEMP event, any current flow would not penetrate into the body. Generally E1 HEMP is considered harmless for human bodies.

10.7　Protecting infrastructure

In 2013, the US House of Representatives considered the "Secure High-voltage Infrastructure for Electricity from Lethal Damage Act" that would provide surge protection for some 300 large transformers around the country.[44]

The problem of protecting civilian infrastructure from electromagnetic pulse has also been intensively studied throughout the European Union, and in particular by the United Kingdom.[45][46]

10.8　In fiction and popular culture

Main article: Electromagnetic pulse in fiction and popular culture

Especially since the 1980s, Nuclear EMP weapons have gained a significant presence in fiction and popular culture.

The popular media often depict EMP effects incorrectly, causing misunderstandings among the public and even professionals, and official efforts have been made in the United States to set the record straight.[43] See, for example, the Oak Ridge quotation in the above section of this article on "Common Misconceptions." Also, the United States Space Command commissioned science educator Bill Nye to produce a video called "Hollywood vs. EMP" so that Hollywood fiction would not confuse those who must deal with real EMP events.[47] The U.S. Space Command video is not available to the general public.

10.9 See also

- Electromagnetic compatibility (EMC)
- Electromagnetic environment
- Electromagnetic hypersensitivity
- Electromagnetic pulse in fiction and popular culture
- Electromagnetic weapon
- Electromagnetism
- Electronic warfare
- Explosively pumped flux compression generator
- Faraday's law of induction
- Gamma ray burst
- Geomagnetic storm
- High-altitude nuclear explosion
- High-power microwave
- Marx generator
- Operation Fishbowl
- Pulsed power
- Soviet Project K nuclear tests
- Starfish Prime
- Ultrashort pulse

10.10 References

[1] Broad, William J. "Nuclear Pulse (I): Awakening to the Chaos Factor", Science. 29 May 1981 212: 1009–1012

[2] Bainbridge, K.T., (Report LA-6300-H), Los Alamos Scientific Laboratory. May 1976. p. 53 *Trinity*

[3] Baum, Carl E., IEEE Transactions on Electromagnetic Compatibility. Vol. 49, No. 2. pp. 211–218. May 2007. *Reminiscences of High-Power Electromagnetics*

[4] Baum, Carl E., Proceedings of the IEEE, Vol.80, No. 6, pp. 789–817. June 1992 *From the Electromagnetic Pulse to High-Power Electromagnetics*

[5] Defense Atomic Support Agency. 23 September 1959. Operation Hardtack Preliminary Report. Technical Summary of Military Effects. Report ADA369152. pp. 346-350.

[6] Broad, William J. "Nuclear Pulse (II): Ensuring Delivery of the Doomsday Signal", Science. 5 June 1981 212: 1116–1120

[7] Broad, William J. "Nuclear Pulse (III): Playing a Wild Card", Science. 12 June 1981 212: 1248–1251

[8] Vittitoe, Charles N., "Did High-Altitude EMP Cause the Hawaiian Streetlight Incident?" Sandia National Laboratories. June 1989.

[9] Longmire, Conrad L., NBC Report, Fall/Winter, 2004. pp. 47–51. U.S. Army Nuclear and Chemical Agency "Fifty Odd Years of EMP"

[10] Theoretical Notes - Note 353 - March 1985 - EMP on Honolulu from the Starfish Event* Conrad L. Longmire - Mission Research Corporation

[11] Rabinowitz, Mario (1987) "Effect of the Fast Nuclear Electromagnetic Pulse on the Electric Power Grid Nationwide: A Different View". IEEE Trans. Power Delivery, PWRD-2, 1199–1222 arXiv:physics/0307127

[12] Zak, Anatoly "The K Project: Soviet Nuclear Tests in Space", The Nonproliferation Review, Volume 13, Issue 1 March 2006 , pp. 143–150

[13] SUBJECT: US-Russian meeting – HEMP effects on national power grid & telecommunications From: Howard Seguine. 17 Feb. 1995 MEMORANDUM FOR RECORD

[14] Pfeffer, Robert and Shaeffer, D. Lynn. Combating WMD Journal, (2009) Issue 3. pp. 33-38. "A Russian Assessment of Several USSR and US HEMP Tests"

[15] Greetsai, Vasily N., et al. IEEE Transactions on Electromagnetic Compatibility, Vol. 40, No. 4, November 1998, "Response of Long Lines to Nuclear High-Altitude Electromagnetic Pulse (HEMP)"

[16] Loborev, Vladimir M. "Up to Date State of the NEMP Problems and Topical Research Directions", Electromagnetic Environments and Consequences: Proceedings of the EUROEM 94 International Symposium, Bordeaux, France, 30 May – 3 June 1994, pp. 15–21

[17] Metatech Corporation (January 2010). *The Early-Time (E1) High-Altitude Electromagnetic Pulse (HEMP) and Its Impact on the U.S. Power Grid." Section 3 – E1 HEMP History* (PDF). Report Meta-R-320. Oak Ridge National Laboratory.

[18] Electromagnetic compatibility (EMC) - Part 2: Environment - Section 9: Description of HEMP environment - Radiated disturbance. Basic EMC publication, IEC 61000-2-9

[19] U.S. Army White Sands Missile Range, *Nuclear Environment Survivability*. Report ADA278230. Page D-7. 15 April 1994.

[20] Longmire, Conrad L. LLNL-9323905, Lawrence Livermore National Laboratory. June 1986 "Justification and Verification of High-Altitude EMP Theory, Part 1" (Retrieved 2010-15-12)

[21] March 8, 2005 "Statement, Dr. Peter Vincent Pry, EMP Commission Staff, before the United States Senate Subcommittee on Terrorism, Technology and Homeland Security"

[22] Report of the Commission to Assess the Threat to the United States from Electromagnetic Pulse (EMP) Attack. Volume 1. Executive Report. 2004. p. 6.

[23] High-Altitude Electromagnetic Pulse (HEMP): A Threat to Our Way of Life, 09.07, By William A. Radasky, Ph.D., P.E. - IEEE

[24] Report of the Commission to Assess the Threat to the United States from Electromagnetic Pulse (EMP) Attack

[25] Report Meta-R-321: "The Late-Time (E3) High-Altitude Electromagnetic Pulse (HEMP) and Its Impact on the U.S. Power Grid" January 2010. Written by Metatech Corporation for Oak Ridge National Laboratory.

[26] "EMPACT America, Inc. - Solar EMP". Web.archive.org. 2011-07-26. Archived from the original on July 26, 2011. Retrieved 2013-05-21.

[27] Louis W. Seiler, Jr. *A Calculational Model for High Altitude EMP*. Air Force Institute of Technology. Report ADA009208. pp. 33 and 36. March 1975

[28] Glasstone, Samuel and Dolan, Philip J., The Effects of Nuclear Weapons. Chapter 11. 1977. United States Department of Defense.

[29] Federation of American Scientists. *Nuclear Weapon EMP Effects*

[30] Hess, Wilmot N. (September 1964). "The Effects of High Altitude Explosions" (PDF). National Aeronautics and Space Administration. NASA TN D-2402. Retrieved 2015-05-13.

[31] U.S. Congressional hearing Transcript H.S.N.C No. 105–18, p. 39

[32] U.S. Congressional hearing Transcript H.A.S.C. No. 106–31, p. 48

[33] Glasstone, Samuel (March 29, 2006). "EMP radiation from nuclear space bursts in 1962". Subsequent tests with lower yield devices [410 kt *Kingfish* at 95 km altitude, 410 kt *Bluegill* at 48 km altitude, and 7 kt *Checkmate* at 147 km] produced electronic upsets on an instrumentation aircraft [presumably the KC-135 that filmed the tests from above the clouds?] that was approximately 300 kilometers away from the detonations.

[34] EMP Commission Critical National Infrastructures Report

[35] Glasstone & Dolan 1977, Chapter 11, section 11.73.

[36] Glasstone & Dolan 1977, Chapter 11, section 11.09.

[37] Seregelyi, J.S, et al. Report ADA266412 "EMP Hardening Investigation of the PRC-77 Radio Set" Retrieved 2009-25-11

[38] Miller, Colin R., Major, USAF "Electromagnetic Pulse Threats in 2010" Air War College, Air University, United States Air Force, November 2005

[39] Commission to Assess the Threat to the United States from Electromagnetic Pulse (EMP) Attack

[40] Report of the Commission to Assess the Threat to the United States from Electromagnetic Pulse (EMP) Attack Volume 1: Executive Report 2004

[41] Ross, Lenard H., Jr. and Mihelic, F. Matthew, "Healthcare Vulnerabilities to Electromagnetic Pulse" American Journal of Disaster Medicine, Vol. 3, No. 6, pp. 321–325. November/December 2008.

[42] "Survivability of Systems and Assets to Electromagnetic Pulse (EMP)"

[43] Report Meta-R-320: "The Early-Time (E1) High-Altitude Electromagnetic Pulse (HEMP) and Its Impact on the U.S. Power Grid" January 2010. Written by Metatech Corporation for Oak Ridge National Laboratory. Appendix: E1 HEMP Myths

[44] McCormack, John. "Lights out: House plan would protect nation's electricity from solar flare, nuclear bomb". WashingtonExaminer.com. Retrieved 2013-06-18.

[45] House of Commons Defence Committee, "Developing Threats: Electro-Magnetic Pulses (EMP)" Tenth Report of Session 2010–12.

[46] Extreme Electromagnetics – The Triple Threat to Infrastructure, 14 January 2013 (Proceedings of a seminar)

[47] 2009 Telly Award Winners, (Manitou Motion Picture Company, Ltd.)

• This article incorporates public domain material from the General Services Administration document "Federal Standard 1037C" (in support of MIL-STD-188).

10.11   Further reading

• ISBN 978-1-59-248389-1 A 21st Century Complete Guide to Electromagnetic Pulse (EMP) Attack Threats, Report of the Commission to Assess the Threat to the United States from Electromagnetic ... High-Altitude Nuclear Weapon EMP Attacks (CD-ROM)

- ISBN 978-0-16-056127-6 Threat posed by electromagnetic pulse (EMP) to U.S. military systems and civil infrastructure: Hearing before the Military Research and Development Subcommittee - first session, hearing held July 16, 1997 (Unknown Binding)

- ISBN 978-0-471-01403-4 Electromagnetic Pulse Radiation and Protective Techniques

- ISBN 978-0-16-080927-9 Report of the Commission to Assess the Threat to the United States from Electromagnetic Pulse (EMP) Attack

10.12   External links

- Glasstone, Samuel; Dolan, Philip J. (1977). "The Effects of Nuclear Weapons". United States Department of Defense.

- GlobalSecurity.org – Electromagnetic Pulse: From chaos to a manageable solution

- Electromagnetic Pulse (EMP) and Tempest Protection for Facilities – U.S. Army Corps of Engineers

- EMP data from *Starfish* nuclear test measured by Richard Wakefield of LANL, and review of evidence pertaining to the effects 1,300 km away in Hawaii, also review of Russian EMP tests of 1962

- Read Congressional Research Service (CRS) Reports regarding HEMP

- MIL-STD-188-125-1

- Electromagnetic Pulse Risks & Terrorism

- How E-Bombs Work

- Commission to Assess the Threat to the United States from Electromagnetic Pulse (EMP) Attack

- NEMP and Nuclear plant

Chapter 11

Nuclear weapons delivery

Nuclear weapons delivery is the technology and systems used to place a nuclear weapon at the position of detonation, on or near its target. Several methods have been developed to carry out this task.

Strategic nuclear weapons are used primarily as part of a doctrine of deterrence by threatening large targets, such as cities. Weapons meant for use in limited military maneuvers, such as destroying specific military, communications, or infrastructure targets, are known as *tactical* nuclear weapons. In terms of explosive yields, nowadays the former have much larger yield than the latter, even though it is not a rule. The bombs that destroyed Hiroshima and Nagasaki in 1945 (with TNT equivalents between 15 and 22 kilotons) were weaker than many of today's tactical weapons, yet they achieved the desired effect when used strategically.

The "Little Boy" and the "Fat Man" devices were large and cumbersome gravity bombs.

11.1 Nuclear triad

Main article: Nuclear triad

A **nuclear triad** refers to a strategic nuclear arsenal which consists of three components, traditionally strategic bombers, intercontinental ballistic missiles (ICBMs), and submarine-launched ballistic missiles (SLBMs). The purpose of having a three-branched nuclear capability is to significantly reduce the possibility that an enemy could destroy all of a nation's nuclear forces in a first-strike attack; this, in turn, ensures a credible threat of a second strike, and thus increases a nation's nuclear deterrence.[1][2][3]

11.2 Main delivery mechanisms

11.2.1 Gravity bomb

Historically, the first method of delivery, and the method used in the only two nuclear weapons actually used in warfare, was a gravity bomb dropped by a bomber. In the years leading up to the development and deployment of nuclear-armed missiles, nuclear bombs represented the most practical means of nuclear weapons delivery; even today, and especially with the decommissioning of nuclear missiles, aerial bombing remains the primary means of offensive nuclear weapons delivery, and the majority of U.S. nuclear warheads are represented in bombs, although some are in the form of missiles.

Gravity bombs are designed to be dropped from planes, which requires that the weapon be able to withstand vibrations and changes in air temperature and pressure during the course of a flight. Early weapons often had a removable core for safety, known as in flight insertion (IFI) cores, being inserted or assembled by the air crew during flight. They had to meet safety conditions, to prevent accidental detonation or dropping. A variety of types also had to have a fuse to initiate detonation. US nuclear weapons that met these criteria are designated by the letter "B" followed, without a hyphen, by the sequential number of the "physics package" it contains. The "B61", for example, was the primary bomb in the US arsenal for decades.

Various air-dropping techniques exist, including toss bombing, parachute-retarded delivery, and laydown modes, intended to give the dropping aircraft time to escape the ensuing blast.

The early gravity nuclear bombs could only be carried by the B-29 Superfortress. The next generation of weapons were still so big and heavy that they could only be carried by bombers such as the B-52 Stratofortress and V bombers, but by the mid-1950s smaller weapons had been developed that could be carried and deployed by fighter-bombers.

11.2.2 Ballistic missile

Main article: Ballistic missile

Missiles using a ballistic trajectory usually deliver a

Trident II SLBM launched by Royal Navy Vanguard class submarine.

warhead over the horizon, at distances of thousands of kilometers, as in the case of intercontinental ballistic missiles (ICBMs) and submarine-launched ballistic missiles (SLBMs). Most ballistic missiles exit the Earth's atmosphere and re-enter it in their sub-orbital spaceflight.

Placement of nuclear missiles on the low Earth orbit has been banned by the Outer Space Treaty as early as 1967. Also, the eventual Soviet Fractional Orbital Bombardment System (FOBS) that served a similar purpose—it was just deliberately designed to deorbit before completing a full circle—was phased out in January 1983 in compliance with the SALT II treaty.

An ICBM is more than 20 times as fast as a bomber and

more than 10 times as fast as a fighter plane, and also flying at a much higher altitude, and therefore more difficult to defend against. ICBMs can also be fired quickly in the event of a surprise attack.

Early ballistic missiles carried a single warhead, often of megaton-range yield. Because of the limited accuracy of the missiles, this kind of high yield was considered necessary in order to ensure a particular target's destruction. Since the 1970s modern ballistic weapons have seen the development of far more accurate targeting technologies, particularly due to improvements in inertial guidance systems. This set the stage for smaller warheads in the hundreds-of-kilotons-range yield, and consequently for ICBMs having multiple independently targetable reentry vehicles (MIRV). Advances in technology have enabled a single missile to launch a payload containing several warheads. The number of independent warheads capable of deployment from ballistic missiles depends on the weapons platform the missile is launched from. For example, one D5 Trident missile carried by an Ohio-class submarine is capable of launching eight independent warheads,[4] while a Typhoon has missiles capable of deploying 10 warheads at a time.[5][6] MIRV has a number of advantages over a missile with a single warhead. With small additional costs, it allows a single missile to strike multiple targets, or to inflict maximum damage on a single target by attacking it with multiple warheads. It makes anti-ballistic missile defense even more difficult, and even less economically viable, than before.

Missile warheads in the American arsenal are indicated by the letter "W"; for example, the W61 missile warhead would have the same physics package as the B61 gravity bomb described above, but it would have different environmental requirements, and different safety requirements since it would not be crew-tended after launch and remain atop a missile for a great length of time.[7]

11.2.3 Cruise missile

Main article: Cruise missile
See also: Supersonic Low Altitude Missile

A cruise missile is a jet engine or rocket-propelled missile that flies at low altitude using an automated guidance system (usually inertial navigation, sometimes supplemented by either GPS or mid-course updates from friendly forces) to make them harder to detect or intercept. Cruise missiles can carry a nuclear warhead. They have a shorter range and smaller payloads than ballistic missiles, so their warheads are smaller and less powerful.

The AGM-86 ALCM is the US Air Force's current nuclear-armed air-launched cruise missile. The ACM is only carried on the B-52 Stratofortress which can carry 20 missiles. Thus the cruise missiles themselves can be compared

Cruise missiles have a shorter range than ICBMs. U/RGM-109E Tomahawk pictured (not nuclear capable anymore).

The Davy Crockett artillery shell is the smallest known nuclear weapon developed by the USA.

with MIRV warheads. The BGM/UGM-109 Tomahawk submarine-launched cruise missile is capable of carrying nuclear warheads, but all nuclear warheads were removed.

Cruise missiles may also be launched from mobile launchers on the ground, and from naval ships.

There is no letter change in the US arsenal to distinguish the warheads of cruise missiles from those for ballistic missiles.

Cruise missiles, even with their lower payload, have a number of advantages over ballistic missiles for the purposes of delivering nuclear strikes:

- Launch of a cruise missile is difficult to detect early from satellites and other long-range means, contributing to a surprise factor of attack.

- That, coupled with the ability to actively maneuver in flight, allows for penetration of strategic anti-missile systems aimed at intercepting ballistic missiles on calculated trajectory of flight.

Partially for those reasons, nuclear-armed cruise missiles are amongst the least deployed of all nuclear weapons, as their deployment is restricted by treaties such as SALT II.

11.2.4 Other delivery systems

Other delivery methods included artillery shells, mines such as the Medium Atomic Demolition Munition and the novel Blue Peacock, nuclear depth charges, and nuclear torpedoes. An 'Atomic Bazooka' was also fielded, designed to be used against large formations of tanks.

In the 1950s the U.S. developed small nuclear warheads for air defense use, such as the Nike Hercules. From the 1950s to the 1980s, the United States and Canada fielded a

The Mk-17 was an early U.S. thermonuclear weapon and weighed around 21 short tons (19,000 kg).

low-yield nuclear-tipped air-to-air rocket, the AIR-2 Genie. Further developments of this concept, some with much larger warheads, led to the early anti-ballistic missiles. The United States have largely taken nuclear air-defense weapons out of service with the fall of the Soviet Union in the early 1990s. Russia updated its nuclear tipped Soviet era anti-ballistic missile(ABM) system, known as the A-135 anti-ballistic missile system in 1995. It is believed that the, in development(2013) successor to the nuclear A-135, the A-235 Samolet-M will dispense with nuclear interception warheads and instead rely on a conventional *hit-to-kill* capability to destroy its target.[8]

Small, two-man portable tactical weapons (erroneously referred to as suitcase bombs), such as the Special Atomic Demolition Munition, have been developed, although the difficulty to combine sufficient yield with portability limits their military utility.

11.3 Costs

See also: Space race, Global Positioning System, Corningware, and WD-40

According to an audit by the Brookings Institution, between 1940 and 1996, the U.S. spent $8.78 trillion in present-day terms[9] on nuclear weapons programs. 57 percent of which was spent on building delivery mechanisms for nuclear weapons. 6.3 percent of the total, $551 billion in present-day terms, was spent on weapon nuclear waste management, for example, cleaning up the Hanford site with environmental remediation, and 7 percent of the total, $617 billion was spent on the manufacturing of nuclear weapons themselves.[10]

11.3.1 Technology spin-offs

Edward White during the first US "Spacewalk" Extravehicular activity (EVA), Project Gemini 4, June 1965.

Strictly speaking however not all this 57 percent was spent solely on "weapons programs" delivery systems.

Launch vehicles

For example, two such delivery mechanisms, the Atlas ICBM and Titan II, were re-purposed as human launch vehicles for manned spaceflight, both were used in the civilian Project Mercury and Project Gemini programs respectively, which are regarded as stepping stones in the evolution of US manned spaceflight.[11][12][13] The Atlas vehicle sent John Glenn, the first American into orbit. Similarly in the Soviet Union it was the R-7 ICBM/launch vehicle that placed the first artificial satellite in space, Sputnik, on 4 October 1957, and the first human spaceflight in history was accomplished on a derivative of the R-7, the Vostok, on 12 April 1961, by cosmonaut Yuri Gagarin. A modernized version of the R-7 is still in use as the launch vehicle for the Russian Federation, in the form of the Soyuz spacecraft.

Weather satellites

The first true weather satellite, the TIROS-1 was launched on the Thor-Able launch vehicle April 1, 1960.[14] The PGM-17 Thor was the first operational IRBM(intermediate ballistic missile) deployed by the U.S. Air Force (USAF). The Soviet Union's first fully operational weather satellite, the Meteor 1 was launched 26 March 1969 on the Vostok rocket,[15] a derivative of the R-7 ICBM.

Lubricants

WD-40 was first used by Convair to protect the outer skin, and more importantly, the paper thin "balloon tanks" of the Atlas missile from rust and corrosion.[16][17] These stainless steel fuel tanks were so thin that, when empty, they had to be kept inflated with nitrogen gas to prevent their collapse.

Thermal isolation

In 1953, Dr. S. Donald Stookey of the Corning Research and Development Division invented Pyroceram, a white glass-ceramic material capable of withstanding a thermal shock (sudden temperature change) of up to 450 °C (840 °F). It evolved from materials originally developed for a U.S. ballistic missile program, and Stookey's research involved heat-resistant material for nose cones.[18]

Satellite assisted positioning

Precise navigation would enable United States submarines to get an accurate fix of their positions before they launched their SLBMs, this spurred development of triangulation methods that ultimately culminated in GPS.[19] The motivation for having accurate launch position fixes, and missile velocities,[20] is twofold. It results in a tighter target impact circular error probable and therefore by extension, reduces the need for the earlier generation of heavy multi-megaton nuclear warheads, such as the W53 to ensure the target is destroyed. With increased target accuracy, a greater number of lighter, multi-kiloton range warheads can be packed on a given missile, giving a higher number of separate targets that can be hit per missile.

Global positioning system   Main article: Global Positioning System

During a Labor Day weekend in 1973, a meeting of about twelve military officers at the Pentagon discussed the creation of a *Defense Navigation Satellite System (DNSS)*. It was at this meeting that "the real synthesis that became GPS was created." Later that year, the DNSS program was named *Navstar*, or Navigation System Using Timing and Ranging.[21]

During the development of the submarine-launched Polaris missile, a requirement to accurately know the submarine's location was needed to ensure a high circular error probable warhead target accuracy. This led the US to develop the Transit system.[22] In 1959, ARPA (renamed DARPA in 1972) also played a role in Transit.[23][24][25]

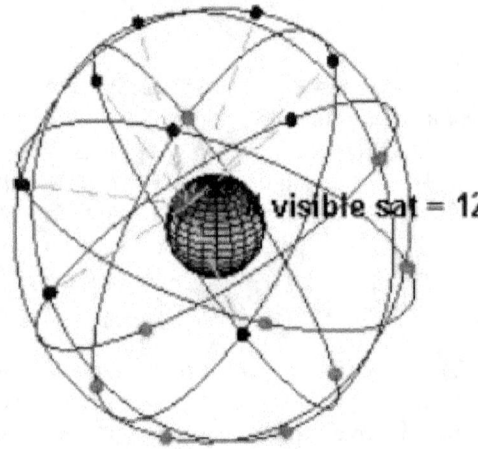

A visual example of a 24 satellite GPS constellation in motion with the Earth rotating. Notice how the number of satellites in view from a given point on the Earth's surface, in this example at 45°N, changes with time. GPS was initially developed to increase Ballistic Missile Circular Error Probable accuracy, accuracy which is vital in a counterforce attack.[26][27][28]

The first satellite navigation system, Transit, used by the United States Navy, was first successfully tested in 1960. It used a constellation of five satellites and could provide a navigational fix approximately once per hour. In 1967, the U.S. Navy developed the Timation satellite that proved the ability to place accurate clocks in space, a technology required by the latter Global Positioning System. In the 1970s, the ground-based Omega Navigation System, based on phase comparison of signal transmission from pairs of

stations,[29] became the first worldwide radio navigation system. Limitations of these systems drove the need for a more universal navigation solution with greater accuracy.

While there were wide needs for accurate navigation in military and civilian sectors, almost none of those was seen as justification for the billions of dollars it would cost in research, development, deployment, and operation for a constellation of navigation satellites. During the Cold War arms race, the nuclear threat to the existence of the United States was the one need that did justify this cost in the view of the United States Congress. This deterrent effect is why GPS was funded. The nuclear triad consisted of the United States Navy's submarine-launched ballistic missiles (SLBMs) along with United States Air Force (USAF) strategic bombers and intercontinental ballistic missiles (ICBMs). Considered vital to the nuclear-deterrence posture, accurate determination of the SLBM launch position was a force multiplier.

Precise navigation would enable United States submarines to get an accurate fix of their positions before they launched their SLBMs.[19] The USAF, with two thirds of the nuclear triad, also had requirements for a more accurate and reliable navigation system. The Navy and Air Force were developing their own technologies in parallel to solve what was essentially the same problem. To increase the survivability of ICBMs, there was a proposal to use mobile launch platforms (such as Russian SS-24 and SS-25) and so the need to fix the launch position had similarity to the SLBM situation.

In 1960, the Air Force proposed a radio-navigation system called MOSAIC (MObile System for Accurate ICBM Control) that was essentially a 3-D LORAN. A follow-on study, Project 57, was worked in 1963 and it was "in this study that the GPS concept was born". That same year, the concept was pursued as Project 621B, which had "many of the attributes that you now see in GPS"[30] and promised increased accuracy for Air Force bombers as well as ICBMs. Updates from the Navy Transit system were too slow for the high speeds of Air Force operation. The Navy Research Laboratory continued advancements with their Timation (Time Navigation) satellites, first launched in 1967, and with the third one in 1974 carrying the first atomic clock into orbit.[31]

Another important predecessor to GPS came from a different branch of the United States military. In 1964, the United States Army orbited its first Sequential Collation of Range (SECOR) satellite used for geodetic surveying. The SECOR system included three ground-based transmitters from known locations that would send signals to the satellite transponder in orbit. A fourth ground-based station, at an undetermined position, could then use those signals to fix its location precisely. The last SECOR satellite was launched in 1969.[32] Decades later, during the early years of GPS, civilian surveying became one of the first fields to make use of the new technology, because surveyors could reap benefits of signals from the less-than-complete GPS constellation years before it was declared operational. GPS can be thought of as an evolution of the SECOR system where the ground-based transmitters have been migrated into orbit.

11.4 See also

- History of nuclear weapons
- List of nuclear weapons
- Mutual assured destruction doctrine
- National missile defense of the United States
- Nuclear explosion
- Nuclear strategy
- Nuclear weapon design
- Nuclear terrorism

11.5 Notes

[1] John Barry (2009-12-12). "Do We Still Need a Nuclear 'Triad'?". *Newsweek*. Retrieved 2010-10-08.

[2] Office for the Deputy Assistant to the Secretary of Defense for Nuclear Matters. "Nuclear Stockpile". US Department of Defense. Retrieved 2010-10-08.

[3] "Toning Up the Nuclear Triad". *Time*. 1985-09-23. Retrieved 2010-10-08.

[4] "SSBN", *CNO* (Navy) (87).

[5] "Red October no more: Russia scraps Cold War era Typhoon submarine", *The Telegraph* (UK).

[6] *The World's biggest nuclear submarine is also one of the sneakiest*, Gizmodo.

[7] *Nav Air*, Navy.

[8] Honkova, Jana (Apr 13, 2013). "Current Developments in Russia's Ballistic Missile Defense" (PDF). George C. Marshall Institute.

[9] Federal Reserve Bank of Minneapolis Community Development Project. "Consumer Price Index (estimate) 1800–". Federal Reserve Bank of Minneapolis. Retrieved November 10, 2015.

[10] *Estimated Minimum Incurred Costs of U.S. Nuclear Weapons Programs, 1940–1996*, Brookings Institution.

[11] "Titan", *Military launch program*, FAS, The Titan II ICBM was converted into the Titan/Gemini space launch vehicle (SLV) by man-rating critical systems. It served as a significant stepping stone in the evolution of the US manned spaceflight program using expendable launch vehicles, culminating in the Apollo program. Twelve successful Gemini launches occurred between April 1964 and November 1966.

[12] "Evolution of US expendable launch vehicles", *What, when, how*.

[13] "Titan History", *Space flight now*.

[14] Darling, David, "Tiros", *Encyclopedia*.

[15] *Soviet Weather Satellite Falls in Antarctica*, RIA Novosti, 27 March 2012, retrieved 28 March 2012.

[16] "Our History". WD-40.

[17] Martin, Douglas. "John S. Barry, Main Force Behind WD-40, Dies at 84". *The New York Times*, July 22, 2009.

[18] "Annual Report: 10-K" (Securities and Exchange Commission filing). WKI. 2001-04-13. Retrieved 2007-03-26.

[19] "Why Did the Department of Defense Develop GPS?". Trimble Navigation. Archived from the original on October 18, 2007. Retrieved January 13, 2010.

[20] Caston, Lauren; et al. "The Future of the U.S. Intercontinental Ballistic Missile Force" (PDF). The RAND corporation.

[21] "MX Deployment Reconsidered", *Air chronicles* (Air force), May–Jun 1981, retrieved 7 June 2013.

[22] Johnson, Steven (2010), *Where good ideas come from, the natural history of innovation*, New York: Riverhead Books

[23] Worth, Helen E; Warren, Mame (2009). *Transit to Tomorrow. Fifty Years of Space Research* (PDF). The Johns Hopkins University Applied Physics Laboratory.

[24] Alexandrow, Catherine (Apr 2008). "The Story of GPS". Darpa.

[25] "50 Years of Bridging the Gap", *History*, DARPA, Apr 2008

[26] "Counterforce issues for the US strategic nuclear forces" (PDF). CBO. 1978.

[27] Forden, Geoffrey. "Strategic uses for China's Bei Dou satellite system" (PDF). MIT.

[28] Scott, Logan. "Circular Error Probable (CEP) mathematics". Earth link.

[29] Proc, Jerry. "Omega". CA: Jerry Proc. Retrieved December 8, 2009.

[30] "Charting a Course Toward Global Navigation". The Aerospace Corporation. Summer 2002. Retrieved January 14, 2010.

[31] "GPS Timeline". *A Guide to the Global Positioning System (GPS)*. Radio Shack. Retrieved January 14, 2010.

[32] Wade, Mark. "SECOR Chronology". *Encyclopedia Astronautica*. Astronautix. Retrieved January 19, 2010.

11.6   References

- Annotated bibliography for nuclear weapon delivery systems from the Alsos Digital Library for Nuclear Issues

11.7   External links

- Nuclear Missile Research Centre

Chapter 12

Nuclear triad

A **nuclear triad** refers to the nuclear weapons delivery of a strategic nuclear arsenal which consists of three components: traditionally strategic bombers, intercontinental ballistic missiles (ICBMs), and submarine-launched ballistic missiles (SLBMs). The purpose of having a three-branched nuclear capability is to significantly reduce the possibility that an enemy could destroy all of a nation's nuclear forces in a first-strike attack; this, in turn, ensures a credible threat of a second strike, and thus increases a nation's nuclear deterrence.[1][2][3]

12.1 Traditional components

While traditional nuclear strategy holds that a nuclear triad provides the best level of deterrence from attack, in reality, most nuclear powers do not have the military budget to sustain a full triad. Only the United States and Russia have maintained nuclear triads for most of the nuclear age.[3] Both the US and the Soviet Union composed their triads along the same lines, including the following components:

1. Bomber aircraft capable of delivering nuclear bombs (carrier-based or land-based; usually armed with long-range missiles).[1]

2. Land-based missiles (MRBMs or ICBMs).[1][3]

3. Ballistic missile submarines (SSBNs). Nuclear missiles launched from ships or submarines.[1][3] Although in early years the US Navy sea leg was carrier aircraft based with a very short period using sub launched cruise missiles such as the Regulus before SLBMs were ready to be deployed.

The triad also gives the commander in chief the flexibility to use different types of weapons for the appropriate strike while also preserving a reserve of nuclear armaments theoretically safe from a counter-force strike:

- ICBMs allow for a long-range strike launched from a controlled or friendly environment at a lower cost per

delivered warhead and easiest targeting from a surveyed geographic location.[4] If launched from a fixed position, such as a missile silo, they are vulnerable to a first strike, though their interception once aloft is substantially difficult.[1][3] Some ICBMs are either rail or road mobile. Medium-range ballistic missiles and ground-launched cruise missiles were also assigned for strategic targets based in nations closer to the potential confrontation, but were eventually forbidden by arms control treaty to the US and Russia.

- SLBMs, launched from submarines, allow for a greater chance of survival from a first strike, giving the commander a second-strike capability.[1][3] Some long-range submarine-launched cruise missiles are counted towards triad status; this was the first type of submarine-launched strategic second-strike nuclear weapon before ballistic missile submarines became available. A SLBM is the most difficult to get accurate targeting for as it requires obtaining an accurate geographical fix to program targeting data to the missile, the total cost of a SLBM is increased by the cost of the submarine force, large crews and deterrence patrols.[4]

- Strategic bombers have greater flexibility in their deployment and weaponry. They can serve as both a first- and second-strike weapon. A bomber armed with AGM-129 ACM missiles, for example, could be classified as a first-strike weapon. A number of bombers often with aerial refueling aircraft kept at safe points would constitute a second-strike weapon.[1][3] In some strategic contexts either with nearby potential enemies or with forward basing lighter aircraft can be used on the strategic level as either a first-strike weapon or if dispersed at small airfields or aboard an aircraft carrier can reasonably avoid a counterstrike giving them regional second-strike capacity, aircraft such as the Mirage 2000, F-15E, A-5 Vigilante, Sea Harrier, or FB-111 are or were tasked part or full-time with land or sea-based strategic nuclear attack missions. An aerial refueling fleet supports intercontinental strategic operations both for heavy bombers and smaller aircraft; it also makes possible around the clock airborne standby of bombers and command aircraft making these airborne assets nearly impossible to eliminate in a first strike. Bomber airborne alert patrols are very expensive in terms of fuel and aircraft maintenance, even non-airborne alert basing requires both crew training hours and aircraft upkeep.[4]

Tactical nuclear weapons are used in air, land and sea warfare. Air-to-air missiles and rockets, surface-to-air missiles, and small air-to-ground rockets, bombs, and precision munitions have been developed and deployed with nuclear warheads. Ground forces have included tactical nuclear artillery shells, surface-to-surface rockets, land mines, medium and small man-packable nuclear engineering demolition charges, even man-carried or vehicle-mounted recoilless rifles. Naval forces have carried nuclear-armed naval rocket-assisted and standard depth charges and torpedoes, and naval gunnery shells. Tactical nuclear weapons and the doctrine for their use is primarily for use in a non-strategic warfighting role destroying military forces in the battle area; they are not counted toward triad status despite the possibility of many of these systems being usable as strategic weapons depending on the target.

12.2   Triad powers

The following nations are considered fully established triad nuclear powers, they have robust capability to launch a worldwide second strike in all three legs and can disperse their air forces and their sea forces on deterrent patrols. They possess nuclear forces consisting of land-based missiles, ballistic or long-range cruise missile submarines, and strategic bombers or long-range tactical aircraft.

12.2.1   Russia

Main articles: Strategic Missile Troops, Long Range Aviation, Borei-class submarine, Typhoon-class submarine, and Delta-class submarine

Also a nuclear power,[5] Russia inherited the arsenal of all of the former Soviet states; this consists of silo-based as well as rail and road mobile ICBMs, sea-based SLBMs, strategic bombers, strategic aerial refueling aircraft, and long-range tactical aircraft capable of carrying gravity bombs, standoff missiles, and cruise missiles. The Russian Strategic Rocket Forces have ICBMs capable of delivering nuclear warheads, silo-based R-36M2 (SS-18), silo-based UR-100N (SS-19), mobile RT-2PM "Topol" (SS-25), silo-based RT-2UTTH "Topol M" (SS-27), mobile RT-2UTTH "Topol M" (SS-

27), mobile RS-24 "Yars" (SS-29) (*Future replacement for R-36 & UR-100N missiles*). Russian strategic nuclear submarine forces are equipped with the following SLBM's, R-29R "Vysota", NATO name SS-N-18 "Stingray", RSM-54 R-29RMU "Sineva", NATO name SS-N-23 "Skiff" and the R-29RMU2.1 "Liner" are in use with the Delta-class submarine, but the RSM-56 R-30 "Bulava", NATO name SS-NX-32 is under development for the Borei-class submarine. The Russian Long Range Aviation operates supersonic Tupolev Tu-22M, and Tupolev Tu-160 bombers and the long range turboprop powered Tupolev Tu-95, they are all mostly armed with strategic stand off missiles or cruise missiles such as the KH-15 and the KH-55/Kh-102. These bombers and nuclear capable strike aircraft such as the Sukhoi Su-24 are supported by Ilyushin Il-78 aerial refuelling aircraft. The USSR was required to destroy its stock of IRBMs in accordance with the INF treaty.

12.2.2   United States

Main articles: Air Force Global Strike Command, Boeing B-52 Stratofortress, Northrop Grumman B-2 Spirit, LGM-30 Minuteman, and Ohio-class submarine

The United States operates Minuteman ICBMs from underground hardened silos, Trident SLBMs carried by *Ohio*-class submarines, it also operates B-52, B-2 strategic bombers, as well as land-based tactical aircraft, some capable of carrying strategic and tactical B61 and large strategic B83 gravity bombs, and AGM-86 ALCMs. While the US no longer keeps nuclear armed bombers on airborne alert, it has the ability to do so, along with the airborne nuclear command and control aircraft with its fleet of KC-10 and KC-135 aerial refueling planes. Previous to development of submarine-launched ballistic missiles, the US Navy strategic nuclear role was provided by aircraft carrier–based bombers and, for a short time, submarine-launched cruise missiles. With the end of the cold war, the US never deployed the rail-mobile version of the Peacekeeper ICBM or the road mobile Midgetman small ICBM. The US destroyed its stock of road-mobile Pershing II IRBMs and ground-launched cruise missiles in accordance with the INF treaty. The US also has shared strategic nuclear weapons and still deploys shared tactical nuclear weapons to some NATO countries.[1][3][6]

12.2.3   China

Main articles: Second Artillery Corps, Xian H-6, Type 094 submarine, and Type 092 submarine

Unlike the United States and Russia where strategic nuclear

forces are enumerated by treaty limits and subject to verification, China, a nuclear power since 1964, is not subject to these requirements but currently has a triad structure smaller in size compared to Russia and the United States. China's nuclear force is much smaller than the US or Russia and is closer in number and capability to that of France or the United Kingdom. This force is mainly land-based missiles including ICBMs, IRBMs, and tactical ballistic missiles as well as cruise missiles. Unlike the US and Russia, China stores many of its missiles in huge underground tunnel complexes; U.S. Representative Michael Turner[7] referring to 2009 Chinese media reports said "This network of tunnels could be in excess of 5,000 kilometers (3,110 miles), and is used to transport nuclear weapons and forces,"[8] the Chinese Army newsletter calls this tunnel system an Underground Great Wall of China.[9]

Currently China has one Type 092 submarine that is currently active with JL-1 SLBM according to Office of Naval Intelligence.[10][11] In addition, the PLAN has deployed 4 newer Type 094 submarines and plan to deploy up to 8 of these Jin-class SSBN by the end of 2020.[12][13] The new Type 094 fleet uses the newer JL-2 SLBM. China carried out a series of successful JL-2 launches in 2009,[14] 2012[15][16] and 2015.[17] The United States expect the 094 SSBN to carry out its first deterrent patrol by 2015 with the JL-2 missile active.[12] There is an aged albeit upgraded bomber force consisting of Xian H-6s with an unclear nuclear delivery role. The PLAAF has a limited capability fleet of H-6 bombers modified for aerial refuelling as well as forthcoming Russian Ilyushin Il-78 aerial refuelling tankers.[18] China also introduced a newer and modernized H-6 variant the H-6K with enhanced capabilities such as launching long ranged cruise missile the CJ-10. In addition to the H-6 bomber, there are numerous tactical fighter and fighter bombers such as the: J-16, J-10, JH-7A and Su-30 which all capable of carrying nuclear weapons.

12.3 Emerging triad powers

An emerging triad power is one where one or more legs have only limited known or suspected near term nuclear strategic capability in the context of its international relations with land based missiles, submarine launched missiles, and air launched weaponry. An emerging power may have one or more experimental vessels or may not yet be able to disperse or sortie out in force for a deterrence patrol in its strategic context.

12.3.1 India

Main articles: Strategic Forces Command, Andaman and Nicobar Command, Agni (missile), K Missile family, and Sukhoi Su-30MKI

India has recently become a nuclear triad with INS Arihant, which is expected to be commissioned in 2016.[19][20][21][22] INS Arihant will either carry 12 K-15 missiles with a range of 750 km[23] or 4 K-4 missiles with an extended range of 3500 km.[24][25] India maintains a no first use nuclear policy and has been developing a nuclear triad capability as a part of its credible minimum deterrence doctrine.[26] India's nuclear-weapons program possesses surface-to-surface missiles such as the Agni II and Agni III. In addition, the 5,000–8000 km range Agni-V ICBM was also successfully tested for third time on 31 January 2015 [27] and is expected to enter service by 2016.[28] India has nuclear-capable fighter aircraft such as the Dassault Mirage 2000H, Sukhoi Su-30 MKI, MIG-29 and SEPECAT Jaguar. Land and air strike capabilities are already in place under the control of Strategic Forces Command which is a part of Nuclear Command Authority (India).

12.4 Former triad powers

12.4.1 France

A former triad power, the French Force de frappe possesses sea-based and air-based nuclear forces through the *Triomphant*-class ballistic missile submarines deployed with M45 intercontinental SLBMs armed with multiple warheads, nuclear capable Dassault Rafale F3 and Dassault Mirage 2000N fighter aircraft (armed with Air-Sol Moyenne Portée) which replaced the long-range Dassault Mirage IV supersonic nuclear bomber and KC-135 aerial refuelling tankers in its inventory. France had S2 and then S3 silo based strategic nuclear IRBMs, the S3 with a 3,500 km range, but these have been phased out of service since the dissolution of the USSR. France operates aircraft with a nuclear strike role from its aircraft carrier.

12.5 Non-triad powers

Non-triad powers are nuclear armed nations which have never developed a strategic nuclear delivery triad.

12.5.1 United Kingdom

The UK never rolled out its own land based missile nuclear delivery system. It only possesses sea-based nuclear forces through its Royal Navy *Vanguard*-class ballistic missile submarines, deployed with Trident II intercontinental SLBMs armed with multiple warheads. The Royal Air Force used to operate V bomber strategic bombers throughout the Cold War and continued airborne delivery using Tornado and Jaguar aircraft until the late 1990s. The planned UK silo-based IRBM, the Blue Streak missile, was cancelled as it was not seen as a credible deterrent, considering the population density of areas in the UK geologically suited for missile silos. The tactical Corporal surface-to-surface missile was operated by the British Army. The American made intermediate range Thor missile aimed at Soviet targets was operated briefly by the RAF but before the arrival of the Polaris SLBM. Previously having a nuclear strike mission for carrier-based Buccaneer attack aircraft and later Sea Harriers, the UK no longer deploys nuclear weapons for delivery by carrier-based naval aircraft or any other means other than the Vanguard submarine-launched Trident SLBM.

12.5.2 Pakistan

Pakistan does not have an active nuclear triad. Its nuclear weapons are primarily land-based. The Minimum Credible Deterrence (MCD) is a defense and strategic principle on which the atomic weapons program of Pakistan is based.[29] This doctrine is not a part of the nuclear doctrine, which is designed for the use of the atomic weapons in a full-scale declared war if the conditions of the doctrine are surpassed.[30] Instead, the MCD policy falls under minimal deterrence as an inverse to Mutually Assured Destruction (MAD).[31] In August 2012, *The Economist* magazine wrote an article stating that Pakistan was an emerging nuclear triad state. Pakistani plans of responding to any capture or pre-emptive destruction of their nuclear defences seems to be one reason why they are determined to develop a third leg, after air- and land-based delivery systems, to Pakistan's nuclear triad, consisting of nuclear-armed ships and submarines. As Iskander Rehman of the Carnegie Endowment, a think-tank, observes in a recent paper, Pakistani nuclear expansion and methods of delivery is drifting "from the dusty plains of the Punjab into the world's most congested shipping lanes... It is only a matter of time before Pakistan formally brings nuclear weapons into its own fleet."[32]

Pakistan possesses several ballistic missiles such as the Shaheen-1A and the Shaheen-II, missiles having ranges of 900 km and 2000 km respectively. They also contain systems said to be capable of carrying several nuclear warheads as well as being designed to evade missile-defense systems.[33][34] Pakistan also possesses the Babur cruise missile with a range up to 700 km. These land-based missiles are controlled by Army Strategic Forces Command of the Pakistan Army.

The PAF has two dedicated units (the No. 16 *Black Panthers* and the No. 26 *Black Spiders*) operating 18 aircraft in each squadron of the JF-17 Thunder, believed to be the preferred vehicle for delivery of nuclear weapons.[35] These units are a major part of the Air Force Strategic Command, a command responsible for nuclear response. The PAF also operates a fleet of F-16 fighters, of which 18 were delivered in 2012 and, as confirmed by General Ashfaq Parvez Kayani, are capable of carrying nuclear weapons.[36] The PAF also possesses the Ra'ad air-launched cruise missile which has a range of 350 km and can carry a nuclear warhead with a yield of between 10 kilotons to 35 kilotons.[37]

In 2004, the Pakistan Navy established the Naval Strategic Forces Command and made it responsible for countering and battling naval-based weapons of mass destruction. It is believed by most experts that Pakistan is developing a sea-based variant of the Hatf VII Babur, which is a nuclear-capable ground-launched cruise missile.[38]

12.5.3 North Korea

North Korea has claimed to have indigenous nuclear weapons technology since a large underground explosion was detected in 2006. The DPRK has both aircraft and missiles which may be tasked to deliver nuclear weapons. The North Korean missile program is largely based on domestically produced variants of the Soviet Scud missile, some of which are sufficiently powerful to attempt satellite launch. The DPRK also has short-range ballistic missiles and cruise missiles. Western researchers believe the current generation of the DPRK's suspected nuclear weapons are too large to be fitted to the country's existing missile stock.[39]

12.6 Suspected triad powers

Main articles: Jericho (missile), Popeye Turbo, and F-15I

Israel has been reported in congressional testimony by the US Department of Defense of having aircraft-delivered nuclear weapons as early as the mid-1960s, a demonstrated missile-based force since the mid-1960s, an IRBM in the mid-1980s, an ICBM in the early 2000s[40] and the suspected second-strike capability arrived with the *Dolphin*-class submarine and Popeye Turbo submarine-launched cruise missile. Israel is suspected of using their inventory of nuclear-capable fighter aircraft such as the long-range

F-15E Strike Eagle, F-16 and formerly the F-4 Phantom, Dassault Mirage III, A-4 Skyhawk and Nesher. Israel has appreciable and growing numbers of long-range tanker aircraft and aerial refueling capacity on its long-range fighter-bomber aircraft, this capacity was used in the 1985 long-range conventional strike against the PLO in Tunisia.[41] *Jane's Defence Weekly* reports that the Israeli *Dolphin*-class submarines are widely believed to be nuclear armed, offering Israel a second-strike capability with a demonstrated range of at least 1500 km in a 2002 test.[42][43] According to an official report which was submitted to the American congress in 2004,[40] it may be that with a payload of 1,000 kg the Jericho 3 gives Israel nuclear strike capabilities within the entire Middle East, Africa, Europe, Asia and almost all parts of North America, as well as within large parts of South America and North Oceania, Israel also has the regional reach of its Jericho 2 IRBM force. The existence of a nuclear force is often hinted at blatantly and evidence of an advanced weapons program including miniaturized and thermonuclear devices has been presented, especially the extensive photographic evidence given by former Israeli nuclear weapons assembler Mordechai Vanunu. There have been incidents where Israel has been suspected of testing, but so far Israel for diplomatic reasons has not openly admitted to having operational nuclear weapons, and so is only a suspect triad state.

12.7 Other nuclear delivery systems

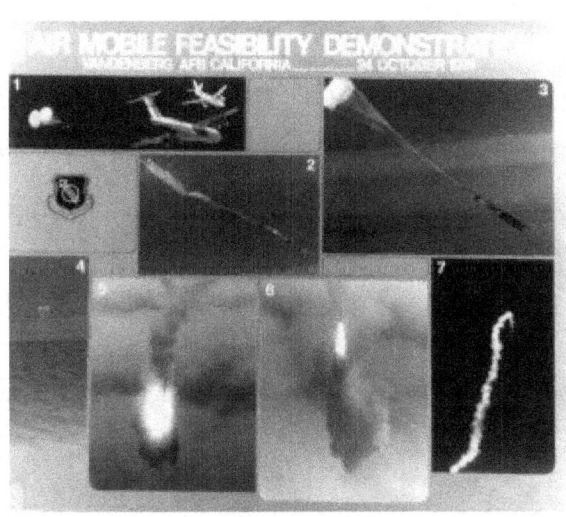

Air Mobile ICBM Feasibility Demonstration—24 October 1974

There is nothing in nuclear strategy to mandate only these three delivery systems. For example, orbital weapons or spacecraft for purposes of orbital bombardment using nuclear devices have been developed and silo deployed by the USSR from 1969 to 1983, these would not fit into the categories listed above. However, actual space-based weapon systems used for weapons of mass destruction have been banned under the Outer Space Treaty and launch ready deployment for the US and former USSR by the SALT II treaty. Another example is the US, UK, and France do or have previously included a strategic nuclear strike mission for carrier-based aircraft, which especially in the past were far harder to track and target with ICBMs or strategic nuclear bombers than fixed bomber or missile bases, permitting some second-strike flexibility; this was the first sea-based deterrent before the SLBM. The US and UK jointly explored an air-launched strategic ballistic nuclear missile, the Skybolt, but canceled the program in favor of submarine-based missiles. In 1974 a Lockheed C-5 Galaxy successfully tested an air launch of a Minuteman ICBM; this system was not deployed, but was used as a bargaining point in the SALT treaty negotiations with the USSR.

12.8 See also

- Fail-deadly

- Mutual Assured Destruction

- Nuclear weapons and the United States

- Nuclear weapons of Russia

- Nuclear weapons of People's Republic of China

- Nuclear weapons of India

12.9 References

[1] Barry, John (12 December 2009). "Do We Still Need a Nuclear 'Triad'?". *Newsweek*. Retrieved 8 October 2010.

[2] Office for the Deputy Assistant to the Secretary of Defense for Nuclear Matters. "Nuclear Stockpile". US Department of Defense. Retrieved 8 October 2010.

[3] "Toning Up the Nuclear Triad". *Time'*. 23 September 1985. Retrieved 8 October 2010.

[4] "Time to Modernize and Revitalize the Nuclear Triad". *The Heritage Foundation*. Retrieved 18 December 2015.

[5] "Russia continues to modernize its nuclear triad". RIA Novosti. 18 November 2009. Retrieved 8 October 2010.

[6] "WMD411 - Case Studies: The New Triad". Nuclear Threat Initiative. 6 April 2010. Retrieved 8 October 2010.

[7] "U.S. Lawmaker Warns of China's Nuclear Strategy". *China Digital Times*. Retrieved 18 December 2015.

[8] http://www.straitstimes.com/BreakingNews/Asia/Story/STIStory_723617.html

[9] "China Builds Underground 'Great Wall' Against Nuke Attack". *The Chosun Ilbo* (English ed.). Retrieved 18 December 2015.

[10] https://fas.org/irp/agency/oni/pla-navy.pdf

[11] http://www.oni.navy.mil/Intelligence_Community/china_media/2015_PLA_NAVY_PUB_Print.pdf

[12] "US upgrades assessment of China's Type 094 SSBN fleet l IHS Jane's 360". *IHS Jane's 360*. Retrieved 18 December 2015.

[13] "Home Security Systems : My Home Security". *GlobalSecurityNewswire.org*. Retrieved 18 December 2015.

[14] "JL-2 (CSS-NX-14)". GlobalSecurity.org. Retrieved 28 October 2014.

[15] Taylor, Marcus; Tamerlani, Eric (3 June 2013). "Pentagon Sees China Progressing on SLBM". Arms Control Association. Retrieved 28 October 2015.

[16] Gertz, Bill (21 August 2012). "Ready To Launch: China conducts rare flight test of new submarine-launched missile". *The Washington Free Beacon*. Retrieved 15 January 2013.

[17] Gertz, Bill (18 February 2015). "China conducts JL-2 sub missile test". *The Washington Times*. Retrieved 10 March 2015.

[18] "HY-6 (Hongzhaji You-6) Aerial Refueling Tanker". GlobalSecurity.org. Retrieved 18 December 2015.

[19] "India's first Nuclear submarine INS Arihant ready for operation, passes deep sea test".

[20] "Nuclear triad weapons ready for deployment: DRDO".

[21] "India close to attaining nuclear triad status". *Deccan Chronicle*. 26 March 2014. Retrieved 26 March 2014.

[22] "After missile test, India inches closer to N-Triad". *Free Press Journal*. 26 March 2014. Retrieved 26 March 2014.

[23] "India tests new underwater nuclear missile". *The Times of India*. Retrieved 3 March 2016.

[24] "India successfully test-fires underwater missile". *The Hindu*. 27 January 2013. ISSN 0971-751X. Retrieved 3 March 2016.

[25] "India tests 3,000 km range n-missile in secret". *Sunday Guardian*. London. Retrieved 3 March 2016.

[26] Nair, Vijai K. "No More Ambiguity: India's Nuclear Policy" (PDF). Archived from the original (PDF) on 27 September 2007. Retrieved 7 June 2007.

[27]

[28] "Agni-V, India's first ICBM test-fired successfully". *Times of India*. Retrieved 26 June 2012.

[29] Farah Zahra, PhD (Political Science) (12 August 2011). "Credible minimum nuclear deterrence". *Daily Times*. Retrieved 19 July 2012. The nuclear arms race in South Asia is not purely a quantitative matter; it encompasses a qualitative dimension where the nuclear weapons and delivery systems on both sides are improving in quality as well ... Dr. Farah Zahra

[30] IISS. "Nuclear policy, doctrine and planning Rationales for nuclear weapons". International Institute for Strategic Studies. Retrieved 19 July 2012.

[31] Paul K. Kerr; Mary Beth Nikitin (10 May 2012) [first published on 30 November 2011]. "Pakistan and Nuclear weapons" (PDF). United States Congress: Congressional Research Services. p. 1. Retrieved 19 July 2012.

[32] "Nuclear profusion". *The Economist*. ISSN 0013-0613. Retrieved 18 December 2015.

[33] Masood, Salman (25 April 2012). "Pakistan Says It Test-Fires Nuclear-Capable Missile". *The New York Times*. Retrieved 26 April 2012.

[34] "Design Characteristics of Pakistan's Ballistic Missiles". NTI. Retrieved 4 July 2012.

[35] "Associated Press Of Pakistan (Pakistan's Premier NEWS Agency) - First Squadron of JF-17 Thunder inducted in PAF". *www.app.com.pk*. Retrieved 18 December 2015.

[36] "Boosting air defence: F-16s replace Americans at Jacobabad airbase". *The Express Tribune*. Retrieved 18 December 2015.

[37] http://www.missilethreat.com/cruise/id.144/cruise_detail.asp

[38] NTI, Nuclear Threat Initiatives (April 2015). "Pakistan's Naval capabilities: Submarine system". *Research: Submarine Proliferation by countries*. NTI: Research: Submarine Proliferation by countries. Retrieved 28 October 2015.

[39] Postol, Theodore (6 May 2009). "A Technical Assessment of Iran's Ballistic Missile Program" (PDF). Institute of Technology.

[40] Feickert, Andrew (5 March 2004). Missile Survey: Ballistic and Cruise Missiles of Foreign Countries (PDF) (Report). Congressional Research Service. RL30427. Retrieved 21 June 2010.

[41] "Israel Air Force, Israel". GlobalSecurity.org. Retrieved 18 December 2015.

[42] "Popeye Turbo". Federation of American Scientists. 20 June 2000.

[43] Ben-David, Alon (1 October 2009). "Israel seeks sixth *Dolphin* in light of Iranian 'threat'". *Jane's Defence Weekly*. Retrieved 3 November 2009.

12.10 External links

- Pros and Cons of Bombers, Missiles, and Submarines

Chapter 13

Strategic bomber

Contemporary U.S. strategic bombers, top to bottom: the B-52 Stratofortress, B-1 Lancer and B-2 Spirit.

A **strategic bomber** is a medium to long range penetration bomber designed to drop large amounts of air-to-ground weaponry onto a distant target for the purposes of debilitating the enemy's capacity to wage war. Unlike tactical bombers, penetrators, fighter-bombers, and attack aircraft, which are used in air interdiction operations to attack enemy combatants and military equipment, strategic bombers are designed to fly into enemy territory to destroy strategic targets (e.g., infrastructure, logistics, military installations, factories, and cities). In addition to strategic bombing, strategic bombers can be used for tactical missions. The United States, Russia, and China maintain strategic bombers.[1]

The modern strategic bomber role appeared after strategic bombing was widely employed, and atomic bombs were first used in combat during World War II. Nuclear strike missions (i.e., delivering nuclear-armed missiles or bombs) can potentially be carried out by most modern fighter-bombers and strike fighters, even at intercontinental range, with the use of **aerial refueling**, so any nation possessing this combination of equipment and techniques theoretically has such capability. Primary delivery aircraft for a modern strategic bombing mission need not always necessarily be a heavy

bomber type, and any modern aircraft capable of nuclear strikes at long range is equally able to carry out tactical missions with conventional weapons. An example is France's Mirage IV, a small strategic bomber replaced in service by the ASMP-equipped Mirage 2000N fighter-bomber and Rafale multirole fighter.

13.1 History

13.1.1 First and Second World Wars

Sikorsky Ilya Muromets was designed by Igor Sikorsky as the first ever airliner, but was turned into a bomber for the Imperial Russian Air Force.

The first strategic bombing efforts took place during World War I (1914–18), by the Russians with their Sikorsky Ilya Muromets bomber (the first heavy four-engine aircraft), and by the Germans using Zeppelins or long-range multi-engine Gotha aircraft. Zeppelins reached England on bombing raids by 1915, forcing the British to create extensive defense systems including some of the first anti-aircraft guns which were often used with searchlights to highlight the enemy machines overhead. Late in the war, American fliers under the command of Brig. Gen. Billy Mitchell were developing multi-aircraft "mass" bombing missions behind German lines, although the Armistice ended full realization of what was being planned.

Study of strategic bombing continued in the interwar years. Many books and articles predicted a fearful prospect for any future war, paced by political fears such as those expressed by British Prime Minister Stanley Baldwin who told the House of Commons early in the 1930s that "the bomber will always get through" no matter what defensive systems were undertaken. It was widely believed by the late 1930s that strategic "terror" bombing of cities in any war would quickly result in devastating losses and might decide a conflict in a matter of days or weeks. But theory far exceeded what most air forces could actually put into the air. Germany focused on short-range tactical bombers. Britain's Royal Air Force began developing four-engine long-range bombers only in the late 1930s. The U.S. Army Air Corps (*Army Air Forces* as of mid-1941) was severely limited by small budgets in the late 1930s, and only barely saved the **B-17** bomber that would soon be vital. The equally important **B-24** first flew in 1939. Both aircraft would constitute the bulk of the American bomber force that made the Allied daylight bombing of Nazi Germany possible in 1943–45.

At the start of World War II, so-called "strategic" bombing was initially carried out by medium bomber aircraft which were typically twin-engined, armed with several defensive guns, but only possessed limited bomb-carrying capacity and range. Both Britain and the U.S. were developing larger two- and four-engined designs, which began to replace or supplement the smaller aircraft by 1941–42. After American entry into the war, late, in 1941, the U.S. 8th Air Force began to develop a daylight bombing capacity using improved B-17 and B-24 four-engine aircraft. The RAF concentrated its efforts on night bombing. But neither force was able to develop adequate bombsights or tactics to allow for often-bragged "pinpoint" accuracy. The post-war U.S. Strategic Bombing Survey studies supported the overall notion of strategic bombing, but underlined many of its shortcomings as well. Attempts to create pioneering examples of "smart bombs" resulted in the Azon ordnance, deployed in the European Theater and CBI Theater from B-24s.

Following the untimely death of the top German advocate for strategic bombing, *General* Walther Wever in early June 1936, the focus of Nazi Germany's Luftwaffe bomber forces, the so-named *Kampfgeschwader* (bomber wings) became the battlefield support of the *Wehrmacht Heer* as part of the general Blitzkrieg form of warfare, carried out with both medium bombers such as the Heinkel He 111, and *Schnellbombers* such as the Junkers Ju 88A. *General* Wever's support of the Ural bomber project before WW II's start dwindled after his passing, with the only aircraft design that could closely match the Allied bomber force's

The only operational strategic bomber with the Luftwaffe in World War II was the troubled Heinkel He 177.

own aircraft - the early November 1937-origin Heinkel He 177A, deployed in its initial form in 1941–42, hampered by a RLM requirement for the He 177A to also perform medium-angle dive bombing, not rescinded until September 1942 - unable to perform either function properly, with a powerplant selection and particular powerplant installation design features on the 30-meter wingspan *Greif*, that led to endless problems with engine fires. The March 1942-origin, trans-Atlantic ranged *Amerika Bomber* program sought to ameliorate the lack of a seriously long-ranged bomber for the *Luftwaffe*, but resulted with only three Messerschmitt-built and a pair of Junkers-built prototypes ever flown, and no operational "heavy bombers" for strategic use for the Third Reich, outside of the roughly one thousand examples of the He 177 that were built.

By the end of the Second World War in 1945, the "heavy" bomber, epitomized by the British Avro Lancaster and American Boeing B-29 Superfortress used in the Pacific Theater, showed what could be accomplished by area bombing of Japan's cities and the often small and dispersed factories within them. Under Major General Curtis LeMay, the U.S. 20th Air Force, based in the Mariana Islands, undertook low-level incendiary bombing missions, results of which were soon measured in the number of square miles destroyed. The air raids on Japan had withered the nation's ability to continue fighting, although the Japanese government resisted every means to surrender, resulting the atomic bombs dropped on Hiroshima and Nagasaki in August 1945.

13.1.2 The Cold War and its aftermath

During the Cold War, the United States and United Kingdom on one side and the Soviet Union on the other kept strategic bombers ready to take off on short notice as part of the deterrent strategy of mutual assured destruction (MAD). Most strategic bombers of the two superpowers were designed to deliver nuclear weapons. For a time, some squadrons of Boeing B-52 Stratofortress bombers were kept in the air around the clock, orbiting some distance away from their fail-safe points near the Soviet border.

The Royal Air Force's British-produced "V bombers" were designed and designated to be able to deliver British-made nuclear bombs to targets in European Russia. These bombers could have been able to reach and destroy cities like Kiev or Moscow before American strategic bombers.

The Soviet Union produced hundreds of unlicensed, reverse-engineered copies of the American Boeing B-29 Superfortress, which the Soviet Air Forces called the Tupolev Tu-4. The Soviets later developed the jet-powered Tupolev Tu-16 "Badger".

The People's Republic of China produced a version of Tupolev Tu-16 on license from the Soviet Union in the 1960s which they named the Xian H-6; it remains in service today.

During the 1960s France produced its Dassault Mirage IV nuclear-armed bomber for the French Air Force as a part of its independent nuclear strike force, the *Force de Frappe*, using French-made bombers and IRBMs to deliver French-made nuclear weapons. Mirage IVs served until mid-1996 in the bomber role, and to 2005 as a reconnaissance aircraft.

Today the French Republic has limited its strategic armaments to a squadron of four nuclear-powered ballistic missile submarines, with 16 SLBM tubes apiece. France also maintains an active force of supersonic fighter-bombers carrying stand-off nuclear missiles such as the ASMP, with Mach 3 speed and a range of 500 kilometers. These missiles can be delivered by the Dassault Mirage 2000N and Rafale fighter-bombers; the Rafale is also capable of refueling others in flight using a buddy refueling pod.

Newer strategic bombers such as the Rockwell International B-1B Lancer, the Tupolev Tu-160, and the Northrop Grumman B-2 Spirit designs incorporate various levels of stealth technology in an effort to avoid detection, especially by radar networks. Despite these advances earlier strategic bombers, for example the B-52 (last produced in 1962) or the Tupolev Tu-95 remain in service and can also deploy the latest air-launched cruise missiles and other "stand-off" or precision guided weapons such as the JASSM and the JDAM.

Indeed, it is likely that the USAF's B-52 fleet will, with continual upgrades, outlive its squadrons of B-1Bs; but the USAF has also recently launched a program to study possible future production of a new strategic bomber to complement the current fleet. It is likely that this bomber would also serve as a replacement for both the B-52 and B-1.

The Russian Air Force's new Tu-160 strategic bombers are expected to be delivered on a regular basis over the course

USAF Rockwell B-1 Lancer supersonic, swing wing strategic bomber.

of 10 to 20 years. In addition, the current Tu-95 and Tu-160 bombers will be periodically updated, as was done during the 1990s with the Tu-22M bombers.

Strategic bombers of the Cold War were primarily armed with nuclear weapons. During the post-1940s Indochina Wars, and also since the end of the Cold War, modern bombers originally intended for strategic use have been exclusively employed using non-nuclear, high explosive weapons. During the Vietnam War, Operation Menu, Operation Freedom Deal, Gulf War, military action in Afghanistan, and the 2003 invasion of Iraq, American B-52s and B-1s were mostly employed in tactical roles. During the Soviet-Afghan war in 1979–88, Soviet Air Forces Tu-95s carried out several mass air raids in various regions of Afghanistan.

13.2   Notable strategic bombers

13.2.1   Nomenclature

Bombers listed below were used in the main or represented a shift in long-range bomber design. (Maximum bomb load) in parenthesis. In practice, bomb loads carried are dependent on factors such as the distance to target and the individual type, size or weight of bombs used.

Nomenclature for size classification of aircraft types used in strategic bombing varies, particularly since the time of World War II due to sequential technological advancements and changes in aerial warfare strategy and tactics. The B-29, for example was a benchmark aircraft of the heavy bomber type at end of World War II due to its size, range and load carrying ability; as the Cold War began, it became an intercontinental range strategic bomber with the development of new techniques, such as aerial refueling (which also greatly extended the range of other medium- to long-range bombers, fighter-bombers and attack aircraft).

During the 1950s the U.S. Strategic Air Command also briefly brought back the outdated term **"medium bomber"** to distinguish its Boeing B-47 Stratojets from somewhat larger contemporary Boeing B-52 Stratofortress **"heavy bombers"** in bombardment wings; older B-29 and B-50 heavy bombers were also redesignated as "medium" during this period.[2][3][4] SAC's nomenclature here was purely semantic and bureaucratic, however as both the B-47 and B-52 strategic bombers were much larger and had far greater performance and load-carrying ability than any of the World War II-era heavy or medium bombers.

Other aircraft such as the twin-engine U.S. FB-111, A-3 Skywarrior and France's Mirage IV had nominal warloads of less than 20,000 lb (9,100 kg), and were significantly smaller in size and gross weight compared with their strategic bomber contemporaries, based on which they might be classified as **medium bombers**. In the nuclear strike role, France would replace its Mirage IVs beginning in the late 1980s with the even smaller, single-engine Mirage 2000N fighter-bomber, a further example of advancing technologies and changing tactics in military aviation and aircraft design. France's newer twin-engine Rafale multirole fighter also has nuclear strike capability.

13.2.2   World War I

- Caproni Ca.1
- Caproni Ca.3 (1,700 lb)
- Gotha G.IV (1,100 lb)
- Handley Page Type O (2,000 lb)
- Handley Page V/1500 (7,500 lb)
- Sikorsky Ilya Muromets (1,100 lb)
- Zeppelin Staaken R.VI (4,400 lb)
- Zeppelin (about 5,000 lb)

13.2.3   World War II

- Martin B-10 (successors B-17 and B-24 therefore and theoredically the first strategic bomber of the USAAF at that time despite being a medium bomber.)
- Boeing B-17 Flying Fortress (8,000 lb) (theoretical maximum: 17,410 lb)
- Consolidated B-24 Liberator (8,000 lb)

- Boeing B-29 Superfortress (20,000 lb)
- Consolidated B-32 Dominator (20,000 lb)
- Handley Page Halifax (13,000 lb)[5]
- Avro Lancaster (22,000 lb)
- Short Stirling (18,000 lb)
- Farman F.220 (9,240 lb)
- Heinkel He 177 (15,870 lb)
- Petlyakov Pe-8 (11,000 lb)
- Piaggio P.108 (7,700 lb)

13.2.4   Cold War

Weapons loads can include nuclear-armed missiles as well as aerial bombs

Boeing B-52 Stratofortress

Avro Vulcan

Tupolev Tu-160

- **Reciprocating/Turbine engine**

  - 🇬🇧 Avro Lincoln (22,000 lb)

  - 🇺🇸 Lockheed P-2 Neptune – small number converted as carrier-launched nuclear-armed bombers which would have to ditch/recover at land bases

  - 🇺🇸 Boeing B-50 Superfortress (28,000 lb)

  - 🇷🇺 Tupolev Tu-4 – reverse-engineered version of B-29 Superfortress

  - 🇺🇸 Convair B-36 Peacemaker (72,000 lb)

  - 🇷🇺 Tupolev Tu-95 (55,000 lb)

- **Jet engine**

  - 🇺🇸 North American B-45 Tornado (22,000 lb)

  - 🇺🇸 Boeing B-47 Stratojet (25,000 lb)

  - 🇷🇺 Myasishchev M-4 (52,910 lb)

  - 🇷🇺 Tupolev Tu-16 (20,000 lb)

    - 🇨🇳 Xian H-6 (20,000 lb)

  - 🇺🇸 Boeing B-52 Stratofortress (70,000 lb)

  - 🇬🇧 Vickers Valiant (21,000 lb)

  - 🇬🇧 Avro Vulcan (21,000 lb)

  - 🇺🇸 Douglas A-3 Skywarrior – nuclear-armed, carrier-based

  - 🇬🇧 Handley Page Victor (35,000 lb)

- **Supersonic**

  - 🇺🇸 Convair B-58 Hustler (19,450 lb)

  - 🇫🇷 Dassault Mirage IV (16,000 lb)

  - 🇫🇷 Dassault Mirage 2000N/2000D (nuclear strike variant)

- 🇺🇸 General Dynamics FB-111A – strategic bomber version of the F-111 swing wing strike aircraft

- 🇷🇺 Tupolev Tu-22 Blinder (20,000 lb)

- 🇷🇺 Tupolev Tu-22M Backfire (46,300 lb)

- 🇺🇸 Rockwell B-1 Lancer (75,000 lb – use of external hardpoints restricted by START I)

- 🇷🇺 Tupolev Tu-160 Blackjack (88,200 lb)

- others designed and built which did not enter operational service:

  - 🇷🇺 Myasishchev M-50 Bounder

  - 🇺🇸 North American XB-70 Valkyrie

  - 🇷🇺 Sukhoi T-4 Sotka

  - 🇬🇧 BAC TSR-2

13.2.5   Post Cold War

Northrop Grumman B-2 Spirit

- 🇺🇸 Northrop Grumman B-2 Spirit (40,000 lb)

13.2.6   Future

- 🇺🇸 Northrop Grumman B-21. A proposed stealth bomber, with a goal of supplanting a portion or all of the current B-52 and B-1. Planned to be deployed in the 2020s.[6][7]

- 🇺🇸 2037 Bomber. A stealth, supersonic, long-range, heavy-payload, strategic bomber project to replace the B-52 Stratofortress, with a deployment time frame goal of 2037.[8][9]

- 🇷🇺 PAK DA

13.3 See also

- Strategic bombing

- Carpet bombing

- High level bombing

- Tactical bombing

13.4 References

[1] Paul, T. V.; Wirtz, James J.; Fortmann, Michael. *Balance of power: theory and practice in the 21st century*, Stanford University Press, 2004, p. 332. ISBN 0-8047-5017-3

[2] Boeing RB-47H Stratojet: U.S. Air Force Fact Sheet

[3] Eighth Air Force History: U.S. Air Force Fact Sheet

[4] Strategic-Air-Command.com, 509th Composite Group, 509th Bombardment Wing

[5] for the Mark III

[6] New Long-Range Bomber On Horizon For 2018

[7] https://www.flightglobal.com/news/articles/usaf-reveals-northrops-b-21-long-range-strike-bombe-422459/

[8] Air Force Assoc. Feb. 2007, p. 11.

[9] Tirpak, John A. "The Bomber Roadmap", Air Force Magazine, June 1999.

- Brown, Michael E. Flying Blind: The Politics of the U.S. Strategic Bomber Program. Ithaca, NY: Cornell University Press, 1992.

- Cross, Robin. The Bombers: The Illustrated Story of Offensive Strategy and Tactics in the Twentieth Century. New York: Macmillan, 1987.

- Green, William. Famous Bombers of the Second World War. New York: Doubleday, 1959, 1960 (two vols).

- Green, William. Warplanes of the Third Reich. New York: Doubleday, 1970.

- Haddow, G. W., and Peter M. Grosz The German Giants: The German R-Planes 1914-1918. London: Putnam, 1969 (2nd ed.)

- Hastings, Max. Bomber Command. New York: Dial Press, 1979

- Jones, Lloyd S. U.S. Bombers 1926 to 1980s. Fallbrook, CA: Aero Publishers, 1980 (3rd ed.)

- Neillands, Robin. The Bomber War: The Allied Offensive Against Nazi Germany. Woodstock, NY: Overlook, 2001.

- Robinson, Douglas H. The Zeppelin in Combat: A History of the German Naval Airship Division, 1912-1918. Atglen, PA: Schiffer, 1994.

- United States Strategic Bombing Survey. Over-all Report (European War). Washington: Government Printing Office, September 30, 1945.

Chapter 14

Intercontinental ballistic missile

"ICBM" redirects here. For the geotag, see ICBM address. For the Institute, see Institute for Chemistry and Biology of the Marine Environment.

An **intercontinental ballistic missile (ICBM)** is a guided ballistic missile with a minimum range of 5,500 kilometres (3,400 mi)[1] primarily designed for nuclear weapons delivery (delivering one or more thermonuclear warheads). Similarly, conventional, chemical, and biological weapons can also be delivered with varying effectiveness, but have never been deployed on ICBMs. Most modern designs support multiple independently targetable reentry vehicles (MIRVs), allowing a single missile to carry several warheads, each of which can strike a different target.

Early ICBMs had limited precision (circular error probable) that allowed them to be used only against the largest targets such as cities. They were seen as a "safe" basing option, one that would keep the deterrent force close to home where it would be difficult to attack. Attacks against (especially hardened) military targets, if desired, still demanded the use of a more precise manned bomber. This is due to the inverse-square law, which predicts that the amount of energy dispersed from a single point release of energy (such as a thermonuclear blast) dissipates by the inverse of the square of the distance from the single point of release. The result is that the power of a nuclear explosion to rupture hardened structures is greatly decreased by the distance from the impact point of the nuclear weapon. So a near-direct hit is generally necessary, as only diminishing returns are gained by increasing bomb yield.

Second- and third-generation designs (e.g. the LGM-118 Peacekeeper) dramatically improved accuracy to the point where even the smallest point targets can be successfully attacked.

ICBMs are differentiated by having greater range and speed than other ballistic missiles: intermediate-range ballistic missiles (IRBMs), medium-range ballistic missiles (MRBMs), and short-range ballistic missiles (SRBMs). Such shorter-range ballistic missiles are known collectively as theatre ballistic missiles.

14.1 History

14.1.1 World War II

The development of the world's first practical design for an ICBM, A9/10, intended for use in bombing New York and other American cities, was undertaken in Nazi Germany by the team of Wernher von Braun under *Projekt Amerika*. The ICBM A9/A10 rocket initially was intended to be guided by radio, but was changed to be a piloted craft after the failure of Operation Elster. The second stage of the A9/A10 rocket was tested a few times in January and February 1945. The progenitor of the A9/A10 was the German V-2 rocket, also designed by von Braun and widely used at the end of World War II to bomb British and Belgian cities. All of these rockets used liquid propellants. Following the war, von Braun and other leading German scientists were relocated to the United States to work directly for the US Army through Operation Paperclip, developing the IRBMs, ICBMs, and launchers.

This technology was also predicted by US Army General Hap Arnold, who wrote in 1943:

> Someday, not too distant, there can come streaking out of somewhere – we won't be able to hear it, it will come so fast – some kind of gadget with an explosive so powerful that one projectile will be able to wipe out completely this city of Washington.[2][3]

14.1.2 Cold War

In the immediate post-war era, the US and USSR both started rocket research programs based on the German wartime designs, especially the V-2. In the US, each branch of the military started its own programs, leading to considerable duplication of effort. In the USSR, rocket research was centrally organized, although several teams worked on different designs. Early designs from both countries

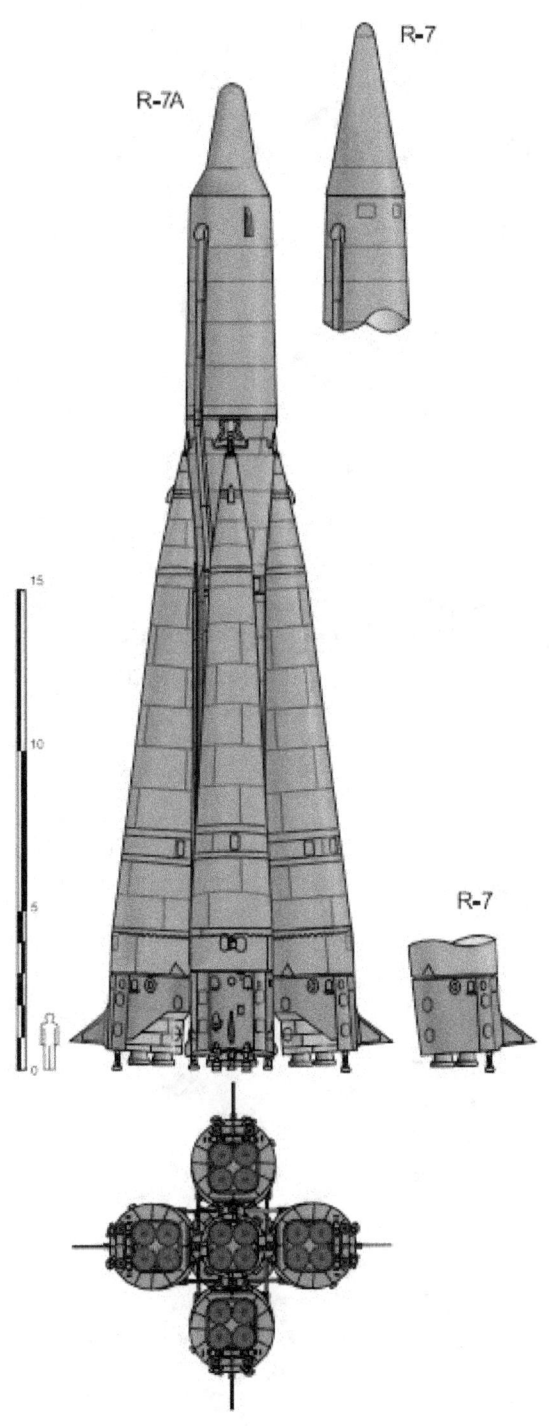

R-7A

R-7

R-7

R-7

15

10

5

0

The R-7 Semyorka was the world's first ICBM and satellite launch vehicle

The SM-65 Atlas was the first US ICBM. First launch 1957, flew full range 1958

were short-range missiles, like the V-2, but improvements quickly followed.

In the USSR early development was focused on missiles able to attack European targets. This changed in 1953

when Sergei Korolyov was directed to start development of a true ICBM able to deliver newly developed hydrogen bombs. Given steady funding throughout, the R-7 developed with some speed. The first launch took place on 15 May 1957 and led to an unintended crash 400 km (250 mi) from the site. The first successful test followed on 21 August 1957; the R-7 flew over 6,000 km (3,700 mi) and became the world's first ICBM.[4] The first strategic-missile unit became operational on 9 February 1959 at Plesetsk in north-west Russia.[5]

It was the same R-7 launch vehicle that placed the first artificial satellite in space, Sputnik, on 4 October 1957. The first human spaceflight in history was accomplished on a derivative of R-7, Vostok, on 12 April 1961, by Soviet cosmonaut Yuri Gagarin. A heavily modernized version of the R-7 is still used as the launch vehicle for the Soviet/Russian Soyuz spacecraft, marking more than 50 years of operational history of Sergei Korolyov's original rocket design.

The U.S. initiated ICBM research in 1946 with the RTV-A-2 Hiroc project. This was a three-stage effort with the

Test launch of an LGM-25C Titan II ICBM from an underground silo at Vandenberg AFB during the mid-1970s

The Soviet R-36M (SS-18 Satan) is the largest ICBM in history, with a throw weight of 8,800 kg, twice that of the Peacekeeper.

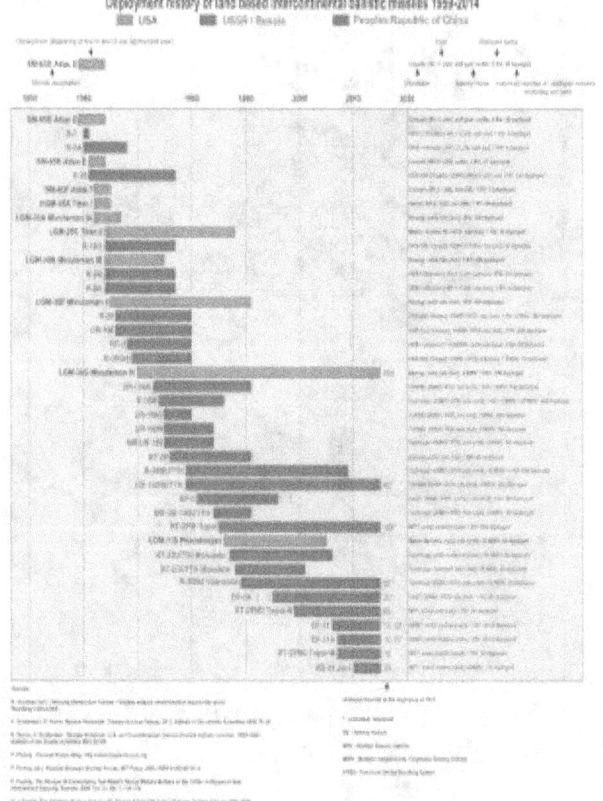

Deployment history of land based ICBM 1959-2014

A Minuteman III ICBM test launch from Vandenberg Air Force Base, CA, United States

ICBM development not starting until the third stage. However, funding was cut after only three partially successful launches in 1948 of the second stage design, used to test

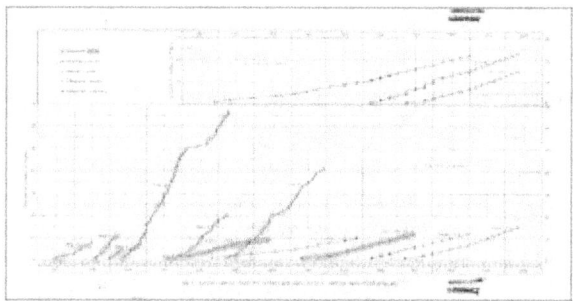

1965 graph of USAF Atlas and Titan ICBM launches, cumulative by month with failures highlighted (pink). This clearly shows how NASA use of ICBM boosters for Projects Mercury and Gemini (blue) served as a highly visible demonstration of confidence in reliability at a time when failure rates had been substantial (Apollo-Saturn history and projections shown as well.)

U.S. Peacekeeper missile after silo launch.

variations on the V-2 design. With overwhelming air superiority and truly intercontinental bombers, the newly forming US Air Force did not take the problem of ICBM development seriously. Things changed in 1953 with the Soviet testing of their first thermonuclear weapon, but it was not until 1954 that the Atlas missile program was given the highest national priority. The Atlas A first flew on 11 June 1957; the flight lasted only about 24 seconds before the rocket blew up. The first successful flight of an Atlas missile to full range occurred 28 November 1958.[6] The first armed version of the Atlas, the Atlas D, was declared operational in January 1959 at Vandenberg, although it had not yet flown. The first test flight was carried out on 9 July 1959,[7][8] and the missile was accepted for service on 1 September.

The R-7 and Atlas each required a large launch facility, making them vulnerable to attack, and could not be kept in a ready state. Failure rates were very high throughout the early years of ICBM technology. Human spaceflight programs (Vostok, Mercury, Voskhod, Gemini, etc.) served as a highly visible means of demonstrating confidence in reliability, with successes translating directly to national defense implications. The US was well behind the Soviet Union in the Space Race, so U.S. President John F. Kennedy increased the stakes with the Apollo program, which used Saturn rocket technology that had been funded by President Dwight D. Eisenhower.

These early ICBMs also formed the basis of many space launch systems. Examples include R-7, Atlas, Redstone, Titan, and Proton, which was derived from the earlier ICBMs but never deployed as an ICBM. The Eisenhower administration supported the development of solid-fueled missiles such as the LGM-30 Minuteman, Polaris and Skybolt. Modern ICBMs tend to be smaller than their ancestors, due to increased accuracy and smaller and lighter warheads, and use solid fuels, making them less useful as orbital launch vehicles.

The Western view of the deployment of these systems was governed by the strategic theory of Mutual Assured Destruction. In the 1950s and 1960s, development began on Anti-Ballistic Missile systems by both the U.S. and USSR; these systems were restricted by the 1972 ABM treaty. The first successful ABM test were conducted by the USSR in 1961, that later deployed a fully operating system defending Moscow in the 1970s (see Moscow ABM system).

The 1972 SALT treaty froze the number of ICBM launchers of both the U.S. and the USSR at existing levels, and allowed new submarine-based SLBM launchers only if an equal number of land-based ICBM launchers were dismantled. Subsequent talks, called SALT II, were held from 1972 to 1979 and actually reduced the number of nuclear warheads held by the U.S. and USSR. SALT II was never ratified by the United States Senate, but its terms were nevertheless honored by both sides until 1986, when the Reagan administration "withdrew" after accusing the USSR of violating the pact.

In the 1980s, President Ronald Reagan launched the Strategic Defense Initiative as well as the MX and Midgetman ICBM programs.

China developed a minimal independent nuclear deterrent entering its own cold war after an ideological split with the Soviet Union beginning in the early 1960s. After first testing a domestic built nuclear weapon in 1964, it went on to develop various warheads and missiles. Beginning in the early 1970s, the liquid fuelled DF-5 ICBM was developed and used as a satellite launch vehicle in 1975. The DF-5, with range of 10,000 to 12,000 km (6,200 to 7,500 mi) long enough to strike the western US and the USSR, was silo deployed with the first pair in service by 1981 with possibly twenty missiles in service by the late 1990s.[9] China also deployed the JL-1 Medium-range ballistic missile with a reach of 1,700 kilometres (1,100 mi) aboard the ultimately unsuccessful type 92 submarine.[10]

14.1.3 Post-Cold War

In 1991, the United States and the Soviet Union agreed in the START I treaty to reduce their deployed ICBMs and attributed warheads.

As of 2009, all five of the nations with permanent seats on the United Nations Security Council have operational long-range ballistic missile systems: all except China have operational submarine-launched missiles, and Russia, the United States and China also have land-based ICBMs (the US missiles are silo-based, while China and Russia have both silo and road-mobile (DF-31, RT-2PM2 Topol-M) missiles).

Israel is believed to have deployed a road mobile nuclear ICBM, the Jericho III, which entered service in 2008; an upgraded version is in development.[11][12]

India successfully test fired Agni V, with a strike range of more than 5,000 km (3,100 mi) on 19 April 2012, claiming entry into the ICBM club.[13] The missile's actual range is speculated by foreign researchers to be up to 8,000 km (5,000 mi) with India having downplayed its capabilities to avoid causing concern to other countries.[14]

It is speculated by some intelligence agencies that North Korea is developing an ICBM.[15] North Korea successfully put a satellite into space on 12 December 2012 using the 32-metre-tall (105 ft) Unha-3 rocket. The United States claimed that the launch was in fact a way to test an ICBM.[16] (See Timeline of first orbital launches by country)

In July 2014 China announced the development of its newest generation of ICBM, the Dongfeng-41 (DF-41), which has a range of 12,000 kilometres (7,500 miles), capable of reaching the United States, and which analysts believe is capable of being outfitted with MIRV technology.[17]

Most countries in the early stages of developing ICBMs have used liquid propellants, with the known exceptions being the Indian Agni-V, the planned but cancelled[18] South African RSA-4 ICBM, and the now in service Israeli Jericho III.[19]

14.2 Flight phases

The following flight phases can be distinguished:

- boost phase: 3 to 5 minutes; it is shorter for a solid-fuel rocket than for a liquid-propellant rocket; depending on the trajectory chosen, typical burnout speed is 4 km/s (2.5 mi/s), up to 7.8 km/s (4.8 mi/s); altitude at the end of this phase is typically 150 to 400 km (93 to 249 mi).

- midcourse phase: approx. 25 minutes—sub-orbital spaceflight with a flightpath being a part of an ellipse with a vertical major axis; the apogee (halfway through the midcourse phase) is at an altitude of approximately 1,200 km (750 mi); the semi-major axis is between 3,186 and 6,372 km (1,980 and 3,959 mi); the projection of the flightpath on the Earth's surface is close to a great circle, slightly displaced due to earth rotation during the time of flight; the missile may release several independent warheads and penetration aids, such as metallic-coated balloons, aluminum chaff, and full-scale warhead decoys.

- reentry phase (starting at an altitude of 100 km, 62 mi): 2 minutes – impact is at a speed of up to 7 km/s (4.3 mi/s) (for early ICBMs less than 1 km/s (0.62 mi/s)); see also maneuverable reentry vehicle.

ICBMs usually use the trajectory which optimizes range for a given amount of payload (the *minimum-energy trajectory*); an alternative is a depressed trajectory, which allows less payload, shorter flight time, and has a much lower apogee.[20]

14.3 Modern ICBMs

Modern ICBMs typically carry multiple independently targetable reentry vehicles (*MIRVs*), each of which carries a separate nuclear warhead, allowing a single missile to hit multiple targets. MIRV was an outgrowth of the rapidly shrinking size and weight of modern warheads and the Strategic Arms Limitation Treaties which imposed limitations on the number of launch vehicles (SALT I and SALT II). It has also proved to be an "easy answer" to proposed deployments of ABM systems—it is far less expensive to add more warheads to an existing missile system than to build an ABM system capable of shooting down the additional warheads; hence, most ABM system proposals have been judged to be impractical. The first operational ABM systems were deployed in the U.S. during the 1970s. Safeguard ABM facility was located in North Dakota and was operational from 1975 to 1976. The USSR deployed its ABM-

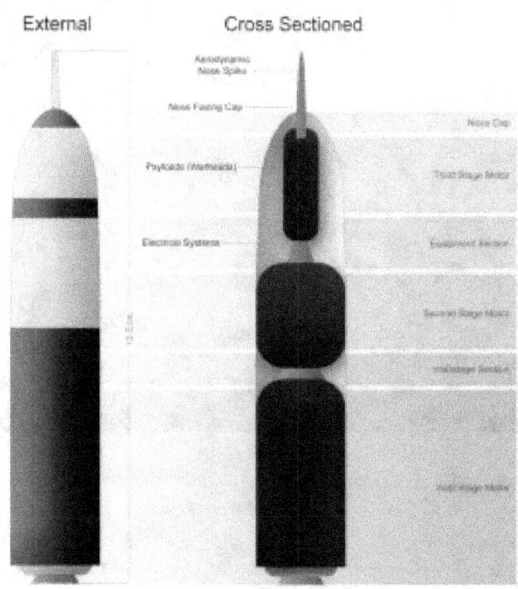

External | Cross Sectioned

External and cross sectional views of a Trident II D5 nuclear missile system. It is a submarine-launched missile capable of carrying multiple nuclear warheads up to 8,000 km (5,000 mi). Trident missiles are carried by fourteen active US Navy Ohio-class and four Royal Navy Vanguard-class submarines.

1 Galosh system around Moscow in the 1970s, which remains in service. Israel deployed a national ABM system based on the Arrow missile in 1998,[21] but it is mainly designed to intercept shorter-ranged theater ballistic missiles, not ICBMs. The Alaska-based United States national missile defense system attained initial operational capability in 2004.[22]

ICBMs can be deployed from TELs such as the Russian RT-2PM2 Topol-M.

ICBMs can be deployed from multiple platforms:

- in missile silos, which offer some protection from military attack (including, the designers hope, some protection from a nuclear first strike)

- on submarines: submarine-launched ballistic missiles (SLBMs); most or all SLBMs have the long range of ICBMs (as opposed to IRBMs)

- on heavy trucks; this applies to one version of the Topol which may be deployed from a self-propelled mobile launcher, capable of moving through roadless terrain, and launching a missile from any point along its route

- mobile launchers on rails; this applies, for example, to РТ−23УТТХ "Молодец" (RT-23UTTH "Molodets"—SS-24 "Scalpel")

The last three kinds are mobile and therefore hard to find. During storage, one of the most important features of the missile is its serviceability. One of the key features of the first computer-controlled ICBM, the Minuteman missile, was that it could quickly and easily use its computer to test itself.

In flight, a booster pushes the warhead and then falls away. Most modern boosters are solid-fueled rocket motors, which can be stored easily for long periods of time. Early missiles used liquid-fueled rocket motors. Many liquid-fueled ICBMs could not be kept fuelled all the time as the cryogenic fuel liquid oxygen boiled off and caused ice formation, and therefore fueling the rocket was necessary before launch. This procedure was a source of significant operational delay, and might allow the missiles to be destroyed by enemy counterparts before they could be used. To resolve this problem the United Kingdom invented the missile silo that protected the missile from a first strike and also hid fuelling operations underground.

Once the booster falls away, the warhead continues on an unpowered ballistic trajectory, much like an artillery shell or cannonball. The warhead is encased in a cone-shaped reentry vehicle and is difficult to detect in this phase of flight as there is no rocket exhaust or other emissions to mark its position to defenders. The high speeds of the warheads make them difficult to intercept and allow for little warning, striking targets many thousands of kilometers away from the launch site (and due to the possible locations of the submarines: anywhere in the world) within approximately 30 minutes.

Many authorities say that missiles also release aluminized balloons, electronic noise-makers, and other items intended to confuse interception devices and radars.

As the nuclear warhead reenters the Earth's atmosphere its high speed causes compression of the air, leading to a dramatic rise in temperature which would destroy it if it were not shielded in some way. As a result, warhead components are contained within an aluminium honeycomb sub-structure, sheathed in a pyrolytic carbon-epoxy synthetic resin composite material heat shield. Warheads are also often radiation-hardened (to protect against nuclear-tipped ABMs or the nearby detonation of friendly warheads), one neutron-resistant material developed for this purpose in the UK is three-dimensional quartz phenolic.

Circular error probable is crucial, because halving the circular error probable decreases the needed warhead energy by a factor of four. Accuracy is limited by the accuracy of the navigation system and the available geodetic information.

Strategic missile systems are thought to use custom integrated circuits designed to calculate navigational differential equations thousands to millions of FLOPS in order to reduce navigational errors caused by calculation alone. These circuits are usually a network of binary addition circuits that continually recalculate the missile's position. The inputs to the navigation circuit are set by a general purpose computer according to a navigational input schedule loaded into the missile before launch.

One particular weapon developed by the Soviet Union - the Fractional Orbital Bombardment System - had a partial orbital trajectory, and unlike most ICBMs its target could not be deduced from its orbital flight path. It was decommissioned in compliance with arms control agreements, which address the maximum range of ICBMs and prohibit orbital or fractional-orbital weapons. However, according to reports, Russia is working on the new Sarmat ICBM which leverages Fractional Orbital Bombardment concepts to use a Southern polar approach instead of flying over the Northern polar regions. Using this approach, it is theorized, avoids the US missile defense batteries in California and Alaska.

14.4 Specific ICBMs

Main articles: Comparison of ICBMs and List of ICBMs

14.4.1 Land-based ICBMs

- Minuteman (13,000km) (United States) 🇺🇸 U.S.

- R-36M2 (SS-18) (10,000+ km) (Soviet Union, Russia) ▬ Russia

- UR-100N (SS-19) (10,000+ km) (Soviet Union, Russia) ▬ Russia

- RT-2PM "Topol" (SS-25) (10,000+ km) (Soviet Union, Russia) ▬ Russia

Testing of the Peacekeeper re-entry vehicles at the Kwajalein Atoll. All eight fired from only one missile. Each line, if its warhead were live, represents the potential explosive power of about 300 kilotons of TNT, about nineteen times larger than the detonation of the atomic bomb in Hiroshima.

Artist's concept of SS-24 deployed on railway.

- RT-2UTTH "Topol M" (SS-27) (11,000 km) (Russia) ▬ Russia

- RS-24 "Yars" (SS-29) (11,000 km) (Russia) ▬ Russia

- RS-26 Rubezh(Russia) (6,000-12,600 km) ▬ Russia

- Sarmat heavy ICBM (SS-18 R-36M replacement)(Russia) ▬ Russia

- DF-4 (~5,500-7,000 km) (China) 🇨🇳 China

- DF-31 (7,200-11,200 km) (China) 🇨🇳 China

- DF-5 (12,000-15,000 km) (China) 🇨🇳 China

- DF-41 (12,000-15,000+ km) (China) ⬜ China

- Agni-V (5,000 to 8,000 km) (India) 🟧 India

- Jericho 3 (5,000 to 11,500 km) (Israel) ✡ Israel

Russia, the United States, China and India are the only countries currently known to possess land-based ICBMs. Israel has also tested ICBMs but is not open about actual deployment.[23][24]

The United States currently operates 450 ICBMs in three USAF bases. The only model deployed is LGM-30G Minuteman-III.

All previous USAF Minuteman II missiles have been destroyed in accordance with START, and their launch silos have been sealed or sold to the public. To comply with the START II most U.S. multiple independently targetable reentry vehicles, or MIRVs, have been eliminated and replaced with single warhead missiles. The powerful MIRV-capable Peacekeeper missiles were phased out in 2005.[25] However, since the abandonment of the START II treaty, the U.S. is said to be considering retaining 800 warheads on an existing 450 missiles.[26]

The Russian Strategic Rocket Forces have 369 ICBMs able to deliver 1,247 nuclear warheads, 58 silo-based R-36M2 (SS-18), 70 silo-based UR-100N (SS-19), 171 mobile RT-2PM "Topol" (SS-25), 52 silo-based RT-2UTTH "Topol M" (SS-27), 18 mobile RT-2UTTH "Topol M" (SS-27), 6 (15 in December 2011[27]) mobile RS-24 "Yars" (SS-29) *(Future replacement for R-36 & UR-100N missiles)*

China has developed several long range ICBMs, like the DF-31. The Dongfeng 5 or DF-5 is a 3-stage liquid fuel ICBM and has an estimated range of 13,000 kilometers. The DF-5 had its first flight in 1971 and was in operational service 10 years later. One of the downsides of the missile was that it took between 30 and 60 minutes to fuel. The Dong Feng 31 (a.k.a. CSS-10) is a medium-range, three-stage, solid-propellant intercontinental ballistic missile, and is a land-based variant of the submarine-launched JL-2.

The DF-41 or CSS-X-10 can carry up to 10 nuclear warheads, which are MIRVs and has a range of approximately 12,000–14,000 km (7,500–8,700 mi).[28][29][30][31] The DF-41 deployed in underground Xinjiang, Qinghai,Gansu and Inner Mongolia area. The mysterious underground subway ICBM carrier systems they called "Underground Great Wall Project[32]".

Israel is believed to have deployed a road mobile nuclear ICBM, the Jericho III, which entered service in 2008. It is possible for the missile to be equipped with a single 750 kg (1,650 lb) nuclear warhead or up to three MIRV warheads. It is believed to be based on the Shavit space launch vehicle and is estimated to have a range of 4,800 to 11,500 km

(3,000 to 7,100 mi).[11] In November 2011 Israel tested an ICBM believed to be an upgraded version of the Jericho III.[12]

India has a series of ballistic missiles called Agni, On 19 April 2012, India successfully test fired its first Agni-V, a three-stage solid fueled missile, with a strike range of more than 7,500 km (4,700 mi). The missile was test-fired for the second time on 15 September 2013.[13] On 31 January 2015, India conducted a third successful test flight of the Agni-V from the Wheeler Island facility. The test used a canisterised version of the missile, mounted over a Tatra truck.[33]

14.4.2 Submarine-launched

Main article: Submarine-launched ballistic missile

All current designs of submarine launched ballistic missiles have intercontinental range except the current generation of short range Indian SLBMs.[34] Current operators of such missiles are the United States, Russia, United Kingdom, France, India and the People's Republic of China.[35]

14.5 Missile defense

Main articles: Anti-ballistic missile and Missile defense

An anti-ballistic missile is a missile which can be deployed to counter an incoming nuclear or non-nuclear ICBM. ICBMs can be intercepted in three regions of their trajectory: boost phase, mid-course phase or terminal phase. Currently the US, Russia, France, India, and Israel have developed anti-ballistic missile systems, of which the Russian A-135 anti-ballistic missile system, US Ground-Based Midcourse Defense, and Indian Prithvi and Advanced Air Defence Systems have the capability to intercept ICBMs carrying nuclear, chemical, biological, or conventional warheads.

14.6 See also

- Air Force Global Strike Command

- Anti-Ballistic Missile Treaty

- Atmospheric reentry

- Countermeasure

- Dense Pack

- Emergency Action Message

- Fractional Orbital Bombardment System

- France and weapons of mass destruction

- General Bernard Adolph Schriever

- Heavy ICBM

- High-alert nuclear weapon

- India and weapons of mass destruction

- Israel and weapons of mass destruction

- ICBM address

- List of ICBMs

- Missile Defense Agency

- Nuclear disarmament

- Nuclear navy

- Nuclear warfare

- Nuclear weapon

- People's Republic of China and weapons of mass destruction

- Russia and weapons of mass destruction

- SLBM

- Strike Force (France)

- Submarine

- Throw-weight

- United Kingdom and weapons of mass destruction

- United States and weapons of mass destruction

- Underground Great Wall of China

14.7   References

[1] "Intercontinental Ballistic Missiles".   *Special Weapons Primer*. Federation of American Scientists. Retrieved 2012-12-14.

[2] Dolman, Everett C.; Cooper, Henry F., Jr. "19: Increasing the Military Uses of Space". *Toward a Theory of Space Power*. NDU Press. Retrieved 2012-04-19.

[3] Correll, John T. (July 2005). "How the Air Force got the ICBM". *Airforce Magazine*. Retrieved 2012-04-19.

[4] Wade, Mark. "R-7". *Encyclopedia Astronautica*. Retrieved 2011-07-04.

[5] "This Week in EUCOM History: February 6–12, 1959". EUCOM. 6 February 2012. Retrieved 2012-02-08.

[6] "Atlas". *The Exploration of Space*. Century of Flight. Retrieved 2012-12-14.

[7] "Atlas D". Missile Threat. Archived from the original on 10 February 2012. Retrieved 2012-04-19.

[8] "Atlas". *Encyclopedia Astronautica*. Astronautix. Retrieved 2012-04-19.

[9] "DF-5". *Weapons of Mass Destruction / WMD Around the World*. Federation of American Scientists. Retrieved 2012-12-14.

[10] "Type 92 Xia". *Weapons of Mass Destruction Around the World*. Federation of American Scientists. Retrieved 2012-12-14.

[11] Feickert, Andrew (5 March 2004). Missile Survey: Ballistic and Cruise Missiles of Foreign Countries (PDF). *Congressional Research Service* (Report).   Library of Congress. RL30427. Retrieved 2010-06-21.

[12] Pfeffer, Anshel (2 November 2011). "IDF test-fires ballistic missile in central Israel". *Haaretz*. Reuters. Retrieved 2011-11-03.

[13] Mallikarjun, Y; Subramanian, TS (19 April 2012). "Agni-V successfully test-fired". *The Hindu*. Retrieved 2012-04-19.

[14] "India downplayed Agni-V's capacity: Chinese experts". Beijing, China: The Hindustan Times. Indo-Asian News Service. April 20, 2012. Retrieved 13 July 2014.

[15] "North Korea's Taepodong and Unha Missiles". *Programs*. Federation of American Scientists. Retrieved 2012-04-19.

[16] "North Korea says it successfully launched satellite into orbit". *NBC News*. 12 December 2012. Retrieved 2013-04-13.

[17] "China 'confirms new generation long range missiles'". *Telegraph.co.uk*. 1 August 2014. Retrieved 1 April 2015.

[18] "South Africa". astronautix.com. Retrieved 2016-07-08.

[19] "Jericho". *Encyclopedia Astronautica*. Astronautix. Retrieved 2012-12-14.

[20] Science & Global Security, 1992, Volume 3, pp.101-159 Depressed Trajectory SLBMs: A Technical Evaluation and Arms Control Possibilities

[21] "Israeli Arrow ABM System is Operational as War Butts Darken". *Israel High-Tech & Investment Report*. November 2002. Retrieved 2012-04-19.

[22] "Fort Greely". *Systems*. Missile Threat. 8 December 1998. Archived from the original on 30 January 2012. Retrieved 2012-04-19.

[23] "ICBM". *Encyclopedia Britannica.* Retrieved 2012-04-19.

[24] *India test launches Agni-V long-range missile,* UK: BBC News, April 19, 2012, retrieved 2016-03-11.

[25] Edwards, Joshua S. (20 September 2005). "Peacekeeper missile mission ends during ceremony". US: Air force. Archived from the original on 2012-10-18. Retrieved 2016-04-28.

[26] Norris, Robert S; Kristensen, Hans (January–February 2009). "Nuclear Notebook: U.S. and Soviet/Russian intercontinental ballistic missiles, 1959–2008" (PDF). *Bulletin of the Atomic Scientists.* doi:10.2968/065001008. Retrieved 2012-12-14.

[27] "Second RS-24 regiment begins combat duty". *Russian strategic nuclear forces.* Russian forces. Retrieved 2012-12-14.

[28] "Five types of missiles to debut on National Day". *Xinhua.* 2 September 2009. Retrieved 2010-04-06.

[29] "DF-41, CSS-X-10". *Weapons of Mass Destruction.* Global security. Retrieved 6 April 2010.

[30] "DF-41 (CSS-X-10; China)". *Jane's Strategic Weapon Systems.* Jane's Information Group. 2 July 2009. Retrieved 2010-04-06.

[31] "DF-41 (CSS-X-10)". Missile Threat. Archived from the original on 8 April 2016. Retrieved 2015-01-26.

[32] Zhang, Hui. "China's underground Great Wall: subterranean ballistic missile". *Power & Policy.* Power and Policy, Belfer Center for Science and International Affairs, Kennedy School of Government, Harvard University. Retrieved 14 June 2015.

[33] "Agni 5, India's Longest Range Ballistic Missile, Successfully Test-Fired". *NDTV.com.* Retrieved 2016-02-08.

[34] "Going ballistic: India looks to join elite missile club". *The Times of India.* Retrieved 1 April 2015.

[35] "Type 094 (Jin Class) Nuclear-Powered Missile Submarine". Sino defence. Retrieved 2012-12-14.

14.8 External links

- *Missile Defense Agency,* US: Department of Defense.

- *Estimated Strategic Nuclear Weapons Inventories,* Rice, September 2004.

- *The 10 longest range Intercontinental Ballistic Missiles.*

- *Intercontinental Ballistic and Cruise Missiles* (guide), US: FAS

- Hawes, Kingdon R. "King", Lt Col, USAF (Ret.), *A Tale of Two Airplanes,* RC135.

- ICBM Modernization: Approaches to Basing Options and Interoperable Warhead Designs Need Better Planning and Synchronization: Report to Congressional Committees Government Accountability Office

14.9 Literature

- «Korolev: Facts and myths» - J. K. Golovanov, M: Nauka, 1994, - ISBN 5-02-000822-2

- "Rockets and people" – B. E. Chertok, M: "mechanical engineering", 1999. ISBN 5-217-02942-0 (Russian);

- "Testing of rocket and space technology - the business of my life" Events and facts - A.I. Ostashev, Korolyov, 2001.

- "Nesterenko" series Lives of great people - Authors: Gregory Sukhina A., Ivkin, Vladimir Ivanovich, publishing house "Young guard" in 2015, ISBN 978-5-235-03801-1

Chapter 15

Submarine-launched ballistic missile

A UGM-96 Trident I clears the water after launch from a US Navy submarine in 1984

A **submarine-launched ballistic missile** (**SLBM**) is a ballistic missile capable of being launched from submarines. Modern variants usually deliver multiple independently targetable reentry vehicles (MIRVs) each of which carries a nuclear warhead and allows a single launched missile to strike several targets. Submarine-launched ballistic missiles operate in a different way from submarine-launched cruise missiles.

Modern submarine-launched ballistic missiles are closely related to intercontinental ballistic missiles (range of over 5,500 kilometres (3,000 nmi)), and in many cases SLBMs and ICBMs may be part of the same family of weapons.

15.1 History

15.1.1 Origins

The first practical design of a submarine-based launch platform was developed by the Germans near the end of World War II involving a launch tube which contained a V-2 ballistic missile variant and was towed behind a submarine, known by the code-name *Prüfstand XII*. The war ended before it could be tested, but the engineers who had worked on it went on to work for the USA and USSR on their SLBM programs. These and other early SLBM systems required vessels to be surfaced when they fired missiles, but launch systems eventually were adapted to allow underwater launching in the 1950-1960s. A converted Project 611 (Zulu-IV class) submarine launched the world's first SLBM, an R-11FM (SS-N-1 Scud-A, naval variant of the SS-1 Scud) on 16 September 1955.[1] Five additional Project V611 and AV611 (Zulu-V class) submarines became the world's first operational ballistic missile submarines (SSBNs) with two R-11FM missiles each, entering service in 1956-57.[2]

The United States Navy initially worked on a sea-based variant of the US Army Jupiter intermediate-range ballistic missile, projecting four of the large, liquid-fueled missiles per submarine.[3] Rear Admiral W. F. "Red" Raborn headed a Special Project Office to develop Jupiter for the Navy, beginning in late 1955.[4][5] However, at the Project Nobska submarine warfare conference in 1956, physicist Edward Teller stated that a physically small one-megaton warhead could be produced for the relatively small, solid-fueled Polaris missile,[6] and this prompted the Navy to leave the Jupiter program in December of that year. Soon Chief of Naval Operations Admiral Arleigh Burke concentrated all Navy strategic research on Polaris, still under Admiral Raborn's Special Project Office.[7] All US SLBMs have been solid-fueled while all Soviet and Russian SLBMs have been liquid-fueled except for the Russian RSM-56 Bulava, which entered service in 2014.

The world's first operational SSBN was USS *George Wash-*

Polaris A-1

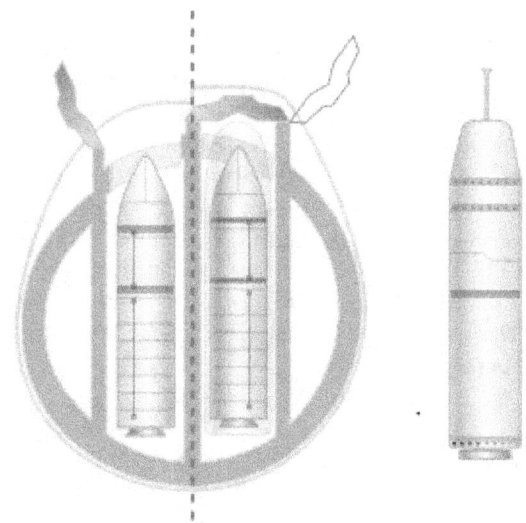

French M45 SLBM and M51 SLBM in cross-section of a submarine.

16 missiles was the Project 667A (Yankee class), which first entered service in 1967 with 32 boats completed by 1974.[14][15] By the time the first Yankee was commissioned the US had built 41 SSBNs, nicknamed the "41 for Freedom".[16][17]

ington (SSBN-598) with 16 Polaris A-1 missiles, which entered service in December 1959 and conducted the first SSBN deterrent patrol November 1960-January 1961.[8] *George Washington* also conducted the first successful submerged SLBM launch with a Polaris A-1 on 20 July 1960.[9] Forty days later, the Soviet Union made its first successful underwater launch of a submarine ballistic missile in the White Sea, on 10 September 1960 from the same converted Project 611 (NATO reporting name Zulu-IV class) submarine that first launched the R-11FM.[10] The Soviets were only a year behind the US with their first SSBN, the ill-fated K-19 of Project 658 (Hotel class), commissioned in November 1960. However, the Hotel class carried only three R-13 missiles (NATO reporting name SS-N-4) each and had to surface and raise the missile to launch.[11] Submerged launch was not an operational capability for the Soviets until 1963, when the R-21 missile (SS-N-5) was first backfitted to Project 658 (Hotel class) and Project 629 (Golf class) submarines.[12] The Soviet Union was able to beat the U.S. in launching and testing the first SLBM with a live nuclear warhead, an R-13 that detonated in the Novaya Zemlya Test Range in the Arctic Ocean, doing so on 20 October 1961,[13] just ten days before the gigantic 50 Mt Tsar Bomba's detonation in the same general area. The United States eventually conducted a similar test in the Pacific Ocean on 6 May 1962, with a Polaris A-2 launched from USS *Ethan Allen* (SSBN-608) as part of the nuclear test series Operation Dominic. The first Soviet SSBN with

15.1.2 Deployment and further development

The short range of the early SLBMs dictated basing and deployment locations. By the late 1960s the Polaris A-3 was deployed on all US SSBNs with a range of 4,600 kilometres (2,500 nmi), a great improvement on the 1,900 kilometres (1,000 nmi) range of Polaris A-1. The A-3 also had three warheads that landed in a pattern around a single target.[18][19] The Yankee class was initially equipped with the R-27 Zyb missile (SS-N-6) with a range of 2,400 kilometres (1,300 nmi). The US was much more fortunate in its basing arrangements than the Soviets. Thanks to NATO and the US possession of Guam, US SSBNs were permanently forward deployed at Advanced Refit Sites in Holy Loch, Scotland, Rota, Spain, and Guam by the middle 1960s, resulting in short transit times to patrol areas near the Soviet Union. The SSBN facilities at the Advanced Refit Sites were austere, with only a submarine tender and floating dry dock. Converted merchant ships designated T-AKs (Military Sealift Command cargo ships) were provided to ferry missiles and supplies to the sites. With two rotating crews per boat, about one-third of the total US force could be in a patrol area at any time. The Soviet bases, in the Murmansk area for the Atlantic and the Petropavlovsk-Kamchatsky area for the Pacific, required their SSBNs to make a long transit (through NATO-monitored waters in the Atlantic) to their mid-ocean patrol areas to hold the continental United States (CONUS) at risk. This resulted in only a small percentage of the Soviet force occupying patrol areas at any time, and was a great motivation for longer-range Soviet SLBMs, which would allow them to patrol close to their bases, in areas sometimes referred to as "deep bastions". These missiles were the R-29 Vysota series (SS-N-8, SS-N-18, SS-N-23), equipped on Projects 667B, 667BD, 667BDR, and 667BDRM (Delta-I through Delta-IV classes).[20] The SS-N-8, with a range of 7,700 kilometres (4,200 nmi), entered service on the first Delta-I boat in 1972, before the Yankee class was even completed. A total of 43 Delta-class boats of all types entered service 1972-90, with the SS-N-18 on the Delta III class and the R-29RM Shtil (SS-N-23) on the Delta IV class.[21][22][23][24] The new missiles had increased range and eventually Multiple Independently-targeted Re-entry Vehicles (MIRV), multiple warheads that could each hit a different target.[25]

15.1.3   Poseidon and Trident I

Although the US did not commission any new SSBNs from 1967 through 1981, they did introduce two new SLBMs. Thirty-one of the 41 original US SSBNs were built with larger diameter launch tubes with future missiles in mind. In the early 1970s the Poseidon (C-3) missile entered service, and those 31 SSBNs were backfitted with it.[26] Poseidon offered a massive MIRV capability of up to 14 warheads per missile.[27] Like the Soviets, the US also desired a longer-range missile that would allow SSBNs to be based in CONUS. In the late 1970s the Trident I (C-4) missile with a range of 7,400 kilometres (4,000 nmi) and eight MIRV warheads was backfitted to 12 of the Poseidon-equipped submarines.[28][29][30] The SSBN facilities (primarily a submarine tender and floating dry dock) of the base at Rota, Spain were disestablished and the Naval Submarine Base King's Bay in Georgia was built for the Trident I-equipped force.

15.1.4   Trident and Typhoon submarines

Both the United States and the Soviet Union commissioned larger SSBNs designed for new missiles in 1981. The American large SSBN was the *Ohio* class, also called the "Trident submarine", with the largest SSBN armament ever of 24 missiles, initially Trident I but built with much larger tubes for the Trident II (D-5) missile, which entered service in 1990.[31][32] The entire class was converted to use Trident II by the early 2000s. Trident II offered a range of over 8,000 kilometres (4,300 nmi) with eight larger MIRV warheads than Trident I. When the USS *Ohio* (SSBN-726) commenced sea trials in 1980, two of the first ten US

A Trident II missile just after launch.

SSBNs had their missiles removed to comply with SALT treaty requirements; the remaining eight were converted to attack submarines (SSN) by the end of 1982. These were all in the Pacific, and the Guam SSBN base was disestablished; the first several *Ohio*-class boats used new Trident facilities at Naval Submarine Base Bangor, Washington. Eighteen *Ohio*-class boats were commissioned by 1997,[33] four of which were converted as cruise missile submarines (SSGN) in the 2000s to comply with START I treaty requirements. The Soviet large SSBN was the Project 941 Akula, famous as the Typhoon-class (and not to be confused with the Project 971 Shchuka attack submarine, called "Akula" by NATO). The Typhoons were the largest submarines ever built at 48,000 tons submerged. They were armed with 20 of the new R-39 Rif (SS-N-20) missiles with a range of 8,300 kilometres (4,500 nmi) and 10 MIRV warheads. Six Typhoons were commissioned 1981-89.[34]

15.1.5   Post-Cold War

New SSBN construction terminated for over 10 years in Russia and slowed in the US with the collapse of the Soviet Union and the end of the Cold War in 1991. The US rapidly decommissioned its remaining 31 older SSBNs, with a few converted to other roles, and the base at Holy Loch was disestablished. Most of the former Soviet SSBN force was gradually scrapped under the provisions of the Nunn–Lugar Cooperative Threat Reduction agreement through 2012.[35]

By that time the Russian SSBN force stood at six Delta-IVs, three Delta-IIIs, and a lone Typhoon used as a testbed for new missiles (the R-39s unique to the Typhoons were reportedly scrapped in 2012). Upgraded missiles such as the R-29RMU Sineva (SS-N-23 Sineva) were developed for the Deltas. In 2013 the Russians commissioned the first Borei-class submarine, also called the *Dolgorukiy* class after the lead vessel. By 2015 two others had entered service. This class is intended to replace the aging Deltas, and carries 16 solid-fuel RSM-56 Bulava missiles, with a reported range of 10,000 kilometres (5,400 nmi) and six MIRV warheads. The US is designing a replacement for the *Ohio* class; however, as of early 2015 none have been laid down.

Ballistic missile submarines have been of great strategic importance for the USA, Russia, and other nuclear powers since they entered service in the Cold War, as they can hide from reconnaissance satellites and fire their nuclear weapons with virtual impunity. This makes them immune to a first strike directed against nuclear forces, allowing each side to maintain the capability to launch a devastating retaliatory strike, even if all land-based missiles have been destroyed. This relieves each side of the necessity to adopt a launch on warning posture, with its grave attendant risk of accidental nuclear war. Additionally, the deployment of highly accurate missiles on ultra-quiet submarines allows an attacker to sneak up close to the enemy coast and launch a missile on a depressed trajectory (a non-optimal ballistic trajectory which trades off reduced throw-weight for a faster and lower path, effectively reducing the time between launch and impact), thus opening the possibility of a decapitation strike.

15.2 Types of SLBMs

Montage of the launch of a Trident I C-4 SLBM and the paths of its reentry vehicles

Specific types of SLBMs (current, past and under development) include:

United States of America (also known as *Fleet Ballistic Missiles*)

- UGM-27 Polaris (A-1 through A-3) – Decommissioned

- UGM-73 Poseidon (C-3) – Decommissioned

- UGM-96 Trident I (C-4) – Decommissioned

- UGM-133 Trident II (D-5) – Operational

Soviet Union / Russian Federation

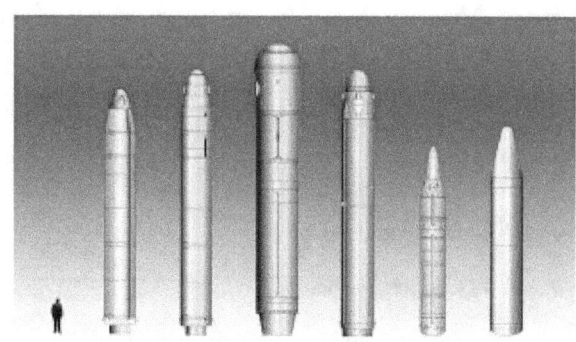

Selected Russian and Chinese SLBMs. L to R: R-29 Vysota (SS-N-8), R-29R (SS-N-18), R-39 (SS-N-20), R-29RM (SS-N-23), JL-1, JL-2

- R-13 NATO name SS-N-4 – Decommissioned

- R-21 NATO name SS-N-5 – Decommissioned

- RSM-25[36] R-27 NATO name SS-N-6 – Decommissioned

- RSM-40[36] R-29 "Vysota", NATO name SS-N-8 "Sawfly" – Decommissioned

- R-27K, NATO name SS-NX-13 – designed for anti-ship use, never operational[37]

- RSM-45 R-31 NATO name SS-N-17 "Snipe"[36] – Decommissioned

- RSM-50[36] R-29R "Vysota", NATO name SS-N-18 "Stingray" – Decommissioned

- RSM-52[36] R-39 "Rif", NATO name SS-N-20 "Sturgeon" – Decommissioned

- RSM-54 R-29RM "Shtil", NATO name SS-N-23 "Skiff" – Decommissioned (last ship is now under rebuild to R-29RMU "Sineva")[38]

- RSM-54 R-29RMU "Sineva", NATO name SS-N-23 "Skiff" – Operational

- RSM-54 R-29RMU2 "Layner" – Operational

- RSM-56 R-30 "Bulava", SS-NX-32[39] – Operational

United Kingdom

- UGM-27 Polaris (A-3) and Chevaline – Decommissioned

- UGM-133 Trident II (D-5) – Operational

France

- M1 – Decommissioned

- M2 – Decommissioned

- M20 – Decommissioned

- M4 – Decommissioned

- M45 – Operational

- M51 – Operational

People's Liberation Army Navy

- JL-1[40] Decommissioned

- JL-2 – Operational

India

- K-15 Sagarika -Operational on the Arihant class submarine.

- K-4 Test fired from Arihant class submarine.[41]

- K-5 – Under development

North Korea

- KN-11 – Under development

15.3   Non-military use

Some former Russian SLBMs have been converted into Volna and Shtil' launch vehicles to launch satellites – either from a submarine or from a launch site on land.

15.4   See also

- ICBM

- Nuclear warfare

- Vertical launching system

- Ballistic missile submarine

15.5   References

[1] Wade, Mark. "R-11". Encyclopedia Astronautica. Retrieved 23 April 2011.

[2] List of Project 611 submarines

[3] Friedman, pp. 192-195

[4] Friedman, pp. 192-195

[5] History of the Jupiter Missile, pp. 23-35

[6] Teller, Edward (2001). *Memoirs: A Twentieth Century Journey in Science and Politics*. Cambridge, Massachusetts: Perseus Publishing. pp. 420–421. ISBN 0-7382-0532-X.

[7] History of the Jupiter Missile, pp. 23-35

[8] Friedman, p. 196

[9] "Missiles 1963", *Flight International*: 752, 7 November 1963

[10] Dygalo, V.A. "Start razreshaju (in Russian)". Nauka i Zhizn'. Retrieved 23 April 2011.

[11] List of Project 658 submarines

[12] Gardiner and Chumbley, pp. 355-357

[13] Polmar, Norman; White, Michael (2010). *Project Azorian: The CIA and the Raising of the K-129*. Naval Institute Press. p. 21. ISBN 978-1-59114-690-2.

[14] Gardiner and Chumbley, p. 403

[15] List of Project 667A (Yankee class) submarines

[16] Gardiner and Chumbley, pp. 610-613

[17] Polmar American Submarine, p. 133

[18] Friedman, pp. 199-200

[19] Polmar American Submarine, pp. 131-133

[20] Gardiner and Chumbley, pp. 355-357

[21] List of Project 667B (Delta I class) submarines

[22] List of Project 667BD (Delta II class) submarines

[23] List of Project 667BDR (Delta III class) submarines

[24] List of Project 667BDRM (Delta IV class) submarines

[25] Gardiner and Chumbley, pp. 355-357

[26] Friedman, p. 201

[27] Polmar American Submarine, p. 133

[28] Gardiner and Chumbley, pp. 553-554

[29] Friedman, p. 206

[30] Polmar American Submarine, pp. 133-135

[31] Friedman, pp. 206-207

[32] Gardiner and Chumbley, p. 554

[33] Gardiner and Chumbley, p. 613

[34] List of Project 941 Typhoon-class submarines

[35] DTRA verification page

[36] Korabli VMF SSSR, Vol. 1, Part 1, Yu. Apalkov, Sankt Peterburg, 2003, ISBN 5-8172-0069-4

[37] SS-NX-13 SLBM System (U), Defense Intelligence Agency, D5T-1020S-4l7-75, 1 October 1975

[38] "SSBN K-51 Verkhoturye arrived to Zvezdochka for repairs today". Rusnavy.com. 23 August 2010. Retrieved 8 October 2010.

[39] NASIC-1031-0985-09

[40] "JL-1 [CSS-N-3] – China Nuclear Forces". Fas.org. Retrieved 10 February 2012.

[41] http://www.newindianexpress.com/nation/ EXPRESS-EXCLUSIVE-Maiden-Test-of-Undersea-K-4-Missile-From-Arihant-Submarine/ 2016/04/09/article3370608.ece

- Friedman, Norman (1994). *U.S. Submarines Since 1945: An Illustrated Design History*. Annapolis, Maryland: United States Naval Institute. ISBN 1-55750-260-9.

- Gardiner, Robert; Chumbley, Stephen (1995). *Conway's All the World's Fighting Ships 1947-1995*. London: Conway Maritime Press. ISBN 1-55750-132-7.

- Polmar, Norman (1981). *The American Submarine*. Annapolis, Maryland: Nautical and Aviation Publishing. pp. 123–136. ISBN 0-933852-14-2.

15.6 External links

- Navweaps.com US naval missiles index page

- Navweaps.com Soviet and Russian naval missiles index page

- Navweaps.com UK naval missiles index page

- Video showing the launch of a Trident SLBM.

- Estimated Strategic Nuclear Weapons Inventories (September 2004)

- R-11 SLBM

- Trident Submarines Are Killing Machines Unparalleled In Human History.

- NavSource.org SSBN photo gallery index

Chapter 16

Multiple independently targetable reentry vehicle

For the band, see M.I.R.V.

A **multiple independently targetable reentry vehicle**

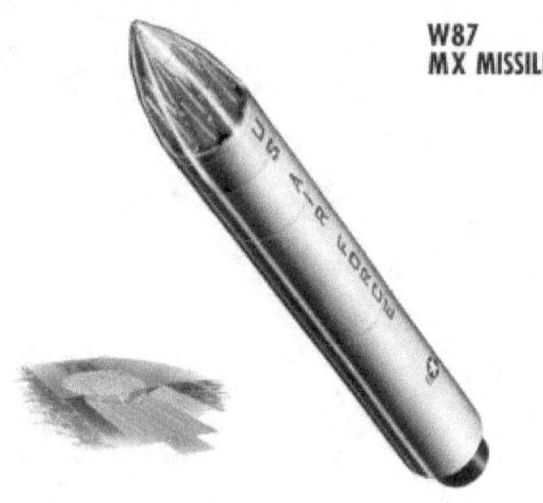

**W87
MX MISSILE**

The MIRV U.S. Peacekeeper missile, with the reentry vehicles highlighted in red.

Technicians secure a number of Mk21 reentry vehicles on a Peacekeeper MIRV bus.

(**MIRV**) is a ballistic missile payload containing several warheads, each capable of being aimed to hit one of a group of targets. By contrast a unitary warhead is a single warhead on a single missile. An intermediate case is the multiple reentry vehicle (MRV) missile which carries several warheads which are dispersed but not individually aimed. Only the United States, Russia, France, and China are known to have developed MIRV missiles.

16.1   Purpose

The military purpose of a MIRV is fourfold:

- Enhance first-strike proficiency for strategic forces.[2]

- Providing   greater   target   damage   for   a   given

thermonuclear weapon payload.   Several small warheads cause much more target damage area than a single warhead alone. This in turn reduces the number of missiles and launch facilities required for a given destruction level - much the same as the purpose of a cluster munition.[3]

- With single warhead missiles, one missile must be launched for each target. By contrast with a MIRV warhead, the post-boost (or bus) stage can dispense the warheads against multiple targets across a broad area.

- Reduces the effectiveness of an anti-ballistic missile system that relies on intercepting individual warheads.[4] While a MIRV attacking missile can have multiple warheads (3–12 on United States and Russian missiles), interceptors may have only one warhead per missile. Thus, in both a military and an economic sense, MIRVs render ABM systems less effective, as the costs of maintaining a workable defense

A Trident II missile, operated exclusively by the US Navy and Royal Navy. Each missile can carry up to 14 warheads.[1]

against MIRVs would greatly increase, requiring multiple defensive missiles for each offensive one. Decoy reentry vehicles can be used alongside actual warheads to minimize the chances of the actual warheads being intercepted before they reach their targets. A system that destroys the missile earlier in its trajectory (before MIRV separation) is not affected by this but is more difficult, and thus more expensive to implement.

MIRV land-based ICBMs were considered destabilizing because they tended to put a premium on striking first. The world's first MIRV—US Minuteman III missile of 1970—threatened to rapidly increase the US's deployable nuclear arsenal and thus the possibility that it would have enough bombs to destroy virtually all of the Soviet Union's nuclear weapons and negate any significant retaliation. Later on the US feared the Soviet's MIRVs because Soviet missiles had a greater throw-weight and could thus put more warheads on each missile than the US could. For example, the US MIRVs might have increased their warhead per missile count by a factor of 6 while the Soviets increased theirs by a factor of 10. Furthermore, the US had a much smaller proportion of its nuclear arsenal in ICBMs than the Soviets. Bombers could not be outfitted with MIRVs so their capacity would not be multiplied. Thus the US did not seem to have as much potential for MIRV usage as the Soviets. However, the US had a larger number of Submarine-launched ballistic missiles, which could be outfitted with

MIRVs, and helped offset the ICBM disadvantage. It is because of this that this type of weapon was banned under the START II agreement. However, START II was never ratified by the Russian Duma due to disagreements about the ABM treaty.

16.2 Mode of operation

In a MIRV, the main rocket motor (or booster) pushes a "bus" (see illustration) into a free-flight suborbital ballistic flight path. After the boost phase the bus maneuvers using small on-board rocket motors and a computerised inertial guidance system. It takes up a ballistic trajectory that will deliver a reentry vehicle containing a warhead to a target, and then releases a warhead on that trajectory. It then maneuvers to a different trajectory, releasing another warhead, and repeats the process for all warheads.

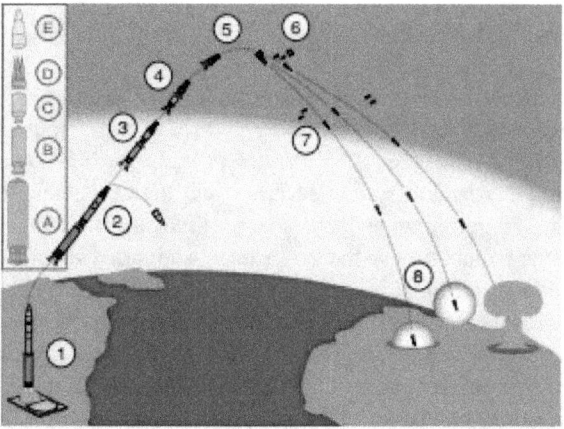

Minuteman III MIRV launch sequence: 1. The missile launches out of its silo by firing its first-stage boost motor (A). 2. About 60 seconds after launch, the 1st stage drops off and the second-stage motor (B) ignites. The missile shroud (E) is ejected. 3. About 120 seconds after launch, the third-stage motor (C) ignites and separates from the 2nd stage. 4. About 180 seconds after launch, third-stage thrust terminates and the post-boost vehicle (D) separates from the rocket. 5. The post-boost vehicle maneuvers itself and prepares for reentry vehicle (RV) deployment. 6. While the post-boost vehicle backs away, the RVs, decoys, and chaff are deployed (this may occur during ascent). 7. The RVs and chaff reenter the atmosphere at high speeds and are armed in flight. 8. The nuclear warheads detonate, either as air bursts or ground bursts.

The precise technical details are closely guarded military secrets, to hinder any development of enemy countermeasures. The bus' on-board propellant limits the distances between targets of individual warheads to perhaps a few hundred kilometers.[5] Some warheads may use small hypersonic airfoils during the descent to gain additional cross-range distance. Additionally, some buses (e.g. the

British Chevaline system) can release decoys to confuse interception devices and radars, such as aluminized balloons or electronic noisemakers.

Testing of the Peacekeeper reentry vehicles, all eight (ten capable) fired from only one missile. Each line represents the path of an individual warhead.

Accuracy is crucial, because doubling the accuracy decreases the needed warhead energy by a factor of four for radiation damage and by a factor of eight for blast damage. Navigation system accuracy and the available geophysical information limits the warhead target accuracy. Some writers believe that government-supported geophysical mapping initiatives and ocean satellite altitude systems such as Seasat may have a covert purpose to map mass concentrations and determine local gravity anomalies, in order to improve accuracies of ballistic missiles. Accuracy is expressed as circular error probable (CEP). This is simply the radius of the circle that the warhead has a 50 percent chance of falling into when aimed at the center. CEP is about 90–100 m for the Trident II and Peacekeeper missiles.[6]

16.3 MRV

A multiple reentry vehicle payload for a ballistic missile deploys multiple warheads in a pattern against a single target (as opposed to multiple independently targetable reentry vehicle, which deploys multiple warheads against multiple targets). The advantage of an MRV over a single warhead is that the damage produced in the center of the pattern is far greater than the damage possible from any single warhead in the MRV cluster, this makes for an efficient area attack weapon. The number of warheads makes interception by anti-ballistic missiles unlikely.

Improved warhead designs allow smaller warheads for a given yield, while better electronics and guidance systems allowed greater accuracy. As a result, MIRV technology has proven more attractive than MRV for advanced nations. Because of the larger amount of nuclear material consumed by MRVs and MIRVs, single warhead missiles are more attractive for nations with less advanced technology. The United States deployed an MRV payload on the Polaris A-3. The Soviet Union deployed MRVs on the R-36 Mod 4 ICBM. Refer to atmospheric reentry for more details.

16.4 See also

16.5 References

Notes

[1] Parsch, Andreas. "UGM-133". Directory of U.S. Military Rockets and Missiles. Retrieved 2014-06-13.

[2] "Multiple Independently Targetable Reentry Vehicles (MIRVs)". Retrieved 14 June 2014.

[3] The best overall printed sources on nuclear weapons design are: Hansen, Chuck. *U.S. Nuclear Weapons: The Secret History*. San Antonio, TX: Aerofax, 1988; and the more-updated Hansen, Chuck, "Swords of Armageddon: U.S. Nuclear Weapons Development since 1945" (CD-ROM & download available). PDF. 2,600 pages, Sunnyvale, California, Chukelea Publications, 1995, 2007. ISBN 978-0-9791915-0-3 (2nd Ed.)

[4] Robert C. Aldridge (1983). *First Strike!: The Pentagon's Strategy for Nuclear War*. South End Press. pp. 65–. ISBN 978-0-89608-154-3. Retrieved 26 February 2013.

[5] Question Re Mirv Warheads — Military Forum | Airliners.net

[6] Cimbala, Stephen J. (2010). *Military Persuasion: Deterrence and Provocation in Crisis and War*. Penn State Press. p. 86. ISBN 978-0-271-04126-1. Retrieved 3 May 2013.

16.6 External links

- "MIRV: A BRIEF HISTORY OF MINUTEMAN and MULTIPLE REENTRY VEHICLES" by Daniel Buchonnet, Lawrence Livermore Laboratory, February 1976.

- Operation 1964

- The Defense of the United States, 1981 CBS Five-Part TV Series from Google Video

Chapter 17

Nuclear strategy

Nuclear strategy involves the development of doctrines and strategies for the production and use of nuclear weapons.

As a sub-branch of military strategy, nuclear strategy attempts to match nuclear weapons as means to political ends. In addition to the actual use of nuclear weapons whether in the battlefield or strategically, a large part of nuclear strategy involves their use as a bargaining tool.

Some of the issues considered within nuclear strategy include:

- Under what conditions does it serve a nation's interest to develop nuclear weapons?

- What types of nuclear weapons should be developed?

- When and how should such weapons be used?

Many strategists argue that nuclear strategy differs from other forms of military strategy. The immense and terrifying power of the weapons makes their use, in seeking victory in a traditional military sense, impossible.

Perhaps counterintuitively, an important focus of nuclear strategy has been determining how to prevent and deter their use, a crucial part of mutual assured destruction.

In the context of nuclear proliferation and maintaining the balance of power, states also seek to prevent other states from acquiring nuclear weapons as part of nuclear strategy.

17.1 Nuclear deterrent composition

The doctrine of mutual assured destruction (MAD) assumes that a nuclear deterrent force must be credible and survivable. That is, each deterrent force must survive a first strike with sufficient capability to effectively destroy the other country in a second strike. Therefore, a first strike would be suicidal for the launching country.

In the late 1940s and 1950s as the Cold War developed, the United States and Soviet Union pursued multiple delivery methods and platforms to deliver nuclear weapons. Three types of platforms proved most successful and are collectively called a "nuclear triad". These are air-delivered weapons (bombs or missiles), ballistic missile submarines (usually nuclear-powered and called SSBNs), and intercontinental ballistic missiles (ICBMs), usually deployed in land-based hardened missile silos or on vehicles.

Although not considered part of the deterrent forces, all of the nuclear powers deployed large numbers of tactical nuclear weapons in the Cold War. These could be delivered by virtually all platforms capable of delivering large conventional weapons.

During the 1970s there was growing concern that the combined conventional forces of the Soviet Union and the Warsaw Pact could overwhelm the forces of NATO. It seemed unthinkable to respond to a Soviet/Warsaw Pact incursion into Western Europe with strategic nuclear weapons, inviting a catastrophic exchange. Thus, technologies were developed to greatly reduce collateral damage while being effective against advancing conventional military forces. Some of these were low-yield neutron bombs, which were lethal to tank crews, especially with tanks massed in tight formation, while producing relatively little blast, thermal radiation, or radioactive fallout. Other technologies were so-called "suppressed radiation devices," which produced mostly blast with little radioactivity, making them much like conventional explosives, but with much more energy.[1]

17.2 See also

- Military strategy

- Counterforce, Countervalue

- Decapitation strike

- Deterrence

- Doctrine for Joint Nuclear Operations

- Fail-deadly

- Force de frappe

- First strike, Second strike

- Game theory & wargaming

- Madman theory

- Massive retaliation

- Minimal deterrence

- Mutual assured destruction (MAD)

- Assured destruction

- No first use

- National Security Strategy of the United States

- Nuclear blackmail

- Nuclear proliferation

- Nuclear utilization target selection (NUTS)

- Nuclear weapons debate

- Single Integrated Operational Plan (SIOP)

- Strategic bombing

- Tactical nuclear weapons

- Bernard Brodie

- Herman Kahn

- Stanley Kubrick's *Dr. Strangelove* (1964), a film satirizing nuclear strategy.

- Thomas Schelling

17.3   Bibliography

17.3.1   Early texts

- Brodie, Bernard. *The Absolute Weapon*. Freeport, N.Y.: Books for Libraries Press, 1946.

- Brodie, Bernard. *Strategy in the Missile Age*. Princeton: Princeton University Press, 1959.

- Dunn, Lewis A. *dead link]Deterrence Today – Roles, Challenges, and Responses* Paris: IFRI Proliferation Papers n° 19, 2007.

- Kahn, Herman. *On Thermonuclear War*. 2nd ed. Princeton, N.J.: Princeton University Press, 1961.

- Kissinger, Henry A. *Nuclear Weapons and Foreign Policy*. New York: Harper, 1957.

- Schelling, Thomas C. *Arms and Influence*. New Haven: Yale University Press, 1966.

- Wohlstetter, Albert. "The Delicate Balance of Terror." *Foreign Affairs* 37, 211 (1958): 211–233.

17.3.2   Secondary literature

- Baylis, John, and John Garnett. *Makers of Nuclear Strategy*. London: Pinter, 1991. ISBN 1-85567-025-9.

- Buzan, Barry, and Herring, Eric. "The Arms Dynamic in World Politics". London: Lynne Rienner Publishers, 1998. ISBN 1-55587-596-3.

- Freedman, Lawrence. *The Evolution of Nuclear Strategy*. 2nd ed. New York: St. Martin's Press, 1989. ISBN 0-333-97239-2 .

- Heuser, Beatrice. *NATO, Britain, France and the FRG: Nuclear Strategies and Forces for Europe, 1949–2000* (London: Macmillan, hardback 1997, paperback 1999), 256p., ISBN 0-333-67365-4

- Heuser, Beatrice. *Nuclear Mentalities? Strategies and Belief Systems in Britain, France and the FRG* (London: Macmillan, July 1998), 277p., Index, Tables. ISBN 0-333-69389-2

- Heuser, Beatrice. "Victory in a Nuclear War? A Comparison of NATO and WTO War Aims and Strategies", *Contemporary European History* Vol. 7 Part 3 (November 1998), pp. 311–328.

- Heuser, Beatrice. "Warsaw Pact Military Doctrines in the 70s and 80s: Findings in the East German Archives", *Comparative Strategy* Vol. 12 No. 4 (Oct.–Dec. 1993), pp. 437–457.

- Kaplan, Fred M. *The Wizards of Armageddon*. New York: Simon and Schuster, 1983. ISBN 0-671-42444-0.

- Rai Chowdhuri, Satyabrata. *Nuclear Politics: Towards A Safer World*, Ilford: New Dawn Press, 2004.

- Rosenberg, David. "The Origins of Overkill: Nuclear Weapons and American Strategy, 1945–1960." *International Security* 7, 4 (Spring, 1983): 3–71.

- Schelling, Thomas C. *The Strategy of Conflict.* Cambridge: Harvard University Press, 1960.

- Smoke, Richard. National Security and the Nuclear Dilemma. 3rd ed. New York: McGraw–Hill, 1993. ISBN 0-07-059352-3.

17.4   References

[1] Solem, J. C. (1974). "Tactical nuclear deterrence". *Los Alamos Scientific Laboratory Report LA-74-1362.*

Chapter 18

Nuclear warfare

"Nuclear War" redirects here. For other uses, see Nuclear War (disambiguation).
"Atomic war" redirects here. It is not to be confused with Atomic Wars.

Nuclear warfare (sometimes **atomic warfare** or **ther-**

The Titan II Intercontinental ballistic missile (ICBM) carried a 9 Mt W53 warhead, one of the most powerful nuclear weapons fielded by the United States during the Cold War.

monuclear warfare) is a military conflict or political strategy in which nuclear weaponry is used to inflict damage on the enemy. Compared to conventional warfare, nuclear warfare can be vastly more destructive in range and extent of damage, and in a much shorter time. A major nuclear exchange would have long-term effects, primarily from the fallout released, and could also lead to a "nuclear winter"

that could last for decades, centuries, or even millennia after the initial attack.[1][2] Some analysts claim that with this potential nuclear winter side-effect of a nuclear war almost every human on Earth could starve to death.[3][4] Other analysts, who dismiss the nuclear winter hypothesis, calculate that with nuclear weapon stockpiles at Cold War highs, in a surprise countervalue global nuclear war, billions of casualties would have resulted but billions of people would nevertheless have survived.[5][6][7][8]

So far, two nuclear weapons have been used in the course of warfare, both by the United States near the end of World War II. On August 6, 1945, a uranium gun-type device (code name "Little Boy") was detonated over the Japanese city of Hiroshima. Three days later, on August 9, a plutonium implosion-type device (code name "Fat Man") was detonated over the Japanese city of Nagasaki. These two bombings resulted in the deaths of approximately 120,000 people.

After World War II, nuclear weapons were also developed by the Soviet Union (1949), the United Kingdom (1952), France (1960), and the People's Republic of China (1964), which contributed to the state of conflict and extreme tension that became known as the Cold War. In 1974, India, and in 1998, Pakistan, two countries that were openly hostile toward each other, developed nuclear weapons. Israel (1960s) and North Korea (2006) are also thought to have developed stocks of nuclear weapons, though it is not known how many. The Israeli government has never admitted to having nuclear weapons, although it is known to have constructed the reactor and reprocessing plant necessary for building nuclear weapons.[9] South Africa also manufactured several complete nuclear weapons in the 1980s, but subsequently became the first country to voluntarily destroy their domestically made weapons stocks and abandon further production (1990s).[10] Nuclear weapons have been detonated on over 2,000 occasions for testing purposes and demonstrations.[11][12]

After the collapse of the Soviet Union in 1991 and the resultant end of the Cold War, the threat of a major nuclear war between the two nuclear superpowers was generally

156

thought to have declined. Since then, concern over nuclear weapons has shifted to the prevention of localized nuclear conflicts resulting from nuclear proliferation, and the threat of nuclear terrorism.

18.1 Types of nuclear warfare

The possibility of using nuclear weapons in war is usually divided into two subgroups, each with different effects and potentially fought with different types of nuclear armaments.

The first, *limited nuclear war* [13] (sometimes *attack* or *exchange*), refers to a small-scale use of nuclear weapons by two (or more) belligerents. A "limited nuclear war" could include targeting military facilities—either as an attempt to pre-emptively cripple the enemy's ability to attack as a defensive measure, or as a prelude to an invasion by conventional forces, as an offensive measure. This term could apply to *any* small-scale use of nuclear weapons that may involve military or civilian targets (or both).

The second, *full-scale nuclear war*, could consist of large numbers of nuclear weapons used in an attack aimed at an entire country, including military, economic, and civilian targets. Such an attack would almost certainly destroy the entire economic, social, and military infrastructure of the target nation, and would probably have a devastating effect on Earth's biosphere.

Some Cold War strategists such as Henry Kissinger[14] argued that a limited nuclear war *could* be possible between two heavily armed superpowers (such as the United States and the Soviet Union). Some predict, however, that a limited war could potentially "escalate" into a full-scale nuclear war. Others have called limited nuclear war "global nuclear holocaust in slow motion", arguing that—once such a war took place—others would be sure to follow over a period of decades, effectively rendering the planet uninhabitable in the same way that a "full-scale nuclear war" between superpowers would, only taking a much longer (and arguably more agonizing) path to the same result.

Even the most optimistic predictions of the effects of a major nuclear exchange foresee the death of many millions of victims within a very short period of time. More pessimistic predictions argue that a full-scale nuclear war could potentially bring about the extinction of the human race, or at least its *near* extinction, with only a relatively small number of survivors (mainly in remote areas) and a reduced quality of life and life expectancy for centuries afterward. However, such predictions, assuming total war with nuclear arsenals at Cold War highs, have not been without criticism.[5] Such a horrific catastrophe as global nuclear warfare would almost certainly cause permanent damage to most complex life on the planet, its ecosystems, and the global climate. If predictions about the production of a nuclear winter are accurate, it would also change the balance of global power, with countries such as Australia, New Zealand, India, China, Argentina and Brazil predicted to become world superpowers if the Cold War ever led to a large-scale nuclear attack.[6]

A study presented at the annual meeting of the American Geophysical Union in December 2006 asserted that even a small-scale regional nuclear war could produce as many direct fatalities as all of World War II and disrupt the global climate for a decade or more. In a regional nuclear conflict scenario in which two opposing nations in the subtropics each used 50 Hiroshima-sized nuclear weapons (c. 15 kiloton each) on major population centers, the researchers predicted fatalities ranging from 2.6 million to 16.7 million per country. The authors of the study estimated that as much as five million tons of soot could be released, producing a cooling of several degrees over large areas of North America and Eurasia (including most of the grain-growing regions). The cooling would last for years and could be "catastrophic", according to the researchers.[15]

Either a limited or full-scale nuclear exchange could occur during an *accidental nuclear war*, in which the use of nuclear weapons is triggered unintentionally. Postulated triggers for this scenario have included malfunctioning early warning devices and/or targeting computers, deliberate malfeasance by rogue military commanders, consequences of an accidental straying of warplanes into enemy airspace, reactions to unannounced missile tests during tense diplomatic periods, reactions to military exercises, mistranslated or miscommunicated messages, and others. A number of these scenarios actually occurred during the Cold War, though none resulted in the use of nuclear weapons.[16] Many such scenarios have been depicted in popular culture, such as in the 1962 novel *Fail-Safe* (released as a film in 1964), the film *WarGames*, released in 1983 and the film *Dr. Strangelove or: How I Learned to Stop Worrying and Love the Bomb*, also released in 1964.

18.2 History

Main articles: History of nuclear weapons and Timeline of nuclear weapons development

18.2.1 1940s

Atomic bombings of Hiroshima and Nagasaki

Main article: Atomic bombings of Hiroshima and Nagasaki
During the final stages of World War II in 1945, the United

Mushroom cloud from the atomic explosion over Nagasaki rising 18,000 m (59,000 ft) into the air on the morning of August 9, 1945.

States conducted atomic raids on the Japanese cities of Hiroshima and Nagasaki, the first on August 6, 1945, and the second on August 9, 1945. These two events were the only times nuclear weapons have been used in combat.[17]

For six months before the atomic bombings, the U.S. 20th Air Force under General Curtis LeMay executed low-level incendiary raids against Japanese cities. The worst air raid to occur during the process was not the nuclear attacks, but the *Operation Meetinghouse* raid on Tokyo. On the night of March 9–10, 1945, *Operation Meetinghouse* commenced and 334 Boeing B-29 Superfortress bombers took off to raid, with 279 of them dropping 1,665 tons of incendiaries and explosives on Tokyo. The bombing was meant to burn wooden buildings and indeed the bombing caused fire that created a 50 m/s wind, which is comparable to tornadoes. Each bomber carried 6 tons of bombs. A total of 381,300 bombs, which amount to 1,783 tons of bombs, were used in the bombing. Within a few hours of the raid, it had killed an estimated 100,000 people and destroyed 41 km^2 (16 sq mi) of the city and 267,000 buildings in a single night — the deadliest bombing raid in military aviation history other than the atomic raids on Hiroshima and Nagasaki.[18][19][20][21] By early August 1945, an esti-

mated 450,000 people had died as the U.S. had intensely firebombed a total of 67 Japanese cities.

In late June 1945, as the U.S. wrapped up the two-and-a-half-month Battle of Okinawa (which cost the lives of 260,000 people, including 150,000 civilians),[22][23] it was faced with the prospect of invading the Japanese home islands in an operation codenamed Operation Downfall. Based on the U.S. casualties from the preceding island-hopping campaigns, American commanders estimated that between 50,000 and 500,000 U.S. troops would die and at least 600,000-1,000,000 others would be injured while invading the Japanese home islands. The U.S. manufacture of 500,000 Purple Hearts from the anticipated high level of casualties during the U.S. invasion of Japan gave a demonstration of how deadly and costly it would be. President Harry S. Truman realized he could not afford such a horrendous casualty rate, especially since over 400,000 American combatants had already died fighting in both the European and the Pacific theaters of the war.[24]

On July 26, 1945, the United States, the United Kingdom, and the Republic of China issued a Potsdam Declaration that called for the unconditional surrender of Japan. It stated that if Japan did not surrender, it would face "prompt and utter destruction."[25][26] The Japanese government ignored this ultimatum, sending a message that they were not going to surrender. In response to the rejection, President Truman authorized the dropping of the atomic bombs. At the time of its use, there were only two atomic bombs available, and despite the fact that more were in production back in mainland U.S., the third bomb wouldn't be available for combat until September.[27][28]

On August 6, 1945, the uranium-type nuclear weapon codenamed "Little Boy" was detonated over the Japanese city of Hiroshima with an energy of about 15 kilotons of TNT (63,000 gigajoules), destroying nearly 50,000 buildings (including the headquarters of the 2nd General Army and Fifth Division) and killing approximately 70,000 people, including 20,000 Japanese combatants and 20,000 Korean slave laborers.[29][30] Three days later, on August 9, a plutonium-type nuclear weapon codenamed "Fat Man" was used against the Japanese city of Nagasaki, with the explosion equivalent to about 20 kilotons of TNT (84,000 gigajoules), destroying 60% of the city and killing approximately 35,000 people, including 23,200-28,200 Japanese munitions workers, 2,000 Korean slave laborers, and 150 Japanese combatants.[31] The industrial damage in Nagasaki was high, partly owing to the inadvertent targeting of the industrial zone, leaving 68-80 percent of the non-dock industrial production destroyed.[32]

Six days after the detonation over Nagasaki, Japan announced its surrender to the Allied Powers on August 15, 1945, signing the Instrument of Surrender on September

2, 1945, officially ending the Pacific War and, therefore, World War II, as Germany had already signed its Instrument of Surrender on May 7, 1945, ending the war in Europe. The two atomic bombings led, in part, to post-war Japan's adopting of the Three Non-Nuclear Principles, which forbade the nation from developing nuclear armaments.[33]

Immediately after the Japan bombings

Immediately after the atomic bombings of Japan, the status of atomic weapons in international and military relations was unclear. Presumably, the United States hoped atomic weapons could offset the Soviet Union's larger conventional ground forces in Eastern Europe, and possibly be used to pressure Soviet leader Joseph Stalin into making concessions. Under Stalin, the Soviet Union pursued its own atomic capabilities through a combination of scientific research and espionage directed against the American program. The Soviets believed that the Americans, with their limited nuclear arsenal, were unlikely to engage in any new world wars, while the Americans were not confident they could prevent a Soviet takeover of Europe, despite their atomic advantage.

Within the United States the authority to produce and develop nuclear weapons was removed from military control and put instead under the civilian control of the United States Atomic Energy Commission. This decision reflected an understanding that nuclear weapons had unique risks and benefits that were separate from other military technology known at the time.

Convair B-36 bomber

For several years after World War II, the United States developed and maintained a strategic force based on the Convair B-36 bomber that would be able to attack any potential enemy from bomber bases in the United States. It deployed atomic bombs around the world for potential use in conflicts. Over a period of a few years, many in the American defense community became increasingly convinced of the invincibility of the United States to a nuclear attack. Indeed, it became generally believed that the threat of nuclear

war would deter any strike against the United States.

Many proposals were suggested to put all American nuclear weapons under international control (by the newly formed United Nations, for example) as an effort to deter both their usage and an arms race. However, no terms could be arrived at that would be agreed upon by both the United States and the Soviet Union.

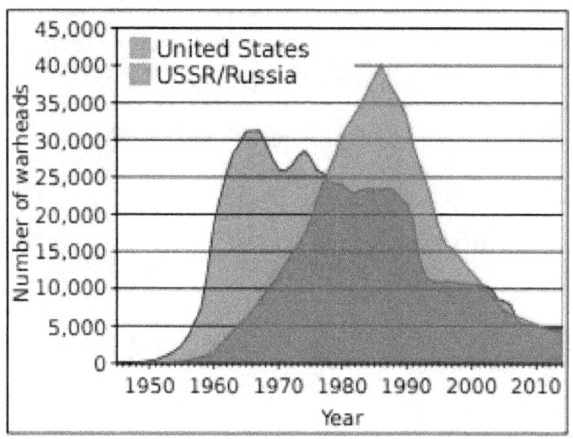

American and Soviet nuclear stockpiles.

On August 29, 1949, the Soviet Union tested its first nuclear weapon at Semipalatinsk in Kazakhstan (see also Soviet atomic bomb project). Scientists in the United States from the Manhattan Project had warned that, in time, the Soviet Union would certainly develop nuclear capabilities of its own. Nevertheless, the effect upon military thinking and planning in the United States was dramatic, primarily because American military strategists had not anticipated the Soviets would "catch up" so soon. However, at this time, they had not discovered that the Soviets had conducted significant nuclear espionage of the project from spies at Los Alamos, the most significant of which was done by the theoretical physicist Klaus Fuchs. The first Soviet bomb was more or less a deliberate copy of the Fat Man plutonium device.

With the monopoly over nuclear technology broken, worldwide nuclear proliferation accelerated. The United Kingdom tested its first independent atomic bomb in 1952, followed by France in 1960 and then China in 1964. While much smaller than the arsenals of the United States and the Soviet Union, Western Europe's nuclear reserves were nevertheless a significant factor in strategic planning during the Cold War. A top-secret White Paper, compiled by the Royal Air Force and produced for the British Government in 1959, estimated that British bombers carrying nuclear weapons were capable of destroying key cities and military targets in the Soviet Union, with an estimated 16 million deaths in the Soviet Union (half of whom were estimated to be killed on impact and the rest fatally injured) *before* bomber aircraft from the U.S. Strategic Air Command reached their targets.

After the successful Trinity nuclear test July 16, 1945, which was the very first nuclear detonation, the Manhattan project lead manager J. Robert Oppenheimer recalled:

> We knew the world would not be the same. A few people laughed, a few people cried, most people were silent. I remembered the line from the Hindu scripture the *Bhagavad Gita*. Vishnu is trying to persuade the prince that he should do his duty and to impress him takes on his multiarmed form and says, "Now, I am become Death, the destroyer of worlds." I suppose we all thought that one way or another."
> — J. Robert Oppenheimer, *The Decision To Drop The Bomb*[34]

18.2.2   1950s

Although the Soviet Union had nuclear weapon capabilities in the beginning of the Cold War, the United States still had an advantage in terms of bombers and weapons. In any exchange of hostilities, the United States would have been capable of bombing the Soviet Union, whereas the Soviet Union would have more difficulty carrying out the reverse mission.

The widespread introduction of jet-powered interceptor aircraft upset this imbalance somewhat by reducing the effectiveness of the American bomber fleet. In 1949 Curtis LeMay was placed in command of the Strategic Air Command and instituted a program to update the bomber fleet to one that was all-jet. During the early 1950s the B-47 and B-52 were introduced, providing the ability to bomb the Soviet Union more easily. Before the development of a capable strategic missile force in the Soviet Union, much of the war-fighting doctrine held by western nations revolved around using a large number of smaller nuclear weapons used in a tactical role. It is debatable whether such use could be considered "limited" however, because it was believed that the United States would use its own strategic weapons (mainly bombers at the time) should the Soviet Union deploy any kind of nuclear weapon against civilian targets. Douglas MacArthur, an American general, was fired by President Harry Truman, partially because he persistently requested permission to use his own discretion in deciding whether to use atomic weapons on the People's Republic of China in 1951 during the Korean War.[35] Mao Zedong, China's communist leader, gave the impression that he would welcome a nuclear war with the capitalists because it would annihilate what he viewed as their "imperialist"

system.[36][37]

> Let us imagine how many people would die if war breaks out. There are 2.7 billion people in the world, and a third could be lost. If it is a little higher it could be half ... I say that if the worst came to the worst and one-half dies, there will still be one-half left, but imperialism would be razed to the ground and the whole world would become socialist. After a few years there would be 2.7 billion people again.
> — Mao Zedong, 1957 [38]

The U.S. and USSR conducted hundreds of nuclear tests, including the Desert Rock exercises at the Nevada Test Site, USA, pictured above during the Korean War.

The concept of a "Fortress North America" emerged during the Second World War and persisted into the Cold War to refer to the option of defending Canada and the United States against their enemies if the rest of the world were lost to them. This option was rejected with the formation of NATO and the decision to permanently station troops in Europe.

In the summer of 1951 Project Vista started, in which project analysts such as Robert F. Christy looked at how to defend Western Europe from a Soviet invasion. The emerging development of tactical nuclear weapons were looked upon as a means to give Western forces a qualitative advantage over the Soviet numerical supremacy in conventional weapons.[39]

Several scares about the increasing ability of the Soviet Union's strategic bomber forces surfaced during the 1950s. The defensive response by the United States was to deploy a fairly strong "layered defense" consisting of interceptor aircraft and anti-aircraft missiles, like the Nike, and guns, like the Skysweeper, near larger cities. However, this was a small response compared to the construction of a huge fleet of nuclear bombers. The principal nuclear strategy was to massively penetrate the Soviet Union. Because such a large area could not be defended against this overwhelming attack in any credible way, the Soviet Union would lose any exchange.

This logic became ingrained in American nuclear doctrine and persisted for much of the duration of the Cold War. As long as the strategic American nuclear forces could overwhelm their Soviet counterparts, a Soviet pre-emptive strike could be averted. Moreover, the Soviet Union could not afford to build any reasonable counterforce, as the economic output of the United States was far larger than that of the Soviets, and they would be unable to achieve "nuclear parity".

Soviet nuclear doctrine, however, did not match American nuclear doctrine.[40][41] Soviet military planners assumed they could win a nuclear war.[40][42][43] Therefore, they *expected* a large-scale nuclear exchange, followed by a "conventional war" which itself would involve heavy use of tactical nuclear weapons. American doctrine rather assumed that Soviet doctrine was similar, with the *mutual* in Mutually Assured Destruction necessarily requiring that the other side see things in much the same way, rather than believing—as the Soviets did—that they could fight a large-scale, "combined nuclear and conventional" war.

In accordance with their doctrine, the Soviet Union conducted large-scale military exercises to explore the possibility of defensive and offensive warfare during a nuclear war. The exercise, under the code name of "Snowball", involved the detonation of a nuclear bomb about twice as powerful as that which fell on Nagasaki and an army of approximately 45,000 soldiers on maneuvers through the hypocenter immediately after the blast.[44] The exercise was conducted on September 14, 1954, under command of Marshal Georgy Zhukov to the north of Totskoye village in Orenburg Oblast, Russia.

A revolution in nuclear strategic thought occurred with the introduction of the intercontinental ballistic missile (ICBM), which the Soviet Union first successfully tested in August 1957. In order to deliver a warhead to a target, a missile was much faster and more cost-effective than a bomber, and enjoyed a higher survivability due to the enormous difficulty of interception of the ICBMs (due to their high altitude and extreme speed). The Soviet Union could now afford to achieve nuclear parity with the United States in raw numbers, although for a time, they appeared to have chosen not to.

Photos of Soviet missile sites set off a wave of panic in the U.S. military, something the launch of Sputnik would do for the American public a few months later. Politicians,

notably then-U.S. Senator John F. Kennedy suggested that a "missile gap" existed between the Soviet Union and the United States. The US military gave missile development programs the highest national priority, and several spy aircraft and reconnaissance satellites were designed and deployed to observe Soviet progress.

Early ICBMs and bombers were relatively inaccurate, which led to the concept of countervalue strikes — attacks directly on the enemy population, which would theoretically lead to a collapse of the enemy's will to fight. During the Cold War, the Soviet Union invested in extensive protected civilian infrastructure, such as large "nuclear-proof" bunkers and non-perishable food stores. By comparison, smaller scale civil defense programs were instituted in the United States starting in the 1950s, where schools and other public buildings had basements stocked with non-perishable food supplies, canned water, first aid, and dosimeter and Geiger counter radiation-measuring devices. Many of the locations were given "Fallout shelter" designation signs. CONELRAD radio information systems were adopted, whereby the commercial radio sector (later supplemented by the National Emergency Alarm Repeaters) would broadcast on two AM frequencies in the event of a Civil Defense (CD) emergency. These two frequencies, 640 and 1240 kHz, were marked with small CD triangles on the tuning dial of radios of the period, as can still be seen on 1950s-vintage radios on online auction sites and museums. A few backyard fallout shelters were built by private individuals.

Henry Kissinger's view on tactical nuclear war in his controversial 1957 book Nuclear Weapons and Foreign Policy was that any nuclear weapon exploded in air burst mode that was below 500 kiloton in yield and thus averting serious fallout, may be more decisive and less costly in human lives than a protracted conventional war.

A list of targets made by the U.S.A. was released sometime during December 2015 by the U.S. National Archives and Records Administration. The language used to describe targets is "designated ground zeros". The list was released after a request was made during 2006 by William Burr who belongs to a research group at George Washington University, and belongs to a previously top-secret 800-page document. The list is entitled "Atomic Weapons Requirements Study for 1959" and was produced by U.S. Strategic Air Command during the year 1956.[45]

18.2.3   1960s

In 1960, the United States developed its first Single Integrated Operational Plan, a range of targeting options, and described launch procedures and target sets against which nuclear weapons would be launched, variants of which were

RF-101 Voodoo reconnaissance photograph of the MRBM launch site in San Cristóbal, Cuba (1962)

in use from 1961 to 2003. That year also saw the start of the Missile Defense Alarm System, an American system of 12 early-warning satellites that provided limited notice of Soviet intercontinental ballistic missile launches between 1960 and 1966. The Ballistic Missile Early Warning System was completed in 1964.

A complex and worrisome situation developed in 1962, in what is called the Cuban Missile Crisis. The Soviet Union placed medium-range ballistic missiles 90 miles (140 km) from the United States, possibly as a direct response to American Jupiter missiles placed in Turkey. After intense negotiations, the Soviets ended up removing the missiles from Cuba and decided to institute a massive weapons-building program of their own. In exchange, the United States dismantled its launch sites in Turkey, although this was done secretly and not publicly revealed for over two decades. Khrushchev did not even reveal this part of the agreement when he came under fire by political opponents for mishandling the crisis. Communication delays during the crisis led to the establishment of the Moscow–Washington hotline to allow reliable, direct communications between the two nuclear powers.

By the late 1960s, the number of ICBMs and warheads was so high on both sides that it was believed that both the United States and the Soviet Union were capable of completely destroying the infrastructure and a large proportion of the population of the other country. Thus, by some western game theorists, a balance of power system known as mutually assured destruction (or *MAD*) came into being. It was thought that no full-scale exchange between the powers would result in an outright winner, with at best one side emerging the pyrrhic victor. Thus both sides were deterred from risking the initiation of a direct confrontation, instead being forced to engage in lower intensity proxy wars.

During this decade the People's Republic of China began to build subterranean infrastructure such as the Underground Project 131 following the Sino-Soviet split.

One drawback of the MAD doctrine was the possibility of a nuclear war occurring without either side intentionally striking first. Early Warning Systems (EWS) were notoriously error-prone. For example, on 78 occasions in 1979 alone, a "missile display conference" was called to evaluate detections that were "potentially threatening to the North American continent". Some of these were trivial errors and were spotted quickly, but several went to more serious levels. On September 26, 1983, Stanislav Petrov received convincing indications of an American first strike launch against the Soviet Union, but positively identified the warning as a false alarm. Though it is unclear what role Petrov's actions played in preventing a nuclear war during this incident, he has been honored by the United Nations for his actions.

Similar incidents happened many times in the United States, due to failed computer chips,[46] misidentifications of large flights of geese, test programs, and bureaucratic failures to notify early warning military personnel of legitimate launches of test or weather missiles. For many years, the U.S. Air Force's strategic bombers were kept airborne on a daily rotating basis "around the clock" (see Operation Chrome Dome), until the number and severity of accidents, the 1968 Thule Air Base B-52 crash in particular, persuaded policymakers it was not worthwhile.

18.2.4   1970s

Israel responded to the Arab Yom Kippur War attack on 6 October 1973 by assembling 13 nuclear weapons in a tunnel under the Negev desert when Syrian tanks were sweeping in across the Golan Heights. On 8 October 1973, Israeli Prime Minister Mrs Golda Meir authorized Defense Minister Moshe Dayan to activate the 13 Israeli nuclear warheads and distribute them to Israeli air force units, with the intent that they be used if Israel began to be overrun.[47]

On 24 October 1973 when US President Nixon was preoccupied with the Watergate scandal Henry Kissinger ordered a DEFCON−3 alert preparing American B-52 nuclear bombers for war, after intelligence reports indicated that the USSR was preparing to defend Egypt in its Yom Kippur war with Israel. It had become apparent that if Israel had dropped nuclear weapons on Egypt or Syria, as it prepared to do, then the USSR would have retaliated against Israel, with the US then committed to providing Israeli assistance, possibly escalating to a general nuclear war.[47]

By the late 1970s, people in both the United States and the Soviet Union, along with the rest of the world, had been living with the concept of mutual assured destruction (MAD) for about a decade, and it became deeply ingrained into the psyche and popular culture of those countries.

On May 18, 1974, India conducted its first nuclear test in the Pokhran test range. The name of the operation was Smiling Buddha, and India termed the test as a "peaceful nuclear explosion".

The Soviet Duga-3 early warning over-the-horizon radar system was made operational in 1976. The extremely powerful radio transmissions needed for such a system led to much disruption of civilian shortwave broadcasts, earning it the nickname "Russian Woodpecker".

The idea that any nuclear conflict would eventually escalate was a challenge for military strategists. This challenge was particularly severe for the United States and its NATO allies because it was believed (until the 1970s) that a Soviet tank invasion of Western Europe would quickly overwhelm NATO conventional forces, leading to the necessity of the West escalating to the use of tactical nuclear weapons, one of which was the W-70.

This strategy had one major (and possibly critical) flaw, which was soon realized by military analysts but highly underplayed by the U.S. military: conventional NATO forces in the European theatre of war were far outnumbered by similar Soviet and Warsaw Pact forces, and it was assumed that in case of a major Soviet attack (commonly envisioned as the "Red tanks rolling towards the North Sea" scenario) that NATO—in the face of quick conventional defeat—would soon have no other choice but to resort to tactical nuclear strikes against these forces. Most analysts agreed that once the first nuclear exchange had occurred, escalation to global nuclear war would likely become inevitable. The Soviet bloc's vision of an atomic war between NATO and Warsaw Pact forces was simulated in the top secret exercise Seven Days to the River Rhine in 1979. The British government exercised their vision of Soviet nuclear attack with Square Leg in early 1980.

Large hardened nuclear weapon storage areas were built across European countries in anticipation of local US and European forces falling back as the conventional NATO defense from the Soviet Union, named REFORGER, was believed to only be capable of stalling the Soviets for a short time.

18.2.5   1980s

In the late 1970s and, particularly, during the early 1980s under U.S. President Ronald Reagan, the United States renewed its commitment to a more powerful military, which

Montage of the launch of a Trident C4 SLBM and the paths of its reentry vehicles.

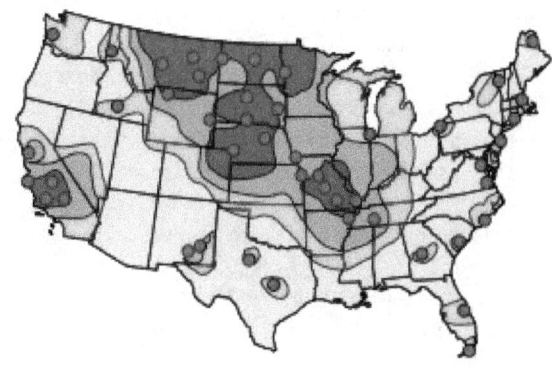

FEMA-estimated primary counterforce targets for Soviet ICBMs in 1990. The resulting fall-out is indicated with the darkest considered as lethal to lesser fall-out yellow zones.[48]

required a large increase in spending on U.S. military programs. These programs, which were originally part of the defense budget of U.S. President Jimmy Carter, included spending on conventional and nuclear weapons systems. Under Reagan, defensive systems like the Strategic Defense Initiative were emphasized as well.

Another major shift in nuclear doctrine was the development and the improvement of the submarine-launched, nuclear-armed, ballistic missile, or SLBM. It was hailed by many military theorists as a weapon that would make nuclear war less likely. SLBMs—which can move with "stealth" (greatly lessened detectability) virtually anywhere in the world—give a nation a "second strike" capability (i.e., after absorbing a "first strike"). Before the advent of the SLBM, thinkers feared that a nation might be tempted to initiate a first strike if it felt confident that such a strike would incapacitate the nuclear arsenal of its enemy, making retaliation impossible. With the advent of SLBMs, no nation could be certain that a first strike would incapacitate its enemy's entire nuclear arsenal. To the contrary, it would have to fear a near certain retaliatory second strike from SLBMs. Thus, a first strike was a much less feasible (or desirable) option, and a deliberately initiated nuclear war was thought to be less likely to start.

However, it was soon realized that submarines could approach enemy coastlines undetected and decrease the warning time (the time between detection of the missile launch and the impact of the missile) from as much as half an hour to possibly under three minutes. This effect was especially significant to the United States, Britain and China, whose capitals all lay within 100 miles (160 km) of their coasts. Moscow was much more secure from this type of threat, due to its considerable distance from the sea. This greatly increased the credibility of a "surprise first strike" by one faction and (theoretically) made it possible to knock out or disrupt the chain of command of a target nation before any counterstrike could be ordered (known as a "decapitation strike"). It strengthened the notion that a nuclear war could possibly be "won", resulting not only in greatly increased tensions and increasing calls for fail-deadly control systems, but also in a dramatic increase in military spending. The submarines and their missile systems were very expensive, and one fully equipped nuclear-powered and nuclear-armed missile submarine could cost more than the entire GNP of a developing country.[49] It was also calculated, however, that the greatest cost came in the development of *both* sea- and land-based anti-submarine defenses and in improving and strengthening the "chain of command", and as a result, military spending skyrocketed.

South Africa developed a nuclear weapon capability during the 1970s and early 1980s. It was operational for a brief period before being dismantled in the early 1990s.

According to the 1980 United Nations report *General and Complete Disarmament: Comprehensive Study on Nuclear Weapons: Report of the Secretary-General*, it was estimated that there were a total of about 40,000 nuclear warheads in existence at that time, with a potential combined explosive yield of approximately 13,000 megatons. By comparison, when the volcano Mount Tambora erupted in 1815—turning 1816 into the Year Without A Summer due to the levels of global dimming sulfate aerosols and ash expelled—it exploded with a force of roughly 800 to 1,000 megatons, and ejected 160 km³ (38 cu mi) of mostly rock/tephra,[50] that included 120 million tonnes of sulfur dioxide as an upper estimate.[51] A larger eruption, approximately 74,000 years ago, in Mount Toba produced 2,800 km³ (670 cu mi) of tephra, forming lake Toba,[52] and produced an estimated 6,000 million tonnes (6.6×10^9 short tons) of sulfur dioxide.[53][54] The explosive energy of the eruption may have been as high as equivalent to 20,000,000 megatons (Mt) of TNT,[55] while the asteroid created Chicxulub im-

pact, that is connected with the extinction of the dinosaurs corresponds to at least 70,000,000 Mt of energy, which is roughly 7000 times the maximum arsenal of the US and Soviet Union.[55]

However, comparisons with supervolcanos are more misleading than helpful due to the different aerosols released, the likely air burst fuzing height of nuclear weapons and the globally scattered location of these potential nuclear detonations all being in contrast to the singular and subterranean nature of a supervolcanic eruption.[7] Moreover, assuming the entire world stockpile of weapons were grouped together, it would be difficult, due to the nuclear fratricide effect, to ensure the individual weapons would go off all at once. Nonetheless, many people believe that a full-scale nuclear war would result, through the nuclear winter effect, in the extinction of the human species, though not all analysts agree on the assumptions that underpin these nuclear winter models.[5]

On Sept. 1, 1983, Korean Air Lines Flight 007 was shot down by Soviet jet fighters. On the 26th, a Soviet early warning station under the command of Stanislav Petrov falsely detected 5 inbound intercontinental ballistic missiles from the US. Petrov correctly assessed the situation as a false alarm, and hence did not report his finding to his superiors. It is quite possible that his actions prevented "World War III", as the Soviet policy at that time was immediate nuclear response upon discovering inbound ballistic missiles.

The world came unusually close to nuclear war when the Soviet Union thought that the NATO military exercise Able Archer 83 was a ruse or "cover up" to begin a nuclear first strike. The Soviets responded by raising readiness and preparing their nuclear arsenal for immediate use. Soviet fears of an attack ceased once the exercise concluded without incident.

18.2.6   Post–Cold War

Although the dissolution of the Soviet Union ended the Cold War and greatly reduced tensions between the United States and the Russian Federation, the Soviet Union's formal successor state, both countries remained in a "nuclear stand-off" due to the continuing presence of a very large number of deliverable nuclear warheads on both sides. Additionally, the end of the Cold War led the United States to become increasingly concerned with the development of nuclear technology by other nations outside of the former Soviet Union. In 1995, a branch of the U.S. Strategic Command produced an outline of forward-thinking strategies in the document "Essentials of Post–Cold War Deterrence".

In 1996, a Russian continuity of government facility, Kosvinsky Mountain, which is believed to be a counterpart to the US Cheyenne Mountain Complex, was completed.[56][57][58] It was designed to resist US earth-penetrating nuclear warheads,[56] and is believed to host the Russian Strategic Rocket Forces alternate command post, a post for the general staff built to compensate for the vulnerability of older Soviet era command posts in the Moscow region. In spite of this, the primary command posts for the Strategic Rocket Forces remains Kuntsevo in Moscow and the secondary is the Kosvinsky Mountain in the Urals.[59] The timing of the Kosvinsky facilities completion date is regarded as one explanation for U.S. interest in a new nuclear bunker buster/Earth penetrating warhead and the declaration of the deployment of the B-61 mod 11 in 1997, Kosvinksy is protected by about 1000 feet of granite.[58]

As a consequence of the 9/11 attacks on the USA, American forces immediately increased their readiness to the highest level in 28 years, closing the blast doors of the NORAD's Cheyenne Mountain Operations Center for the first time due to a non-exercise event. But unlike similar increases during the Cold War, Russia immediately decided to stand down a large military exercise in the Arctic region, in order to minimize the risk of incidents, rather than following suit.[60]

The former chair of the United Nations disarmament committee stated that there are more than 16,000 strategic and tactical nuclear weapons ready for deployment and another 14,000 in storage, with the U.S. having nearly 7,000 ready for use and 3,000 in storage, and Russia having about 8,500 ready for use and 11,000 in storage. In addition, China is thought to possess about 400 nuclear weapons, Britain about 200, France about 350, India about 80-100, and Pakistan 100-110. North Korea is confirmed as having nuclear weapons, though it is not known how many, with most estimates between 1 and 10. Israel is also widely believed to possess usable nuclear weapons. NATO has stationed about 480 American nuclear weapons in Belgium, the Netherlands, Italy, Germany, and Turkey, and several other nations are thought to be in pursuit of an arsenal of their own.[61]

A key development in nuclear warfare throughout the 2000s and early 2010s is the proliferation of nuclear weapons to the developing world, with India and Pakistan both publicly testing several nuclear devices, and North Korea conducting an underground nuclear test on October 9, 2006. The U.S. Geological Survey measured a 4.2 magnitude earthquake in the area where the North Korean test is said to have occurred. A further test was announced by the North Korean government on May 25, 2009.[62] Iran, meanwhile, has embarked on a nuclear program which, while officially for civilian purposes, has come under close scrutiny by the United Nations and many individual states.

Recent studies undertaken by the CIA cite the enduring India-Pakistan conflict as the one "flash point" most likely

to escalate into a nuclear war. During the Kargil War in 1999, Pakistan came close to using its nuclear weapons in case the conventional military situation underwent further deterioration.[63] Pakistan's foreign minister had even warned that it would "use any weapon in our arsenal", hinting at a nuclear strike against India.[64] The statement was condemned by the international community, with Pakistan denying it later on. This conflict remains the only war (of any sort) between two declared nuclear powers. The 2001-2002 India-Pakistan standoff again stoked fears of nuclear war between the two countries. Despite these very serious and relatively recent threats, relations between India and Pakistan have been improving somewhat over the last few years. A bus line directly linking Indian- and Pakistani-administered Kashmir has recently been established. However, with the November 26, 2008 Mumbai terror attacks, India currently will not rule out war with Pakistan.

Another potential geopolitical issue which is considered particularly worrisome by military analysts is a possible conflict between the United States and the People's Republic of China over Taiwan. Although economic forces are thought to have reduced the possibility of a military conflict, there remains concern about the increasing military buildup of China (China is rapidly increasing its naval capacity), and that any move toward Taiwan independence could potentially spin out of control.

Israel is thought to possess somewhere between one hundred and four hundred nuclear warheads. It has been asserted that the submarines which Israel received from Germany have been adapted to carry missiles with nuclear warheads, so as to give Israel a second strike capability.[65] Israel has been involved in wars with its neighbors in the Middle East (and with other "non-state actors") on numerous prior occasions, and its small geographic size and population could mean that, in the event of future wars, the Israeli military might have very little time to react to an invasion or other major threat. Such a situation could escalate to nuclear warfare very quickly in some scenarios.

On March 7, 2013, North Korea threatened the United States with a pre-emptive nuclear strike.[66] On April 9, North Korea urged foreigners to leave South Korea, stating that both countries were on the verge of nuclear war.[67] On April 12, North Korea stated that a nuclear war was unavoidable. The country declared Japan as its first target.[68]

18.2.7 Sub-strategic use

See also: Nuclear bunker buster and Edward_Teller § Decision to drop the bombs

The above examples envisage nuclear warfare at a strategic level, i.e., total war. However, nuclear powers have the ability to undertake more limited engagements.

"Sub-strategic use" includes the use of either "low-yield" tactical nuclear weapons, or of variable yield strategic nuclear weapons in a very limited role, as compared to battlefield exchanges of larger-yield strategic nuclear weapons. This was described by the UK Parliamentary Defence Select Committee as "the launch of one or a limited number of missiles against an adversary as a means of conveying a political message, warning or demonstration of resolve".[69] It is believed that all current nuclear weapons states possess tactical nuclear weapons, with the exception of the United Kingdom, which decommissioned its tactical warheads in 1998. However, the UK does possess scalable-yield strategic warheads, and this technology tends to blur the difference between "strategic", "sub-strategic", and "tactical" use or weapons. American, French and British nuclear submarines are believed to carry at least *some* missiles with dial-a-yield warheads for this purpose, potentially allowing a strike as low as one kiloton (or less) against a single target. Only the People's Republic of China and the Republic of India have declarative, unqualified, unconditional "no first use" nuclear weapons policies. India and Pakistan maintain only a credible minimum deterrence.

Commodore Tim Hare, former Director of Nuclear Policy at the British Ministry of Defence, has described "sub-strategic use" as offering the Government "an extra option in the escalatory process before it goes for an all-out strategic strike which would deliver unacceptable damage".[70] However, this sub-strategic capacity has been criticized as potentially increasing the "acceptability" of using nuclear weapons. Combined with the trend in the reduction in the worldwide nuclear arsenal as of 2007 is the warhead miniaturization and modernization of the remaining strategic weapons that is presently occurring in all the declared nuclear weapon states, into more "useable" configurations. The Stockholm International Peace Research Institute suggests that this is creating a culture where use of these weapons is more acceptable and therefore is increasing the risk of war, as these modern weapons do not possess the same psychological deterrent value as the large Cold-War era, multi-megaton warheads.[71]

In many ways, this present change in the balance of terror can be seen as the complete embracement of the switch from the 1950s Eisenhower doctrine of "massive retaliation"[72] to one of "flexible response", which has been growing in importance in the US nuclear war fighting plan/SIOP every decade since.

For example, the United States adopted a policy in 1996 of allowing the targeting of its nuclear weapons at non-state actors ("terrorists") armed with weapons of mass destruction.[73]

Another dimension to the tactical use of nuclear weapons is that of such weapons deployed at sea for use against surface and submarine vessels. Until 1992, vessels of the United States Navy (and their aircraft) deployed various such weapons as bombs, rockets (guided and unguided), torpedoes, and depth charges. Such tactical naval nuclear weapons were considered more acceptable to use early in a conflict because there would be few civilian casualties. It was feared by many planners that such use would probably quickly have escalated into large-scale nuclear war.[74] This situation was particularly exacerbated by the fact that such weapons at sea were not constrained by the safeguards provided by the Permissive Action Link attached to U.S. Air Force and Army nuclear weapons. It is unknown if the navies of the other nuclear powers yet today deploy tactical nuclear weapons at sea.

18.3   Nuclear terrorism

Main article: Nuclear terrorism

Nuclear terrorism by non-state organizations or actors (even individuals) is a largely unknown and understudied factor in nuclear deterrence thinking, as states possessing nuclear weapons are susceptible to retaliation in kind, while sub- or trans-state actors may be less so. The collapse of the Soviet Union has given rise to the possibility that former Soviet nuclear weapons might become available on the black market (so-called 'loose nukes').

A number of other concerns have been expressed about the security of nuclear weapons in newer nuclear powers with relatively less stable governments, such as Pakistan, but in each case, the fears have been addressed to some extent by statements and evidence provided by those nations, as well as cooperative programs between nations. Worry remains, however, in many circles that a relative decrease in security of nuclear weapons has emerged in recent years, and that terrorists or others may attempt to exert control over (or use) nuclear weapons, militarily applicable technology, or nuclear materials and fuel.

Another possible nuclear terrorism threat are devices designed to disperse radioactive materials over a large area using conventional explosives, called dirty bombs. The detonation of a "dirty bomb" would not cause a nuclear explosion, nor would it release enough radiation to kill or injure a large number of people. However, it could cause severe disruption and require potentially very costly decontamination procedures and increased spending on security measures.[75]

18.4 Survival

The predictions of the effects of a major countervalue nuclear exchange include millions of city dweller deaths within a short period of time. Some predictions argue that a full-scale nuclear war could eventually bring about the extinction of the human race.[76] Such predictions, sometimes but not always based on total war with nuclear arsenals at Cold War highs, received contemporary criticism.[5] A number of Cold War publications advocated preparations that could purportedly enable a large proportion of civilians to survive even a total nuclear war. Among the most famous of these is Nuclear War Survival Skills.[77]

To avoid injury and death from a nuclear weapons heat flash and blast effects, the two most far ranging prompt effects of nuclear weapons, schoolchildren were taught to duck and cover by the early Cold War film of the same name. Such advice is once again being given in case of nuclear terrorist attacks.[78]

Prussian blue, or "Radiogardase", is stockpiled in the US, along with potassium iodide and DPTA as pharmaceuticals useful in treating internal exposure to harmful radioisotopes in fallout.[79]

Publications on adapting to a changing diet and supplying nutritional food sources following a nuclear war, with particular focus on agricultural radioecology, include *Nutrition in the postattack environment* by the RAND corporation.[80]

The British government developed a public alert system for use during nuclear attack with the expectation of a four-minute warning before detonation. The United States expected a warning time of anywhere from half an hour (for land based missiles) to less than three minutes (for submarine based weapons). Many countries maintain plans for continuity of government and continuity of operations following a nuclear attack or similar disasters. These range from a designated survivor, intended to ensure survival of some form of government leadership, to the Soviet Dead Hand system, which allows for retaliation even if all Soviet leadership were destroyed. Nuclear submarines are given letters of last resort; orders on what action to take in the event that an enemy nuclear strike has destroyed the government.

A number of other countries around the world have taken significant efforts to maximize their survival prospects in the event of large calamities, both natural and manmade. For example, metro stations in Pyongyang, North Korea, were constructed 110 metres (360 ft) below ground, and were designed to serve as nuclear shelters in the event of war, with each station entrance built with thick steel blast doors.[81][82] An example of a privately funded fallout shelters is the Ark Two Shelter in Ontario, Canada, and autonomous shelters constructed with an emphasis on post-war networking and reconstruction.[83] In Switzerland, the majority of homes have an underground blast and fallout shelter. The country has an overcapacity of such shelters and can accommodate slightly more than the nation's population size.[84][85]

18.5 In fiction

Main article: List of nuclear holocaust fiction
See also: Nuclear weapons in fiction, film, and theater

Nuclear warfare and weapons are staple elements of speculative fiction.

18.6 See also

- World War III
- Broken-Backed War Theory
- Transition to war
- Nuclear briefcase
- Permissive Action Link
- Mount Yamantaw
- People's Liberation Army Rocket Force
- Air Force Global Strike Command
- Doomsday Clock
- Nuclear weapons debate
- List of states with nuclear weapons
- Nuclear weapons and the United States

18.7 References

[1] "7 Possible Toxic Environments Following a Nuclear War - The Medical Implications of Nuclear War - The National Academies Press".

[2] "nuclear winter". *Encyclopedia Britannica*.

[3] "Overview of the Doomsday Clock". *Bulletin of the Atomic Scientists*.

[4] The Nuclear Winter: The World After Nuclear War, Sagan, Carl et al., Sidgwick & Jackson, 1985

[5] http://www.bmartin.cc/pubs/82jpr.html Critique of Nuclear Extinction - Brian Martin 1982

[6] "The Effects of a Global Thermonuclear War".

[7] http://www.bmartin.cc/pubs/82cab/index.html the global health effects of nuclear war

[8] http://trove.nla.gov.au/work/21437545?selectedversion= NBD238850 Long-term worldwide effects of multiple nuclear-weapons detonations. Assembly of Mathematical and Physical Sciences, National Research Council.

[9] Hersh, Seymour (1991). *The Samson Option*. Random House. p. 130. ISBN 0-394-57006-5.

[10] John Pike. "Nuclear Weapons Program".

[11] ""1945-1998" by Isao Hashimoto".

[12] "The Nuclear Testing Tally - Arms Control Association".

[13] N.N. Sokov (2015). *Why Russia calls a limited nuclear strike "de-escalation"*. Bulletin of the Atomic Scientists. Retrieved 2015-12-28.

[14] "Henry Kissinger: The Mike Wallace Interview".

[15] ScienceDaily - Regional Nuclear War Could Devastate Global Climate

[16] Alan F. Philips, 20 Mishaps That Might Have Started Accidental Nuclear War.

[17] Hakim, Joy (1995). *A History of Us: War, Peace and all that Jazz*. New York: Oxford University Press. ISBN 0-19-509514-6.

[18] "1945 Tokyo Firebombing Left Legacy of Terror, Pain". *Common Dreams*.

[19] "March 9, 1945: Burning the Heart Out of the Enemy". *Wired*. Condé Nast Digital. 9 March 2011. Retrieved June 8, 2014.

[20] Wolk (2010), p. 125

[21] "Firebombing of Tokyo". *HISTORY.com*.

[22] Japan's Battle of Okinawa, March–June 1945, Command and General Staff College at the Wayback Machine (archived February 14, 2009)

[23] *Ways of Forgetting, Ways of Remembering: Japan in the Modern World*. New Press. February 4, 2014. p. 273.

[24] "United States Dept. of the Army, Army Battle Casualties and Non Battle Deaths in World War II". Cgsc.cdmhost.com. Retrieved 2011-06-15.

[25] "Potsdam Declaration: Proclamation Defining Terms for Japanese Surrender Issued, at Potsdam, July 26, 1945". National Science Digital Library.

[26] "Milestones: 1937-1945 / The Potsdam Conference, 1945". United States Department of State, Office of the Historian.

[27] Newman, Robert P. (1995). *Truman and the Hiroshima Cult*. p. 86.

[28] "The Atomic Bomb and the End of World War II, A Collection of Primary Sources" (PDF). *National Security Archive Electronic Briefing Book No. 162*. George Washington University. August 13, 1945.

[29] Adams, S. & Crawford, A.. 2000. *World War II*. First edition. Printed in association with the Imperial War Museum. Eyewitness Books series. New York, Doring Kindersley Limited

[30] Alan Axelrod (May 6, 2008). *The Real History of World War II: A New Look at the Past*. Sterling. p. 350.

[31] *Nuke-Rebuke: Writers & Artists Against Nuclear Energy & Weapons (The Contemporary anthology series)*. The Spirit That Moves Us Press. May 1, 1984. pp. 22–29.

[32] Robert Hull (October 11, 2011). *Welcome To Planet Earth - 2050 - Population Zero*. AuthorHouse. p. 215. ISBN 1-4634-2604-6.

[33] Koizumi, Junichiro (August 6, 2005). "Address by Prime Minister Junichiro Koizumi at the Hiroshima Memorial Service for the Hiroshima Peace Memorial Ceremony". Prime Minister of Japan and His Cabinet. Retrieved Nov 28, 2007.

[34] Pontin, Jason (November–December 2007). "Oppenheimer's Ghost". *Technology Review*.

[35] Nuclear Chronology 1945-1959 Archived October 15, 2008, at the Wayback Machine.

[36] "Instant Wisdom: Beyond the Little Red Book". TIME. September 20, 1976.

[37] Robert Service. *Comrades!: A History of World Communism*. Harvard University Press, 2007. p. 321. ISBN 0-674-02530-X

[38] Dikötter, Frank. *Mao's Great Famine: The History of China's Most Devastating Catastrophe, 1958–62*. Walker & Company, 2010. p.13. ISBN 0-8027-7768-6

[39] Project Vista, Caltech, and the dilemmas of Lee DuBridge

[40] Military Planning for European Theatre Conflict During the Cold War

[41] "Nuclear Strategy differences in Soviet and American thinking".

[42] Why the Soviet Union thinks it can fight and win a Nuclear War, Richard Pipes, Professor of History Harvard University 1977

[43] "Candid Interviews with Former Soviet Officials Reveal U.S. Strategic Intelligence Failure Over Decades".

[44] Viktor Suvorov, *Shadow of Victory* (Тень победы), Donetsk, 2003, ISBN 966-696-022-2, pages 353-375.

[45] S. Shane - 1950s U.S. Nuclear Target List Offers Chilling Insight, *The New York Times*, Retrieved 2015-12-23

[46] June 80: Faulty Computer Chip, *20 Mishaps that Might Have Started Accidental Nuclear War*, by Alan F. Phillips, M.D., January 1998, Nuclear Age Peace Foundation

[47] Nuclear Weapons in the Cold War, Bernard Brodie

[48] Continental US Fallout Pattern for Prevailing Winds (FEMA-196/September 1990) Archived March 15, 2011, at the Wayback Machine.

[49] "The Cost of Submarines". *Fast Attacks and Boomers*. Retrieved 2008-07-13.

[50] Stothers, Richard B. (1984). "The Great Tambora Eruption in 1815 and Its Aftermath". *Science*. **224** (4654): 1191–1198. Bibcode:1984Sci...224.1191S. doi:10.1126/science.224.4654.1191. PMID 17819476.

[51] Oppenheimer, Clive (2003). "Climatic, environmental and human consequences of the largest known historic eruption: Tambora volcano (Indonesia) 1815". *Progress in Physical Geography*. **27** (2): 230–259. doi:10.1191/0309133303pp379ra.

[52] "Supersized eruptions are all the rage!". USGS. April 28, 2005.

[53] Robock, A.; C.M. Ammann; L. Oman; D. Shindell; S. Levis; G. Stenchikov (2009). "Did the Toba volcanic eruption of ~74k BP produce widespread glaciation?". *Journal of Geophysical Research*. **114**: D10107. Bibcode:2009JGRD..11410107R. doi:10.1029/2008JD011652.

[54] Huang, C.Y.; Zhao, M.X.; Wang, C.C.; Wei, G.J. (2001). "Cooling of the South China Sea by the Toba Eruption and correlation with other climate proxies ~71,000 years ago". *Geophysical Research Letters*. **28** (20): 3915–3918. Bibcode:2001GeoRL..28.3915H. doi:10.1029/2000GL006113.

[55] https://web.archive.org/web/20141010114324/http://ocw.nd.edu/physics/nuclear-warfare/notes/lecture-18. Archived from the original on October 10, 2014. Retrieved September 23, 2014. Missing or empty |title= (help)

[56] "WINDOW ON HEARTLAND Geopolitical notes on Eastern Europe, the Caucasus and Central Asia". Archived from the original on April 24, 2013.

[57] "Moscow builds bunkers against nuclear attack", by Bill Gertz, *Washington Times*, April 1, 1997

[58] "Global Security.org Kosvinsky Mountain, Kos'vinskiy Kamen'. Gora, MT 59°31'00"N 59°04'00"E".

[59] globalsecurity.org, Strategic C3I Facilities, accessed October 2007

[60] "Context of '(After 11:15 a.m.) September 11, 2001: Russian President Putin Speaks with President Bush'".

[61] London Free Press - Disarmament expert warns of nuclear threat Non functioning link

[62] "BBC NEWS - Asia-Pacific - N Korea faces new UN resolution".

[63] "Pakistan 'prepared nuclear strike'". BBC News.

[64] "Pakistan May Use Any Weapon," The News, Islamabad, May 31, 1999

[65] "Israel buys 2 nuclear-capable submarines from Germany".

[66] "North Korea threatens nuclear strike, U.N. expands sanctions". Reuters. 2013-03-07. Retrieved 2013-05-07.

[67] "North Korea urges foreigners to leave South Korea". CBC. 2013-04-09. Retrieved 2013-05-07.

[68] Meredith, Charlotte (2013-04-12). "North Korea states 'nuclear war is unavoidable' as it declares first target will be Japan". Express. Retrieved 2013-05-07.

[69] UK Parliament, House of Commons, "Select Committee on Defence, Eighth Report", , 20 June 2006. Fetched from URL on 23 December 2012.

[70] "House of Commons - Defence - Eighth Report - THE 1998 STRATEGIC DEFENCE REVIEW".

[71] Risk of nuclear warfare rising (AP) Updated: 2007-06-12 08:36

[72] Jones, Matthew (2008). "Targeting China: U.S. Nuclear Planning and 'Massive Retaliation' in East Asia, 1953–1955". *Journal of Cold War Studies*. **10** (4): 37–65. doi:10.1162/jcws.2008.10.4.37.

[73] Daniel Plesch & Stephen Young, "Senseless policy", *Bulletin of the Atomic Scientists*, November/December 1998, page 4. Fetched from URL on 18 April 2011.

[74] "Declassified: Nuclear Weapons at Sea, Conclusions and Recommendations". *Federation of American Scientists*. Retrieved 2016-04-19.

[75] US Nuclear Regulatory Commission (May 2007). "Backgrounder on Dirty Bombs". Retrieved 2010-04-26.

[76] Ehrlich, P. R., J. Harte, M. A. Harwell, P. H. Raven, C. Sagan, G. M. Woodwell, J. Berry, E. S. Ayensu, A. H. Ehrlich, T. Eisner, S. J. Gould, H. D. Grover, R. Herrera, R. M. May, E. Mayr, C. P. McKay, H. A. Mooney, N. Myers, D. Pimentel, and J. M. Teal (1983). "Long-term biological consequences of nuclear war". *Science*. **222**: 1293–1300. doi:10.1126/science.6658451.

[77] Kearny, Cresson H (1986). *Nuclear War Survival Skills*. Oak Ridge, TN: Oak Ridge National Laboratory. pp. 6–11. ISBN 0-942487-01-X.

[78] Glenn Harlan Reynolds. "The Unexpected Return of 'Duck and Cover'". *The Atlantic*.

[79] "Strategic National Stockpile (SNS)".

[80] "Nutrition in the postattack environment".

[81] Robinson, Martin; Bartlett, Ray; Whyte Rob (2007). *Korea*. Lonely Planet. p. 364. ISBN 978-1-74104-558-1.

[82] Springer, Chris (2003). *Pyongyang: the hidden history of the North Korean capital*. Entente Bt. p. 125. ISBN 978-963-00-8104-7.

[83] "Doomsday Preppers: It's Gonna Get Worse Tonight at 9P et/pt". *Nat Geo TV Blogs*.

[84] Ball, Deborah (2011-06-25). "Swiss Renew Push for Bomb Shelters". *The Wall Street Journal*.

[85] Foulkes, Imogen (2007-02-10). "Swiss still braced for nuclear war". *BBC News*.

18.8   Further reading

- "Presidency in the Nuclear Age", conference and forum at the JFK Library, Boston, October 12, 2009. Four panels: "The Race to Build the Bomb and the Decision to Use It", "Cuban Missile Crisis and the First Nuclear Test Ban Treaty", "The Cold War and the Nuclear Arms Race", and "Nuclear Weapons, Terrorism, and the Presidency".

- "Possibility of Nuclear War in Asia: An Indian Perspective", a project of United Service Institution of India, USI, Discusses the possibility of a nuclear war in Asia from the Indian point of view.

18.9   External links

- History of Nuclear Warfare World History Database

- Fallout: After a Nuclear Attack - slideshow by *Life magazine*

- *The Effects of Nuclear War* (1979) — handbook produced by the United States Office of Technology Assessment (hosted by the Federation of American Scientists)

- *Nuclear Attack Planning Base - 1990* (1987) — assessment of the effects of a major Soviet attack on the United States produced by the Federal Emergency Management Agency (hosted by the Federation of American Scientists)

- *Nuclear War Survival Skills* (1979/1987) — handbook produced by Oak Ridge National Laboratory (use menu at left to navigate)

- Nuclear Warfare OpenCourseWare from the University of Notre Dame

- Nuclear News at HavenWorks.com

- Ground Zero: A Javascript simulation of the effects of a nuclear explosion in a city

- British RAF manual on the effects of nuclear explosions dated 1955

- 20 Mishaps That Might Have Started Accidental Nuclear War by Alan F. Philips, M.D.

- US Doctrine for Joint Nuclear Operations

- Nuclear Files.org Interactive Timeline of the Nuclear Age

- Annotated bibliography on nuclear warfare from the Alsos Digital Library for Nuclear Issues

- DeVolpi, Alexander, Vladimir E. Minkov, Vadim A. Simonenko, and George S. Stanford. 2004. *Nuclear Shadowboxing: Contemporary Threats from Cold War Weaponry*, Vols. 1 and 2. Fidlar Doubleday.

- Air Weapons for the Cold War An in depth history of American air weapons and nuclear bombs from the reference book *American Combat Planes of the 20th Century* by Ray Wagner

- Nuclear Emergency and Radiation Resources

- **NUKEMAP3D** - a 3D nuclear weapons effects simulator powerd by Google Maps.

Chapter 19

On Thermonuclear War

On Thermonuclear War is a book by Herman Kahn, a military strategist at the RAND Corporation, although it was written only a year before he left RAND to form the Hudson Institute. It is a controversial treatise on the nature and theory of war in the thermonuclear weapon age. In it, Kahn addresses the strategic doctrines of nuclear war and its effect on the international balance of power.

Kahn introduced the Doomsday Machine as a rhetorical device to show the limits of John von Neumann's strategy of mutual assured destruction or MAD. The book helped popularize the term megadeath, which Kahn coined in 1953.

Kahn's stated purpose in writing the book was "avoiding disaster and buying time, without specifying the use of this time." The title of the book was inspired by the classic volume *On War*, by Carl von Clausewitz.

Widely read on both sides of the Iron Curtain—the book sold 30,000 copies in hardcover[1]—it is noteworthy for its views on the lack of credibility of a purely thermonuclear deterrent and how a country could "win" a nuclear war.

19.1 Reception

Of the book, Hubert H. Humphrey said: "New thoughts, particularly those which contradict current assumptions, are always painful for the human mind to contemplate. *On Thermonuclear War* is filled with such thoughts."

19.2 In popular culture

Lines from the character General Buck Turgidson in the Stanley Kubrick's 1964 film Dr. Strangelove directly mimic passages from this book,[1] such as Turgidson's phrase "two admittedly regrettable, but nevertheless, distinguishable post-war environments" which reflects a chart from this book labeled "Tragic but Distinguishable Postwar States" (also discussed in the related article Megadeath).

19.3 Publication

First published in 1960 by the Princeton University Press (ISBN 0-313-20060-2), it was republished as a paperback by Transaction Publishers in 2007 (ISBN 978-1412806640).

19.4 Sources

- Herman Kahn (1960), *On Thermonuclear War*. Princeton University Press ISBN 0-313-20060-2

19.5 References

[1] Kaplan, Fred (10 October 2004). "Truth Stranger Than 'Strangelove'". *The New York Times*. Retrieved 11 January 2015. In 1960, Mr. Kahn published a 652-page tome called "On Thermonuclear War," which sold 30,000 copies in hardcover.

19.6 External links

- Essays about and by Herman Kahn

- RAND Corporation unclassified papers by Herman Kahn, 1948–59

- "Fat Man: Herman Kahn and the nuclear age" -(*The New Yorker*)

Chapter 20

Treaty on the Non-Proliferation of Nuclear Weapons

The **Treaty on the Non-Proliferation of Nuclear Weapons**, commonly known as the **Non-Proliferation Treaty** or **NPT**, is an international treaty whose objective is to prevent the spread of nuclear weapons and weapons technology, to promote cooperation in the peaceful uses of nuclear energy, and to further the goal of achieving nuclear disarmament and general and complete disarmament.[1]

Opened for signature in 1968, the treaty entered into force in 1970. As required by the text, after twenty-five years, NPT Parties met in May 1995 and agreed to extend the treaty indefinitely.[2] More countries have adhered to the NPT than any other arms limitation and disarmament agreement, a testament to the treaty's significance.[1] As of August 2016, 191 states have adhered to the treaty, though North Korea, which acceded in 1985 but never came into compliance, announced its withdrawal from the NPT in 2003, following detonation of nuclear devices in violation of core obligations.[3] Four UN member states have never accepted the NPT, three of which are thought to possess nuclear weapons: India, Israel, and Pakistan. In addition, South Sudan, founded in 2011, has not joined.

The treaty recognizes five states as nuclear-weapon states: the United States, Russia, the United Kingdom, France, and China (also the five permanent members of the United Nations Security Council). Four other states are known or believed to possess nuclear weapons: India, Pakistan and North Korea have openly tested and declared that they possess nuclear weapons, Israel is deliberately ambiguous regarding its nuclear weapons status.

The NPT is often seen to be based on a central bargain: "the NPT non-nuclear-weapon states agree never to acquire nuclear weapons and the NPT nuclear-weapon states in exchange agree to share the benefits of peaceful nuclear technology and to pursue nuclear disarmament aimed at the ultimate elimination of their nuclear arsenals".[4] The treaty is reviewed every five years in meetings called Review Conferences of the Parties to the Treaty of Non-Proliferation

of Nuclear Weapons. Even though the treaty was originally conceived with a limited duration of 25 years, the signing parties decided, by consensus, to extend the treaty indefinitely and without conditions during the Review Conference in New York City on 11 May 1995, culminating successful U.S. government efforts led by Ambassador Thomas Graham Jr.

At the time the NPT was proposed, there were predictions of 25–30 nuclear weapon states within 20 years. Instead, over forty years later, five states are not parties to the NPT, and they include the only four additional states believed to possess nuclear weapons.[4] Several additional measures have been adopted to strengthen the NPT and the broader nuclear nonproliferation regime and make it difficult for states to acquire the capability to produce nuclear weapons, including the export controls of the Nuclear Suppliers Group and the enhanced verification measures of the IAEA Additional Protocol.

Critics argue that the NPT cannot stop the proliferation of nuclear weapons or the motivation to acquire them. They express disappointment with the limited progress on nuclear disarmament, where the five authorized nuclear weapons states still have 22,000 warheads in their combined stockpile and have shown a reluctance to disarm further. Several high-ranking officials within the United Nations have said that they can do little to stop states using nuclear reactors to produce nuclear weapons.[5][6]

20.1 Treaty structure

The NPT consists of a preamble and eleven articles. Although the concept of "pillars" is not expressed anywhere in the NPT, the treaty is nevertheless sometimes interpreted as a *three-pillar* system,[7] with an implicit balance among them:

1. *non-proliferation*,

2. *disarmament*, and

3. *the right to peacefully use nuclear technology.*[8]

The "pillars" concept has been questioned by some who believe that the NPT is, as its name suggests, principally about nonproliferation, and who worry that "three pillars" language misleadingly implies that the three elements have equivalent importance.[9]

20.1.1 First pillar: non-proliferation

Five states are recognized by NPT as nuclear weapon states (NWS): China (signed 1992), France (1992), the Soviet Union (1968; obligations and rights now assumed by the Russian Federation), the United Kingdom (1968), and the United States (1968) (The United States, UK, and the Soviet Union – the World War II's "Big Three" — were the only states openly possessing such weapons among the original ratifiers of the treaty, which entered into force in 1970). These five nations are also the five permanent members of the United Nations Security Council.

These five NWS agree not to transfer "nuclear weapons or other nuclear explosive devices" and "not in any way to assist, encourage, or induce" a non-nuclear weapon state (NNWS) to acquire nuclear weapons (Article I). NNWS parties to the NPT agree not to "receive," "manufacture" or "acquire" nuclear weapons or to "seek or receive any assistance in the manufacture of nuclear weapons" (Article II). NNWS parties also agree to accept safeguards by the **International Atomic Energy Agency** (IAEA) to verify that they are not diverting nuclear energy from peaceful uses to nuclear weapons or other nuclear explosive devices (Article III).

The five NWS parties have made undertakings not to use their nuclear weapons against a non-NWS party except in response to a nuclear attack, or a conventional attack in alliance with a Nuclear Weapons State. However, these undertakings have not been incorporated formally into the treaty, and the exact details have varied over time. The U.S. also had nuclear warheads targeted at North Korea, a non-NWS, from 1959 until 1991. The previous United Kingdom Secretary of State for Defence, Geoff Hoon, has also explicitly invoked the possibility of the use of the country's nuclear weapons in response to a non-conventional attack by "rogue states".[10] In January 2006, President Jacques Chirac of France indicated that an incident of state-sponsored terrorism on France could trigger a small-scale nuclear retaliation aimed at destroying the "rogue state's" power centers.[11][12]

20.1.2 Second pillar: disarmament

Article VI of the NPT represents the only binding commitment in a multilateral treaty to the goal of disarmament by the nuclear-weapon states. The NPT's preamble contains language affirming the desire of treaty signatories to ease international tension and strengthen international trust so as to create someday the conditions for a halt to the production of nuclear weapons, and treaty on general and complete disarmament that liquidates, in particular, nuclear weapons and their delivery vehicles from national arsenals.

The wording of the NPT's Article VI arguably imposes only a vague obligation on all NPT signatories to move in the general direction of nuclear and total disarmament, saying, "Each of the Parties to the Treaty undertakes to pursue negotiations in good faith on effective measures relating to cessation of the nuclear arms race at an early date and to nuclear disarmament, and on a treaty on general and complete disarmament."[13] Under this interpretation, Article VI does not strictly require all signatories to actually conclude a disarmament treaty. Rather, it only requires them "to negotiate in good faith."[14]

On the other hand, some governments, especially non-nuclear-weapon states belonging to the Non-Aligned Movement, have interpreted Article VI's language as being anything but vague. In their view, Article VI constitutes a formal and specific obligation on the NPT-recognized nuclear-weapon states to disarm themselves of nuclear weapons, and argue that these states have failed to meet their obligation. The International Court of Justice (ICJ), in its advisory opinion on the Legality of the Threat or Use of Nuclear Weapons, issued 8 July 1996, unanimously interprets the text of Article VI as implying that

> "There exists an obligation to pursue in good faith and bring to a conclusion negotiations leading to nuclear disarmament in all its aspects under strict and effective international control."

The ICJ opinion notes that this obligation involves all NPT parties (not just the nuclear weapon states) and does not suggest a specific time frame for nuclear disarmament.[15]

Critics of the NPT-recognized nuclear-weapon states (the United States, Russia, China, France, and the United Kingdom) sometimes argue that what they view as the failure of the NPT-recognized nuclear weapon states to disarm themselves of nuclear weapons, especially in the post–Cold War era, has angered some non-nuclear-weapon NPT signatories of the NPT. Such failure, these critics add, provides justification for the non-nuclear-weapon signatories to quit the NPT and develop their own nuclear arsenals.[16]

Other observers have suggested that the linkage between proliferation and disarmament may also work the other way, i.e., that the failure to resolve proliferation threats in Iran and North Korea, for instance, will cripple the prospects for disarmament. No current nuclear weapons state, the argument goes, would seriously consider eliminating its last nuclear weapons without high confidence that other countries would not acquire them. Some observers have even suggested that the very progress of disarmament by the superpowers—which has led to the elimination of thousands of weapons and delivery systems[17]—could eventually make the possession of nuclear weapons more attractive by increasing the perceived strategic value of a small arsenal. As one U.S. official and NPT expert warned in 2007, "logic suggests that as the number of nuclear weapons decreases, the 'marginal utility' of a nuclear weapon as an instrument of military power increases. At the extreme, which it is precisely disarmament's hope to create, the strategic utility of even one or two nuclear weapons would be huge."[18]

20.1.3 Third pillar: peaceful use of nuclear energy

The third pillar allows for and agrees upon the transfer of nuclear technology and materials to NPT signatory countries for the development of civilian nuclear energy programs in those countries, as long as they can demonstrate that their nuclear programs are not being used for the development of nuclear weapons.[19]

Since very few of the states with nuclear energy programs are willing to abandon the use of nuclear energy, the third pillar of the NPT under Article IV provides other states with the possibility to do the same, but under conditions intended to make it difficult to develop nuclear weapons.[20]

The treaty recognizes the inalienable right of sovereign states to use nuclear energy for peaceful purposes, but restricts this right for NPT parties to be exercised "in conformity with Articles I and II" (the basic nonproliferation obligations that constitute the "first pillar" of the treaty). As the commercially popular light water reactor nuclear power station uses enriched uranium fuel, it follows that states must be able either to enrich uranium or purchase it on an international market. Mohamed ElBaradei, then Director General of the International Atomic Energy Agency, has called the spread of enrichment and reprocessing capabilities the "Achilles' heel" of the nuclear nonproliferation regime. As of 2007 13 states have an enrichment capability.[21]

Because the availability of fissile material has long been considered the principal obstacle to, and "pacing element" for, a country's nuclear weapons development effort, it was declared a major emphasis of U.S. policy in 2004 to prevent the further spread of uranium enrichment and plutonium reprocessing (a.k.a. "ENR") technology.[22] Countries pos-

sessing ENR capabilities, it is feared, have what is in effect the option of using this capability to produce fissile material for weapons use on demand, thus giving them what has been termed a "virtual" nuclear weapons program.[23] The degree to which NPT members have a "right" to ENR technology notwithstanding its potentially grave proliferation implications, therefore, is at the cutting edge of policy and legal debates surrounding the meaning of Article IV and its relation to Articles I, II, and III of the treaty.

Countries that have signed the treaty as Non-Nuclear Weapons States and maintained that status have an unbroken record of not building nuclear weapons. However, Iraq was cited by the IAEA with punitive sanctions enacted against it by the UN Security Council for violating its NPT safeguards obligations; North Korea never came into compliance with its NPT safeguards agreement and was cited repeatedly for these violations,[24] and later withdrew from the NPT and tested multiple nuclear devices; Iran was found in non-compliance with its NPT safeguards obligations in an unusual non-consensus decision because it "failed in a number of instances over an extended period of time" to report aspects of its enrichment program;[25][26] and Libya pursued a clandestine nuclear weapons program before abandoning it in December 2003.

In 1991, Romania reported previously undeclared nuclear activities by the former regime and the IAEA reported this non-compliance to the Security Council for information only. In some regions, the fact that all neighbors are verifiably free of nuclear weapons reduces any pressure individual states might feel to build those weapons themselves, even if neighbors are known to have peaceful nuclear energy programs that might otherwise be suspicious. In this, the treaty works as designed.

In 2004, Mohamed ElBaradei said that by some estimates thirty-five to forty states could have the knowledge to develop nuclear weapons.[27]

20.2   Key articles

Article I:[28] Each nuclear-weapons state (NWS) undertakes not to transfer, to any recipient, nuclear weapons, or other nuclear explosive devices, and not to assist any non-nuclear weapon state to manufacture or acquire such weapons or devices.

Article II: Each non-NWS party undertakes not to receive, from any source, nuclear weapons, or other nuclear explosive devices; not to manufacture or acquire such weapons or devices; and not to receive any assistance in their manufacture.

Article III: Each non-NWS party undertakes to conclude an

agreement with the IAEA for the application of its safeguards to all nuclear material in all of the state's peaceful nuclear activities and to prevent diversion of such material to nuclear weapons or other nuclear explosive devices.

Article IV: 1. Nothing in this Treaty shall be interpreted as affecting the inalienable right of all the Parties to the Treaty to develop research, production and use of nuclear energy for peaceful purposes without discrimination and in conformity with Articles I and II of this Treaty.

2. All the Parties to the Treaty undertake to facilitate, and have the right to participate in, the fullest possible exchange of equipment, materials and scientific and technological information for the peaceful uses of nuclear energy. Parties to the Treaty in a position to do so shall also co-operate in contributing alone or together with other States or international organizations to the further development of the applications of nuclear energy for peaceful purposes, especially in the territories of non-nuclear-weapon States Party to the Treaty, with due consideration for the needs of the developing areas of the world.

Article VI: Each party "undertakes to pursue negotiations in good faith on effective measures relating to cessation of the nuclear arms race at an early date and to nuclear disarmament, and on a Treaty on general and complete disarmament under strict and effective international control".

Article X. Establishes the right to withdraw from the Treaty giving 3 months' notice. It also establishes the duration of the Treaty (25 years before 1995 Extension Initiative).

20.3   History

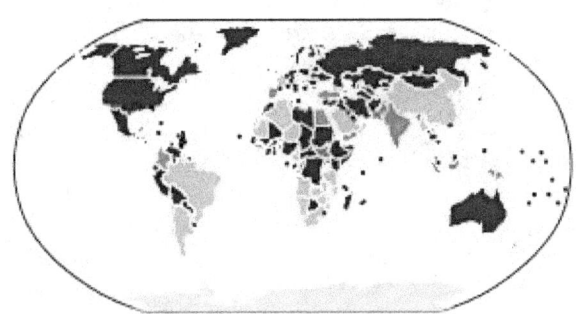

Date NPT first effective (including USSR, YU, CS of that time)
1st decade: ratified or acceded 1968–1977
2nd decade: ratified or acceded 1978–1987
3rd decade: ratified or acceded since 1988
Never signed (India, Israel, Pakistan, South Sudan)

See also: Nuclear proliferation

The impetus behind the NPT was concern for the safety of a world with many nuclear weapon states. It was recognized that the cold war deterrent relationship between just the United States and Soviet Union was fragile. Having more nuclear-weapon states would reduce security for all, multiplying the risks of miscalculation, accidents, unauthorized use of weapons, or from escalation in tensions, nuclear conflict.

The NPT process was launched by Frank Aiken, Irish Minister for External Affairs, in 1958. It was opened for signature in 1968, with Finland the first State to sign. Accession became nearly universal after the end of the Cold War and of South African apartheid. In 1992 China and France acceded to the NPT, the last of the five nuclear powers recognized by the treaty to do so. In 1995 the treaty was extended indefinitely. After Brazil acceded to the NPT in 1998 the only remaining non-nuclear-weapons state which had not signed was Cuba, which joined NPT (and the Treaty of Tlatelolco NWFZ) in 2002.

Several NPT signatories have given up nuclear weapons or nuclear weapons programs. South Africa undertook a nuclear weapons program, but has since renounced it and signed the treaty in 1991 after destroying its small nuclear arsenal; after this, the remaining African countries signed the treaty. The former Soviet Republics where nuclear weapons had been based, namely Ukraine, Belarus and Kazakhstan, transferred those weapons to Russia and joined NPT by 1994 following the signature of the Budapest Memorandum on Security Assurances.

Successor states from the breakups of Yugoslavia and Czechoslovakia also joined the treaty soon after their independence. Montenegro and East Timor were the last countries to sign the treaty on their independence in 2006 and 2003; the only other country to sign in the 21st century was Cuba in 2002. The three Micronesian countries in Compact of Free Association with the USA joined NPT in 1995, along with Vanuatu.

Major South American countries Argentina, Chile, and Brazil joined in 1995 and 1998. Arabian Peninsula countries included Saudi Arabia and Bahrain in 1988, Qatar and Kuwait in 1989, UAE in 1995, and Oman in 1997. The tiny European states of Monaco and Andorra joined in 1995-6. Also signing in the 1990s were Myanmar in 1992 and Guyana in 1993.

20.3.1   United   States-NATO   nuclear
weapons sharing

Main article: Nuclear sharing

At the time the treaty was being negotiated, NATO had in place secret nuclear weapons sharing agreements whereby

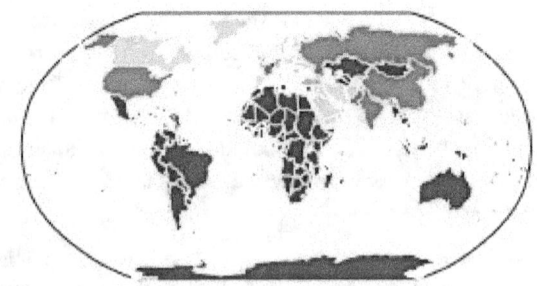

Nuclear-Weapon-Free Zones
Nuclear weapons states
Nuclear sharing
Neither, but NPT

the United States provided nuclear weapons to be deployed by, and stored in, other NATO states. Some argue this is an act of proliferation violating Articles I and II of the treaty. A counter-argument is that the U.S. controlled the weapons in storage within the NATO states, and that no transfer of the weapons or control over them was intended "unless and until a decision were made to go to war, at which the treaty would no longer be controlling", so there is no breach of the NPT.[29] These agreements were disclosed to a few of the states, including the Soviet Union, negotiating the treaty, but most of the states that signed the NPT in 1968 would not have known about these agreements and interpretations at that time.[30]

As of 2005, it is estimated that the United States still provides about 180 tactical B61 nuclear bombs for use by Belgium, Germany, Italy, the Netherlands and Turkey under these NATO agreements.[31] Many states, and the Non-Aligned Movement, now argue this violates Articles I and II of the treaty, and are applying diplomatic pressure to terminate these agreements. They point out that the pilots and other staff of the "non-nuclear" NATO states practice handling and delivering the U.S. nuclear bombs, and non-U.S. warplanes have been adapted to deliver U.S. nuclear bombs which must have involved the transfer of some technical nuclear weapons information. NATO believes its "nuclear forces continue to play an essential role in war prevention, but their role is now more fundamentally political".[32]

U.S. nuclear sharing policies were originally designed to help prevent the proliferation of nuclear weapons—not least by persuading the then West Germany not to develop an independent nuclear capability by assuring it that West Germany would be able, in the event of war with the Warsaw Pact, to wield (U.S.) nuclear weapons in self-defense. (Until that point of all-out war, however, the weapons themselves would remain in U.S. hands.) The point was to limit the spread of countries having their own nuclear weapons programs, helping ensure that NATO allies would not choose to go down the proliferation route.[33] (West Germany was dis-

cussed in U.S. intelligence estimates for a number of years as being a country with the potential to develop nuclear weapons capabilities of its own if officials in Bonn were not convinced that their defense against the Soviet Union and its allies could otherwise be met.[34])

20.3.2   Non-signatories

Four states—India, Israel, Pakistan, and South Sudan—have never signed the treaty. India and Pakistan have publicly disclosed their nuclear weapon programs, and Israel has a long-standing policy of deliberate ambiguity with regards to its nuclear program (see List of countries with nuclear weapons).

India

See also: India and weapons of mass destruction

India has detonated nuclear devices, first in 1974 and again in 1998.[35] India is estimated to have enough fissile material for more than 150 warheads.[36] India was among the few countries to have a no first use policy, a pledge not to use nuclear weapons unless first attacked by an adversary using nuclear weapons, however India's former NSA Shivshankar Menon signaled *a significant shift from "no first use" to "no first use against non-nuclear weapon states"* in a speech on the occasion of Golden Jubilee celebrations of the National Defence College in New Delhi on 21 October 2010, a doctrine Menon said reflected India's "strategic culture, with its emphasis on minimal deterrence".[37][38]

India argues that the NPT creates a club of "nuclear haves" and a larger group of "nuclear have-nots" by restricting the legal possession of nuclear weapons to those states that tested them before 1967, but the treaty never explains on what ethical grounds such a distinction is valid. India's then External Affairs Minister Pranab Mukherjee said during a visit to Tokyo in 2007: *"If India did not sign the NPT, it is not because of its lack of commitment for non-proliferation, but because we consider NPT as a flawed treaty and it did not recognize the need for universal, non-discriminatory verification and treatment."*[39] Although there have been unofficial discussions on creating a South Asian nuclear weapons free zone, including India and Pakistan, this is considered to be highly unlikely for the foreseeable future.[40]

In early March 2006, India and the United States finalized an agreement, in the face of criticism in both countries, to restart cooperation on civilian nuclear technology. Under the deal India has committed to classify 14 of its 22 nuclear power plants as being for civilian use and to place them under IAEA safeguards. Mohamed ElBaradei, then Director

General of the IAEA, welcomed the deal by calling India "an important partner in the non-proliferation regime."[41]

In December 2006, United States Congress approved the United States-India Peaceful Atomic Energy Cooperation Act, endorsing a deal that was forged during Prime Minister Singh's visit to the United States in July 2005 and cemented during President Bush's visit to India earlier in 2006. The legislation allows for the transfer of civilian nuclear material to India. Despite its status outside the Nuclear Non-Proliferation Treaty, nuclear cooperation with India was permitted on the basis of its clean non-proliferation record, and India's need for energy fueled by its rapid industrialization and a billion-plus population.[42]

On 1 August 2008, the IAEA approved the India Safeguards Agreement[43] and on 6 September 2008, India was granted the waiver at the Nuclear Suppliers Group (NSG) meeting held in Vienna, Austria. The consensus was arrived after overcoming misgivings expressed by Austria, Ireland and New Zealand and is an unprecedented step in giving exemption to a country, which has not signed the NPT and the Comprehensive Test Ban Treaty (CTBT).[44][45] While India could commence nuclear trade with other willing countries.[46] The U.S. Congress approved this agreement and President Bush signed it on 8 October 2008.[47]

When China announced expanded nuclear cooperation with Pakistan in 2010, proponents of arms control denounced both the deals, claiming that they weakened the NPT by facilitating nuclear programmes in states which are not parties to the NPT.[48]

As of January 2011, Australia, a top three producer and home to worlds largest known reserves, had continued its refusal to export Uranium to India despite diplomatic pressure from India.[49] In November 2011 the Australian Prime Minister announced a desire to allow exports to India,[50] a policy change which was authorized by her party's national conference in December.[51] On 4 December 2011, Prime Minister Julia Gillard overturned Australia's long-standing ban on exporting uranium to India.[52] She further said "We should take a decision in the national interest, a decision about strengthening our strategic partnership with India in this the Asian century," and said that any agreement to sell uranium to India would include strict safeguards to ensure it would only be used for civilian purposes, and not end up in nuclear weapons.[52] On Sep 5, 2014; Australian Prime Minister Tony Abbott sealed a civil nuclear deal to sell uranium to India. "We signed a nuclear cooperation agreement because Australia trusts India to do the right thing in this area, as it has been doing in other areas," Abbott told reporters after he and Indian Prime Minister Narendra Modi signed a pact to sell uranium for peaceful power generation.[53]

Pakistan

See also: Pakistan and weapons of mass destruction

In May 1998, following India's nuclear tests earlier that month, Pakistan conducted two sets of nuclear tests, the Chagai-I and Chagai-II. Although there is little confirmed information in public, as of 2015, Pakistan was estimated as many as 120 warheads.[36][54] According to analyses of the Carnegie Endowment for International Peace and the Stimson Center, Pakistan has enough fissile material for 350 warheads.[55]

Like India, Pakistani officials argue that the NPT is discriminatory. When asked at a briefing in 2015 whether Islamabad would sign the NPT if Washington requested it, Foreign Secretary Aizaz Ahmad Chaudhry was quoted as responding "It is a discriminatory treaty. Pakistan has the right to defend itself, so Pakistan will not sign the NPT. Why should we?." [56] Until 2010, Pakistan had always maintained the position that it would sign the NPT if India did so. In 2010, Pakistan abandoned this historic position and stated that it would join the NPT only as a recognized nuclear-weapon state.[57]

The NSG Guidelines currently rule out nuclear exports by all major suppliers to Pakistan, with very narrow exceptions, since it does not have full-scope IAEA safeguards (i.e. safeguards on all its nuclear activities). Pakistan has sought to reach an agreement similar to that with India,[58] but these efforts have been rebuffed by the United States and other NSG members, on the grounds that Pakistan's track record as a nuclear proliferator makes it impossible for it to have any sort of nuclear deal in the near future.

By 2010, China reportedly signed a civil nuclear agreement with Pakistan, using the justification that the deal was "peaceful."[59] The British government criticized this, on the grounds that 'the time is not yet right for a civil nuclear deal with Pakistan'.[48] China did not seek formal approval from the nuclear suppliers group, and claimed instead that its cooperation with Pakistan was "grandfathered" when China joined the NSG, a claim that was disputed by other NSG members.[60] Pakistan applied for membership on May 19, 2016,[61] supported by Turkey and China.[62][63] However, many NSG members opposed Pakistan's membership bid due to its track record, including the illicit procurement network of Pakistani scientist A.Q. Khan, which aided the nuclear programs of Iran, Libya and North Korea.[64][65] Pakistani officials reiterated the request in August 2016.[66]

See also: Israel and weapons of mass destruction

Israel

See also: Israel and weapons of mass destruction

Israel has a long-standing policy of deliberate ambiguity with regards to its nuclear program (see List of countries with nuclear weapons). Israel has been developing nuclear technology at its Dimona site in the Negev since 1958, and some nonproliferation analysts estimate that Israel may have stockpiled between 100 and 200 warheads using reprocessed plutonium. The posiition on the NPT is explained in terms of "Israeli exceptionality", a term coined by Professor Gerald M. Steinberg, in reference to the perception that the country's small size, overall vulnerability, as well as the history of deep hostility and large-scale attacks by neighboring states, require a deterrent capability.[67][68]

The Israeli government refuses to confirm or deny possession of nuclear weapons, although this is now regarded as an open secret after Israeli low level nuclear technician Mordechai Vanunu—subsequently arrested and sentenced for treason by Israel—published evidence about the program to the British *Sunday Times* in 1986.

On 18 September 2009 the General Conference of the International Atomic Energy Agency called on Israel to open its nuclear facilities to IAEA inspection and adhere to the non-proliferation treaty as part of a resolution on "Israeli nuclear capabilities," which passed by a narrow margin of 49–45 with 16 abstentions. The chief Israeli delegate stated that "Israel will not co-operate in any matter with this resolution."[69] However, similar resolutions were defeated in 2010, 2013 and 2014. As with Pakistan, the NSG Guidelines currently rule out nuclear exports by all major suppliers to Israel.

20.3.3   North Korea

See also: North Korea and weapons of mass destruction, 2006 North Korean nuclear test, and Six-party talks

North Korea ratified the treaty on 12 December 1985, but gave notice of withdrawal from the treaty on 10 January 2003 following U.S. allegations that it had started an illegal enriched uranium weapons program, and the U.S. subsequently stopping fuel oil shipments under the Agreed Framework[70] which had resolved plutonium weapons issues in 1994.[71] The withdrawal became effective 10 April 2003 making North Korea the first state ever to withdraw from the treaty.[72] North Korea had once before announced withdrawal, on 12 March 1993, but suspended that notice before it came into effect.[73]

On 10 February 2005, North Korea publicly declared that it

possessed nuclear weapons and pulled out of the six-party talks hosted by China to find a diplomatic solution to the issue. "We had already taken the resolute action of pulling out of the Nuclear Non-Proliferation Treaty and have manufactured nuclear arms for self-defence to cope with the Bush administration's evermore undisguised policy to isolate and stifle the DPRK [Democratic People's Republic of Korea]," a North Korean Foreign Ministry statement said regarding the issue.[74] Six-party talks resumed in July 2005.

On 19 September 2005, North Korea announced that it would agree to a preliminary accord. Under the accord, North Korea would scrap all of its existing nuclear weapons and nuclear production facilities, rejoin the NPT, and readmit IAEA inspectors. The difficult issue of the supply of light water reactors to replace North Korea's indigenous nuclear power plant program, as per the 1994 Agreed Framework, was left to be resolved in future discussions.[75] On the next day North Korea reiterated its known view that until it is supplied with a light water reactor it will not dismantle its nuclear arsenal or rejoin the NPT.[76]

On 2 October 2006, the North Korean foreign minister announced that his country was planning to conduct a nuclear test "in the future", although it did not state when.[77] On Monday, 9 October 2006 at 01:35:28 (UTC) the United States Geological Survey detected a magnitude 4.3 seismic event 70 km (43 mi) north of Kimchaek, North Korea indicating a nuclear test.[78] The North Korean government announced shortly afterward that they had completed a successful underground test of a nuclear fission device.

In 2007, reports from Washington suggested that the 2002 CIA reports stating that North Korea was developing an enriched uranium weapons program, which led to North Korea leaving the NPT, had overstated or misread the intelligence.[79][80][81][82] On the other hand, even apart from these press allegations—which some critics worry could have been planted in order to justify the United States giving up trying to verify the dismantlement of Pyongyang's uranium program in the face of North Korean intransigence—there remains some information in the public record indicating the existence of a uranium effort. Quite apart from the fact that North Korean First Vice Minister Kang Sok Ju at one point admitted the existence of a uranium enrichment program, Pakistan's then-President Musharraf revealed that the A.Q. Khan proliferation network had provided North Korea with a number of gas centrifuges designed for uranium enrichment. Additionally, press reports have cited U.S. officials to the effect that evidence obtained in dismantling Libya's WMD programs points toward North Korea as the source for Libya's uranium hexafluoride (UF_6) – which, if true, would mean that North Korea has a uranium conversion facility for producing feedstock for centrifuge enrichment.[83]

20.3.4 Iran

Main articles: Iran and weapons of mass destruction and Nuclear program of Iran
See also: Joint Comprehensive Plan of Action

Iran is a party to the NPT but was found in non-compliance with its NPT safeguards agreement and the status of its nuclear program remains in dispute. In November 2003 IAEA Director General Mohamed ElBaradei reported that Iran had repeatedly and over an extended period failed to meet its safeguards obligations, including by failing to declare its uranium enrichment program.[25] After about two years of EU3-led diplomatic efforts and Iran temporarily suspending its enrichment program,[84] the IAEA Board of Governors, acting under Article XII.C of the IAEA Statute, found in a rare non-consensus decision with 12 abstentions that these failures constituted non-compliance with the IAEA safeguards agreement.[26] This was reported to the UN Security Council in 2006,[85] after which the Security Council passed a resolution demanding that Iran suspend its enrichment.[86] Instead, Iran resumed its enrichment program.[87]

The IAEA has been able to verify the non-diversion of declared nuclear material in Iran, and is continuing its work on verifying the absence of undeclared activities.[88] In February 2008, the IAEA also reported that it was working to address "alleged studies" of weaponization, based on documents provided by certain Member States, which those states claimed originated from Iran. Iran rejected the allegations as "baseless" and the documents as "fabrications."[89] In June 2009, the IAEA reported that Iran had not "cooperated with the Agency in connection with the remaining issues ... which need to be clarified to exclude the possibility of military dimensions to Iran's nuclear program."[90]

The United States concluded that Iran violated its Article III NPT safeguards obligations, and further argued based on circumstantial evidence that Iran's enrichment program was for weapons purposes and therefore violated Iran's Article II nonproliferation obligations.[91] The November 2007 US National Intelligence Estimate (NIE) later concluded that Iran had halted an active nuclear weapons program in the fall of 2003 and that it had remained halted as of mid-2007. The NIE's "Key Judgments," however, also made clear that what Iran had actually stopped in 2003 was only "nuclear weapon design and weaponization work and covert uranium conversion-related and uranium enrichment-related work"—namely, those aspects of Iran's nuclear weapons effort that had not by that point already been leaked to the press and become the subject of IAEA investigations.[92]

Since Iran's uranium enrichment program at Natanz—and its continuing work on a heavy water reactor at Arak that would be ideal for plutonium production—began secretly years before in conjunction with the very weaponization work the NIE discussed and for the purpose of developing nuclear weapons, many observers find Iran's continued development of fissile material production capabilities distinctly worrying. Particularly because fissile material availability has long been understood to be the principal obstacle to nuclear weapons development and the primary "pacing element" for a weapons program, the fact that Iran has reportedly suspended weaponization work may not mean very much.[93] As U.S. Director of National Intelligence Mike McConnell has put it, the aspects of its work that Iran allegedly suspended were thus "probably the least significant part of the program."[94]

Iran states it has a legal right to enrich uranium for peaceful purposes under the NPT, and further says that it "has constantly complied with its obligations under the NPT and the Statute of the International Atomic Energy Agency".[95] Iran also states that its enrichment program is part of its civilian nuclear energy program, which is allowed under Article IV of the NPT. The Non-Aligned Movement has welcomed the continuing cooperation of Iran with the IAEA and reaffirmed Iran's right to the peaceful uses of nuclear technology.[96] UN Secretary General Ban Ki-moon has welcomed the continued dialogue between Iran and the IAEA, and has called for a peaceful resolution to the issue.[97]

In April 2010, during the signing of the U.S.-Russia New START Treaty, President Obama said that the United States, Russia, and other nations are demanding that Iran face consequences for failing to fulfill their obligations under the Nuclear Non-Proliferation Treaty, and that "we will not tolerate actions that flout the NPT, risk an arms race in a vital region, and threaten the credibility of the international community and our collective security."[98]

20.3.5 South Africa

See also: South Africa and weapons of mass destruction

South Africa is the only country that developed nuclear weapons by itself and later dismantled them – unlike the former Soviet states Ukraine, Belarus and Kazakhstan, which inherited nuclear weapons from the former USSR and also acceded to the NPT as non-nuclear weapon states.

During the days of apartheid, the South African government developed a deep fear of both a black uprising and the threat of communism. This led to the development of a secret nuclear weapons program as an ultimate deterrent. South Africa has a large supply of uranium, which is mined in the country's gold mines. The government built a nuclear research facility at Pelindaba near Pretoria where uranium was enriched to fuel grade for the Koeberg Nuclear Power Station as well as weapon grade for bomb production.

In 1991, after international pressure and when a change of government was imminent, South African Ambassador to the United States Harry Schwarz signed the Nuclear Non-Proliferation Treaty. In 1993, the then president Frederik Willem de Klerk openly admitted that the country had developed a limited nuclear weapon capability. These weapons were subsequently dismantled before South Africa acceded to the NPT and opened itself up to IAEA inspection. In 1994, the IAEA completed its work and declared that the country had fully dismantled its nuclear weapons program.

20.3.6   Libya

See also: Libya and weapons of mass destruction

Libya had signed and ratified the Nuclear Non-Proliferation Treaty and was subject to IAEA nuclear safeguards inspections, but undertook a secret nuclear weapons development program in violation of its NPT obligations, using material and technology provided by the A.Q. Khan proliferation network[99]—including actual nuclear weapons designs allegedly originating in China. Libya began secret negotiations with the United States and the United Kingdom in March 2003 over potentially eliminating its WMD programs. In October 2003, Libya was embarrassed by the interdiction of a shipment of Pakistani-designed centrifuge parts sent from Malaysia, also as part of A. Q. Khan's proliferation ring.[100]

In December 2003, Libya announced that it had agreed to eliminate all its WMD programs, and permitted U.S. and British teams (as well as IAEA inspectors) into the country to assist this process and verify its completion. The nuclear weapons designs, gas centrifuges for uranium enrichment, and other equipment—including prototypes for improved SCUD ballistic missiles—were removed from Libya by the United States. (Libyan chemical weapons stocks and chemical bombs were also destroyed on site with international verification, with Libya joining the Chemical Weapons Convention.) Libya's non-compliance with its IAEA safeguards was reported to the U.N. Security Council, but with no action taken, as Libya's return to compliance with safeguards and Article II of the NPT was welcomed.[101]

In 2011 the Libyan government was overthrown in the Libyan Civil War with the assistance of a military intervention by NATO forces acting under the auspices of United Nations Security Council Resolution 1973.[102][103] It was speculated in the media (especially in the Middle Eastern media) that NATO's intervention in Libya shortly after the nation agreed to nuclear and chemical weapons disarmament would make other countries such as North Korea more reluctant to give up nuclear programs due to the risk of being weakened as a result.[104]

20.4   Leaving the treaty

Article X allows a state to leave the treaty if "extraordinary events, related to the subject matter of this Treaty, have jeopardized the supreme interests of its country", giving three months' (ninety days') notice. The state is required to give reasons for leaving the NPT in this notice.

NATO states argue that when there is a state of "general war" the treaty no longer applies, effectively allowing the states involved to leave the treaty with no notice. This is a necessary argument to support the NATO nuclear weapons sharing policy, but a troubling one for the logic of the treaty. NATO's argument is based on the phrase "the consequent need to make every effort to avert the danger of such a war" in the treaty preamble, inserted at the behest of U.S. diplomats, arguing that the treaty would at that point have failed to fulfill its function of prohibiting a general war and thus no longer be binding.[30] Many states do not accept this argument. See United States-NATO nuclear weapons sharing above.

North Korea has also caused an uproar by its use of this provision of the treaty. Article X.1 only requires a state to give three months' notice in total, and does not provide for other states to question a state's interpretation of "supreme interests of its country". In 1993, North Korea gave notice to withdraw from the NPT. However, after 89 days, North Korea reached agreement with the United States to freeze its nuclear program under the Agreed Framework and "suspended" its withdrawal notice. In October 2002, the United States accused North Korea of violating the Agreed Framework by pursuing a secret uranium enrichment program, and suspended shipments of heavy fuel oil under that agreement. In response, North Korea expelled IAEA inspectors, disabled IAEA equipment, and, on 10 January 2003, announced that it was ending the suspension of its previous NPT withdrawal notification. North Korea said that only one more day's notice was sufficient for withdrawal from the NPT, as it had given 89 days before.[105]

The IAEA Board of Governors rejected this interpretation.[106] Most countries held that a new three-months withdrawal notice was required, and some questioned whether North Korea's notification met the "extraordinary events" and "supreme interests" requirements of the treaty. The Joint Statement of 19 September 2005 at the end of the Fourth Round of the Six-Party Talks called for North Korea to "return" to the NPT, implicitly acknowledging that it had withdrawn.

20.5 Recent and coming events

The main outcome of the 2000 Conference was the adoption by consensus of a comprehensive Final Document,[107] which included among other things "practical steps for the systematic and progressive efforts" to implement the disarmament provisions of the NPT, commonly referred to as the Thirteen Steps.

On 18 July 2005, US President George W. Bush met Indian Prime Minister Manmohan Singh and declared that he would work to change US law and international rules to permit trade in US civilian nuclear technology with India.[108] Some, such as British columnist George Monbiot, argue that the U.S.-India nuclear deal, in combination with US attempts to deny Iran (an NPT signatory) civilian nuclear fuel-making technology, may destroy the NPT regime,[109] while others contend that such a move will likely bring India, an NPT non-signatory, under closer international scrutiny.

In the first half of 2010, it was strongly believed that China had signed a civilian nuclear deal with Pakistan claiming that the deal was "peaceful".[59]

Arms control advocates criticised the reported China-Pakistan deal as they did in case of U.S.-India deal claiming that both the deals violate the NPT by facilitating nuclear programmes in states which are not parties to the NPT.[48] Some reports asserted that the deal was a strategic move by China to balance US influence in South-Asia.[60]

According to a report published by U.S. Department of Defense in 2001, China had provided Pakistan with nuclear materials and has given critical technological assistance in the construction of Pakistan's nuclear weapons development facilities, in violation of the Nuclear Non-Proliferation Treaty, of which China even then was a signatory.[110][111]

At the Seventh Review Conference in May 2005, there were stark differences between the United States, which wanted the conference to focus on non-proliferation, especially on its allegations against Iran, and most other countries, who emphasized the lack of serious nuclear disarmament by the nuclear powers. The non-aligned countries reiterated their position emphasizing the need for nuclear disarmament.[112]

The 2010 Review Conference was held in May 2010 in New York City, and adopted a final document that included a summary by the Review Conference President,

Ambassador Libran Capactulan of the Philippines, and an Action Plan that was adopted by consensus.[113][114] The 2010 conference was generally considered a success because it reached consensus where the previous Review Conference in 2005 ended in disarray, a fact that many attributed to the U.S. President Barack Obama's commitment to nuclear nonproliferation and disarmament. Some have warned that this success raised unrealistically high expectations that could lead to failure at the next Review Conference in 2015.[115]

The "Global Summit on Nuclear Security" took place 12–13 April 2010. The summit was proposed by President Obama in Prague and was intended to strengthen the Nuclear Non-Proliferation Treaty in conjunction with the Proliferation Security Initiative and the Global Initiative to Combat Nuclear Terrorism.[116] Forty seven states and three international organizations took part in the summit,[117] which issued a communiqué[118] and a work plan.[119] For further information see 2010 Nuclear Security Summit.

In a major policy speech at the Brandenburg Gate in Berlin on 19 June 2013, United States President Barack Obama outlined plans to further reduce the number of warheads in the U.S. nuclear arsenal.[120] According to *Foreign Policy*, Obama proposed a "one-third reduction in strategic nuclear warheads - on top of the cuts already required by the New START treaty - bringing the number of deployed warheads to about 1,000."[120] Obama is seeking to "negotiate these reductions with Russia to continue to move beyond Cold War nuclear postures," according to briefing documents provided to *Foreign Policy*.[120] In the same speech, Obama emphasized his administration's efforts to isolate any nuclear weapons capabilities emanating from Iran and North Korea. He also called for a renewed bipartisan effort in the United States Congress to ratify the Comprehensive Nuclear-Test-Ban Treaty and called on countries to negotiate a new treaty to end the production of fissile material for nuclear weapons.

On 24 April 2014, it was announced that the nation of the Marshall Islands has brought suit in The Hague against the United States, the former Soviet Union, the United Kingdom, France, China, India, Pakistan, North Korea and Israel seeking to have the disarmament provisions of the NNPT enforced.[121]

20.6 Criticism and responses

Over the years the NPT has come to be seen by many Third World states as "a conspiracy of the nuclear 'haves' to keep the nuclear 'have-nots' in their place".[122] This argument has roots in Article VI of the treaty which "obligates the nuclear weapons states to liquidate their nuclear stockpiles and pursue complete disarmament. The non-nuclear states see no signs of this happening".[4][6] Some argue that the NWS have not fully complied with their disarmament obligations under Article VI of the NPT.[123] Some countries such as India have criticized the NPT, because it "discriminated against states not possessing nuclear weapons on January 1, 1967," while Iran and numerous Arab states have criticized Israel for not signing the NPT.[124][125] There has been disappointment with the limited progress on nuclear disarmament, where the five authorized nuclear weapons states still have 22,000 warheads among them and have shown a reluctance to disarm further.[5]

As noted above, the International Court of Justice, in its advisory opinion on the Legality of the Threat or Use of Nuclear Weapons, stated that "there exists an obligation to pursue in good faith and bring to a conclusion negotiations leading to nuclear disarmament in all its aspects under strict and effective international control.[15] Such an obligation requires that states actively pursue measures to reduce the numbers of nuclear weapons and the importance of their role in military force structures. Some critics of the nuclear-weapons states contend that they have failed to comply with Article VI by failing to make disarmament the driving force in national planning and policy with respect to nuclear weapons, even while they ask other states to plan for their security without nuclear weapons.[126]

The United States responds to criticism of its disarmament record by pointing out that since the end of the Cold War it has eliminated over 13,000 nuclear weapons and eliminated over 80% of its deployed strategic warheads and 90% of non-strategic warheads deployed to NATO, in the process eliminating whole categories of warheads and delivery systems and reducing its reliance on nuclear weapons. U.S. officials have also pointed out the ongoing U.S. work to dismantle nuclear warheads. When current accelerated dismantlement efforts ordered by President George W. Bush have been completed, the U.S. arsenal will be less than a quarter of its size at the end of the Cold War, and smaller than it has been at any point since the Eisenhower administration, well before the drafting of the NPT.[127]

The United States has also purchased many thousands of weapons' worth of uranium formerly in Soviet nuclear weapons for conversion into reactor fuel.[128] As a consequence of this latter effort, it has been estimated that the equivalent of one lightbulb in every ten in the United States is powered by nuclear fuel removed from warheads previously targeted at the United States and its allies during the Cold War.[129]

The U.S. Special Representative for Nuclear Nonproliferation agreed that nonproliferation and disarmament are linked, noting that they can be mutually reinforcing but also that growing proliferation risks create an environment that makes disarmament more difficult.[130] The United Kingdom,[131] France[132] and Russia[133] likewise defend their nuclear disarmament records, and the five NPT NWS issued a joint statement in 2008 reaffirming their Article VI disarmament commitments.[134]

According to Thomas Reed and Danny Stillman, the "NPT has one giant loophole": Article IV gives each non-nuclear weapon state the 'inalienable right' to pursue nuclear energy for the generation of power.[6] A "number of high-ranking officials, even within the United Nations, have argued that they can do little to stop states using nuclear reactors to produce nuclear weapons".[5] A 2009 United Nations report said that:

> The revival of interest in nuclear power could result in the worldwide dissemination of uranium enrichment and spent fuel reprocessing technologies, which present obvious risks of proliferation as these technologies can produce fissile materials that are directly usable in nuclear weapons.[5]

According to critics, those states which possess nuclear weapons, but are not authorized to do so under the NPT, have not paid a significant price for their pursuit of weapons capabilities. Also, the NPT has been explicitly weakened by a number of bilateral deals made by NPT signatories, notably the United States.[5]

20.7   See also

- 13 steps (an important section in the Final Document of the 2000 Review Conference of the Treaty)

- Comprehensive Test Ban Treaty (CTBT)

- Humanitarian Initiative

- Global Initiative to Combat Nuclear Terrorism (GICNT)

- List of countries with nuclear weapons

- Missile Technology Control Regime (MTCR)

- New Agenda Coalition (NAC)

- Non-Proliferation and Disarmament Initiative (NPDI)

- Nuclear armament

- Nuclear warfare

- Nuclear-weapon-free zone

- Nuclear terrorism

- Proliferation Security Initiative (PSI)

- Renovation of the nuclear weapon arsenal of the United States

- Strategic Arms Limitation Talks (SALT)

- Strategic Offensive Reductions Treaty (SORT)

- Weapon of mass destruction (WMD)

- Treaty for the Prohibition of Nuclear Weapons in Latin America and the Caribbean (Treaty of Tlatelolco)

20.8 References

[1] "UNODA - Non-Proliferation of Nuclear Weapons (NPT)". *un.org*. Retrieved 2016-02-20.

[2]

[3] "Nuclear Non-Proliferation Treaty (NPT)" (PDF). *Defense Treaty Inspection Readiness Program - United States Department of Defense*. Defense Treaty Inspection Readiness Program. Archived from the original (PDF) on 11 March 2013. Retrieved 19 June 2013.

[4] Graham, Jr., Thomas (November 2004). "Avoiding the Tipping Point". Arms Control Association.

[5] Benjamin K. Sovacool (2011). *Contesting the Future of Nuclear Power: A Critical Global Assessment of Atomic Energy*, World Scientific, pp. 187–190.

[6] Thomas C. Reed and Danny B. Stillman (2009). *The Nuclear Express: A Political History of the Bomb and its Proliferation*, Zenith Press, p. 144.

[7] See, for example, the Canadian government's NPT web site The Nuclear Non-Proliferation Treaty.

[8] Ambassador Sudjadnan Parnohadiningrat, 26 April 2004, United Nations, New York, Third Session of the Preparatory Committee for the 2005 Review Conference of the Parties to the Treaty on the Non-Proliferation of Nuclear Weapons, furnished by the Permanent Mission of the Republic of Indonesia to the United Nations (indonesiamission-ny.org) Archived 20 November 2005 at the Wayback Machine.

[9] This view was expressed by Christopher Ford, the U.S. NPT representative at the end of the Bush Administration. See "The 2010 Review Cycle So Far: A View from the United States of America," presented at Wilton Park, United Kingdom, 20 December 2007.

[10] UK 'prepared to use nuclear weapons' BBC article dated 20 March 2002

[11] France 'would use nuclear arms', BBC article dated 19 January 2006

[12] Chirac: Nuclear Response to Terrorism Is Possible, Washington Post article dated 20 January 2006

[13] "Information Circulars" (PDF). *iaea.org*. Archived from the original (PDF) on 7 August 2007.

[14] "U.S. Compliance With Article VI of the NPT". Acronym.org.uk. Retrieved 25 November 2010.

[15] The ICJ (8 July 1996). "Legality of the threat or use of nuclear weapons". Retrieved 6 July 2011.

[16] Mishra, J. "NPT and the Developing Countries", *(Concept Publishing Company, 2008)*.

[17] See, e.g., Disarmament, the United States, and the NPT, Christopher Ford, U.S. Special Representative for Nuclear Nonproliferation, delivered at the Conference on "Preparing for 2010: Getting the Process Right," Annecy, France, 17 March 2007; Nuclear Disarmament Progress and Challenges in the Post-Cold War World, U.S. statement to the Second Session of the Preparatory Committee for the 2010 NPT Review Conference, Geneva (30 April 2008) Archived 26 June 2008 at the Wayback Machine.

[18] U.S. Special Representative for Nuclear Nonproliferation Christopher Ford, "Disarmament and Non-Nuclear Stability in Tomorrow's World," remarks to the Conference on Disarmament and Nonproliferation Issues, Nagasaki, Japan (31 August 2007).

[19] Zaki, Mohammed M. (2011-05-24). *American Global Challenges: The Obama Era*. Palgrave Macmillan. ISBN 9780230119116.

[20] "Treaty on the Non-Proliferation of Nuclear Weapons" (PDF). Retrieved 8 October 2015.

[21] Daniel Dombey (19 February 2007). "Director General's Interview on Iran and DPRK". Financial Times. Archived from the original on 22 February 2007. Retrieved 4 May 2006.

[22] See Remarks by President Bush at the National Defense University (11 February 2004), available at http://georgewbush-whitehouse.archives.gov/news/releases/2004/02/20040211-4.html (announcing initiative to stop spread of ENR technology).

[23] "IAEA predicts more nuclear states". BBC. 16 October 2006. Retrieved 12 March 2016.

[24] (PDF) https://web.archive.org/web/20070607003906/http://www.iaea.org/NewsCenter/Focus/IaeaDprk/dprk.pdf. Archived from the original (PDF) on 7 June 2007. Retrieved 28 March 2007. Missing or empty |title= (help) and http://www.iaea.org/NewsCenter/MediaAdvisory/2003/med-advise_048.shtml

[25] "Implementation of the NPT Safeguards Agreement in the Islamic Republic of Iran" (PDF). IAEA. 10 November 2003. GOV/2003/75. Archived from the original (PDF) on 25 October 2007. Retrieved 25 October 2007.

[26] "Implementation of the NPT Safeguards Agreement in the Islamic Republic of Iran" (PDF). IAEA. 24 September 2005. GOV/2005/77. Archived from the original (PDF) on 25 October 2007. Retrieved 25 October 2007.

[27] Mohamed ElBaradei (2004). "Preserving the Non-Proliferation Treaty" (PDF). Disarmament Forum. Archived from the original (PDF) on 27 November 2007. Retrieved 17 November 2007.

[28] "Nuclear Non-Proliferation Treaty (PDF) – IAEA" (PDF). Archived from the original (PDF) on 2 December 2010. Retrieved 25 November 2010.

[29] Brian Donnelly, Foreign and Commonwealth Office, *The Nuclear Weapons Non-Proliferation Articles I, II and VI of the Treaty on the Non-Proliferation of Nuclear Weapons*, Agency for the Prohibition of Nuclear Weapons in Latin America and the Caribbean, archived from the original on 5 January 2009, retrieved 2009-08-07

[30] Otfried Nassauer, Institute for Energy and Environmental Research (ieer.org), Science for Democratic Action Volume 9 Number 3, May 2001, Nuclear Sharing in NATO: Is it Legal?

[31] Hans M. Kristensen, National Resources Defence Council (nrdc.org), February 2005, U.S. Nuclear Weapons in Europe: A Review of Post-Cold War Policy, Force Levels, and War Planning

[32] NATO (nato.int), NATO's Nuclear Forces in the New Security Environment Archived 29 August 2005 at the Wayback Machine.

[33] See, e.g., U.S. Director of Central Intelligence, Likelihood and Consequences of a Proliferation of Nuclear Weapons Systems, declassified U.S. National Intelligence Estimate, NIE 4–63 (28 June 1963), at p.17, paragraph 40.

[34] See, e.g., U.S. Director of Central Intelligence, Annex to National Intelligence Estimate No. 100-2-58: Development of Nuclear Capabilities by Fourth Countries: Likelihood and Consequences, declassified U.S. National Intelligence Estimate, NIE 100-2-58 (1 July 1958), at p.4, paragraphs 18–19; U.S. Director of Central Intelligence, Likelihood and Consequences of the Development of Nuclear Capabilities by Additional Countries, declassified U.S. National Intelligence Estimate, NIE 100-4-60 (20 September 1960), at p. 2, paragraph 4, & p.8, paragraphs 27–29.

[35] BBC

[36] Tellis, Ashley J. "Atoms for War?" (PDF). *Carnegie Endowment for International Peace*. Retrieved 24 October 2015.

[37] "NSA Shivshankar Menon at NDC (Speech)". india Blooms. Retrieved 30 April 2013.

[38] https://web.archive.org/web/20091205231912/http://www.indianembassy.org/policy/CTBT/nuclear_doctrine_aug_17_1999.html. Archived from the original on 5 December 2009. Retrieved 30 April 2013. Missing or empty |title= (help)

[39] "India seeks Japan's support, calls NPT 'flawed'". Whereincity.com. 24 March 2007. Archived from the original on 12 January 2012. Retrieved 25 November 2010.

[40] Banerjee, Dipankar (1998). Thakur, Ramesh, ed. *The Obstacles to a South Asian Nuclear-Weapons-Free-Zone. Nuclear Weapons-Free Zones*. London: Macmillan. ISBN 0-333-73980-9.

[41] "Remarks With International Atomic Energy Agency Director General Mohamed ElBaradei" (PDF). Retrieved 25 November 2010.

[42] (AFP) – 1 October 2008 (1 October 2008). "AFP: India energised by nuclear pacts". Afp.google.com. Archived from the original on 20 May 2011. Retrieved 25 November 2010.

[43] "IAEA Board Approves India-Safeguards Agreement". Iaea.org. Retrieved 25 November 2010.

[44] "NSG CLEARS NUCLEAR WAIVER FOR INDIA". CNN-IBN. 6 September 2008. Retrieved 6 September 2008.

[45] "INDIA JOINS NUCLEAR CLUB, GETS NSG WAIVER". NDTV.com. 6 September 2008. Archived from the original on 8 September 2008. Retrieved 6 September 2008.

[46] "Yes for an Answer". *The Washington Post*. 12 September 2008. Retrieved 20 May 2010.

[47] "President Bush Signs H.R. 7081, the United States-India Nuclear Cooperation Approval and Nonproliferation Enhancement Act". Georgewbush-whitehouse.archives.gov. 8 October 2008. Retrieved 25 November 2010.

[48] "White House Low-Key on China-Pakistan Nuke Deal". *Global Issues*. 30 June 2010. Retrieved 12 March 2012.

[49] "Rudd resists India's push for uranium deal". ABC Online. 20 January 2011. Retrieved 20 January 2011.

[50] "An opportunity for sound and fury signifying something". The Age. 15 November 2011. Retrieved 15 November 2011.

[51] "PM changes mind on uranium sales to India". ABC Online. 15 November 2011. Retrieved 15 November 2011.

[52] "Australia's Labor party backs uranium sales to India". BBC News. 4 December 2011. Retrieved 4 December 2011.

[53] "India and Australia seal civil nuclear deal for uranium trade". *Reuters India*.

[54] Anwar Iqbal. "Impact of US wargames on Pakistan N-arms 'negative'". *dawn.com*. Archived from the original on 27 September 2013.

[55] Craig, Tim (27 August 2015). "Report: Pakistan's nuclear arsenal could become the world's third-biggest". *The Washington Post*. Retrieved 24 October 2015.

[56] Anwar Iqbal. "Pakistan will not sign NPT, says foreign secretary".

[57] "Pakistan against signing the NPT as a non-nuclear weapons state".

[58] BBC (bbc.co.uk), 2 March 2006, US and India seal nuclear accord

[59] "BBC News – China says Pakistan nuclear deal 'peaceful'". *BBC News*. 17 June 2010. Retrieved 12 March 2012.

[60] "The China-Pakistan Nuclear Deal: A Realpolitique Fait Accompli". *NTI*. Retrieved 12 March 2012.

[61] "NSG to take up Pakistan, India's membership requests this week". retrieved on 2016-06-23. Dawn.

[62] *Pakistan appreciates China's unequivocal support for admission in NSG*, retrieved June 25, 2016, The Indian Express

[63] "Kiwis soft on India's NSG bid, Turkey backs Pak - TOI Mobile | The Times of India Mobile Site". *m.timesofindia.com*. Retrieved 2016-06-20.

[64] "China says more talks needed to build consensus on nuclear export club". 12 June 2016 – via Reuters.

[65] Guardia, By Ahmed Rashid in Lahore and Anton La. "I've sold nuclear secrets to Libya, Iran and N Korea".

[66] "Pakistan launches fresh push for NSG membership". 2016-08-23. Retrieved 2016-08-24.

[67] Steinberg, Gerald (2006). "EXAMINING ISRAEL'S NPT EXCEPTIONALITY: 1998–2005". *Ther NonProliferation Review*. **13** (1).

[68] Cohen, Avner (2010). *The Worst Kept Secret*. New York: Columbia University Press. p. 266.

[69] "Israel pressured on nuclear sites – Middle East". Al Jazeera English. 18 September 2009. Retrieved 25 November 2010.

[70] "Text of Agreed Framework" (PDF). Archived from the original (PDF) on 4 June 2011. Retrieved 25 November 2010.

[71] Korean News Service, Tokyo (kcna.co.jp), 10 January 2003, Statement of DPRK Government on its withdrawal from NPT

[72] Nuclear Age Peace Foundation (wagingpeace.org), 10 April 2003, North Korea's Withdrawal from Nonproliferation Treaty Official Archived 12 April 2006 at the Wayback Machine.

[73] International Atomic Energy Agency (iaea.org), May 2003, Fact Sheet on DPRK Nuclear Safeguards

[74] Korean News Service, Tokyo (kcna.co.jp), February 2005, DPRK FM on Its Stand to Suspend Its Participation in Six-party Talks for Indefinite Period

[75] Khan, Joseph (19 September 2005). "North Korea Says It Will Abandon Nuclear Efforts". *New York Times*.

[76] Agence France Presse, 2006, N. Korea raises stakes on nuclear deal with reactor demand, furnished by Media Corp News (channelnewsasia.com), 20 September 2005

[77] BBC (news.bbc.co.uk), 3 October 2006, N Korea 'to conduct nuclear test'

[78] (English) Magnitude 4.3—North Korea 2006 October 09 01:35:28 UTC (Report). United States Geological Survey (USGS). 9 October 2006. Retrieved 2010-12-02.

[79] Carol Giacomo (10 February 2007). "N.Korean uranium enrichment program fades as issue". Reuters. Retrieved 11 February 2007.

[80] Sanger, David E.; Broad, William J. (1 March 2007). "U.S. Had Doubts on North Korean Uranium Drive". New York Times. Retrieved 1 March 2007.

[81] Kessler, Glenn (1 March 2007). "New Doubts on Nuclear Efforts by North Korea". Washington Post. Retrieved 1 March 2007.

[82] "Another Intelligence Twist". Washington Post. 2 March 2007. Retrieved 10 March 2007.

[83] See generally U.S. Department of State. "Adherence to and Compliance With Arms Control, Nonproliferation, and Disarmament Agreements and Commitments," August 2005, pp. 87–92, www.state.gov/documents/organization/52113.pdf; Anthony Faiola, "N. Korea Declares Itself a Nuclear Power," The Washington Post, 10 February 2005, www.washingtonpost.com/wp-dyn/articles/A12836-2005Feb10.html; "Khan 'Gave N. Korea Centrifuges,'" BBC News, 24 August 2005, http://news.bbc.co.uk/2/hi/south_asia/4180286.stm; "Pakistan and North Korea: Dangerous Counter-Trades," IISS Strategic Comments, Vol. 8, No. 9 (November 2002).

[84] AFP. "EU and Iran Avert Nuclear Deadlock". Dw-world.de. Retrieved 25 November 2010.

[85] Implementation of the NPT Safeguards Agreement in the Islamic Republic of Iran, Resolution adopted on 4 February 2006, Archived 3 December 2011 at the Wayback Machine.

[86] UN Security Council Resolution 1737

[87] ""BBC": Iran 'resumes' nuclear enrichment". BBC News. 13 February 2006. Retrieved 25 November 2010.

[88] "Microsoft Word - gov2007-58.doc" (PDF). Archived from the original (PDF) on 2 December 2010. Retrieved 25 November 2010.

[89] "Microsoft Word - gov2008-4.doc" (PDF). Archived from the original (PDF) on 2 December 2010. Retrieved 25 November 2010.

[90] GOV/2009/35, Implementation of the NPT Safeguards Agreement and relevant provisions of Security Council resolutions 1737 (2006), 1747 (2007), 1803 (2008) and 1835 (2008) in the Islamic Republic of Iran, 5 June 2009. Archived 13 October 2009 at the Wayback Machine.

[91] Adherence to and Compliance With Arms Control, Nonproliferation, and Disarmament Agreements and Commitments, Bureau of Verification and Compliance, U.S. Department of State, 30 August 2005 Archived 14 March 2007 at the Wayback Machine.

[92] "Iran: Nuclear Intentions and Capabilities (National Intelligence Estimate)" (PDF). Archived from the original (PDF) on 22 November 2010. Retrieved 25 November 2010.

[93] See, e.g., U.S. Special Representative for Nuclear Nonproliferation Christopher A. Ford, "The 2020 NPT Review Cycle So Far: A View from the United States of America," remarks at Wilton Park, UK (20 December 2007), http://www.state.gov/t/isn/rls/rm/98382.htm ("Given that possession of the necessary quantity of fissile material is the most difficult challenge in developing a nuclear weapon, the recently-released U.S. National Intelligence Estimate (NIE) hardly alleviates our concerns about Iran's nuclear work.").

[94] Mark Mazzetti (6 February 2008). "Intelligence Chief Cites Qaeda Threat to U.S.". *New York Times*.

[95] "INFCIRC/724 – Communication dated 26 March 2008 received from the Permanent Mission of the Islamic Republic of Iran to the Agency" (PDF). Archived from the original (PDF) on 11 September 2010. Retrieved 25 November 2010.

[96] "XV Ministerial Conference of the Non-Aligned Movement (July 2008): Statement on the Islamic Republic of Iran's Nuclear Issue" (PDF). Archived from the original (PDF) on 11 September 2010. Retrieved 25 November 2010.

[97] OIC (March 2008): UN Secretary-General's address to the 11th Summit of the Organization of the Islamic Conference Archived 18 December 2008 at the Wayback Machine.

[98] "New START Treaty and Protocol". DipNote. 8 April 2010. Retrieved 2 October 2012.

[99] Implementation of the NPT Safeguards Agreement in the Socialist People's Libyan Arab Jamahiriya, International Atomic Energy Agency, GOV/2008/39, 12 September 2008.

[100] Uncovering the Nuclear Black Market: Working Toward Closing Gaps in the International Nonproliferation Regime, by David Albright and Corey Hinderstein, Institute for Science and International Security, 4 June 2004.

[101] See generally Assistant Secretary of State Paula DeSutter, "Libya Renounces Weapons of Mass Destruction, http://usinfo.state.gov/journals/itps/0305/ijpe/desutter.htm; DeSutter, "Completion of Verification Work in Libya," testimony before the Subcommittee on International Terrorism, Nonproliferation, and Human Rights (22 September 2004), http://www.state.gov/t/vci/rls/rm/2004/37220.htm; DeSutter, "U.S. Government's Assistance to Libya in the Elimination of its Weapons of Mass Destruction (WMD)," testimony before the Senate Foreign Relations Committee (26 February 2004), http://www.state.gov/t/vci/rls/rm/2004/29945.htm.

[102] https://web.archive.org/web/20160201201746/http://www.theepochtimes.com/n2/world/odyssey-dawn-phase-one-of-libya-military-intervention-.html. Archived from the original on 1 February 2016. Retrieved 21 November 2015. Missing or empty |title= (help)

[103] "Libya: Nato to take command of no-fly zone". *BBC News*. Retrieved 2016-02-20.

[104] Braut-Hegghammer, Malfrid. "Relinquished Nuclear Powers: A Case Study of Libya" in *To Join or Not to Join the Nuclear Club: How Nations Think about Nuclear Weapons: Two Middle East Case Studies - Libya and Pakistan* (Middle East Studies, April 2013).

[105] "North Korea Profile – Nuclear Overview". Nti.org. Retrieved 25 November 2010.

[106] "Media Advisory 2003/48 – IAEA Board of Governors Adopts Resolution on Safeguards in North Korea – 12 February". Iaea.org. Retrieved 25 November 2010.

[107] "2000 NPT Review Conference". *un.org*. Retrieved 2016-02-20.

[108] The Associated Press, 2005, Bush opens energy door to India, furnished by CNN (cnn.com), 18 July 2005 Archived 19 February 2006 at the Wayback Machine.

[109] George Monbiot, The Guardian (guardian.co.uk), 2 August 2005, The treaty wreckers Archived 3 August 2005 at the Wayback Machine.

[110] "US Report: China gifted nuclear bomb and Pakistan stole the technology". *The world reporter*. 18 November 2009. Retrieved 12 March 2012.

[111] "Nuclear Weapons Proliferation report". *Canadian Security Intelligence Service*. October 2001. Archived from the original on 12 March 2012. Retrieved 12 March 2012.

[112] Syed Hamid Albar, Minister of Foreign Affairs of Malaysia, United Nations (un.org), New York, 2 May 2005, The General Debate of the 2005 Review Conference of the Parties to the Treaty on the Non-Proliferation of Nuclear Weapons Archived 8 May 2005 at the Wayback Machine.

[113] Summary of the 2010 NPT final outcome document, Beatrice Fihn, Reaching Critical Will, 1 June 2010. Archived 16 July 2011 at the Wayback Machine.

[114] Final Document, 2010 Review Conference of the Parties to the Treaty on the Non-Proliferation of Nuclear Weapons, NPT/CONF.2010/50, adopted 28 May 2010. Archived 16 July 2011 at the Wayback Machine.

[115] "Heinrich Böll Stiftung European Union" (PDF). *boell.eu*. Archived from the original (PDF) on 23 August 2011.

[116] Obama Calls for Global Nuclear Summit in March 2010 from America.gov, retrieved on 8 January 2010. Archived 31 January 2010 at the Wayback Machine.

[117] "Key Facts about the Nuclear Security Summit". *Office of the Press Secretary*. The White House. 13 April 2010. Retrieved 23 August 2012.

[118] "Communiqué of the Washington Nuclear Security Summit". *Office of the Press Secretary*. The White House. 13 April 2010. Retrieved 23 August 2012.

[119] "Work Plan of the Washington Nuclear Security Summit". *Office of the Press Secretary*. The White House. 13 April 2012. Retrieved 23 August 2012.

[120] Lubold, Gordon (19 June 2013). "Foreign Policy Situation Report: Obama to outline big nuke cuts today; DOD civilian owes $500k – to DOD; Petraeus to Team Rubicon; Hastings, dead; Say goodbye, Rambo; Tara Sonenshine on "bottom line diplomacy:" and a bit more.". *Foreign Policy*. Retrieved 19 June 2013.

[121] Newman, Scott (24 April 2014). "Pacific Island Nation Sues U.S., Others For Violating Nuclear Treaty". National Public Radio. Retrieved 24 April 2014.

[122] "Insights Secure-2014: Questions On Current Events". Insights on India. Retrieved 24 October 2014.

[123] Rendall, Steve. "Ignoring the U.S.'s "Bad Atoms"". Fair.org. Archived from the original on 24 November 2010. Retrieved 25 November 2010.

[124] Fidler, David P. "India Wants to Join the Non-Proliferation Treaty as a Weapon State". Yale Global. Retrieved 24 October 2014.

[125] Reuters. "Pakistan and India refused to sign the Nuclear Non-Proliferation Treaty". The Nation. Retrieved 24 October 2014.

[126] lwmdreport.org/ndcs/online/ NuclearDisorderPart1Section2.pdf 1

[127] Fact Sheet: Increasing Transparency in the U.S. Nuclear Weapons Stockpile, 3 May 2010. Archived 6 May 2010 at the Wayback Machine.

[128] See. e.g., "Disarmament, the United States, and the NPT," http://www.state.gov/t/isn/rls/other/81946.htm; U.S. Special Representative for Nuclear Nonproliferation Christopher Ford, "Procedure and Substance in the NPT Review Cycle: The Example of Nuclear Disarmament," remarks to the Conference on "Preparing for 2010: Getting the Process Right," Annecy, France (17 March 2007), http://www.state.gov/t/isn/rls/rm/81940.htm; "The United States and Article VI: A Record of Accomplishment," http://geneva.usmission.gov/CD/updates/05-06-08%20Article%20VI%20Briefing.pdf.

[129] Remarks by U.S. National Security Advisor Stephen Hadley at the Center for International Security and Cooperation, Stanford University (8 February 2008), http://www.whitehouse.gov/news/releases/2008/02/20080211-6.html.

[130] Disarmament, the United States, and the NPT Archived 30 May 2009 at the Wayback Machine.

[131] FCO fact sheet on nuclear weapons Archived 12 June 2009 at the Wayback Machine.

[132] "The 2005 NPT Review Conference: A French Perspective". Armscontrol.org. Retrieved 25 November 2010.

[133] "Statement by Russian Deputy Foreign Minister Kislyak at the 2005 NPT Review Conference" (PDF). Archived from the original (PDF) on 26 March 2009. Retrieved 25 November 2010.

[134] "Statement of the P5 to the 2008 NPT PrepCom" (PDF). Archived from the original (PDF) on 4 June 2011. Retrieved 25 November 2010.

20.9 External links

- Nuclear Non-Proliferation Treaty (PDF) – IAEA

- UN Office of Disarmament Affairs NPT section

- Procedural history, related documents and photos on the *Treaty on the Non-Proliferation of Nuclear Weapons (NPT)* in the Historic Archives of the United Nations Audiovisual Library of International Law

- [Abolition 2000 Europe]

- People vs. The Bomb: Showdown at the UN (Video)

- NuclearFiles.org Summary and text from the nuclear NPT

- Membership/Signatories

- The Nonproliferation Policy Education Center (NPEC)- A not-for-profit organization based in Washington, D.C., and founded in 1994 to promote a better understanding of strategic weapons proliferation issues among policymakers, scholars and the media.

- Annotated Bibliography on the NPT from the Alsos Digital Library for Nuclear Issues

- Nuclear Nonproliferation Treaty Turns 40 Today in 2008.

- George Perkovich, "Principles for Reforming the Nuclear Order", *Proliferation Papers*, Paris, Ifri, Fall 2008.

- U.S. Department of State, website compiling speeches and papers relevant to NPT Review Cycle, http://www.state.gov/t/isn/wmd/nnp/.

- Department of Foreign Affairs, Republic of the Philippines, http://dfa.gov.ph/main/index.php/newsroom/npt-information-a-updates

- Annotated bibliography for the Nuclear Nonproliferation Treaty from the Alsos Digital Library for Nuclear Issues

- The American Academy of Arts and Sciences Nuclear Collisions: Discord, Reform & the Nuclear Nonproliferation Regime

Chapter 21

Intermediate-Range Nuclear Forces Treaty

The **Intermediate-Range Nuclear Forces Treaty (INF Treaty)** is the abbreviated name of the **Treaty Between the United States of America and the Union of Soviet Socialist Republics on the Elimination of Their Intermediate-Range and Shorter-Range Missiles**, a 1987 agreement between the United States and the Soviet Union (and later its successor states, in particular the Russian Federation). Signed in Washington, D.C. by President Ronald Reagan and Soviet leader Mikhail Gorbachev on 8 December 1987, the treaty was ratified by the United States Senate on 27 May 1988 and came into force on 1 June 1988.

The INF Treaty eliminated all nuclear and conventional missiles, as well as their launchers, with with ranges of 500–1,000 kilometres (310–620 mi) (short-range) and 1,000–5,500 kilometres (620–3,420 mi) (intermediate-range). The treaty did not cover sea-launched missiles.[2] By May 1991, 2,692 missiles were eliminated, followed by 10 years of on-site verification inspections.[3]

21.1 Background

In early 1977, the Soviet Union first deployed the SS-20 Saber (also known as the RSD-10) in its European territories, a mobile, concealable intermediate-range ballistic missile (IRBM) with a multiple independently targetable reentry vehicle (MIRV) containing three nuclear 150-kiloton warheads. The SS-20's range of 4,700–5,000 kilometres (2,900–3,100 mi) was great enough to reach Western Europe from well within Soviet territory; the range was just below the SALT II minimum range for an intercontinental ballistic missile, 5,500 kilometres (3,400 mi).[4][5][6] The SS-20 replaced aging Soviet systems of the SS-4 Sandal and SS-5 Skean, which were seen to pose a limited threat to Western Europe due to their poor accuracy, limited payload (one warhead), lengthy preparation time, difficulty in being concealed, and immobility (thus exposing them to pre-emptive NATO strikes ahead of a planned attack).[7] Whereas the SS-4 and SS-5 were seen as defen-

sive weapons, the SS-20 was seen as a potential offensive system.[8]

The US, then under President Jimmy Carter, initially considered its strategic nuclear weapons and nuclear-capable aircraft to be adequate counters to the SS-20 and a sufficient deterrent against Soviet aggression. In 1977, however, Chancellor Helmut Schmidt of West Germany argued in a speech that a Western response to the SS-20 deployment should be explored, a call which was echoed by NATO, given a perceived Western disadvantage in European nuclear forces.[6] Leslie H. Gelb, the US Assistant Secretary of State, later recounted that Schmidt's speech pressured the US into developing a response.[9]

SS-20 launchers

On 12 December 1979, following European pressure for a response to the SS-20, Western foreign and defense ministers meeting in Brussels made the NATO Double-Track Decision.[6] The ministers argued that the Warsaw Pact had "developed a large and growing capability in nuclear systems that directly threaten Western Europe": "theater" nuclear systems (i.e., tactical nuclear weapons[10]). In describing this "aggravated" situation, the ministers made direct reference to the SS-20 featuring "significant improvements over previous systems in providing greater accuracy, more mobility, and greater range, as well as having multiple warheads." The ministers also attributed the altered situation to the deployment of the Soviet Tupolev Tu-22M strategic bomber, which they believed to display "much greater performance" than its predecessors. Furthermore, the ministers expressed concern that the Soviet Union had gained an advantage over NATO in "Long-Range Theater Nuclear Forces" (LRTNF), and also significantly increased short-range theater nuclear capacity. To address these developments, the ministers adopted two policy "tracks." One thousand theater nuclear warheads, out of 7,400 such warheads, would be removed from Europe and the US would pursue bilateral negotiations with the Soviet Union intended to limit theater nuclear forces. Should these negotiations fail, NATO would modernize its own LRTNF, or intermediate-range nuclear forces (INF), by replacing US Pershing 1a missiles with 108 Pershing II launchers in West Germany and deploying 464 BGM-109G Ground Launched Cruise Missiles (GLCMs) to Belgium, Italy, the Netherlands, and the United Kingdom beginning in December 1983.[5][11][12][13]

21.2   Negotiations

21.2.1   Early negotiations: 1981–83

Despite dissatisfaction with the deployment of US weapons in Europe, the Soviet Union agreed to open negotiations and preliminary discussions, named the Preliminary Intermediate-Range Nuclear Forces Talks,[5] which began in Geneva in October 1980. Formal talks began on 30 November 1981, with the US then led by President Ronald Reagan and the Soviet Union by Leonid Brezhnev. The core of the US negotiating position reflected the principles put forth under Carter: any limits placed on US INF capabilities, both in terms of "ceilings" and "rights," must be reciprocated with limits on Soviet systems. Additionally, the US insisted that a sufficient verification regime be in place.[14] Paul Nitze, a longtime hand at defense policy who had participated in the Strategic Arms Limitation Talks (SALT), led the US delegation after being recruited by Secretary of State Alexander Haig. Though Nitze had backed the first SALT treaty, he opposed SALT II and had resigned from the US delegation during its negotiation. Nitze was also then a member of the Committee on the Present Danger, a firmly anti-Soviet group composed of neoconservatives and conservative Republicans.[15][9] Yuli Kvitsinsky, the well-respected second-ranking official at the Soviet embassy in West Germany, headed the Soviet delegation.[8][16][17][18]

On 18 November 1981, shortly before the beginning of formal talks, Reagan made the Zero Option proposal (or the "zero-zero" proposal). [19] The plan called for a hold on US deployment of GLCM and Pershing II systems, recip-

Paul Nitze, 1983

rocated by Soviet elimination of its SS-4, SS-5, and SS-20 missiles. There appeared to be little chance of the Zero Option being adopted, but the gesture was well received in the European public. In February 1982, US negotiators put forth a draft treaty containing the Zero Option and a global prohibition on intermediate- and short-range missiles, with compliance ensured via a stringent, though unspecific, verification program.[16]

Opinion within the Reagan administration on the Zero Option was mixed. Richard Perle, then the Assistant Secretary of Defense for Global Strategic Affairs, was the architect of the plan. Secretary of Defense Caspar Weinberger, who supported a continued US nuclear presence in Europe, was skeptical of the plan, though eventually accepted it for its value in putting the Soviet Union "on the defensive in the European propaganda war." Reagan later recounted that the "zero option sprang out of the realities of nuclear politics in Western Europe."[19] The Soviet Union rejected the plan shortly after the US tabled it in February 1982, arguing that both the US and Soviet Union should be able to retain intermediate-range missiles in Europe. Specifically, Soviet negotiators proposed that the number of INF missiles and aircraft deployed in Europe by one side be capped at 600 by 1985 and 300 by 1990. Concerned that this proposal would force the US to withdraw aircraft from Europe and not deploy INF missiles, given US cooperation with existing

British and French deployments, the US proposed "equal rights and limits"—the US would be permitted to match Soviet SS-20 deployments.[16]

Between 1981 and 1983, US and Soviet negotiators gathered for six rounds of talks, each two months in length—a system based on the earlier SALT talks.[16] The US delegation was composed of Nitze, General William F. Burns of the Joint Chiefs of Staff, Thomas Graham of the Arms Control and Disarmament Agency (ACDA), and officials from the US Department of State, Office of the Secretary of Defense, and US National Security Council. Colonel Norman Clyne, a SALT participant, served as Nitze's chief of staff.[8][20]

There was little convergence between the two sides over these two years. A U.S. effort to separate the question of nuclear-capable aircraft from that of intermediate-range missiles successfully focused attention on the latter, but little clear progress on the subject was made. In the summer of 1982, Nitze and Kvitsinsky took a "walk in the woods" in the Jura Mountains, away from formal negotiations in Geneva, in an independent attempt to bypass bureaucratic procedures and break the negotiating deadlock.[21][8][22] Nitze later said that his and Kvitsinsky's goal was to agree to certain concessions that would allow for a summit meeting meeting Brezhnev and Reagan later in 1982.[23]

Nitze's offer to Kvitsinsky was that the US would forego deployment of the Pershing II and continue deployment of GLCMs, but limited to 75 missile launchers. The Soviet Union, in return, would also have to limit itself to 75 intermediate-range missile launchers in Europe and 90 in Asia. Due to each GLCM launcher containing four GLCMs and each SS-20 launcher containing three warheads, such an agreement would have resulted in the US having 75 more intermediate-range warheads in Europe than the Soviet Union, though SS-20s were seen as more advanced and maneuverable than GLCMs. While Kvitsinsky was skeptical that the plan would be well received in Moscow, Nitze was optimistic about its chances in Washington.[23] The deal ultimately found little traction in either capital. In the US, the Office of the Secretary of Defense opposed Nitze's proposal, as it opposed any proposal that would allow the Soviet Union to deploy missiles to Europe while blocking US deployments. Nitze's proposal was relayed by Kvitsinsky to Moscow, where it was also rejected. The plan accordingly was never introduced into formal negotiations.[21][8]

Thomas Graham, a US negotiator, later recalled that Nitze's "walk in the woods" proposal was primarily of Nitze's own design and known beforehand only to William F. Burns, another arms control negotiator and representative of the Joint Chiefs of Staff (JCS), and Eugene V. Rostow, the director of the Arms Control and Disarmament Agency. In a National Security Council following the Nitze-Kvitsinsky

Demonstrators protest planned deployments of US missiles, 1982

walk, the proposal was received positively by the JCS and Reagan. Following protests by Richard Perle, working within the Office of the Secretary of Defense, Reagan informed Nitze that he would not back the plan. The State Department, then led by Alexander Haig, also indicated that it would not support Nitze's plan and preferred a return to the Zero Option proposal.[8][22][23] Nitze argued that one positive consequence of the walk in the woods was that the European public, which had doubted US interest in arms control, became convinced that the US was participating in the INF negotiations in good faith.[23]

In early 1983, US negotiators indicated that they would support a plan beyond the Zero Option if the plan established equal rights and limits for the US and Soviet Union, with such limits valid worldwide, and excluded British and French missile systems (as well as those of any other third party). As a temporary measure, the US negotiators also proposed a cap of 450 deployed INF warheads around the world for both the US and Soviet Union. In response, Soviet negotiators expressed that a plan would have to block all US INF deployments in Europe, cover both missiles and aircraft, include third parties, and focus primarily on Europe for it to gain Soviet backing. In the fall of 1983, just ahead of the scheduled deployment of US Pershing IIs and GLCMs, the US lowered its proposed limit on global INF deployments to 420 missiles, while the Soviet Union proposed "equal reductions": if the US cancelled the planned deployment of Pershing II and GLCM systems, the Soviet Union would reduce its own INF deployment by 572 warheads. In November 1983, after the first Pershing IIs arrived in West Germany, the Soviet Union walked out of negotiations, as it had warned it would do should the US missile deployments occur.[24]

21.2.2   Restarted negotiations: 1985–87

In March 1986, negotiations between the US and the Soviet Union resumed, covering not only the INF issue but also separate discussions on strategic weapons (START I) and space issues (Nuclear and Space Talks). In late 1985 both sides were moving towards limiting INF systems in Europe and Asia. On 15 January 1986, Gorbachev announced a Soviet proposal for a ban on all nuclear weapons by 2000, which included INF missiles in Europe. This was dismissed by the US and countered with a phased reduction of INF launchers in Europe and Asia to none by 1989. There would be no constraints on British and French nuclear forces.

A series of meetings in August and September 1986 culminated in the Reykjavík Summit between Reagan and Gorbachev on 11 October 1986. Both agreed in principle to remove INF systems from Europe and to equal global limits of 100 INF missile warheads. Gorbachev also proposed deeper and more fundamental changes in the strategic relationship. More detailed negotiations extended throughout 1987, aided by the decision of West German Chancellor Helmut Kohl in August to unilaterally remove the joint US-West German Pershing 1a systems. The treaty text was finally agreed in September 1987. On 8 December 1987, the Treaty was officially signed by President Reagan and General Secretary Gorbachev at a summit in Washington and ratified the following May in a 93-5 vote by the United States Senate.

21.3   Implementation

By the treaty's deadline of 1 June 1991, a total of 2,692 of such weapons had been destroyed, 846 by the US and 1,846 by the Soviet Union. Under the treaty both nations were allowed to inspect each other's military installations. Each nation was permitted to render inoperative and retain 15 missiles, 15 launch canisters and 15 launchers for static display.

On 13 December 2001, President of the US, George W. Bush gave Russia a 6-month notice of US intent to withdraw from the Anti-Ballistic Missile Treaty so that the US could pursue development of the program at that time known as National Missile Defense (NMD), which was already under way in potential violation of US treaty obligations.[25]

On 10 February 2007, Russian president Vladimir Putin declared that the INF Treaty no longer served Russia's interests. On 14 February, the Information Telegraph Agency of Russia and Interfax quoted General Yuri Baluyevsky, the Chief of General Staff of the Armed Forces of the Russian Federation, as saying that Russia could pull out of the INF, and that the decision would depend on the United States' ac-

A Soviet inspector examines a BGM-109G Gryphon ground-launched cruise missile in 1988 prior to its destruction.

tions with its proposed Ground-Based Midcourse Defense missile defense system, parts of which the U.S. at the time planned to deploy in Poland and the Czech Republic. (Subsequently, the plans were abandoned in favor of different systems based on sea and in Romania; see National missile defense.)

Dan Blumenthal of the American Enterprise Institute wrote that the actual Russian problem with the INF was that China is not bound by it and continued to build up their own Intermediate-Range forces.[26]

In 2012, the US complained about alleged Russian treaty violations.[27] The two systems that appeared to be violations were the R-500, a cruise missile using the 9K720 Iskander launcher, and a short-range ICBM.[28] In July 2014, the United States formally notified Russia of a breach for developing and possessing prohibited weapons, while Russian officials called the treaty unsuitable for Russia and unfair because other countries in Asia had such weapons. Russian officials also Russian officials called the restrictions of the treaty unsuitable for Russia given the strategic situation in Asia.[29]

Russia publicly considered US drones to be a violation of the treaty.[30][31]

21.4   Affected programs

Specific missiles destroyed:

- United States
  - BGM-109G Gryphon
  - Pershing 1b and Pershing II
- Soviet Union (listed by NATO reporting name)
  - SS-4 Sandal

- SS-5 Skean
- SS-12 Scaleboard
- SS-20 Saber
- SS-23 Spider
- SSC-X-4 Slingshot

21.5 See also

- *A Walk in the Woods* – a 1988 play based on Paul Nitze and Yuli Kvitsinsky's "walk in the woods"
- Greenham Common Women's Peace Camp
- Hollanditis
- Woensdrecht

21.6 References

21.6.1 Citations

[1] "Treaty Between the United States of America and the Union of Soviet Socialist Republics on the Elimination of Their Intermediate-Range and Shorter-Range Missiles". Nuclear Threat Initiative. 22 June 2016. Retrieved 16 August 2016.

[2] "INF Treaty". United States Department of State. Retrieved 15 August 2016.

[3] Stockholm International Peace Research Institute (2007). *SIPRI Yearbook 2007: Armaments, Disarmament, and International Security*. New York, NY: Oxford University Press. p. 683.

[4] "RSD-10 MOD 1/-MOD 2 (SS-20)". Missile Threat. 17 October 2012. Retrieved 15 August 2016.

[5] "Intermediate-Range Nuclear Forces [INF] Chronology". Federation of American Scientists. Retrieved 15 August 2016.

[6] Bohlen et al. 2012, p. 7.

[7] Bohlen et al. 2012, pp. 6–7.

[8] "Paul Nitze and A Walk in the Woods — A Failed Attempt at Arms Control". Association for Diplomatic Studies and Training. Retrieved 19 August 2016.

[9] "Interview with Leslie H. Gelb". National Security Archive. 28 February 1999. Retrieved 19 August 2016.

[10] Legge 1983, p. 1.

[11] "Special Meeting of Foreign and Defence Ministers (The "Double-Track" Decision on Theatre Nuclear Forces)". NATO. 12 December 1979. Archived from the original on 27 February 2009. Retrieved 15 August 2016.

[12] Legge 1983, pp. 1–2, 35–37.

[13] Bohlen et al. 2012, pp. 8–9.

[14] Bohlen et al. 2012, pp. 6, 9.

[15] Burr, William; Wampler, Robert (27 October 2004). ""The Master of the Game": Paul H. Nitze and U.S. Cold War Strategy from Truman to Reagan". National Security Archive. Retrieved 19 August 2016.

[16] Bohlen et al. 2012, p. 9.

[17] "Yuli A. Kvitsinsky: Chief Soviet arms control negotiator". *United Press International*. 25 September 1981. Retrieved 19 August 2016.

[18] Freudenheim, Milt; Slavin, Barbara (6 December 1981). "The World in Summary; Arms Negotiators In Geneva Begin To Chip the Ice". *The New York Times*. Retrieved 19 August 2016.

[19] Wittner, Lawrence S. (1 April 2000). "Reagan and Nuclear Disarmament". *Boston Review*. Retrieved 17 August 2016.

[20] "Nomination of William F. Burns To Be Director of the United States Arms Control and Disarmament Agency". Ronald Reagan Presidential Library. 7 January 1988. Retrieved 19 August 2016.

[21] Bohlen et al. 2012, pp. 9–10.

[22] Berger, Marilyn (21 October 2004). "Paul H. Nitze, Missile Treaty Negotiator and Cold War Strategist, Dies at 97". *The New York Times*. Retrieved 20 August 2016.

[23] Nitze, Paul (20 October 1990). *Paul Nitze Interview*. Interview with Academy of Achievement. Washington, D.C. Retrieved 20 August 2016.

[24] Bohlen et al. 2012, p. 10.

[25] Giles & Monaghan 2014.

[26] Mark Stokes and Dan Blumenthal "Can a treaty contain China's missiles?" *Washington Post*, 2 January 2011.

[27] Rogin, Josh (7 December 2013). "US Reluctant to Disclose to All NATO Allies that Russia is Violating INF Treaty". The Atlantic Council. Retrieved 7 December 2013.

[28] Marcus, Jonathan (30 January 2014). "US briefs Nato on Russian 'nuclear treaty breach'". BBC News. Retrieved 31 January 2014.

[29] Luhn, Alec; Borger, Julian (29 July 2014). "Moscow may walk out of nuclear treaty after US accusations of breach". *The Guardian*. Retrieved 29 July 2014.

[30] Adomanis, Mark (31 July 2014). "Russian Nuclear Treaty Violation: The Basics". U.S. Naval Institute. Retrieved 31 July 2014.

[31] "Russia: US claims on nuclear missiles treaty unfounded, we have questions too". Russia Today. 30 July 2014. Retrieved 15 August 2016.

21.6.2 Publications

- Bohlen, Avis; Burns, William; Pifer, Steven; Woodworth, John (2012). The Treaty on Intermediate-Range Nuclear Forces: History and Lessons Learned (PDF) (Report). Washington, D.C.: Brookings Institution. Retrieved 16 August 2016.

- Garthoff, Raymond L. (1983). "The NATO Decision on Theater Nuclear Forces". *Political Science Quarterly*. **98** (2): 197–214.

- Giles, Keir; Monaghan, Andrew (2014). *European Missile Defense and Russia*. Carlisle Barracks, PA: United States Army War College Press. ISBN 1-58487-635-2.

- Haass, Richard (1988). *Beyond the INF Treaty: Arms, Arms Control, and the Atlantic Alliance*. Lanham, MD: University Press of America. ISBN 978-0-819-16942-6.

- Legge, J. Michael (1983). Theater Nuclear Weapons and the NATO Strategy of Flexible Response (PDF) (Report). RAND Corporation. Retrieved 15 August 2016.

- Rhodes, Richard (2008). *Arsenals of Folly: The Making of the Nuclear Arms Race*. New York, NY: Vintage. ISBN 978-0-375-71394-1.

21.7 External links

- Text of the INF Treaty

- Video of a 1986 PBS program on the future of arms control

- Video of a 1986 year-in-review for the Soviet Union

- Statements by Ronald Reagan on INF Treaty negotiations in March, April, June, and December 1987

Chapter 22

Strategic Offensive Reductions Treaty

Not to be confused with Strategic Arms Reduction Treaty or Strategic Arms Limitation Talks.

"SORT" redirects here. For other uses, see SORT (disambiguation).

The **Treaty Between the United States of America and the Russian Federation on Strategic Offensive Reductions (SORT)**, also known as the **Treaty of Moscow**, was a strategic arms reduction treaty between the United States and Russia that was in force from June 2003 until February 2011 when it was superseded by the New START treaty.[1] At the time, SORT was positioned as "represent[ing] an important element of the new strategic relationship" between the two countries[2] with both parties agreeing to limit their nuclear arsenal to between 1,700 and 2,200 operationally deployed warheads each. It was signed in Moscow on 24 May 2002. After ratification by the U.S. Senate and the State Duma, SORT came into force on 1 June 2003. It would have expired on 31 December 2012 if not superseded by New START. Either party could have withdrawn from the treaty upon giving three months written notice to the other.

22.1 Mutual nuclear disarmament

SORT was one in a long line of treaties and negotiations on mutual nuclear disarmament between Russia (and its predecessor, the Soviet Union) and the United States, which includes SALT I (1969–1972), the ABM Treaty (1972), SALT II (1972–1979), the INF Treaty (1987), START I (1991), START II (1993) and New START (2010).

The Moscow Treaty was different from START in that it limited operationally deployed warheads, whereas START I limited warheads through declared attribution to their means of delivery (ICBMs, SLBMs, and Heavy Bombers).[3]

Russian and U.S. delegations met twice a year to discuss the implementation of the Moscow Treaty at the Bilateral Implementation Commission (BIC).

22.2 Ratification

The treaty was submitted for ratification on December 2002. However, the passage of the agreement took about a year because the bill had to be resubmitted after its rejection in committee due to concerns about funding for nuclear forces and about cutting systems that had not yet reached the end of their service lives. Further, the deputies were concerned about the U.S. ability to upload reserve nuclear warheads for a first strike (upload potential).

The ratification was also problematic because the chairman of the foreign affairs committee of the Duma, Dmitry Rogozin, disagreed with his Federation Council counterpart Margelov. Deputy Rogozin argued that the Moscow Treaty should be delayed because of the 2003 U.S. invasion of Iraq. In the end, however, this delay never happened. The final vote was similar to START II with nearly a third of the deputies voting against. The ratification resolution mandated presidential reporting on nuclear force developments and noted that key legislators should be included in interagency planning.

22.3 Implementation

Lawrence Livermore National Laboratory reported that President Bush directed the US military to cut its stockpile of both deployed and reserve nuclear weapons in half by 2012. The goal was achieved in 2007, a reduction of US nuclear warheads to just over 50 percent of the 2001 total. A further proposal by Bush would have brought the total down another 15 percent.[4]

22.4   Criticism

The treaty was criticized for various reasons:

- There were no verification provisions to give confidence, to either the signatories or other parties, that the stated reductions have in fact taken place.

- The arsenal reductions were not required to be permanent; warheads are not required to be destroyed and may therefore be placed in storage and later redeployed.

- The arsenal reductions were required to be completed by 31 December 2012, which is also the day on which the treaty loses all force, unless extended by both parties.

- There was a clause in the treaty which provided that withdrawal can occur upon the giving of three months' notice and since no benchmarks are required in the treaty, either side could feasibly perform no actions in furtherance of the treaty, and then withdraw in September 2012.

22.5   See also

- There have been several other treaties known as the Treaty of Moscow

- Russia and weapons of mass destruction

- United States and weapons of mass destruction

- New START Treaty

22.6   Further reading

- Nuclear Files.org Text of the SORT

22.7   Footnotes

[1] http://www.whitehouse.gov/sites/default/files/2010%20New%20START%20msg%20rel.pdf

[2] Letter of Transmittal: The Moscow Treaty 2002

[3] START1 treaty hypertext US State Dept. Article II

[4] Lawrence Livermore National Laboratories.   Science & Technology Review. *Monitoring a Nuclear Weapon from the Inside: Embedded sensors could help transform stockpile stewardship*

Chapter 23

Comprehensive Nuclear-Test-Ban Treaty

The **Comprehensive Nuclear-Test-Ban Treaty (CTBT)** is a multilateral treaty that bans all nuclear explosions, for both civilian and military purposes, in all environments. It was adopted by the United Nations General Assembly on 10 September 1996 but has not entered into force as eight specific states have not yet ratified the treaty.

23.1 History

The movement for international control of nuclear weapons began in 1945, with a call from Canada and United Kingdom for a conference on the subject.[1] In June 1946, Bernard Baruch, an emissary of President Harry S. Truman, proposed the Baruch Plan before the United Nations Atomic Energy Commission, which called for an international system of controls on the production of atomic energy. The plan, which would serve as the basis for United States nuclear policy into the 1950s, was rejected by the Soviet Union as a US ploy to cement its nuclear dominance.[2][3]

Between the Trinity nuclear test of 16 July 1945 and the signing of the Partial Test Ban Treaty (PTBT) on 5 August 1963, 499 nuclear tests were conducted.[4] Much of the impetus for the PTBT, the precursor to the CTBT, was rising public concern surrounding the size and resulting nuclear fallout from underwater and atmospheric nuclear tests, particularly tests of powerful thermonuclear weapons (hydrogen bombs). The Castle Bravo test of 1 March 1954, in particular, attracted significant attention as the detonation resulted in fallout that spread over inhabited areas and sickened a group of Japanese fishermen.[5][6][7][8][9] Between 1945 and 1963, the US conducted 215 atmospheric tests, the Soviet Union conducted 219, the UK conducted 21, and France conducted three.[10]

In 1954, following the Castle Bravo test, Prime Minister Jawaharlal Nehru of India issued the first appeal for a "standstill agreement" on testing, which was soon echoed by the British Labour Party.[11][12][13] Negotiations on a comprehensive test ban, primarily involved the US, UK, and Soviet Union, began in 1955 following a proposal by Soviet leader Nikita Khrushchev.[14][15] Of primary concern throughout the negotiations, which would stretch with some interruptions to July 1963, was the system of verifying compliance with the test ban and detecting illicit tests. On the Western side, there were concerns that the Soviet Union would be able to circumvent any test ban and secretly leap ahead in the nuclear arms race.[16][17][18] These fears were amplified following the US *Rainier* shot of 19 September 1957, which was the first contained underground test of a nuclear weapon. Though the US held a significant advantage in underground testing capabilities, there was worry that the Soviet Union would be able to covertly conduct underground tests during a test ban, as underground detonations were more difficult to detect than above-ground tests.[19][20] On the Soviet side, conversely, the on-site compliance inspections demanded by the US and UK were seen as amounting to espionage.[21] Disagreement over verification would lead to the Anglo-American and Soviet negotiators abandoning a comprehensive test ban (i.e., a ban on all tests, including those underground) in favor of a partial ban, which would be finalized on 25 July 1963. The PTBT, joined by 123 states following the original three parties, banned detonations for military and civilian purposes underwater, in the atmosphere, and in outer space.[22][23][24]

The PTBT had mixed results. On the one hand, enactment of the treaty was followed by a substantial drop in the atmospheric concentration of radioactive particles.[25][26] On the other hand, nuclear proliferation was not halted entirely (though it may have been slowed) and nuclear testing continued at a rapid clip. Compared to the 499 tests from 1945 to the signing of the PTBT, 436 tests were conducted over the ten years following the PTBT.[27][14] Furthermore, US and Soviet underground testing continued "venting" radioactive gas into the atmosphere.[28] Additionally, though underground testing was generally safer than above-ground testing, underground tests continued to risk the leaking of radionuclides, including plutonium, into the ground.[29][30][31] From 1964 through 1996, the year of the CTBT's adoption, an estimated 1,377 underground nuclear tests were conducted. The final non-underground (atmospheric or underwater) test was conducted by China in 1980.[32][33]

The PTBT has been seen as a step towards the Nuclear Nonproliferation Treaty (NPT) of 1968, which directly made reference to the PTBT.[34] Under the NPT, non-nuclear weapon states were prohibited from possessing, manufacturing, and acquiring nuclear weapons or other nuclear explosive devices. All signatories, including nuclear weapon states, were committed to the goal of total nuclear disarmament. However, India, Pakistan, and Israel have declined to sign the NPT on grounds that such a treaty is fundamentally discriminatory as it places limitations on states that do not have nuclear weapons while making no efforts to curb weapons development by declared nuclear weapons states.

In 1974, a step towards a comprehensive test ban was made with the Threshold Test Ban Treaty (TTBT), ratified by the US and Soviet Union, which banned underground tests with yields above 150 kilotons.[28][35] In April 1976, the two states reached agreement on the Peaceful Nuclear Explosions Treaty (PNET), which concerns nuclear detonations outside the weapons sites discussed in the TTBT. As in the TTBT, the US and Soviet Union agreed to bar peaceful nuclear explosions (PNEs) at these other locations with yields above 150 kilotons, as well as group explosions with total yields in excess of 1,500 kilotons. To verify compliance, the PNET requires that states rely on national technical means of verification, share information on explosions, and grant on-site access to counterparties. The TTBT and PNET did not enter into force for the US and Soviet Union until 11 December 1990.[36]

Reagan and Gorbachev, December 1987

In October 1977, the US, UK, and Soviet Union returned to negotiations over a test ban. The three nuclear powers made notable progress in the late 1970s, agreeing to terms on a ban on all testing, including a temporary prohibition on PNEs, but continued disagreements over the compliance mechanisms led to an end to negotiations ahead of Ronald Reagan's inauguration as President in 1981.[34] In 1985, Soviet leader Mikhail Gorbachev announced a unilateral testing moratorium, and in December 1986, Reagan reaffirmed US commitment to pursue the long-term goal of a comprehensive test ban. In November 1987, negotiations on a test ban restarted, followed by a joint US-Soviet program to research underground-test detection in December 1987.[34][37]

23.2   Negotiations

Given the political situation prevailing in the subsequent decades, little progress was made in nuclear disarmament until the end of the Cold War in 1991. Parties to the PTBT held an amendment conference that year to discuss a proposal to convert the Treaty into an instrument banning all nuclear-weapon tests. With strong support from the UN General Assembly, negotiations for a comprehensive test-ban treaty began in 1993.

23.2.1   Adoption

Intensive efforts were made over the next three years to draft the Treaty text and its two annexes. However, the Conference on Disarmament, in which negotiations were being held, did not succeed in reaching consensus on the adoption of the text. Under the direction of Prime Minister John Howard and Foreign Minister Alexander Downer, Australia then sent the text to the United Nations General Assembly in New York, where it was submitted as a draft resolution.[38] On 10 September 1996, the Comprehensive Test-Ban Treaty (CTBT) was adopted by a large majority, exceeding two-thirds of the General Assembly's Membership.[39]

23.2.2   Obligations

(Article I):[40]

1. Each State Party undertakes not to carry out any nuclear weapon test explosion or any other nuclear explosion, and to prohibit and prevent any such nuclear explosion at any place under its jurisdiction or control.

2. Each State Party undertakes, furthermore, to refrain from causing, encouraging, or in any way participating in the carrying out of any nuclear weapon test explosion or any other nuclear explosion.

23.3 Status

Further information: List of parties to the Comprehensive Nuclear-Test-Ban Treaty

The Treaty was adopted by the United Nations General Assembly on 10 September 1996.[41] It opened for signature in New York on 24 September 1996,[41] when it was signed by 71 States, including five of the eight then nuclear-capable states. As of March 2015, 164 states have ratified the CTBT and another 19 states have signed but not ratified it.[42][43]

The treaty will enter into force 180 days after the 44 states listed in Annex 2 of the treaty have ratified it. These "Annex 2 states" are states that participated in the CTBT's negotiations between 1994 and 1996 and possessed nuclear power reactors or research reactors at that time.[44] As of 2015, eight Annex 2 states have not ratified the treaty: China, Egypt, Iran, Israel and the United States have signed but not ratified the Treaty; India, North Korea and Pakistan have not signed it.[45]

23.4 Monitoring

Geophysical and other technologies are used to monitor for compliance with the Treaty: forensic seismology, hydroacoustics, infrasound, and radionuclide monitoring. The technologies are used to monitor the underground, the waters and the atmosphere for any sign of a nuclear explosion. Statistical theories and methods are integral to CTBT monitoring providing confidence in verification analysis. Once the Treaty enters into force, on site inspection will be provided for where concerns about compliance arise.

The Preparatory Commission for the Comprehensive Test Ban Treaty Organization (CTBTO), an international organization headquartered in Vienna, Austria, was created to build the verification regime, including establishment and provisional operation of the network of monitoring stations, the creation of an international data centre, and development of the On Site Inspection capability.

The monitoring network consists of 337 facilities located all over the globe. As of May 2012, more than 260 facilities have been certified. The monitoring stations register data that is transmitted to the international data centre in Vienna for processing and analysis. The data are sent to states that have signed the Treaty.[46]

23.4.1 Subsequent nuclear testing

Three countries have tested nuclear weapons since the CTBT opened for signature in 1996. India and Pakistan both carried out two sets of tests in 1998. North Korea carried out four announced tests in 2006, 2009 and 2013 and 2016. All four North Korean tests were picked up by the International Monitoring System set up by the Comprehensive Nuclear-Test-Ban Treaty Organization Preparatory Commission. A North Korean test is believed to have taken place in January 2016, evidenced by an "artificial earthquake" measured as a magnitude 5.1 by the U.S. Geological Survey.[47][48][49]

23.5 See also

- List of weapons of mass destruction treaties
- Comprehensive Nuclear-Test-Ban Treaty Organization
- Comprehensive Nuclear-Test-Ban Treaty Organization Preparatory Commission
- National technical means of verification
- Nuclear-free zone

23.6 References

23.6.1 Citations

[1] Polsby 1984, p. 56.

[2] Strode 1990, p. 7.

[3] Polsby 1984, pp. 57–58.

[4] Delcoigne, G.C. "The Test Ban Treaty" (PDF). IAEA. p. 18. Retrieved 11 August 2016.

[5] "Limited or Partial Test Ban Treaty (LTBT/PTBT)". Atomic Heritage Foundation. Retrieved 1 August 2016.

[6] Burr, William; Montford, Hector L. (3 August 2003). "The Making of the Limited Test Ban Treaty, 1958–1963". National Security Archive. Retrieved 7 August 2016.

[7] "Treaty Banning Nuclear Tests in the Atmosphere, in Outer Space and Under Water (Partial Test Ban Treaty) (PTBT)". Nuclear Threat Initiative. 26 October 2011. Retrieved 31 July 2016.

[8] Rhodes 2005, p. 542.

[9] Strode 1990, p. 31.

[10] "Archive of Nuclear Data". Natural Resources Defense Council. Archived from the original on 10 October 2007. Retrieved 6 August 2016.

[11] Burns & Siracusa 2013, p. 247.

[12] Polsby 1984, p. 58.

[13] "1 March 1954 – Castle Bravo". Preparatory Commission for the Comprehensive Nuclear-Test-Ban Treaty Organization. Retrieved 31 July 2016.

[14] Rhodes 2008, p. 72.

[15] Reeves 1993, p. 121.

[16] Burns & Siracusa 2013, p. 305.

[17] Ambrose 1991, pp. 457–458.

[18] Seaborg 1981, pp. 8–9.

[19] Seaborg 1981, p. 9.

[20] Evangelista 1999, pp. 85–86.

[21] Evangelista 1999, p. 79.

[22] Schlesinger 2002, pp. 905–906, 910.

[23] "Comprehensive Nuclear Test Ban Treaty & Partial Test Ban Treaty Membership" (PDF). Nuclear Threat Initiative. 8 June 2015. Retrieved 11 August 2016.

[24] "Treaty Banning Nuclear Weapon Tests in the Atmosphere, in Outer Space and Under Water". United Nations Office for Disarmament Affairs. Retrieved 11 August 2016.

[25] "Radiocarbon Dating". Utrecht University. Retrieved 31 July 2016.

[26] "The Technical Details: The Bomb Spike". National Oceanic and Atmospheric Administration. Retrieved 12 August 2016.

[27] Delcoigne, G.C. "The Test Ban Treaty" (PDF). IAEA. p. 18. Retrieved 11 August 2016.

[28] Burr, William (2 August 2013). "The Limited Test Ban Treaty – 50 Years Later: New Documents Throw Light on Accord Banning Atmospheric Nuclear Testing". National Security Archive. Retrieved 12 August 2016.

[29] "General Overview of the Effects of Nuclear Testing". CTBTO Preparatory Commission. Retrieved 12 August 2016.

[30] "Fallout from Nuclear Weapons". Report on the Health Consequences to the American Population from Nuclear Weapons Tests Conducted by the United States and Other Nations (Report). Centers for Disease Control and Prevention. May 2005. pp. 20–21.

[31] Daryl Kimball and Wade Boese (June 2009). "Limited Test Ban Treaty Turns 40". Arms Control Association. Retrieved 21 May 2012.

[32] Stockholm International Peace Research Institute (2007). Armaments, Disarmament and International Security (PDF). New York, NY: Oxford University Press. pp. 555–556.

[33] "Nuclear testing world overview". Preparatory Commission for the Comprehensive Nuclear-Test-Ban Treaty Organization. 2012. Retrieved 21 May 2012.

[34] "Comprehensive Test Ban Treaty Chronology". Federation of American Scientists. Retrieved 7 August 2016.

[35] "The Flawed Test Ban Treaty". Heritage Foundation. 27 March 1984. Retrieved 12 August 2016.

[36] "Peaceful Nuclear Explosions Treaty (PNET)". United States Department of State. Retrieved 14 August 2016.

[37] Blakeslee, Sandra (18 August 1988). "In Remotest Nevada, a Joint U.S. and Soviet Test". The New York Times. Retrieved 11 August 2016.

[38] "Comprehensive nuclear-test-ban treaty: draft resolution". United Nations. 6 September 1996. Retrieved 3 December 2011.

[39] "Resolution adopted by the general assembly:50/245. Comprehensive Nuclear-Test-Ban Treaty". United Nations. 17 September 1996. Retrieved 3 December 2011.

[40] "Comprehensive Nuclear-Test-Ban Treaty CTBTO" (PDF). CTBTO Preparatory Commission. Retrieved 4 December 2011.

[41] United Nations Treaty Collection (2009). "Comprehensive Nuclear-Test-Ban Treaty". Retrieved 23 August 2009.

[42] "Status of Signature and Ratification". Preparatory Commission for the Comprehensive Nuclear-Test-Ban Treaty Organization. 2011. Retrieved 20 September 2011.

[43] David E. Hoffman (1 November 2011), "Supercomputers offer tools for nuclear testing — and solving nuclear mysteries", The Washington Post; National, retrieved 30 November 2013
In this news article, the number of states ratifying was reported as 154.

[44] "The Russian Federation's support for the Comprehensive Nuclear-Test-Ban Treaty". CTBTO Preparatory Commission. 2008. Retrieved 4 December 2011.

[45] "STATE DEPARTMENT TELEGRAM 012545 TO INTSUM COLLECTIVE, "INTSUM: INDIA: NUCLEAR TEST UNLIKELY"". Nuclear Proliferation International History Project.

[46] "US nuclear security administrator dagostino visits the CTBTO". CTBTO Preparatory Commission. 15 September 2009. Retrieved 4 December 2011.

[47] "Highlight 2007: The CTBT Verification Regime Put to the Test – The Event in the DPRK on 9 October 2006". Preparatory Commission for the Comprehensive Nuclear-Test-Ban Treaty Organization. 2012. Retrieved 21 May 2012.

[48] "Press Release June 2009: Experts Sure About Nature of the DPRK Event". *Preparatory Commission for the Comprehensive Nuclear-Test-Ban Treaty Organization*. 2012. Retrieved 21 May 2012.

[49] McKirdy, Euan (6 January 2016). "North Korea announces it conducted nuclear test". CNN. Retrieved 15 August 2016.

23.6.2 Publications

- Ambrose, Stephen E. (1991). *Eisenhower: Soldier and President*. New York, NY: Simon & Schuster. ISBN 978-0671747589.

- Burns, Richard Dean; Siracusa, Joseph M. (2013). *A Global History of the Nuclear Arms Race: Weapons, Strategy, and Politics – Volume 1*. Santa Barbara, CA: ABC-CLIO. ISBN 978-1440800955.

- Evangelista, Matthew (1999). *Unarmed Forces: The Transnational Movement to End the Cold War*. Ithaca, NY: Cornell University Press. ISBN 978-0801487842.

- Polsby, Nelson W. (1984). *Political Innovation in America: The Politics of Policy Initiation*. New Haven, CT: Yale University Press. ISBN 978-0300034288.

- Strode, Rebecca (1990). "Soviet Policy Toward a Nuclear Test Ban: 1958–1963". In Mandelbaum, Michael. *The Other Side of the Table: The Soviet Approach to Arms Control*. New York and London: Council on Foreign Relations Press. ISBN 978-0876090718.

- Reeves, Richard (1993). *President Kennedy: Profile of Power*. New York, NY: Simon & Schuster. ISBN 978-0671892890.

- Rhodes, Richard (2005). *Dark Sun: The Making of the Hydrogen Bomb*. New York, NY: Simon & Schuster. ISBN 978-0684824147.

- Rhodes, Richard (2008). *Arsenals of Folly: The Making of the Nuclear Arms Race*. New York, NY: Vintage. ISBN 978-0375713941.

- Schlesinger, Arthur Meier Jr. (2002). *A Thousand Days: John F. Kennedy in the White House*. New York, NY: Houghton Mifflin. ISBN 978-0618219278.

- Seaborg, Glenn T. (1981). *Kennedy, Khrushchev, and the Test Ban*. Berkeley and Los Angeles, CA: University of California Press. ISBN 978-0520049611.

23.7 External links

- Full text of the treaty

- CTBTO Preparatory Commission – official news and information

- The Test Ban Test: U.S. Rejection has Scuttled the CTBT

- US conducts subcritical nuclear test ABC News, 24 February 2006

- International Physicians for the Prevention of Nuclear War, 1991

- Daryl Kimball and Christine Kucia, Arms Control Association, 2002

- General John M. Shalikashvili, Special Advisor to the President and the Secretary of State for the Comprehensive Test Ban Treaty

- Christopher Paine, Senior Researcher with NRDC's Nuclear Program, 1999

- Obama or McCain Can Finish Journey to Nuclear Test Ban Treaty

- For the number of nuclear explosions conducted in various parts of the globe from 1954-1998 see – http://blip.tv/file/1662914/

- Introductory note by Thomas Graham, Jr., procedural history note and audiovisual material on the *Comprehensive Nuclear Test Ban Treaty* in the United Nations Audiovisual Library of International Law

- Lecture by Masahiko Asada entitled *Nuclear Weapons and International Law* in the Lecture Series of the United Nations Audiovisual Library of International Law

- Comprehensive Nuclear-Test-Ban Treaty: Background and Current Developments Congressional Research Service

- The Woodrow Wilson Center's Nuclear Proliferation International History Project or NPIHP is a global network of individuals and institutions engaged in the study of international nuclear history through archival documents, oral history interviews and other empirical sources.

Chapter 24

List of states with nuclear weapons

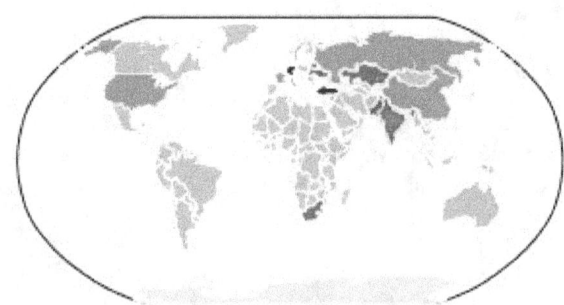

Map of nuclear-armed states of the world.
NPT-designated nuclear weapon states (China, France, Russia, United Kingdom, United States)
Other states with nuclear weapons (India, North Korea, Pakistan)
Other states presumed to have nuclear weapons (Israel)
NATO nuclear weapons sharing states (Belgium, Germany, Italy, Netherlands, Turkey)
States formerly possessing nuclear weapons (Belarus, Kazakhstan, South Africa, Ukraine)

There are eight sovereign states that have successfully detonated nuclear weapons.[1] Five are considered to be "**nuclear-weapon states**" (NWS) under the terms of the Treaty on the Non-Proliferation of Nuclear Weapons (NPT). In order of acquisition of nuclear weapons these are: the United States, the Russian Federation (successor state to the Soviet Union), the United Kingdom, France, and China.

Since the NPT entered into force in 1970, three states that were not parties to the Treaty have conducted nuclear tests, namely India, Pakistan, and North Korea. North Korea had been a party to the NPT but withdrew in 2003. Israel is also widely presumed to have nuclear weapons, though it maintains a policy of deliberate ambiguity regarding this, and is not known definitively to have conducted a nuclear test.[2] According to the Stockholm International Peace Research Institute's SIPRI Yearbook of 2014, Israel is estimated to have approximately 80 nuclear warheads.[3] Furthermore, according to Bulletin of the Atomic Scientists Nuclear Notebook 2014, the total number of nuclear weapons worldwide is estimated at 10,144.[4]

South Africa has the unique status of a nation that developed nuclear weapons but then disassembled its arsenal before joining the NPT. This means that there are three European countries, one country in North America, zero in South America, four Asian countries, zero Oceanian countries and zero African countries that are known to have nuclear weapons.[5] Nations that are known or thought to have nuclear weapons are sometimes referred to informally as the **nuclear club**.

24.1 Statistics and force configuration

The following is a list of states that have admitted the possession of nuclear weapons or are presumed to possess them, the approximate number of warheads under their control, and the year they tested their first weapon and their force configuration. This list is informally known in global politics as the "Nuclear Club."[6] With the exception of Russia and the United States (which have subjected their nuclear forces to independent verification under various treaties) these figures are estimates, in some cases quite unreliable estimates. In particular, under the Strategic Offensive Reductions Treaty thousands of Russian and U.S. nuclear warheads are inactive in stockpiles awaiting processing. The fissile material contained in the warheads can then be recycled for use in nuclear reactors.

From a high of 68,000 active weapons in 1985, as of 2015 there are some 4,000 active nuclear warheads and 10,300 total nuclear warheads in the world.[1] Many of the decommissioned weapons were simply stored or partially dismantled, not destroyed.[7]

It is also noteworthy that since the dawn of the Atomic Age the delivery methods of most states with nuclear weapons has evolved with some achieving a nuclear triad while others have consolidated away from land and air deterrents to submarine based forces.

24.2 Five nuclear-weapon states under the NPT

See also: History of nuclear weapons

These five states are also the UN Security Council's per-

An early stage in the "Trinity" fireball, the first nuclear explosion, 1945

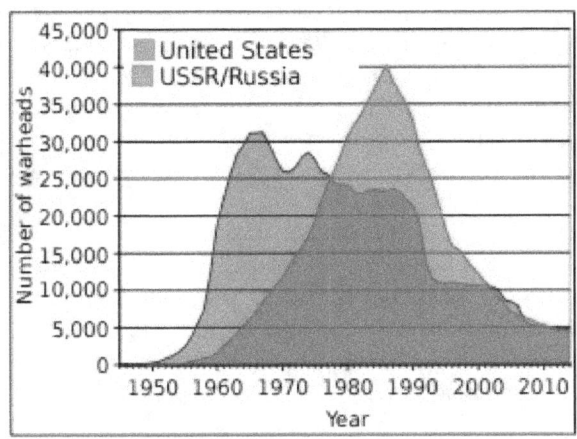

U.S. and USSR/Russian nuclear weapons stockpiles, 1945–2006

manent members with veto power.

A Trident missile launched from a Royal Navy Vanguard *class ballistic missile submarine*

French nuclear-powered aircraft carrier Charles de Gaulle *(right) and the American nuclear-powered carrier USS* Enterprise *(left), each of which carries nuclear-capable warplanes*

24.2.1 United States

Main articles: Nuclear weapons and the United States and United States and weapons of mass destruction

The United States developed the first atomic weapons, during World War II in cooperation with the United Kingdom and Canada as part of the Manhattan Project, out of the fear that Nazi Germany would develop them first. It tested the first nuclear weapon July 16, 1945 ("Trinity") at 5:30 am, and remains the only country to have used nuclear weapons against an enemy state in warfare, devastating the Japanese cities of Hiroshima and Nagasaki. It was the first nation to develop the hydrogen bomb, testing an experimental prototype in 1952 ("Ivy Mike") and a deployable weapon in 1954 ("Castle Bravo"). Throughout the Cold War it continued to modernize and enlarge its nuclear arsenal, but from 1992 on has been involved primarily in a program of Stockpile stewardship.[28][29][30] The U.S. nuclear arsenal contained 31,175 warheads at its Cold War height (in 1966).[31] During the Cold War the United States built approximately 70,000 nuclear warheads, more than all other nuclear-weapon states combined.[32][33]

24.2.2   Russian Federation (formerly Soviet Union)

Main articles: Russia and weapons of mass destruction and Strategic Missile Troops

The Soviet Union tested its first nuclear weapon ("RDS-1") in 1949, in a crash project developed partially with espionage obtained during and after World War II (see: Soviet atomic bomb project). The Soviet Union was the second nation to have developed and tested a nuclear weapon. The direct motivation for Soviet weapons development was to achieve a balance of power during the Cold War. It tested its first megaton-range hydrogen bomb ("RDS-37") in 1955. The Soviet Union also tested the most powerful explosive ever detonated by humans, ("Tsar Bomba"), with a theoretical yield of 100 megatons, intentionally reduced to 50 when detonated. After its dissolution in 1991, the Soviet weapons entered officially into the possession of the Russian Federation.[34] The Soviet nuclear arsenal contained some 45,000 warheads at its peak (in 1986); the Soviet Union built about 55,000 nuclear warheads since 1949.[33]

24.2.3   United Kingdom

Main articles: Nuclear weapons and the United Kingdom and United Kingdom and weapons of mass destruction

The United Kingdom tested its first nuclear weapon ("Hurricane") in 1952. The UK had provided considerable impetus and initial research for the early conception of the atomic bomb, aided by the presence of refugee scientists working in British laboratories who had fled the continent. It collaborated closely with the United States and Canada during the Manhattan Project, but had to develop its own method for manufacturing and detonating a bomb as U.S. secrecy grew after 1945. The United Kingdom was the third country in the world, after the United States and Soviet Union, to develop and test a nuclear weapon. Its programme was motivated to have an independent deterrent against the Soviet Union, while also maintaining its status as a great power. It tested its first hydrogen bomb in 1957 (Operation Grapple), making it the third country to do so after the United States and Soviet Union.[35][36] The UK maintained a fleet of V bomber strategic bombers and ballistic missile submarines (SSBNs) equipped with nuclear weapons during the Cold War. It currently maintains a fleet of four 'Vanguard' class ballistic missile submarines equipped with Trident II missiles. In 2016, the UK House

of Commons voted to renew the British nuclear deterrent, without setting a date for the commencement of service of a replacement to the current system.

24.2.4   France

Main articles: France and weapons of mass destruction and Force de dissuasion

France tested its first nuclear weapon in 1960 ("Gerboise Bleue"), based mostly on its own research. It was motivated by the Suez Crisis diplomatic tension *vis-à-vis* both the Soviet Union and the Free World allies United States and United Kingdom. It was also relevant to retain great power status, alongside the United Kingdom, during the post-colonial Cold War (see: Force de frappe). France tested its first hydrogen bomb in 1968 ("Opération Canopus"). After the Cold War, France has disarmed 175 warheads with the reduction and modernization of its arsenal that has now evolved to a dual system based on submarine-launched ballistic missiles (SLBMs) and medium-range air-to-surface missiles (Rafale fighter-bombers). However new nuclear weapons are in development and reformed nuclear squadrons were trained during Enduring Freedom operations in Afghanistan. France signed the Nuclear Non-Proliferation Treaty in 1992.[37] In January 2006, President Jacques Chirac stated a terrorist act or the use of weapons of mass destruction against France would result in a nuclear counterattack.[38] In February 2015, President Francois Hollande stressed the need for a nuclear deterrent in "a dangerous world". He also detailed the French deterrent as "less than 300" nuclear warheads, three sets of 16 submarine-launched ballistic missiles and 54 medium-range air-to-surface missiles" and urged other states to show similar transparency.[39]

24.2.5   China

Main articles: China and weapons of mass destruction and People's Liberation Army Rocket Force

China tested its first nuclear weapon device ("596") in 1964 at the Lop Nur test site. The weapon was developed as a deterrent against both the United States and the Soviet Union. Two years later, China had a fission bomb capable of being put onto a nuclear missile. It tested its first hydrogen bomb ("Test No. 6") in 1967, a mere 32 months after testing its first nuclear weapon (the shortest fission-to-fusion development known in history).[40] The country is currently thought to have had a stockpile of around 240 warheads, though because of the limited information available, estimates range from 100 to 400.[41][42][43] China is the only NPT nuclear-weapon state to give an unqualified negative security assurance due to its "no first use" policy.[44][45] China signed the Nuclear Non-Proliferation Treaty in 1992.[37] On February 25, 2015 U.S. Vice Admiral Joseph Mulloy stated to the House Armed Services Committee's seapower subcommittee that the U.S. does not believe the PLAN currently deploys SLBMs on their submarine fleet.[46]

24.3 Other states declaring possession of nuclear weapons

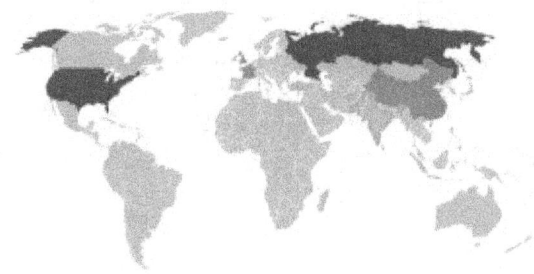

Large stockpile with global range (dark blue), smaller stockpile with global range (medium blue), small stockpile with regional range (light blue)

24.3.1 India

Main article: India and weapons of mass destruction

India is not a party to the Nuclear Non-Proliferation Treaty. India tested what it called a "peaceful nuclear explosive" in 1974 (which became known as "Smiling Buddha"). The test was the first test developed after the creation of the NPT, and created new questions about how civilian nuclear technology could be diverted secretly to weapons purposes (dual-use technology). India's secret development caused great concern and anger particularly from nations, such as Canada, that had supplied its nuclear reactors for peaceful and power generating needs.

Indian officials rejected the NPT in the 1960s on the grounds that it created a world of nuclear "haves" and "have-nots", arguing that it unnecessarily restricted "peaceful activity" (including "peaceful nuclear explosives"), and that India would not accede to international control of their nuclear facilities unless all other countries engaged in unilateral disarmament of their own nuclear weapons. The Indian position has also asserted that the NPT is in many ways a neo-colonial regime designed to deny security to post-colonial powers.[47] Even after its 1974 test, India maintained that its nuclear capability was primar-

ily "peaceful", but between 1988 and 1990 it apparently weaponized two dozen nuclear weapons for delivery by air.[48] In 1998 India tested weaponized nuclear warheads ("Operation Shakti"), including a thermonuclear device.[49]

In July 2005, U.S. President George W. Bush and Indian Prime Minister Manmohan Singh announced plans to conclude an Indo-US civilian nuclear agreement.[50] This came to fruition through a series of steps that included India's announced plan to separate its civil and military nuclear programs in March 2006,[51] the passage of the India–United States Civil Nuclear Agreement by the U.S. Congress in December 2006, the conclusion of a U.S.–India nuclear cooperation agreement in July 2007,[52] approval by the IAEA of an India-specific safeguards agreement,[53] agreement by the Nuclear Suppliers Group to a waiver of export restrictions for India,[54] approval by the U.S. Congress[55] and culminating in the signature of U.S.–India agreement for civil nuclear cooperation[56] in October 2008. The U.S. State Department said it made it "very clear that we will not recognize India as a nuclear-weapon state".[57] The United States is bound by the Hyde Act with India and may cease all cooperation with India if India detonates a nuclear explosive device. The US had further said it is not its intention to assist India in the design, construction or operation of sensitive nuclear technologies through the transfer of dual-use items.[58] In establishing an exemption for India, the Nuclear Suppliers Group reserved the right to consult on any future issues which might trouble it.[59] As of early 2013, India was estimated to have had a stockpile of around 90–110 warheads.[1]

24.3.2 Pakistan

Main article: Pakistan and weapons of mass destruction

Pakistan also is not a party to the Nuclear Non-Proliferation Treaty. Pakistan covertly developed nuclear weapons over decades, beginning in the late 1970s. Pakistan first delved into nuclear power after the establishment of its first nuclear power plant near Karachi with equipment and materials supplied mainly by western nations in the early 1970s. Pakistani President Zulfiqar Ali Bhutto promised in 1971 that if India could build nuclear weapons then Pakistan would too, according to him: "We will develop Nuclear stockpiles, even if we have to eat grass."

It is believed that Pakistan has possessed nuclear weapons since the mid-1980s.[60] The United States continued to certify that Pakistan did not possess such weapons until 1990, when sanctions were imposed under the Pressler Amendment, requiring a cutoff of U.S. economic and military assistance to Pakistan.[61] In 1998, Pakistan conducted its first six nuclear tests at the Ras Koh Hills in response to the five tests conducted by India a few weeks before.

In 2004, the Pakistani metallurgist Abdul Qadeer Khan, a key figure in Pakistan's nuclear weapons program, confessed to heading an international black market ring involved in selling nuclear weapons technology. In particular, Khan had been selling gas centrifuge technology to North Korea, Iran, and Libya. Khan denied complicity by the Pakistani government or Army, but this has been called into question by journalists and IAEA officials, and was later contradicted by statements from Khan himself.[62]

As of early 2013, Pakistan was estimated to have had a stockpile of around 100–120 warheads,[1] and in November 2014 it was projected that by 2020 Pakistan would have enough fissile material for 200 warheads.[63]

24.3.3 North Korea

Main article: North Korea and weapons of mass destruction

North Korea was a party to the Nuclear Non-Proliferation Treaty, but announced a withdrawal on January 10, 2003, after the United States accused it of having a secret uranium enrichment program and cut off energy assistance under the 1994 Agreed Framework. In February 2005, North Korea claimed to possess functional nuclear weapons, though their lack of a test at the time led many experts to doubt the claim. However, in October 2006, North Korea stated that due to growing intimidation by the USA, it would conduct a nuclear test to confirm its nuclear status. North Korea reported a successful nuclear test on October 9, 2006 (see 2006 North Korean nuclear test). Most U.S. intelligence officials believe that North Korea did, in fact, test a nuclear device due to radioactive isotopes detected by U.S. aircraft; however, most agree that the test was probably only partially successful.[64] The yield may have been less than a kiloton, which is much smaller than the first successful tests of other powers; boosted fission weapons may have an unboosted yield in this range, which is sufficient to start deuterium-tritium fusion in the boost gas at the center; the fast neutrons from fusion then ensure a full fission yield. North Korea conducted a second, higher yield test on 25 May 2009 (see 2009 North Korean nuclear test) and a third test with still higher yield on 12 February 2013 (see 2013 North Korean nuclear test). North Korea claimed to have conducted its first H-bomb test on 5 January 2016, though measurements of seismic disturbances indicate that the detonation was not consistent with a hydrogen bomb.[65]

24.4 Other states believed to possess nuclear weapons

24.4.1 Israel

Main articles: Nuclear weapons and Israel and Israel and weapons of mass destruction

Israel is widely believed to have been the sixth country in the world to develop nuclear weapons, with "rudimentary, but deliverable," nuclear weapons available as early as 1967.[66] Israel is not a party to the NPT. Israel engages in strategic ambiguity, saying it would not be the first country to "introduce" nuclear weapons into the region, but refusing to otherwise confirm or deny a nuclear weapons program or arsenal. This policy of "nuclear opacity" has been interpreted as an attempt to get the benefits of deterrence with a minimum political cost.[66][67] In 1968, the Israeli Ambassador to the United States, Yitzhak Rabin, affirmed to the United States State Department that Israel would "not be the first to introduce nuclear weapons into the Middle East." Upon further questioning about what "introduce" meant in this context, however, he said that "he would not consider a weapon that had not been tested as a weapon," and affirmed that he did not believe that "an unadvertised, untested nuclear device" was really "a nuclear weapon." He also agreed, however, that an "advertised but untested" device would be considered "introduction." This has been interpreted to mean that official Israeli policy was that the country could possess a nuclear weapon without technically "introducing" it, so long as it did not test it, and as long as it was "unadvertised".[68][69]

In 1986, a former Dimona technician, Mordechai Vanunu, disclosed extensive information about the nuclear program to the British press, including photographs of the secret areas of the nuclear site, some of which depicted nuclear weapons cores and designs. Vanunu gave detailed descriptions of lithium-6 separation required for the production of tritium, an essential ingredient of fusion-boosted fission bombs, as well as information about the rate of plutonium production. Vanunu's evidence was vetted by experienced technical experts before publication, and is considered to be among the strongest evidence for the advanced state of the Israeli nuclear weapons program.[67][70] Theodore Taylor, a former U.S. nuclear device design expert and physicist leading the field[71] especially in small and efficient nuclear weapons, reviewed the 1986 Vanunu leaks and photographs in detail. Taylor concluded that Israel's thermonuclear weapon designs appeared to be "less complex than those of other nations," and at the time of the 1986 leaks "not capable of producing yields in the megaton or higher range." Nevertheless, "they may produce at least several times the yield of fission weapons with the same quantity of plutonium or highly enriched uranium." In other words, Israel could "boost" the yield of its nuclear fission weapons. According to Taylor, the uncertainties involved in the process of boosting required more than theoretical analysis for full confidence in the weapons' performance. Taylor therefore concluded that Israel had "unequivocally" tested a miniaturized nuclear device. The Institute for Defense Analyses (IDA) concluded after reviewing the evidence given by Vanunu that as of 1987, "the Israelis are roughly where the U.S. was in the fission weapon field in about 1955 to 1960." and would require supercomputers or parallel computing clusters to refine their hydrogen bomb designs for improved yields without testing, though noting in 1987 they were already then developing the computer code base required.[72] Israel was first permitted to import US built supercomputers beginning in November 1995.[72]

In a paper by the USAF Counterproliferation Center researcher Lieutenant Colonel Warner D. Farr wrote that much lateral proliferation happened between pre-nuclear France and Israel stating "the French nuclear test in 1960 made two nuclear powers not one—such was the depth of collaboration" and "the Israelis had unrestricted access to French nuclear test explosion data." minimizing the need for early Israeli testing.[73] West Germany army magazine, *Wehrtechnik* ("military technology"), claimed that western intelligence documented that Israel had conducted an underground test in the Negev in 1963.[74] There is also speculation that Israel may have tested a nuclear weapon along with South Africa in 1979, but this has not been confirmed, and interpretation of the Vela Incident is controversial. The stated purpose of the Negev Nuclear Research Center near Dimona is to advance basic nuclear science and applied research on nuclear energy.[75]

According to the Natural Resources Defense Council and the Federation of American Scientists, Israel likely possesses around 75–200 nuclear weapons.[22][76] The Stockholm International Peace Research Institute estimates that Israel has approximately 80 intact nuclear weapons, of which 50 are for delivery by Jericho II medium-range ballistic missiles and 30 are gravity bombs for delivery by aircraft. SIPRI also reports that there was renewed speculation in 2012 that Israel may also have developed nuclear-capable submarine-launched cruise missiles.[77]

24.5 Nuclear weapons sharing

- **Belgium, Germany, Italy, Netherlands, Turkey**

See also: Germany and weapons of mass destruction and Netherlands and weapons of mass destruction

Under NATO nuclear weapons sharing, the United States has provided nuclear weapons for Belgium, Germany, Italy, the Netherlands, and Turkey to deploy and store.[80] This involves pilots and other staff of the "non-nuclear" NATO states practicing, handling, and delivering the U.S. nuclear bombs, and adapting non-U.S. warplanes to deliver U.S. nuclear bombs. However, since all U.S. nuclear weapons are protected with Permissive Action Links, the host states cannot arm the bombs without authorization codes from the U.S. Department of Defense.[81] Former Italian President Francesco Cossiga acknowledged the presence of U.S. nuclear weapons in Italy.[82] U.S. nuclear weapons were also deployed in Canada until 1984, and in Greece until 2001 for nuclear sharing purposes.[83]

Members of the Non-Aligned Movement have called on all countries to "refrain from nuclear sharing for military purposes under any kind of security arrangements."[84] The Institute of Strategic Studies Islamabad (ISSI) has criticized the arrangement for allegedly violating Articles I and II of the NPT, arguing that "these Articles do not permit the NWS to delegate the control of their nuclear weapons directly or indirectly to others."[85] NATO has argued that the weapons' sharing is compliant with the NPT because "the U.S. nuclear weapons based in Europe are in the sole possession and under constant and complete custody and control of the United States."[86]

24.6 States formerly possessing nuclear weapons

Nuclear weapons have been present in many nations, often as staging grounds under control of other powers. However, in only one instance has a nation given up nuclear weapons after being in control of them; in most cases this has been because of special political circumstances. The fall of the Soviet Union left several former Soviet republics in physical possession of nuclear weapons, though not operational control which was dependent on Russian-controlled electronic Permissive Action Links and the Russian command and control system.[87][88]

24.6.1 South Africa

Main article: South Africa and weapons of mass destruction

South Africa produced six nuclear weapons in the 1980s, but disassembled them in the early 1990s. In 1979, there was a putative detection of a covert nuclear test in the Indian Ocean, called the Vela incident. It has long been speculated that it was possibly a test by Israel, in collaboration with

Alleged Spare bomb casings from South Africa's nuclear weapon programme. Their purpose is disputed.[89]

and support of South Africa, though this has never been confirmed. South Africa could not have constructed such a nuclear bomb until November 1979, two months after the "double flash" incident. South Africa signed the Nuclear Non-Proliferation Treaty in 1991.[90][91]

24.6.2 Former Soviet countries

See also: Nuclear weapons and Ukraine

- **Belarus** had 81 single warhead missiles stationed on its territory after the Soviet Union collapsed in 1991. They were all transferred to Russia by 1996. In May 1992, Belarus acceded to the Nuclear Non-Proliferation Treaty.[92]

- **Kazakhstan** inherited 1,400 nuclear weapons from the Soviet Union, and transferred them all to Russia by 1995. Kazakhstan has since acceded to the Nuclear Non-Proliferation Treaty.[93]

- **Ukraine** has acceded to the Nuclear Non-Proliferation Treaty. Ukraine inherited about 5,000 nuclear weapons when it became independent from the Soviet Union in 1991, making its nuclear arsenal the third-largest in the world.[94] By 1996, Ukraine had agreed to dispose of all nuclear weapons within its territory, with the condition that its borders were respected, as part of the Budapest Memorandum on Security Assurances. The warheads were disassembled in Russia.[95] Despite Russia's subsequent and internationally disputed annexation of Crimea in 2014, Ukraine reaffirmed its 1994 decision to accede to the Nuclear Non-Proliferation Treaty as a non-nuclear-weapon state.[96]

24.7   See also

- Comprehensive Nuclear-Test-Ban Treaty
- Doomsday Clock
- Historical nuclear weapons stockpiles and nuclear tests by country
- Nuclear disarmament
- Nuclear proliferation
- Nuclear war
- Nuclear terrorism
- Nuclear-weapon-free zone
- Nuclear power
- No first use

24.8   Notes

[1] All numbers are estimates from the Federation of American Scientists. The latest update was in September 2015. If differences between active and total stockpile are known, they are given as two figures separated by a forward slash. If specifics are not available (n.a.), only one figure is given. Stockpile number may not contain all intact warheads if a substantial amount of warheads are scheduled for but have not yet gone through dismantlement; not all "active" warheads are deployed at any given time. When a range of weapons is given (e.g., 0–10), it generally indicates that the estimate is being made on the amount of fissile material that has likely been produced, and the amount of fissile material needed per warhead depends on estimates of a country's proficiency at nuclear weapon design.

[2] From the 1960s until the 1990s, the United Kingdom's Royal Air Force maintained the independent capability to deliver nuclear weapons via its V bomber fleet.

[3] France formerly possessed a nuclear triad until 1996 and the retirement of its land-based arsenal.

24.9   References

[1] "Federation of American Scientists: Status of World Nuclear Forces". Fas.org. 2014. Retrieved 2014-05-26.

[2] Harding, Luke (2006-12-12). "Calls for Olmert to resign after nuclear gaffe Israel and the Middle East I Guardian Unlimited". London: Guardian. Retrieved 2009-05-15.

[3] Nuclear Forces, sipri.org

[4] Nuclear Notebook, Bulletin of the Atomic Scientists, the-bulletin.org, 2014

[5] Arms Control and Global Security, Paul R. Viotti - 2010, p 312

[6] "Nuclear club," *Oxford English Dictionary*: "nuclear club n. the nations that possess nuclear weapons." The term's first cited usage is from 1957.

[7] Webster, Paul (July/August 2003). "," *The Guardian*.

[8] "Status of Signature and Ratification of the Comprehensive Test Ban Treaty". Retrieved 13 January 2012.

[9] IISS 2012, pp. 54–55

[10] IISS 2012, p. 192

[11] IISS 2012, p. 169

[12] IISS 2012, p. 111

[13] *The Long Shadow: Nuclear Weapons and Security in 21st Century Asia* by Muthiah Alagappa (NUS Press, 2009), page 169: "China has developed strategic nuclear forces made up of land-based missiles, submarine-launched missiles, and bombers. Within this triad, China has also developed weapons of different ranges, capabilities, and survivability."

[14] IISS 2012, pp. 223-224

[15] IISS 2012, p. 243

[16] Peri, Dinakar (12 June 2014). "India's Nuclear Triad Finally Coming of Age". The Diplomat. Retrieved 10 March 2015.

[17] "Nuclear triad weapons ready for deployment: DRDO".

[18] IISS 2012, p. 272

[19] *Pakistan's Nuclear Weapons*, By Bhumitra Chakma, (Routledge 2012), page 61: "Pakistan possesses two types of nuclear delivery vehicles: aircraft and missiles. Initially in the pre-tests era, Islamabad depended solely on aircraft as its chief means of delivering nuclear weapons. In the early 1990s, Pakistan acquired a few dozen ballistic missiles from China, and subsequently, it developed a number of missile systems which became its mainstay of nuclear delivery".

[20] "U.S.: Test Points to N. Korea Nuke Blast". *The Washington Post*. October 13, 2006.

[21] *Nuclear Proliferation in South Asia: Crisis Behaviour and the Bomb* by Sumit Ganguly, Sumit Ganguly, S. Paul Kapur, (Routledge 2008), page 194-195

[22] There are a wide range of estimates as to the size of the Israeli nuclear arsenal. For a compiled list of estimates, see Avner Cohen, *The Worst-Kept Secret: Israel's bargain with the Bomb* (Columbia University Press, 2010), Table 1, page xxvii and page 82.

[23] Brower, Kenneth S (February 1997), "A Propensity for Conflict: Potential Scenarios and Outcomes of War in the Middle East", *Jane's Intelligence Review* (special report) (14): 14–5.

[24] Farr, Warner D (September 1999), The Third Temple's holy of holies: Israel's nuclear weapons, The Counterproliferation Papers, Future Warfare Series 2, USAF Counterproliferation Center, Air War College, Air University, Maxwell Air Force Base, retrieved July 2, 2006.

[25]   • Hersh, Seymour (1991). *The Samson option: Israel's Nuclear Arsenal and American Foreign Policy*. Random House. ISBN 0-394-57006-5., page 271

[26] *An Atlas of Middle Eastern Affairs* By Ewan W. Anderson, Liam D. Anderson, (Routledge 2013), page 233: "In terms of delivery systems, there is strong evidence that Israel now possesses all three elements of the nuclear triad."

[27] IISS 2012, p. 328

[28] Hansen, Chuck (1988). *U.S. nuclear weapons: The secret history*. Arlington, TX: Aerofax. ISBN 0-517-56740-7.

[29] Hansen, Chuck (1995). *The Swords of Armageddon: U.S. nuclear weapons development since 1945*. Sunnyvale, CA: Chukelea Publications. Retrieved 2016-02-20.

[30] Stephen I. Schwartz, ed., *Atomic Audit: The Costs and Consequences of U.S. Nuclear Weapons Since 1940* (Washington, D.C.: Brookings Institution Press, 1998).

[31] "Fact Sheet: Increasing Transparency in the U.S. Nuclear Weapons Stockpile". U.S. Department of Defense. 3 May 2010. Archived from the original (PDF) on 2015-06-28. Retrieved 31 August 2013.

[32] "Policy Library".

[33] Robert S. Norris and Hans M. Kristensen, "Global nuclear stockpiles, 1945-2006," Bulletin of the Atomic Scientists 62, no. 4 (July/August 2006), 64-66...

[34] Holloway, David (1994). *Stalin and the bomb: The Soviet Union and atomic energy, 1939-1956*. New Haven, CT: Yale University Press. ISBN 0-300-06056-4.

[35] Gowing, Margaret (1974). *Independence and deterrence: Britain and atomic energy, 1945-1952*. London: Macmillan. ISBN 0-333-15781-8.

[36] Arnold, Lorna (2001). *Britain and the H-bomb*. Basingstoke: Palgrave. ISBN 0-312-23518-6.

[37] Treaty on the Non-Proliferation of Nuclear Weapons, United Nations Office for Disarmament Affairs.

[38] France 'would use nuclear arms' (BBC, January 2006)

[39] "Nuclear deterrent important in 'dangerous world', says Hollande". *spacedaily.com*. Retrieved 2016-02-20.

[40] John Wilson Lewis and Xue Litai, *China Builds the Bomb* (Stanford, Calif.: Stanford University Press, 1988). ISBN 0-8047-1452-5

[41]

[42] Norris, Robert S. and Hans M. Kristensen. "Chinese nuclear forces, 2006," *Bulletin of the Atomic Scientists* 62:3 (May/June 2006): 60-63.

[43] Lewis, Jeffery. "The ambiguous arsenal," *Bulletin of the Atomic Scientists* 61:3 (May/June 2005): 52-59.

[44] "No-First-Use (NFU)". *Nuclear Threat Initiative*. Archived from the original on 2010-01-25.

[45] "Statement on security assurances issued on 5 April 1995 by the People's Republic of China". United Nations. 6 April 1995. S/1995/265. Archived from the original on 2007-08-24. Retrieved 20 September 2012.

[46] Reuters Editorial (25 February 2015). "China submarines outnumber U.S. fleet: U.S. admiral". *Reuters*.

[47] George Perkovich, *India's Nuclear Bomb: The Impact on Global Proliferation* (Berkeley: University of California Press, 1999), 120-121, and 7.

[48] George Perkovich, *India's Nuclear Bomb: The Impact on Global Proliferation* (Berkeley: University of California Press, 1999), 293–297.

[49] "India's Nuclear Weapons Program: Operation Shakti". 1998. Retrieved 2006-10-10.

[50] "Joint Statement Between President George W. Bush and Prime Minister Manmohan Singh". Whitehouse.gov. Retrieved 2009-05-15.

[51] Implementation of the India-United States Joint Statement of July 18, 2005: India's Separation Plan

[52] "U.S.- India Civil Nuclear Cooperation Initiative – Bilateral Agreement on Peaceful Nuclear Cooperation".

[53] "IAEA Board Approves India-Safeguards Agreement". Iaea.org. Retrieved 2009-05-15.

[54] Statement on Civil Nuclear Cooperation with India

[55] "Congressional Approval of the U.S.-India Agreement for Cooperation Concerning Peaceful Uses of Nuclear Energy (123 Agreement)".

[56] "Secretary of State Condoleezza Rice and Indian Minister of External Affairs Pranab Mukherjee At the Signing of the U.S.-India Civilian Nuclear Cooperation Agreement".

[57] Interview With Undersecretary of State for Arms Control and International Security Robert Joseph, *Arms Control Today*, May 2006.

[58] Was India misled by America on nuclear deal?, *Indian Express*.

[59] ACA: Final NSG Statement

[60] NTI Pakistan Profile, retrieved 22 April 2012.

[61] "Case Studies in Sanctions and Terrorism: Pakistan". Iie.com. Retrieved 2009-05-15.

[62] See A.Q. Khan: Investigation, dismissal, confession, pardon and aftermath, for citations and details.

[63] "Pakistan to Have 200 Nuke Weapons by 2020: US Think Tank". The Times of india. November 2014. Retrieved 2014-11-28.

[64] "CIA's Hayden: North Korea Nuke Test 'Was a Failure'". Newsmax.com. 2007-03-28. Retrieved 2009-05-15.

[65] "North Korea Test Shows Technical Advance". *The Wall Street Journal*. CCLXVII (5): A6. January 7, 2016.

[66] NTI Israel Profile Retrieved July 12, 2007.

[67] Avner Cohen (2010). *The Worst-Kept Secret: Israel's bargain with the Bomb*. Columbia University Press.

[68] Memcon, "Negotiations with Israel - 4F and Advanced Weapons," November 8, 1968 and Memcon, "Negotiations with Israel - F4 and Advanced Weapons," November 12, 1968, both part of National Security Archive Electronic Briefing Book No. 189, "Israel Crosses the Threshold" (April 28, 2006)

[69] Avner Cohen and William Burr, "The Untold Story of Israel's Bomb," *Washington Post*, April 30, 2006; B01.

[70] "Vanunu 'wanted to avert holocaust'". BBC News. May 29, 2004.

[71]

[72] "Israel's Nuclear Weapon Capability: An Overview".

[73] Farr, Warner D (September 1999). *The Third Temple's holy of holies: Israel's nuclear weapons*. The Counterproliferation Papers, Future Warfare Series. **2**. USAF Counterproliferation Center, Air War College, Air University, Maxwell Air Force Base. Retrieved July 2, 2006.

[74] https://fas.org/nuke/guide/israel/nuke/farr.htm IV. 1974-1999: Bringing the Bomb up the Basement Stairs (paragraph 3)

[75] "Nuclear Research Center NEGEV - NRCN". Israel Atomic Energy Commission. Retrieved February 1, 2012.

[76] Israel's Nuclear Weapons, Federation of American Scientists (August 17, 2000)

[77] "Israel".

[78] Hans M. Kristensen (26 June 2008). "Status of U.S. Nuclear Weapons in Europe" (PDF). Federation of American Scientists. Retrieved 10 November 2012.

[79] "US moves nuclear weapons from Turkey to Romania". *EurActiv.com*. Retrieved 2016-08-18.

[80] "Berlin Information-center for Transatlantic Security: NATO Nuclear Sharing and the N.PT - Questions to be Answered". Bits.de. Retrieved 2009-05-15.

[81] "Nuclear Command and Control". *Security Engineering: A Guide to Building Dependable Distributed Systems* (PDF). Ross Anderson, University of Cambridge Computing Laboratory. Retrieved April 29, 2010.

[82] "Cossiga: "In Italia ci sono bombe atomiche Usa"".

[83] Hans M. Kristensen (February 2005). "U.S. Nuclear Weapons in Europe" (PDF). Natural Resources Defense Council. Retrieved 2006-05-23.

[84] Statement on behalf of the non-aligned state parties to the Treaty on the Non-Proliferation of Nuclear Weapons, 2 May 2005

[85] ISSI - NPT in 2000: Challenges ahead, Zafar Nawaz Jaspal, The Institute of Strategic Studies, Islamabad Archived January 9, 2009, at the Wayback Machine.

[86] NATO's Positions Regarding Nuclear Non-Proliferation, Arms Control and Disarmament and Related Issues, NATO

[87] William C. Martel (1998). "Why Ukraine gave up nuclear weapons : nonproliferation incentives and disincentives". In Barry R. Schneider, William L. Dowdy. *Pulling Back from the Nuclear Brink: Reducing and Countering Nuclear Threats*. Psychology Press. pp. 88–104. ISBN 9780714648569. Retrieved 6 August 2014.

[88] Alexander A. Pikayev (Spring–Summer 1994). "Post-Soviet Russia and Ukraine: Who can push the Button?" (PDF). *The Nonproliferation Review*. **1** (3). doi:10.1080/10736709408436550. Retrieved 6 August 2014.

[89] Lewis, Jeffrey (3 December 2015). "Revisiting Sout Africa's Bomb". *Arms Control Wonk. Leading Voice on Arms Control, Disarmament and Non-Proliferation*. Retrieved 6 December 2015.

[90] Nuclear Weapons Program (South Africa), Federation of American Scientists (May 29, 2000).

[91] Von Wielligh, N. & von Wielligh-Steyn, L. (2015). The Bomb – South Africa's Nuclear Weapons Programme. Pretoria: Litera.

[92] "Belarus Special Weapons". Federation of American Scientists.

[93] "Kazakhstan Special Weapons". Federation of American Scientists.

[94] Ukraine Special Weapons, GlobalSecurity.org

[95] "Ukraine Special Weapons". Federation of American Scientists.

[96] Joint Statement by the United States and Ukraine, March 25, 2014.

24.10 Bibliography

- International Institute for Strategic Studies; Hackett, James (ed.) (7 March 2012). *The Military Balance 2012*. London: Routledge. ISBN 1857436423.

- Farr, Warner D (September 1999), *The Third Temple's holy of holies: Israel's nuclear weapons*, The Counterproliferation Papers, Future Warfare Series, **2**, USAF Counterproliferation Center, Air War College, Air University, Maxwell Air Force Base, retrieved July 2, 2006.

24.11 External links

- Archive of Nuclear Data - List of warheads by country

- Globalsecurity.org – World Special Weapons Guide

- The Nuclear Weapon Archive

- Nuclear Notebook from Bulletin of the Atomic Scientists

- U.S. Nuclear Weapons in Europe: A review of post-Cold War policy, force levels, and war planning NRDC, February 2005

- Online NewsHour with Jim Lehrer:Tracking Nuclear Proliferation

- Stockholm International Peace Research Institute's data on world nuclear forces

- Nuclear Proliferation International History Project For more on the history of nuclear proliferation see the Woodrow Wilson Center's Nuclear Proliferation International History Project website.

- Proliferation Watch: US Intelligence Assessments of Potential Nuclear Powers, 1977–2001

Chapter 25

Nuclear Tipping Point

Nuclear Tipping Point is a 2010 documentary film produced by the Nuclear Threat Initiative. It features interviews with four American government officials who were in office during the Cold War period, but are now advocating for the elimination of nuclear weapons: Henry Kissinger, George Shultz, Sam Nunn, and William Perry.[1] Michael Douglas narrated the film.[2]

These "Four Cold Warriors", who each contributed in important ways to the nuclear arms race, built on classical deterrence theory, now argue that we must eliminate all nuclear weapons or face disaster on an enormous scale. Former Secretary Kissinger puts the new danger this way: "The classical notion of deterrence was that there was some consequences before which aggressors and evildoers would recoil. In a world of suicide bombers, that calculation doesn't operate in any comparable way".[3] Shultz has said, "If you think of the people who are doing suicide attacks, and people like that get a nuclear weapon, they are almost by definition not deterrable".[4]

The film was screened at the White House on April 6, 2010.[5][6]

25.1 See also

- List of films about nuclear issues
  - *Countdown to Zero*
- Nuclear weapons debate

25.2 References

[1] "Documentary Advances Nuclear Free Movement". NPR. Retrieved 2010-06-10.

[2] Pease, Christian (November 10, 2011). "William J. Perry: A cold warrior". *Palo Alto Weekly*. Embarcadero Media. Retrieved August 4, 2015.

[3] Ben Goddard (2010-01-27). "Cold Warriors say no nukes". *The Hill*.

[4] Hugh Gusterson (30 March 2012). "The new abolitionists". *Bulletin of the Atomic Scientists*.

[5] "White House to Host Screening Tonight of Nuclear Tipping Point". PRNewswire-USNewswire. Retrieved 2010-07-09.

[6] "White House to Host Screening Tonight of Nuclear Tipping Point". FOX Business. Retrieved 2010-06-10.

25.3 External links

- *Nuclear Tipping Point* website
- Nuclear Threat Initiative official website
- *Nuclear Tipping Point* at the Internet Movie Database

Chapter 26

Nuclear disarmament

The Campaign for Nuclear Disarmament symbol, designed by Gerald Holtom in 1958.[1]

Nuclear disarmament refers to both the act of reducing or eliminating nuclear weapons and to the end state of a nuclear-weapon-free world, in which nuclear weapons are completely eliminated.

Nuclear disarmament groups include the Campaign for Nuclear Disarmament, Peace Action, Greenpeace, International Physicians for the Prevention of Nuclear War, Mayors for Peace, Global Zero, the International Campaign to Abolish Nuclear Weapons, and the Nuclear Age Peace Foundation. There have been many large anti-nuclear demonstrations and protests. On June 12, 1982, one million people demonstrated in New York City's Central Park against nuclear weapons and for an end to the cold war arms race. It was the largest anti-nuclear protest and the largest political demonstration in American history.[2][3]

In recent years, some U.S. elder statesmen have also advocated nuclear disarmament. Sam Nunn, William Perry, Henry Kissinger, and George Shultz have called upon governments to embrace the vision of a world free of nuclear weapons, and in various op-ed columns have proposed an ambitious program of urgent steps to that end. The four have created the Nuclear Security Project to advance this agenda. Organisations such as Global Zero, an international non-partisan group of 300 world leaders dedicated to achieving nuclear disarmament, have also been established.

Proponents of nuclear disarmament say that it would lessen the probability of nuclear war occurring, especially accidentally. Critics of nuclear disarmament say that it would undermine deterrence.

26.1 History

In 1945 in the New Mexico desert, American scientists conducted "Trinity," the first nuclear weapons test, marking the beginning of the atomic age.[4] Even before the Trinity test, national leaders debated the impact of nuclear weapons on domestic and foreign policy. Also involved in the debate about nuclear weapons policy was the scientific community, through professional associations such as the Federation of Atomic Scientists and the Pugwash Conference on Science and World Affairs.[5]

On August 6, 1945, towards the end of World War II, the Little Boy device was detonated over the Japanese city of Hiroshima. Exploding with a yield equivalent to 12,500 tonnes of TNT, the blast and thermal wave of the bomb destroyed nearly 50,000 buildings (including the headquarters of the 2nd General Army and Fifth Division) and killed approximately 75,000 people, among them 20,000 Japanese soldiers and 20,000 Koreans.[6] Detonation of the Fat Man device exploded over the Japanese city of Nagasaki three days later on 9 August 1945, destroying 60% of the city and killing approximately 35,000 people, among them 23,200-28,200 Japanese civilian munitions workers and 150 Japanese soldiers.[7] Subsequently, the world's nuclear weapons stockpiles grew.[4]

The mushroom cloud over Hiroshima after the dropping of the atomic bomb nicknamed 'Little Boy' (Atomic bombings of Hiroshima and Nagasaki in 1945).

Mushroom-shaped cloud and water column from the underwater nuclear explosion of July 25, 1946, which was part of Operation Crossroads.

Operation Crossroads was a series of nuclear weapon tests conducted by the United States at Bikini Atoll in the Pacific Ocean in the summer of 1946. Its purpose was to test the effect of nuclear weapons on naval ships. Pressure to cancel Operation Crossroads came from scientists and diplomats. Manhattan Project scientists argued that further nuclear testing was unnecessary and environmentally dangerous. A Los Alamos study warned "the water near a recent surface explosion will be a witch's brew" of radioactivity. To prepare the atoll for the nuclear tests, Bikini's native residents were evicted from their homes and resettled on smaller, uninhabited islands where they were unable to sustain themselves.[8]

November 1951 nuclear test at the Nevada Test Site, from Operation Buster, with a yield of 21 kilotons. It was the first U.S. nuclear field exercise conducted on land; troops shown are 6 mi (9.7 km) from the blast.

Radioactive fallout from nuclear weapons testing was first drawn to public attention in 1954 when a Hydrogen bomb test in the Pacific contaminated the crew of the Japanese fishing boat *Lucky Dragon*.[9] One of the fishermen died in Japan seven months later. The incident caused widespread concern around the world and "provided a decisive impetus for the emergence of the anti-nuclear weapons movement in many countries".[9] The anti-nuclear weapons movement grew rapidly because for many people the atomic bomb "encapsulated the very worst direction in which society was moving".[10]

26.2   Nuclear   disarmament   movement

1952 World Peace Council congress in East Berlin showing Picasso's peace dove above the stage

Women Strike for Peace during the Cuban Missile Crisis in 1962.

Demonstration in Lyon, France in the 1980s against nuclear weapons tests

See also: History of the anti-nuclear movement and List of peace activists

Peace movements emerged in Japan and in 1954 they converged to form a unified "Japanese Council Against Atomic and Hydrogen Bombs". Japanese opposition to the Pacific nuclear weapons tests was widespread, and "an estimated 35 million signatures were collected on petitions calling for bans on nuclear weapons".[10] In the United Kingdom, the first Aldermaston March organised by the Direct Action Committee and supported by the Campaign for Nuclear Disarmament took place at Easter 1958, when several thousand people marched for four days from Trafalgar Square, London, to the Atomic Weapons Research Establishment close to Aldermaston in Berkshire, England, to demonstrate

On 12 December 1982, 30,000 women held hands around the 6 miles (9.7 km) perimeter of the RAF Greenham Common base, in protest against the decision to site American cruise missiles there.

their opposition to nuclear weapons.[11][12] CND organised Aldermaston marches into the late 1960s when tens of thousands of people took part in the four-day events.[10]

On November 1, 1961, at the height of the Cold War, about 50,000 women brought together by Women Strike for Peace marched in 60 cities in the United States to demonstrate against nuclear weapons. It was the largest national women's peace protest of the 20th century.[13][14]

In 1958, Linus Pauling and his wife presented the United Nations with the petition signed by more than 11,000 scientists calling for an end to nuclear-weapon testing. The "Baby Tooth Survey," headed by Dr Louise Reiss, demonstrated conclusively in 1961 that above-ground nuclear testing posed significant public health risks in the form of radioactive fallout spread primarily via milk from cows that had ingested contaminated grass.[15][16][17] Public pressure and the research results subsequently led to a moratorium on above-ground nuclear weapons testing, followed by the Partial Test Ban Treaty, signed in 1963 by John F. Kennedy and Nikita Khrushchev.[18] On the day that the treaty went into force, the Nobel Prize Committee awarded Pauling the Nobel Peace Prize, describing him as "Linus Carl Pauling, who ever since 1946 has campaigned ceaselessly, not only against nuclear weapons tests, not only against the spread of these armaments, not only against their very use, but against all warfare as a means of solving international conflicts."[5][19] Pauling started the International League of Humanists in 1974. He was president of the scientific advisory board of the World Union for Protection of Life and also one of the signatories of the Dubrovnik-Philadelphia Statement.

In the 1980s, a popular movement for nuclear disarmament again gained strength in the light of the weapons build-up and aggressive rhetoric of US President Ronald Reagan. Reagan had "a world free of nuclear weapons" as his personal mission,[20][21][22] and was largely scorned for this in Europe.[22] His officials tried to stop such talks but Reagan was able to start discussions on nuclear disarmament with Soviet Union.[22] He changed the name "SALT" (Strategic Arms Limitation Talks) to "START" (Strategic Arms Reduction Talks).[21]

On June 3, 1981, Thomas launched the White House Peace Vigil in Washington, D.C..[23] He was later joined on the vigil by anti-nuclear activists Concepcion Picciotto and Ellen Benjamin.[24]

On June 12, 1982, one million people demonstrated in New York City's Central Park against nuclear weapons and for an end to the cold war arms race. It was the largest anti-nuclear protest and the largest political demonstration in American history.[2][3] International Day of Nuclear Disarmament protests were held on June 20, 1983 at 50 sites across the United States.[25][26] In 1986, hundreds of people walked from Los Angeles to Washington DC in the Great Peace March for Global Nuclear Disarmament.[27] There were many Nevada Desert Experience protests and peace camps at the Nevada Test Site during the 1980s and 1990s.[28][29]

On May 1, 2005, 40,000 anti-nuclear/anti-war protesters marched past the United Nations in New York, 60 years after the atomic bombings of Hiroshima and Nagasaki.[30][31] This was the largest anti-nuclear rally in the U.S. for several decades.[32] In 2008, 2009, and 2010, there have been protests about, and campaigns against, several new nuclear reactor proposals in the United States.[33][34][35]

There is an annual protest against U.S. nuclear weapons research at Lawrence Livermore National Laboratory in California and in the 2007 protest, 64 people were arrested.[36] There have been a series of protests at the Nevada Test Site and in the April 2007 Nevada Desert Experience protest, 39 people were cited by police.[37] There have been anti-nuclear protests at Naval Base Kitsap for many years, and several in 2008.[38][39][40]

26.2.1   World Peace Council

One of the earliest peace organisations to emerge after the Second World War was the World Peace Council, which was directed by the Communist Party of the Soviet Union through the Soviet Peace Committee. Its origins lay in the Communist Information Bureau's (Cominform) doctrine, put forward 1947, that the world was divided between peace-loving progressive forces led by the Soviet Union and warmongering capitalist countries led by the United States. In 1949, Cominform directed that peace "should now become the pivot of the entire activity of the Communist Parties", and most western Communist parties followed this policy.[41] Lawrence Wittner, a historian of the post-war peace movement, argues that the Soviet Union devoted great efforts to the promotion of the WPC in the early post-war years because it feared an American attack and American superiority of arms[42] at a time when the USA possessed the atom bomb but the Soviet Union had not yet developed it.[43]

In 1950, the WPC launched its Stockholm Appeal[44] calling for the absolute prohibition of nuclear weapons. The campaign won popular support, collecting, it is said, 560 million signatures in Europe, most from socialist countries, including 10 million in France (including that of the young Jacques Chirac), and 155 million signatures in the Soviet Union – the entire adult population.[45] Several non-aligned peace groups who had distanced themselves from the WPC advised their supporters not to sign the Appeal.[43]

The WPC had uneasy relations with the non-aligned peace movement and has been described as being caught in contradictions as "it sought to become a broad world movement while being instrumentalized increasingly to serve foreign policy in the Soviet Union and nominally socialist countries."[46] From the 1950s until the late 1980s it tried to use non-aligned peace organizations to spread the Soviet point of view. At first there was limited co-operation between such groups and the WPC, but western delegates who tried to criticize the Soviet Union or the WPC's silence about Russian armaments were often shouted down at WPC conferences[42] and by the early 1960s they had dissociated themselves from the WPC.

26.3   Arms reduction treaties

After the 1986 Reykjavik Summit between U.S. President Ronald Reagan and the new Soviet General Secretary Mikhail Gorbachev, the United States and the Soviet Union concluded two important nuclear arms reduction treaties: the INF Treaty (1987) and START I (1991). After the end of the Cold War, the United States and the Russian Federation concluded the Strategic Offensive Reductions Treaty (2003) and the New START Treaty (2010).

In the Soviet Union (USSR), voices against Soviet nuclear weapons were few and far between since there was no

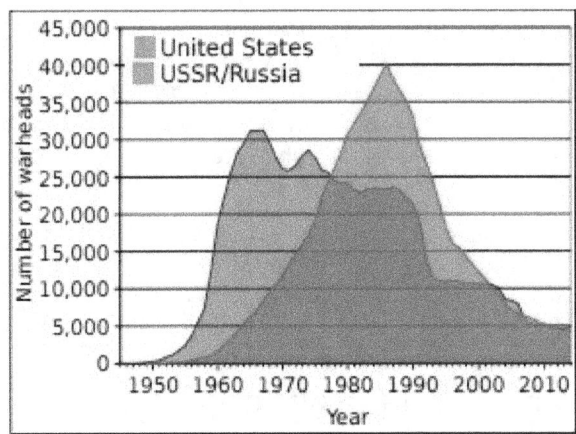

United States and USSR/Russian nuclear weapons stockpiles, 1945-2014. These numbers include warheads not actively deployed, including those on reserve status or scheduled for dismantlement. Stockpile totals do not necessarily reflect nuclear capabilities since they ignore size, range, type, and delivery mode.

widespread Freedom of speech and Freedom of the press as political factors. Certain citizens who had become prominent enough to safely criticize the Soviet government, such as Andrei Sakharov, did speak out against nuclear weapons, but that was to little effect. Dissident movements that emerged in the Soviet bloc in the 1980s drew attention to Soviet armaments, some demonstrating at congresses of the World Peace Council, but they were suppressed.

When the extreme danger intrinsic to nuclear war and the possession of nuclear weapons became apparent to all sides during the Cold War, a series of disarmament and nonproliferation treaties were agreed upon between the United States, the Soviet Union, and several other states throughout the world. Many of these treaties involved years of negotiations, and seemed to result in important steps in arms reductions and reducing the risk of nuclear war.

Key treaties

- Partial Test Ban Treaty (PTBT) 1963: Prohibited all testing of nuclear weapons except underground.

- Nuclear Non-Proliferation Treaty (NPT)—signed 1968, came into force 1970: An international treaty (currently with 189 member states) to limit the spread of nuclear weapons. The treaty has three main pillars: nonproliferation, disarmament, and the right to peacefully use nuclear technology.

- Interim Agreement on Offensive Arms (SALT I) 1972: The Soviet Union and the United States agreed to a freeze in the number of intercontinental ballistic missiles (ICBMs) and submarine-launched ballistic missiles (SLBMs) that they would deploy.

- Anti-Ballistic Missile Treaty (ABM) 1972: The United States and Soviet Union could deploy ABM interceptors at two sites, each with up to 100 ground-based launchers for ABM interceptor missiles. In a 1974 Protocol, the US and Soviet Union agreed to only deploy an ABM system to one site.

- Strategic Arms Limitation Treaty (SALT II) 1979: Replacing SALT I, SALT II limited both the Soviet Union and the United States to an equal number of ICBM launchers, SLBM launchers, and heavy bombers. Also placed limits on Multiple Independent Reentry Vehicles (MIRVS).

- Intermediate-Range Nuclear Forces Treaty (INF) 1987: Created a global ban on short- and long-range nuclear weapons systems, as well as an intrusive verification regime.

- Strategic Arms Reduction Treaty (START I)—signed 1991, ratified 1994: Limited long-range nuclear forces in the United States and the newly independent states of the former Soviet Union to 6,000 attributed warheads on 1,600 ballistic missiles and bombers.

- Strategic Arms Reduction Treaty II (START II)—signed 1993, never put into force: START II was a bilateral agreement between the US and Russia which attempted to commit each side to deploy no more than 3,000 to 3,500 warheads by December 2007 and also included a prohibition against deploying multiple independent reentry vehicles (MIRVs) on intercontinental ballistic missiles (ICBMs)

- Strategic Offensive Reductions Treaty (SORT or Moscow Treaty)—signed 2002, into force 2003: A very loose treaty that is often criticized by arms control advocates for its ambiguity and lack of depth. Russia and the United States agreed to reduce their "strategic nuclear warheads" (a term that remain undefined in the treaty) to between 1,700 and 2,200 by 2012. Was superseded by New Start Treaty in 2010.

- Comprehensive Test Ban Treaty (CTBT)—signed 1996, not yet in force: The CTBT is an international treaty (currently with 181 state signatures and 148 state ratifications) that bans all nuclear explosions in all environments. While the treaty is not in force, Russia has not tested a nuclear weapon since 1990 and the United States has not since 1992.[47]

- New START Treaty—signed 2010, into force in 2011: replaces SORT treaty, reduces deployed nuclear warheads by about half, will remain into force until at least 2021

Only one country has been known to ever dismantle their nuclear arsenal completely—the apartheid government of South Africa apparently developed half a dozen crude fission weapons during the 1980s, but they were dismantled in the early 1990s.

26.4   United Nations

Angela Kane at a 2012 ceremony marking the anniversary of the Hiroshima and Nagasaki atomic bombings

Main article: United Nations Office for Disarmament Affairs

In its landmark resolution 1653 of 1961, "Declaration on the prohibition of the use of nuclear and thermo-nuclear weapons," the UN General Assembly stated that use of nuclear weaponry "would exceed even the scope of war and cause indiscriminate suffering and destruction to mankind and civilization and, as such, is contrary to the rules of international law and to the laws of humanity".[48]

The UN Office for Disarmament Affairs (UNODA) is a department of the United Nations Secretariat established in January 1998 as part of the United Nations Secretary-General Kofi Annan's plan to reform the UN as presented in his report to the General Assembly in July 1997.[49]

Its goal is to promote nuclear disarmament and non-proliferation and the strengthening of the disarmament regimes in respect to other weapons of mass destruction, chemical and biological weapons. It also promotes disarmament efforts in the area of conventional weapons, especially land mines and small arms, which are often the weapons of choice in contemporary conflicts.

Following the retirement of Sergio Duarte in February 2012, Angela Kane was appointed as the new High Representative for Disarmament Affairs.

26.5   U.S. nuclear policy

Despite a general trend toward disarmament in the early 2000s, the George W. Bush administration repeatedly pushed to fund policies that would allegedly make nuclear weapons more usable in the post–Cold War environment.[50][51] To date the U.S. Congress has refused to fund many of these policies. However, some [52] feel that even considering such programs harms the credibility of the United States as a proponent of nonproliferation.

26.5.1   Recent controversial U.S. nuclear policies

- Reliable Replacement Warhead Program (RRW): This program seeks to replace existing warheads with a smaller number of warhead types designed to be easier to maintain without testing. Critics charge that this would lead to a new generation of nuclear weapons and would increase pressures to test. Congress has not funded this program.

- **Complex Transformation**: Complex transformation, formerly known as Complex 2030, is an effort to shrink the U.S. nuclear weapons complex and restore the ability to produce "pits" the fissile cores of the primaries of U.S. thermonuclear weapons. Critics see it as an upgrade to the entire nuclear weapons complex to support the production and maintenance of the new generation of nuclear weapons. Congress has not funded this program.

- Nuclear bunker buster: Formally known as the Robust Nuclear Earth Penetrator (RNEP), this program aimed to modify an existing gravity bomb to penetrate into soil and rock in order to destroy underground targets. Critics argue that this would lower the threshold for use of nuclear weapons. Congress did not fund this proposal, which was later withdrawn.

- Missile Defense: Formerly known as National Missile Defense, this program seeks to build a network of interceptor missiles to protect the United States and its allies from incoming missiles, including nuclear-armed missiles. Critics have argued that this would impede nuclear disarmament and possibly stimulate a nuclear arms race. Elements of missile defense are being deployed in Poland and the Czech Republic, despite Russian opposition.

Former U.S. officials Henry Kissinger, George Shultz, Bill Perry, and Sam Nunn (aka 'The Gang of Four' on Nuclear Deterrence)."[53] proposed in January 2007 that the United States rededicate itself to the goal of eliminating nuclear weapons, concluding: "We endorse setting the goal of a world free of nuclear weapons and working energetically on the actions required to achieve that goal." Arguing a year later that "with nuclear weapons more widely available, deterrence is decreasingly effective and increasingly hazardous," the authors concluded that although "it is tempting and easy to say we can't get there from here, . . . we must chart a course" toward that goal."[54] During his Presidential campaign, U.S. President Elect Barack Obama pledged to "set a goal of a world without nuclear weapons, and pursue it."[55]

26.5.2 U.S. policy options for nuclear terrorism

The United States has taken the lead in ensuring that nuclear materials globally are properly safeguarded. A popular program that has received bipartisan domestic support for over a decade is the Cooperative Threat Reduction Program (CTR). While this program has been deemed a success, many believe that its funding levels need to be increased so as to ensure that all dangerous nuclear materials are secured in the most expeditious manner possible. The CTR program has led to several other innovative and important nonproliferation programs that need to continue to be a budget priority in order to ensure that nuclear weapons do not spread to actors hostile to the United States.

Key programs:

- **Cooperative Threat Reduction** (CTR): The CTR program provides funding to help Russia secure materials that might be used in nuclear or chemical weapons as well as to dismantle weapons of mass destruction and their associated infrastructure in Russia.

- **Global Threat Reduction Initiative** (GTRI): Expanding on the success of the CTR, the GTRI will expand nuclear weapons and material securing and dismantlement activities to states outside of the former Soviet Union.

26.6 Other states

Main article: List of states with nuclear weapons

While the vast majority of states have adhered to the stipulations of the Nuclear Nonproliferation Treaty, a few states have either refused to sign the treaty or have pursued nuclear weapons programs while not being members of the treaty. Many view the pursuit of nuclear weapons by these states as a threat to nonproliferation and world peace, and therefore seek policies to discourage the spread of nuclear weapons to these states, a few of which are often described by the US as "rogue states".

- Declared nuclear weapon states not party to the NPT.[56]

  - Indian nuclear weapons: 80–100 active warheads
  - Pakistani nuclear weapons: 90–110 active warheads
  - North Korean nuclear weapons: <10 active warheads

- Undeclared nuclear weapon states not party to the NPT:

  - Israeli nuclear weapons: 75–200 active warheads[57]

- Nuclear weapon states not party to the NPT that disarmed and joined the NPT as non-nuclear weapons states:

  - South African nuclear weapons: disarmed from 1989–1993

- Former Soviet states that disarmed and joined the NPT as non-nuclear weapons states:

  - Belarus
  - Kazakhstan
  - Ukraine

- Non-nuclear weapon states party to the NPT currently accused of seeking nuclear weapons:

  - Iran

- Non-nuclear weapon states party to the NPT who acknowledged and eliminated past nuclear weapons programs:

  - Libya[58]

26.7 Recent developments

Eliminating nuclear weapons has long been an aim of the pacifist left. But now many mainstream politicians, academic analysts, and retired military leaders also advocate nuclear disarmament. Sam Nunn, William Perry, Henry Kissinger, and George Shultz have called upon governments to embrace the vision of a world free of nuclear weapons, and in three *Wall Street Journal* opeds proposed an ambitious program of urgent steps to that end. The four have created the Nuclear Security Project to advance this agenda. Nunn reinforced that agenda during a speech at the Harvard Kennedy School on October 21, 2008, saying, "I'm much more concerned about a terrorist without a return address that cannot be deterred than I am about deliberate war between nuclear powers. You can't deter a group who is willing to commit suicide. We are in a different era. You have to understand the world has changed."[59] In 2010, the four were featured in a documentary film entitled *Nuclear Tipping Point*. The film is a visual and historical depiction of the ideas laid forth in the Wall Street Journal op-eds and reinforces their commitment to a world without nuclear weapons and the steps that can be taken to reach that goal.[60]

Global Zero is an international non-partisan group of 300 world leaders dedicated to achieving nuclear disarmament.[61] The initiative, launched in December 2008, promotes a phased withdrawal and verification for the destruction of all devices held by official and unofficial members of the nuclear club. The Global Zero campaign works toward building an international consensus and a sustained global movement of leaders and citizens for the elimination of nuclear weapons. Goals include the initiation of United States-Russia bilateral negotiations for reductions to 1,000 total warheads each and commitments from the other key nuclear weapons countries to participate in multilateral negotiations for phased reductions of nuclear arsenals. Global Zero works to expand the diplomatic dialogue with key governments and continue to develop policy proposals on the critical issues related to the elimination of nuclear weapons.

The International Conference on Nuclear Disarmament took place in Oslo in February, 2008, and was organized by The Government of Norway, the Nuclear Threat Initiative and the Hoover Institute. The Conference was entitled *Achieving the Vision of a World Free of Nuclear Weapons* and had the purpose of building consensus between nuclear weapon states and non-nuclear weapon states in relation to the Nuclear Non-proliferation Treaty.[62]

The Tehran International Conference on Disarmament and Non-Proliferation took place in Tehran in April 2010. The conference was held shortly after the signing of the New START, and resulted in a call of action toward eliminating all nuclear weapons. Representatives from 60 countries were invited to the conference. Non-governmental organizations were also present.

Among the prominent figures who have called for the abolition of nuclear weapons are "the philosopher Bertrand Russell, the entertainer Steve Allen, CNN's Ted Turner, former Senator Claiborne Pell, Notre Dame president Theodore Hesburgh, South African Bishop Desmond Tutu and the Dalai Lama".[63]

Others have argued that nuclear weapons have made the world relatively safer, with peace through deterrence and through the stability–instability paradox, including in south Asia.[64][65] Kenneth Waltz has argued that nuclear weapons have created a nuclear peace, and further nuclear weapon proliferation might even help avoid the large scale conventional wars that were so common prior to their invention at the end of World War II.[66] In the July 2012 issue of Foreign Affairs Waltz took issue with the view of most U.S., European, and Israeli, commentators and policymakers that a nuclear-armed Iran would be unacceptable. Instead Waltz argues that it would probably be the best possible outcome, as it would restore stability to the Middle East by balancing Israel's regional monopoly on nuclear weapons.[67] Professor John Mueller of Ohio State University, the author of *Atomic Obsession*,[68] has also dismissed the need to interfere with Iran's nuclear program and expressed that arms control measures are counterproductive.[69] During a 2010 lecture at the University of Missouri, which was broadcast by C-SPAN, Dr. Mueller has also argued that the threat from nuclear weapons, especially nuclear terrorism, has been exaggerated, both in the popular media and by officials.[70]

Former Secretary Kissinger says there is a new danger, which cannot be addressed by deterrence: "The classical notion of deterrence was that there was some consequences before which aggressors and evildoers would recoil. In a world of suicide bombers, that calculation doesn't operate in any comparable way".[71] George Shultz has said, "If you think of the people who are doing suicide attacks, and people like that get a nuclear weapon, they are almost by definition not deterrable".[72]

26.8 See also

- Anti-nuclear organizations

- Comprehensive Test Ban Treaty

- Countdown to Zero

- International Atomic Energy Agency

- List of anti-war organizations

- List of peace activists

- Nuclear-free zone

- Nuclear proliferation

- Nuclear warfare

- Nuclear weapon

- Nuclear weapons convention

- Nuclear-Weapon-Free Zone

- Nuclear weapons and the United States

- Paranuclear

- Seabed Arms Control Treaty

- Strategic Offensive Reductions Treaty (SORT)

- United Nations

- Tehran International Conference on Disarmament and Non-Proliferation, 2010

26.9 References

[1] "BBC NEWS : Magazine : World's best-known protest symbol turns 50". *BBC News*. London. 20 March 2008. Retrieved 2008-05-25.

[2] Jonathan Schell. The Spirit of June 12 *The Nation*, July 2, 2007.

[3] 1982 - a million people march in New York City

[4] Mary Palevsky, Robert Futrell, and Andrew Kirk. Recollections of Nevada's Nuclear Past *UNLV FUSION*, 2005, p. 20.

[5] Jerry Brown and Rinaldo Brutoco (1997). *Profiles in Power: The Anti-nuclear Movement and the Dawn of the Solar Age*. Twayne Publishers, pp. 191-192.

[6] Emsley, John (2001). "Uranium". *Nature's Building Blocks: An A to Z Guide to the Elements*. Oxford: Oxford University Press. p. 478. ISBN 0-19-850340-7.

[7] *Nuke-Rebuke: Writers & Artists Against Nuclear Energy & Weapons (The Contemporary anthology series)*. The Spirit That Moves Us Press. May 1, 1984. pp. 22–29.

[8] Niedenthal, Jack (2008), *A Short History of the People of Bikini Atoll*, retrieved 2009-12-05

[9] Wolfgang Rudig (1990). *Anti-nuclear Movements: A World Survey of Opposition to Nuclear Energy*, Longman, p. 54-55.

[10] Jim Falk (1982). *Global Fission: The Battle Over Nuclear Power*, Oxford University Press, pp. 96-97.

[11] A brief history of CND

[12] "Early defections in march to Aldermaston". Guardian Unlimited. 1958-04-05.

[13] Woo, Elaine (January 30, 2011). "Dagmar Wilson dies at 94; organizer of women's disarmament protesters". *Los Angeles Times*.

[14] Hevesi, Dennis (January 23, 2011). "Dagmar Wilson, Anti-Nuclear Leader, Dies at 94". *The New York Times*.

[15] Louise Zibold Reiss (November 24, 1961). "Strontium-90 Absorption by Deciduous Teeth: Analysis of teeth provides a practicable method of monitoring strontium-90 uptake by human populations" (PDF). Science. Retrieved October 13, 2009.

[16] Thomas Hager (November 29, 2007). "Strontium-90". Oregon State University Libraries Special Collections. Retrieved December 13, 2007.

[17] Thomas Hager (November 29, 2007). "The Right to Petition". Oregon State University Libraries Special Collections. Retrieved December 13, 2007.

[18] Jim Falk (1982). *Global Fission: The Battle Over Nuclear Power*, Oxford University Press, p. 98.

[19] Linus Pauling (October 10, 1963). "Notes by Linus Pauling. October 10, 1963.". Oregon State University Libraries Special Collections. Retrieved December 13, 2007.

[20] Giuliani's Obama-Nuke Critique Defies And Ignores Reagan, Huffington Post 04- 7-10

[21] President Reagan's Legacy and U.S. Nuclear Weapons Policy, Heritage.org, July 20, 2006

[22] "Hyvästi, ydinpommi", Helsingin Sanomat 2010-09-05, p. D1-D2

[23] Colman McCarthy (February 8, 2009). "From Lafayette Square Lookout, He Made His War Protest Permanent". *The Washington Post*.

[24] "The Oracles of Pennsylvania Avenue". *Al Jazeera Documentary Channel*. April 17, 2012.

[25] Harvey Klehr. Far Left of Center: The American Radical Left Today Transaction Publishers, 1988, p. 150.

[26] 1,400 Anti-nuclear protesters arrested *Miami Herald*, June 21, 1983.

[27] Hundreds of Marchers Hit Washington in Finale of Nationwide Peace March *Gainesville Sun*, November 16, 1986.

[28] Robert Lindsey. 438 Protesters are Arrested at Nevada Nuclear Test Site *New York Times*, February 6, 1987.

[29] 493 Arrested at Nevada Nuclear Test Site *New York Times*, April 20, 1992.

[30] Lance Murdoch. Pictures: New York MayDay anti-nuke/war march *IndyMedia*, 2 may 2005.

[31] Anti-Nuke Protests in New York *Fox News*, May 2, 2005.

[32] Lawrence S. Wittner. Nuclear Disarmament Activism in Asia and the Pacific, 1971-1996 *The Asia-Pacific Journal*, Vol. 25-5-09, June 22, 2009.

[33] Protest against nuclear reactor *Chicago Tribune*, October 16, 2008.

[34] Southeast Climate Convergence occupies nuclear facility *Indymedia UK*, August 8, 2008.

[35] Anti-Nuclear Renaissance: A Powerful but Partial and Tentative Victory Over Atomic Energy

[36] Police arrest 64 at California anti-nuclear protest *Reuters*, April 6, 2007.

[37] Anti-nuclear rally held at test site: Martin Sheen among activists cited by police *Las Vegas Review-Journal*, April 2, 2007.

[38] For decades, faith has sustained anti-nuclear movement *Seattle Times*, April 7, 2006.

[39] Bangor Protest Peaceful; 17 Anti-Nuclear Demonstrators Detained and Released *Kitsap Sun*, January 19, 2008.

[40] Twelve Arrests, But No Violence at Bangor Anti-Nuclear Protest *Kitsap Sun*, June 1, 2008.

[41] Deery, P., "The Dove Flies East: Whitehall, Warsaw and the 1950 World Peace Congress", *Australian Journal of Politics and History*, Vol. 48, 2002

[42] Lawrence Wittner, *Resisting the Bomb*, Stanford University Press, 1997

[43] Santi, Rainer, *100 years of peace making: A history of the International Peace Bureau and other international peace movement organisations and networks*, Pax förlag, International Peace Bureau, January 1991

[44] Committee on Un-American Activities, *Report on the Communist "peace" offensive. A campaign to disarm and defeat the United States*, 1951

[45] Y., *Ideas of Peace and Concordance in Soviet Political Propaganda (1950 – 1985*

[46] Wernicke, Günter, "The Communist-Led World Peace Council and the Western Peace Movements: The Fetters of Bipolarity and Some Attempts to Break Them in the Fifties and Early Sixties", Peace & Change, Volume 23, Number 3, July 1998 , pp. 265–311(47)

[47] "Nuclear Disarmament," US Policy World. http://www.uspw.org/

[48] John Burroughs (2012). "Humanitarian Law or Nuclear Weapons: Choose One" (PDF). *Lawyers Committee on Nuclear Policy*.

[49] Renewing the United Nations: A Program for Reform (A/51/950)

[50]

[51]

[52]

[53] The Gang of Four on Nuclear Deterrence

[54] Renewed call from Kissinger, Nunn, Perry and Shultz for Nuclear-Free World

[55] Barack Obama and Joe Biden's Plan to Secure America and Restore our Standing

[56] http://www.sipri.org/yearbook/2011/07 [Yearbook, Ch. 7 -- World Nuclear Forces], Stockholm International Peace Research Institute, June 7, 2011.

[57] Norris, Robert S., William Arkin, Hans M. Kristensen, and Joshua Handler. *Bulletin of the Atomic Scientists* 58:5 (September/October 2002): 73-75. Israeli nuclear forces, 2002

[58] "Nuclear Disarmament". *US Policy World*. Archived from the original on October 19, 2009.

[59] Maclin, Beth (2008-10-20) "A Nuclear weapon-free world is possible, Nunn says", Belfer Center, Harvard University. Retrieved on 2008-10-21.

[60] "The Growing Appeal of Zero". *The Economist*. June 18, 2011. p. 66.

[61] http://www.globalzero.org?name=2.htm&id=2

[62] "International Conference on Nuclear Disarmament". February 2008.

[63] Ernest Lefever (Autumn 1999). "Nuclear Weapons: Instruments of Peace". *Global Dialogue*.

[64] http://www.stimson.org/images/uploads/research-pdfs/ESCCONTROLCHAPTER1.pdf

[65] http://krepon.armscontrolwonk.com/archive/2911/the-stability-instability-paradox

[66] https://www.mtholyoke.edu/acad/intrel/waltz1.htm   Kenneth Waltz, "The Spread of Nuclear Weapons: More May be Better,"

[67] Waltz, Kenneth (July–August 2012). "Why Iran Should Get the Bomb: Nuclear Balancing Would Mean Stability". *Foreign Affairs*. Retrieved 25 August 2012.

[68] http://www.oup.com/us/catalog/general/subject/Politics/InternationalStudies/InternationalSecurityStrategicSt/?view=usa&ci=9780195381368

[69] http://bloggingheads.tv/videos/2333 From 19:00 to 26:00 minutes

[70] http://www.c-spanvideo.org/program/AtomicO: John Mueller, "Atomic Obsession"

[71] Ben Goddard (2010-01-27). "Cold Warriors say no nukes". *The Hill.*

[72] Hugh Gusterson (30 March 2012). "The new abolitionists". *Bulletin of the Atomic Scientists.*

26.10 External links

- New Video: A World Without Nuclear Weapons

- Nuclear Files.org—Arms Control and Disarmament

- Annotated bibliography for nuclear arms control from the Alsos Digital Library for Nuclear Issues

- The Woodrow Wilson Center's Nuclear Proliferation International History Project or NPIHP is a global network of individuals and institutions engaged in the study of international nuclear history through archival documents, oral history interviews and other empirical sources.

- Council for a Livable World

- Center for Arms Control and Non-Proliferation

- People v The Bomb: Showdown at the UN—TV documentary report on 2005 NPT Review crisis

- William Walker, "President-elect Obama and Nuclear Disarmament. Between Elimination and Restraint.", *Proliferation Papers*, Paris, Ifri, Winter 2009

- Robert S. Norris & Hans M. Kristensen, "Nuclear U.S. and Soviet/Russian intercontinental ballistic missiles, 1959-2008"

- Robert S. Norris & Hans M. Kristensen, "U.S. nuclear forces, 2009", Nuclear Notebook, "Bulletin of the Atomic Scientists"

- Seiitsu Tachibana, "Bush administration's nuclear weapons policy : New obstacles to nuclear disarmament" *Hiroshima Peace Science*, Vol. 24, pages 105-133 (2002)

- Nuclear Disarmament at the United Nations Office for Disarmament Affairs

- Stockholm International Peace Research Institute's Research on Nuclear weapons

Chapter 27

History of the anti-nuclear movement

Main article: Anti-nuclear movement

The application of nuclear technology, both as a source

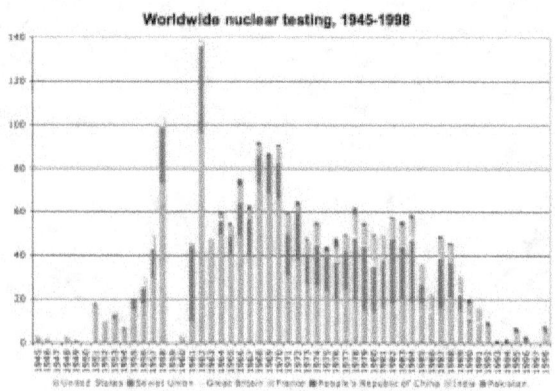

Worldwide nuclear testing totals, 1945-1998.

of energy and as an instrument of war, has been controversial.[1][2][3][4][5]

Scientists and diplomats have debated nuclear weapons policy since before the atomic bombing of Hiroshima in 1945.[6] The public became concerned about nuclear weapons testing from about 1954, following extensive nuclear testing in the Pacific. In 1961, at the height of the Cold War, about 50,000 women brought together by Women Strike for Peace marched in 60 cities in the United States to demonstrate against nuclear weapons.[7][8] In 1963, many countries ratified the Partial Test Ban Treaty which prohibited atmospheric nuclear testing.[9]

Some local opposition to nuclear power emerged in the early 1960s,[10] and in the late 1960s some members of the scientific community began to express their concerns.[11] In the early 1970s, there were large protests about a proposed nuclear power plant in Wyhl, Germany. The project was cancelled in 1975 and anti-nuclear success at Wyhl inspired opposition to nuclear power in other parts of Europe and North America.[12][13] Nuclear power became an issue of major public protest in the 1970s.[14]

27.1 Early years

The 1945 Trinity explosion, 0.016 seconds after detonation. The fireball is about 200 meters (600 ft) wide. Trees may be seen as black objects in the foreground.

In 1945 in the New Mexico desert, American scientists conducted "Trinity," the first nuclear weapons test, marking the beginning of the atomic age.[15] Even before the Trinity test, national leaders debated the impact of nuclear weapons on domestic and foreign policy. Also involved in the debate about nuclear weapons policy was the scientific community, through professional associations such as the Federation of Atomic Scientists and the Pugwash Conference on Science and World Affairs.[6]

On August 6, 1945, towards the end of World War II, the Little Boy device was detonated over the Japanese military city of Hiroshima. Exploding with a yield equivalent to 12,500 tonnes of TNT, the blast and thermal wave of the bomb destroyed nearly 50,000 buildings (including the headquarters of the 2nd General Army and Fifth Division) and killed approximately 75,000 people, among them 20,000 Japanese soldiers and 20,000 Korean slave laborers.[16] Detonation of the Fat Man device exploded over the Japanese industrial city of Nagasaki three days later after Hiroshima, destroying 60% of the city and killing approximately 35,000 people, among them 23,200-28,200

The mushroom cloud over Hiroshima after the dropping of the atomic bomb nicknamed 'Little Boy' (1945).

Operation Crossroads Test Able, a 23-kiloton air-deployed nuclear weapon detonated on July 1, 1946. This bomb used, and consumed, the infamous Demon core that took the lives of two scientists in two separate criticality accidents.

Mushroom-shaped cloud and water column from the underwater nuclear explosion of July 25, 1946, which was part of Operation Crossroads.

November 1951 nuclear test at the Nevada Test Site, from Operation Buster, with a yield of 21 kilotons. It was the first U.S. nuclear field exercise conducted on land; troops shown are 6 mi (9.7 km) from the blast.

Japanese munitions workers, 2,000 Korean slave laborers, and 150 Japanese soldiers.[17] The two bombings remains the only events where nuclear weapons have been used in combat. Subsequently, the world's nuclear weapons stockpiles grew.[15]

Operation Crossroads was a series of nuclear weapon tests conducted by the United States at Bikini Atoll in the Pacific Ocean in the summer of 1946. Its purpose was to test the effect of nuclear weapons on naval ships. Pressure to cancel Operation Crossroads came from scientists and diplomats. Manhattan Project scientists argued that further nu-

clear testing was unnecessary and environmentally dangerous. A Los Alamos study warned "the water near a recent surface explosion will be a witch's brew" of radioactivity. To prepare the atoll for the nuclear tests, Bikini's native residents were evicted from their homes and resettled on smaller, uninhabited islands where they were unable to sustain themselves.[18]

Radioactive fallout from nuclear weapons testing was first drawn to public attention in 1954 when a Hydrogen bomb test in the Pacific contaminated the crew of the Japanese fishing boat Lucky Dragon.[9] One of the fishermen died in Japan seven months later. The incident caused widespread concern around the world and "provided a decisive impetus for the emergence of the anti-nuclear weapons movement in many countries".[9] The anti-nuclear weapons movement grew rapidly because for many people the atomic bomb "encapsulated the very worst direction in which society was moving".[19]

Women Strike for Peace during the Cuban Missile Crisis in 1962.

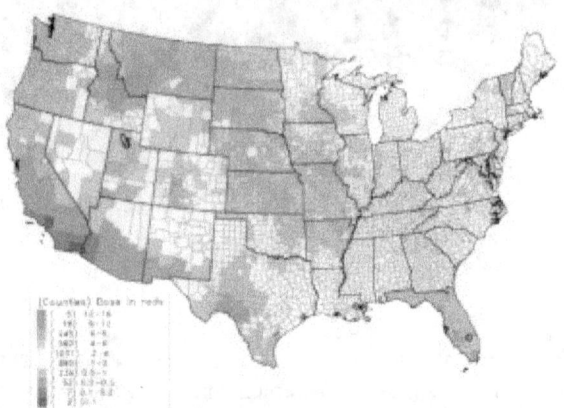

Because of concerns about worldwide fallout levels, the Partial Test Ban Treaty was signed in 1963. Above are the per capita thyroid doses (in rads) in the continental United States resulting from all exposure routes from all atmospheric nuclear tests conducted at the Nevada Test Site from 1951–1962.

Peace movements emerged in Japan and in 1954 they converged to form a unified "Japanese Council Against Atomic and Hydrogen Bombs". Japanese opposition to the Pacific nuclear weapons tests was widespread, and "an estimated 35 million signatures were collected on petitions calling for bans on nuclear weapons".[19]

German publications of the 1950s and 1960s contained criticism of some features of nuclear power including its safety.

The Phoenix of Hiroshima (foreground) in Hong Kong Harbor in 1967, was involved in several famous protest voyages against nuclear testing in the Pacific.

Nuclear waste disposal was widely recognized as a major problem, with concern publicly expressed as early as 1954. In 1964, one author went so far as to state "that the dangers and costs of the necessary final disposal of nuclear waste could possibly make it necessary to forego the development of nuclear energy".[20]

The Russell–Einstein Manifesto was issued in London on July 9, 1955 by Bertrand Russell in the midst of the Cold War. It highlighted the dangers posed by nuclear weapons and called for world leaders to seek peaceful resolutions to international conflict. The signatories included eleven preeminent intellectuals and scientists, including Albert Einstein, who signed it just days before his death on April 18, 1955. A few days after the release, philanthropist Cyrus S. Eaton offered to sponsor a conference—called for in the manifesto—in Pugwash, Nova Scotia, Eaton's birthplace. This conference was to be the first of the Pugwash Conferences on Science and World Affairs, held in July 1957.

In the United Kingdom, the first Aldermaston March organised by the Campaign for Nuclear Disarmament took place at Easter 1958, when several thousand people marched for four days from Trafalgar Square, London, to the Atomic Weapons Research Establishment close to Aldermaston in Berkshire, England, to demonstrate their opposition to nuclear weapons.[21][22] The Aldermaston marches continued into the late 1960s when tens of thousands of people took part in the four-day marches.[19]

In 1959, a letter in the *Bulletin of Atomic Scientists* was the start of a successful campaign to stop the Atomic Energy Commission dumping radioactive waste in the sea 19 kilometres from Boston.[23]

On November 1, 1961, at the height of the Cold War, about 50,000 women brought together by Women Strike for Peace marched in 60 cities in the United States to demonstrate against nuclear weapons. It was the largest national women's peace protest of the 20th century.[7][8]

In 1958, Linus Pauling and his wife presented the United Nations with the petition signed by more than 11,000 scientists calling for an end to nuclear-weapon testing. The "Baby Tooth Survey," headed by Dr Louise Reiss, demonstrated conclusively in 1961 that above-ground nuclear testing posed significant public health risks in the form of radioactive fallout spread primarily via milk from cows that had ingested contaminated grass.[24][25][26] Public pressure and the research results subsequently led to a moratorium on above-ground nuclear weapons testing, followed by the Partial Test Ban Treaty, signed in 1963 by John F. Kennedy and Nikita Khrushchev.[27] On the day that the treaty went into force, the Nobel Prize Committee awarded Pauling the Nobel Peace Prize, describing him as "Linus Carl Pauling, who ever since 1946 has campaigned ceaselessly, not only against nuclear weapons tests, not only against the spread of these armaments, not only against their very use, but against all warfare as a means of solving international conflicts."[6][28]

Pauling started the International League of Humanists in 1974. He was president of the scientific advisory board of the World Union for Protection of Life and also one of the signatories of the Dubrovnik-Philadelphia Statement.

27.2 After the Partial Test Ban Treaty

The Shippingport Atomic Power Station was the first full-scale PWR nuclear power plant in the United States. The reactor went online December 2, 1957, and was in operation until October, 1982.

In the United States, the first commercially viable nuclear power plant was to be built at Bodega Bay, north of San Francisco, but the proposal was controversial and conflict with local citizens began in 1958.[10] The proposed plant site was close to the San Andreas Fault and close to the

Radioactive materials were accidentally released from the 1970 Baneberry Nuclear Test at the Nevada Test Site.

The 18,000 km² expanse of the Semipalatinsk Test Site (indicated in red), which covers an area the size of Wales. The Soviet Union conducted 456 nuclear tests at Semipalatinsk from 1949 until 1989 with little regard for their effect on the local people or environment. The full impact of radiation exposure was hidden for many years by Soviet authorities and has only come to light since the test site closed in 1991.[29]

region's environmentally sensitive fishing and dairy industries. The Sierra Club became actively involved.[31] The conflict ended in 1964, with the forced abandonment of plans for the power plant. Historian Thomas Wellock traces the birth of the anti-nuclear movement to the controversy over Bodega Bay.[10] Attempts to build a nuclear power plant in Malibu were similar to those at Bodega Bay and were also abandoned.[10]

In 1966, Larry Bogart founded the Citizens Energy Coun-

120,000 people attended an anti-nuclear protest in Bonn, Germany, on October 14, 1979, following the Three Mile Island accident.[30]

cil, a coalition of environmental groups that published the newsletters "Radiation Perils," "Watch on the A.E.C." and "Nuclear Opponents". These publications argued that "nuclear power plants were too complex, too expensive and so inherently unsafe they would one day prove to be a financial disaster and a health hazard".[32][33]

The emergence of the anti-nuclear power movement was "closely associated with the general rise in environmental consciousness which had started to materialize in the USA in the 1960s and quickly spread to other Western industrialized countries".[11] Some nuclear experts began to voice dissenting views about nuclear power in 1969, and this was a necessary precondition for broad public concern about nuclear power to emerge.[11] These scientists included Ernest Sternglass from Pittsburg, Henry Kendall from the Massachusetts Institute of Technology, Nobel laureate George Wald and radiation specialist Rosalie Bertell. These members of the scientific community "by expressing their concern over nuclear power, played a crucial role in demystifying the issue for other citizens", and nuclear power became an issue of major public protest in the 1970s.[11][34]

In 1971, 15,000 people demonstrated against French plans to locate the first light-water reactor power plant in Bugey. This was the first of a series of mass protests organized at nearly every planned nuclear site in France.[35]

Also in 1971, the town of Wyhl, in Germany, was a proposed site for a nuclear power station. In the years that followed, public opposition steadily mounted, and there were large protests. Television coverage of police dragging away farmers and their wives helped to turn nuclear power into a major issue. In 1975, an administrative court withdrew the construction licence for the plant,[12][13][36] but the Wyhl occupation generated ongoing debate. This initially centred on the state government's handling of the affair and associated police behaviour, but interest in nuclear issues was also stimulated. The Wyhl experience encouraged the formation of citizen action groups near other planned nuclear

sites.[12] Many other anti-nuclear groups formed elsewhere, in support of these local struggles, and some existing citizen action groups widened their aims to include the nuclear issue.[12] Anti-nuclear success at Wyhl also inspired nuclear opposition in the rest of Europe and North America.[13]

In 1972, the anti-nuclear weapons movement maintained a presence in the Pacific, largely in response to French nuclear testing there. Activists, including David McTaggart from Greenpeace, defied the French government by sailing small vessels into the test zone and interrupting the testing program.[37][38] In Australia, thousands joined protest marches in Adelaide, Melbourne, Brisbane, and Sydney.[38] Scientists issued statements demanding an end to the tests; unions refused to load French ships, service French planes, or carry French mail; and consumers boycotted French products. In Fiji, activists formed an Against Testing on Mururoa organization.[38]

In Spain, in response to a surge in nuclear power plant proposals in the 1960s, a strong anti-nuclear movement emerged in 1973, which ultimately impeded the realisation of most of the projects.[39]

In 1974, organic farmer Sam Lovejoy took a crowbar to the weather-monitoring tower which had been erected at the Montague Nuclear Power Plant site. Lovejoy felled the tower and then took himself to the local police station, where he took full responsibility for the action. Lovejoy's action galvanized local public opinion against the plant.[40][41] The Montague project was canceled in 1980,[42] after $29 million was spent on the project.[40]

By the mid-1970s anti-nuclear activism had moved beyond local protests and politics to gain a wider appeal and influence. Although it lacked a single co-ordinating organization, and did not have uniform goals, the movement's efforts gained a great deal of attention.[4] Jim Falk has suggested that popular opposition to nuclear power quickly grew into an effective anti-nuclear power movement in the 1970s.[43] In some countries, the nuclear power conflict "reached an intensity unprecedented in the history of technology controversies".[44]

In France, between 1975 and 1977, some 175,000 people protested against nuclear power in ten demonstrations.[30]

In West Germany, between February 1975 and April 1979, some 280,000 people were involved in seven demonstrations at nuclear sites. Several site occupations were also attempted. In the aftermath of the Three Mile Island accident in 1979, some 120,000 people attended a demonstration against nuclear power in Bonn.[30]

In May 1979, an estimated 70,000 people, including the governor of California, attended a march and rally against nuclear power in Washington, D.C.[45][46]

On June 12, 1982, one million people demonstrated in New

York City's Central Park against nuclear weapons and for an end to the cold war arms race. It was the largest anti-nuclear protest and the largest political demonstration in American history.[47][48] International Day of Nuclear Disarmament protests were held on June 20, 1983 at 50 sites across the United States.[49][50] In 1986, hundreds of people walked from Los Angeles to Washington DC in the Great Peace March for Global Nuclear Disarmament.[51] There were many Nevada Desert Experience protests and peace camps at the Nevada Test Site during the 1980s and 1990s.[52][53]

On May 1, 2005, 40,000 anti-nuclear/anti-war protesters marched past the United Nations in New York, 60 years after the atomic bombings of Hiroshima and Nagasaki.[54] This was the largest anti-nuclear rally in the U.S. for several decades.[55] In Britain, there were many protests about the government's proposal to replace the aging Trident weapons system with a newer model. The largest protest had 100,000 participants and, according to polls, 59 percent of the public opposed the move.[55]

The International Conference on Nuclear Disarmament took place in Oslo in February 2008, and was organized by The Government of Norway, the Nuclear Threat Initiative and the Hoover Institute. The Conference was entitled *Achieving the Vision of a World Free of Nuclear Weapons* and had the purpose of building consensus between nuclear weapon states and non-nuclear weapon states in relation to the Nuclear Non-proliferation Treaty.[56]

In May 2010, some 25,000 people, including members of peace organizations and 1945 atomic bomb survivors, marched for about two kilometers from downtown New York to the United Nations headquarters, calling for the elimination of nuclear weapons.[57]

27.3   Other issues

Early anti-nuclear advocates expressed the view that affluent lifestyles on a global scale strain the viability of the natural environment and that nuclear energy would enable those lifestyles. Examples of such expressions are:

"We can and should seize upon the energy crisis as a good excuse and a great opportunity for making some very fundamental changes that we should be making anyhow for other reasons."
— Russell E. Train, 1974.[58]

"In fact, giving society cheap, abundant energy at this point would be the moral equivalent of giving an idiot child a machine gun."
— Paul R. Ehrlich, 1975.[59]

"If you ask me, it'd be little short of disastrous for us to discover a source of clean, cheap, abundant energy because of what we would do with it. We ought to be looking for energy sources that are adequate for our needs, but that won't give us the excesses of concentrated energy with which we could do mischief to the earth or to each other."
— Amory Lovins, 1977.[60]

"Let's face it. We don't want safe nuclear power plants. We want NO nuclear power plants."
— Spokesman for the Government Accountability Project, 1985.[61]

"... we also thought that as you provide societies with more energy it enables them to do more environmental destruction. The idea of tying us to the natural forces of the wind and the sun was very appealing in that it would limit and constrain human development"
— Robert Stone (director) (of both anti-nuclear weapons and, recently, pro-nuclear power films), 2014.[62]

27.4   See also

- Debate over the atomic bombings of Hiroshima and Nagasaki

- Nuclear power debate

- Nuclear weapons debate

- The Bomb (film)

- Uranium mining debate

27.5   References

[1] "Sunday Dialogue: Nuclear Energy, Pro and Con". *New York Times*. February 25, 2012.

[2] Robert Benford. The Anti-nuclear Movement (book review) *American Journal of Sociology*, Vol. 89, No. 6, (May 1984), pp. 1456-1458.

[3] James J. MacKenzie. Review of The Nuclear Power Controversy by Arthur W. Murphy *The Quarterly Review of Biology*, Vol. 52, No. 4 (Dec., 1977), pp. 467-468.

[4] Walker, J. Samuel (2004). *Three Mile Island: A Nuclear Crisis in Historical Perspective* (Berkeley: University of California Press), pp. 10-11.

[5] Jim Falk (1982). *Global Fission: The Battle Over Nuclear Power*, Oxford University Press.

[6] Jerry Brown and Rinaldo Brutoco (1997). *Profiles in Power: The Anti-nuclear Movement and the Dawn of the Solar Age*, Twayne Publishers, pp. 191-192.

[7] Woo, Elaine (January 30, 2011). "Dagmar Wilson dies at 94; organizer of women's disarmament protesters". *Los Angeles Times*.

[8] Hevesi, Dennis (January 23, 2011). "Dagmar Wilson, Anti-Nuclear Leader, Dies at 94". *The New York Times*.

[9] Wolfgang Rudig (1990). *Anti-nuclear Movements: A World Survey of Opposition to Nuclear Energy*, Longman, p. 54-55.

[10] Paula Garb. Review of Critical Masses, *Journal of Political Ecology*, Vol 6, 1999.

[11] Wolfgang Rudig (1990). *Anti-nuclear Movements: A World Survey of Opposition to Nuclear Energy*, Longman, p. 52.

[12] Stephen Mills and Roger Williams (1986). Public Acceptance of New Technologies Routledge, pp. 375-376.

[13] Robert Gottlieb (2005). Forcing the Spring: The Transformation of the American Environmental Movement, Revised Edition, Island Press, USA, p. 237.

[14] Jim Falk (1982). *Global Fission: The Battle Over Nuclear Power*, Oxford University Press, pp. 95-96.

[15] Mary Palevsky, Robert Futrell, and Andrew Kirk. Recollections of Nevada's Nuclear Past *UNLV FUSION*, 2005, p. 20.

[16] Emsley, John (2001). "Uranium". *Nature's Building Blocks: An A to Z Guide to the Elements*. Oxford: Oxford University Press. p. 478. ISBN 0-19-850340-7.

[17] *Nuke-Rebuke: Writers & Artists Against Nuclear Energy & Weapons (The Contemporary anthology series)*. The Spirit That Moves Us Press. May 1, 1984. pp. 22–29.

[18] Niedenthal, Jack (2008), *A Short History of the People of Bikini Atoll*, retrieved 2009-12-05

[19] Jim Falk (1982). *Global Fission: The Battle Over Nuclear Power*, Oxford University Press, pp. 96-97.

[20] Wolfgang Rudig (1990). *Anti-nuclear Movements: A World Survey of Opposition to Nuclear Energy*, Longman, p. 63.

[21] A brief history of CND

[22] "Early defections in march to Aldermaston". Guardian Unlimited. 1958-04-05.

[23] Jim Falk (1982). *Global Fission: The Battle Over Nuclear Power*, Oxford University Press, p. 93.

[24] Louise Zibold Reiss (November 24, 1961). "Strontium-90 Absorption by Deciduous Teeth: Analysis of teeth provides a practicable method of monitoring strontium-90 uptake by human populations" (PDF). Science. Retrieved October 13, 2009.

[25] Thomas Hager (November 29, 2007). "Strontium-90". Oregon State University Libraries Special Collections. Retrieved December 13, 2007.

[26] Thomas Hager (November 29, 2007). "The Right to Petition". Oregon State University Libraries Special Collections. Retrieved December 13, 2007.

[27] Jim Falk (1982). *Global Fission: The Battle Over Nuclear Power*, Oxford University Press, p. 98.

[28] Linus Pauling (October 10, 1963). "Notes by Linus Pauling. October 10, 1963.". Oregon State University Libraries Special Collections. Retrieved December 13, 2007.

[29] Togzhan Kassenova (28 September 2009). "The lasting toll of Semipalatinsk's nuclear testing". *Bulletin of the Atomic Scientists*.

[30] Herbert P. Kitschelt. Political Opportunity and Political Protest: Anti-Nuclear Movements in Four Democracies *British Journal of Political Science*, Vol. 16, No. 1, 1986, p. 71.

[31] Thomas Raymond Wellock (1998). Critical Masses: Opposition to Nuclear Power in California, 1958-1978, The University of Wisconsin Press, pp. 27–28.

[32] Keith Schneider. Larry Bogart, an Influential Critic Of Nuclear Power, Is Dead at 77 *The New York Times*, August 20, 1991.

[33] Anna Gyorgy (1980). No Nukes: Everyone's Guide to Nuclear Power South End Press, ISBN 0-89608-006-4, p. 383.

[34] Jim Falk (1982). *Global Fission: The Battle Over Nuclear Power*, Oxford University Press, p. 95.

[35] Dorothy Nelkin and Michael Pollak (1982). *The Atom Besieged: Antinuclear Movements in France and Germany*, ASIN: B0011LXE0A, p. 3.

[36] Nuclear Power in Germany: A Chronology

[37] Paul Lewis. David McTaggart, a Builder of Greenpeace, Dies at 69 *The New York Times*, March 24, 2001.

[38] Lawrence S. Wittner. Nuclear Disarmament Activism in Asia and the Pacific, 1971-1996 *The Asia-Pacific Journal*, Vol. 25-5-09, June 22, 2009.

[39] Lutz Mez, Mycle Schneider and Steve Thomas (Eds.) (2009). *International Perspectives of Energy Policy and the Role of Nuclear Power*, Multi-Science Publishing Co. Ltd, p. 371.

[40] Utilities Drop Nuclear Power Plant Plans *Ocala Star-Banner*, January 4, 1981.

[41] Anna Gyorgy (1980). No Nukes: Everyone's Guide to Nuclear PowerSouth End Press, ISBN 0-89608-006-4, pp. 393-394.

[42] Northeast Utilities System. Some of the Major Events in NU's History Since the 1966 Affiliation

[43] Jim Falk (1982). *Global Fission: The Battle Over Nuclear Power*, Oxford University Press, p. 96.

[44] Herbert P. Kitschelt. Political Opportunity and Political Protest: Anti-Nuclear Movements in Four Democracies *British Journal of Political Science*, Vol. 16, No. 1, 1986, p. 57.

[45] Jon Agnone. Amplifying Public Opinion: The Policy Impact of the U.S. Environmental Movement p. 7.

[46] Social Protest and Policy Change p. 45.

[47] Jonathan Schell. The Spirit of June 12 *The Nation*, July 2, 2007.

[48] 1982 - a million people march in New York City

[49] Harvey Klehr. Far Left of Center: The American Radical Left Today Transaction Publishers, 1988, p. 150.

[50] 1,400 Anti-nuclear protesters arrested *Miami Herald*, June 21, 1983.

[51] Hundreds of Marchers Hit Washington in Finale of Nationwaide Peace March *Gainesville Sun*, November 16, 1986.

[52] Robert Lindsey. 438 Protesters are Arrested at Nevada Nuclear Test Site *New York Times*, February 6, 1987.

[53] 493 Arrested at Nevada Nuclear Test Site *New York Times*, April 20, 1992.

[54] Anti-Nuke Protests in New York *Fox News*, May 2, 2005.

[55] Lawrence S. Wittner. A rebirth of the anti-nuclear weapons movement? Portents of an anti-nuclear upsurge *Bulletin of the Atomic Scientists*, 7 December 2007.

[56] "International Conference on Nuclear Disarmament". February 2008.

[57] A-bomb survivors join 25,000-strong anti-nuclear march through New York *Mainichi Daily News*, May 4, 2010.

[58] Train, R. E. (1974). "The Quality of Growth". *Science*. **184** (4141): 1050–3. doi:10.1126/science.184.4141.1050. PMID 17736183.

[59] "An Ecologist's Perspective on Nuclear Power", Federation of American Scientists Public Issue Report, May-June 1975

[60] Mother Earth News Nov/Dec 1977, p. 22: The Plowboy Interview with Amory Lovins

[61] The American Spectator, Vol 18, No. 11, Nov. 1985

[62] KTH Royal Institute of Technology in Stockholm, Nov. 2014: Interview with Robert Stone

Chapter 28

Effects of nuclear explosions on human health

The **medical effects of the atomic bomb on Hiroshima** upon humans can be put into the four categories below, with the effects of larger thermonuclear weapons producing blast and thermal effects so large that there would be a negligible number of survivors close enough to the center of the blast who would experience prompt/acute radiation effects, which were observed after the 16 kiloton yield Hiroshima bomb, due to its relatively low yield:[1][2]

- Initial stage—the first 1–9 weeks, in which are the greatest number of deaths, with 90% due to thermal injury and/or blast effects and 10% due to super-lethal radiation exposure.

- Intermediate stage—from 10–12 weeks. The deaths in this period are from ionizing radiation in the median lethal range - LD50

- Late period—lasting from 13–20 weeks. This period has some improvement in survivors' condition.

- Delayed period—from 20+ weeks. Characterized by numerous complications, mostly related to healing of thermal and mechanical injuries, and if the individual was exposed to a few hundred to a thousand Millisieverts of radiation, it is coupled with infertility, sub-fertility and blood disorders. Furthermore, ionizing radiation above a dose of around 50-100 Millisievert exposure has been shown to statistically begin increasing ones chance of dying of cancer sometime in their lifetime over the normal unexposed rate of ~25%, in the long term, a heightened rate of cancer, proportional to the dose received, would begin to be observed after ~5+ years, with lesser problems such as eye cataracts and other more minor effects in other organs and tissue also being observed over the long term.

Fallout exposure - Depending on if further afield individuals Shelter in place or evacuate perpendicular to the direction of the wind, and therefore avoid contact with the fallout plume, and stay there for the days and weeks after the nuclear explosion, their exposure to fallout, and therefore their total dose, will vary. With those who do shelter in place, and or evacuate, experiencing a total dose that would be negligible in comparison to someone who just went about their life as normal.[3][4]

Staying indoors until after the most hazardous fallout isotope, I-131 decays away to 0.1% of its initial quantity after ten half lifes - which is represented by 80 days in I-131s case, would make the difference between likely contracting Thyroid cancer or escaping completely from this substance depending on the actions of the individual.

Some scientists estimate that if there were a nuclear war resulting in 100 Hiroshima-size nuclear explosions on cities, it could cause significant loss of life in the tens of millions from long term climatic effects alone. The climatology hypothesis is that *if* each city firestorms, a great deal of soot could be thrown up into the atmosphere which could blanket the earth, cutting out sunlight for years on end, causing the disruption of food chains, in what is termed a Nuclear Winter scenario.[5][6]

28.1 Blast effects - the initial stage

28.1.1 Immediate post-attack period

The main causes of death and disablement in this state are thermal burns and the failure of structures resulting from the blast effect. Injury from the pressure wave is minimal in contrast because the human body can survive up to 2 bar (30 psi) while most buildings can only withstand a 0.8 bar (12 psi) blast. Therefore, the fate of humans is closely related to the survival of the buildings around them.[7]

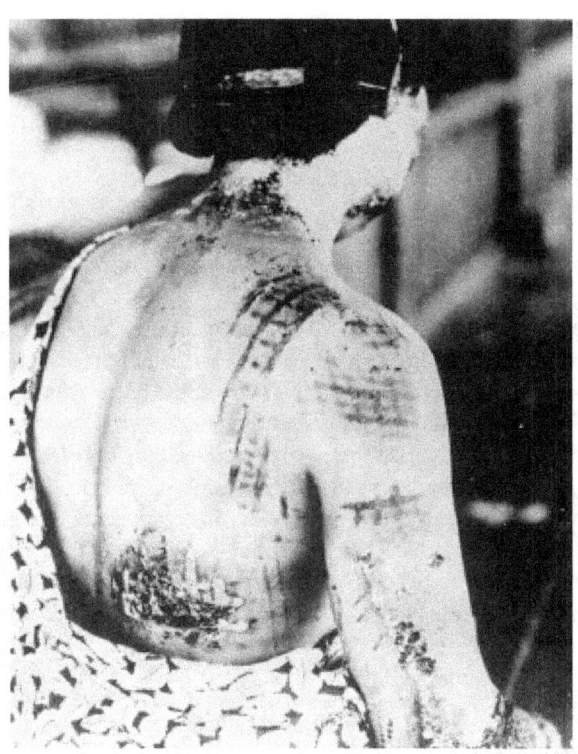

Japanese woman(one of the Hiroshima Maidens) suffering burns from thermal radiation after the United States dropped nuclear bombs on Japan.

Melted and fused pieces of metal (including coins that were in people's pockets) from the Atomic bombings of Japan, the melting of metal like this occurred during the ensuing fires and firestorms, after the bombs had long since exploded.

28.1.2 Fate within certain peak overpressure

- over 0.8 bar (12 psi) - 98% dead, 2% injured

- 0.3 - 0.8 bar (5-12 psi) - 50% dead, 40% injured, 10% safe

- 0.14 - 0.3 bar (2-5 psi) - 5% dead, 45% injured, 50% safe[7]

28.2 Types of radioactive exposure after a nuclear attack

In a nuclear explosion the human body can be irradiated by at least three processes. The first, and most major, cause of burns is due to thermal radiation and not caused by ionizing radiation.

- Thermal burns from infrared heat radiation, these would be the most common burn type experienced by personnel.

- If personnel become in direct contact with fallout, Beta burns from shallow ionizing beta radiation will be experienced, the largest particles (visible to the naked eye) in local fallout would be likely to have very high radioactivity because they would be deposited so soon after detonation; this fraction of the total fallout is called the prompt or local fallout fraction. It is likely that one such particle upon the skin would be able to cause a localized beta burn. This local fallout, termed *Bikini snow* after the Pacific island weapon tests,[8] was experienced by the crew on the deck of the Lucky Dragon fishing ship following the explosion of the 15 megaton *Shrimp* device in the Castle Bravo event. However, these particular decay particles (beta particles) are very weakly penetrating and have a short range, requiring almost direct contact between fallout and personnel to be harmful.

See also Nuclear Fallout

- Rarer still would be personnel who experience radiation burns from highly penetrating gamma radiation. This would likely cause deep gamma penetration within the body, which would result in uniform whole body irradiation rather than only a surface burn. In cases of whole body gamma irradiation (*c.* 10 Gy) due to accidents involving medical product irradiators, some of the human subjects have developed injuries to their skin between the time of irradiation and death.

In the picture above, the normal clothing (a Kimono) that the woman was wearing attenuated the far reaching thermal radiation, the kimono however would have naturally been unable to attenuate any gamma radiation, if she were close enough to the weapon to have experienced any, and it would be likely that any such penetrating radiation effect would be evenly applied to her entire body. Beta burns would be likely all over the body if there was contact with fallout after the explosion, unlike thermal burns, which are only ever on one side of the body, as heat radiation infrared naturally does not penetrate the human body. In addition, the pattern on her clothing has been burnt into the skin by the thermal radiation. This is because white fabric reflects more visible and infrared light than dark fabric. As a result, the skin underneath dark fabric is burned more than the skin covered by white clothing.

There is also the risk of internal radiation poisoning by ingestion of fallout particles, if one is in a fallout zone.

28.3   Radiation poisoning

Radiation poisoning, also called "**radiation sickness**" or a "**creeping dose**", is a form of damage to organ tissue due to excessive exposure to ionizing radiation. The term is generally used to refer to acute problems caused by a large dosage of radiation in a short period, though this also has occurred with long-term exposure to low-level radiation. Many of the symptoms of radiation poisoning occur as ionizing radiation interferes with cell division. There are numerous lethal radiation syndromes, including prodromal syndrome, bone marrow death, central nervous system death and gastrointestinal death.[9]

28.3.1   Prodromal syndrome

The "prodromal syndrome" is not a diagnosis, but the technical term used by mental health professionals to describe a specific group of symptoms that may precede the onset of a mental illness. For example, a fever is "prodromal" to measles, which means that a fever may be a risk factor for developing this illness.

28.3.2   Bone marrow death

Bone marrow death is caused by a dose of radiation between 2 and 10 Gray and is characterized by the part of the bone marrow that makes the blood being broken down. Therefore, production of red and white blood cells and platelets is stopped due to loss of the blood-making stem cells (4.5 Gray kills 95% of stem cells). The loss of platelets greatly increases the chance of fatal hemorrhage, while the lack of white blood cells causes infections; the fall in red blood cells is minimal, and only causes mild anemia.[9]

The exposure to 4.5 Gray of penetrating gamma rays has many effects that occur at different times:

In 24 hours:[9]

- vomiting

- diarrhea

These will usually abate after 6–7 days.

Within 3–4 weeks there is a period of extreme illness.[9]

- severe bloody diarrhea, indicating intestinal disorders causing fluid imbalance

- extensive internal bleeding

- septicemia infections

The peak incidence of acute BM death corresponds to the 30-day nadir in blood cell numbers. The number of deaths then falls progressively until it reaches 0 at 60 days after irradiation. The amount of radiation greatly affects the probability of death. For example, over the range of 2 to 6 Gray the probability of death in untreated adults goes from about 1% to 99%, but these figures are for healthy adults. Therefore, results may differ, because of the thermal and mechanical injuries and infectious conditions.[9]

28.3.3   Gastrointestinal death

Gastrointestinal death is caused by a dose of radiation between 10 and 50 Gray. Whole body doses cause damage to epithelial cells lining the gastrointestinal tract and this combined with the bone marrow damage is fatal. All symptoms become increasingly severe, causing exhaustion and emaciation in a few days and death within 7–14 days from loss of water and electrolytes.[9]

The symptoms of gastrointestinal death are:[9]

- gastrointestinal pain

- anorexia

- nausea

- vomiting

- diarrhea

28.3.4 Central nervous system death

Central nervous system death is the main cause of death in 24–48 hours among those exposed to 50 Gray.[9]

The symptoms are:[9]

- vomiting

- nausea

- diarrhea

- drowsiness

- lethargy

- tremors

- delirium

- frequent seizures

- convulsions

- heat prostration

- coma

- respiratory failure

- death

28.4 Short-term effects (6–8 weeks)

28.4.1 Skin

The skin is susceptible to beta-emitting radioactive fallout. The principal site of damage is the germinal layer, and often the initial response is erythema (reddening) due to blood vessels congestion and edema. Erythema lasting more than 10 days occurs in 50% of people exposed to 5-6 Gray.[9]

Other effects with exposure include:[9]

- 2–3 Gray—temporary hair loss

- 7 Gray—permanent epilation occurs

- 10 Gray—itching and flaking occurs

- 10–20 Gray—weeping blistering and ulceration will occur

28.4.2 Lungs

The lungs are the most radiosensitive organ, and radiation pneumonitis can occur leading to pulmonary insufficiency and death (100% after exposure to 50 Gray of radiation), in a few months.

Radiation pneumonitis is characterized by:[9]

- Loss of epithelial cells

- Edema

- Inflammation

- Occlusions of airways, air sacs and blood vessels

- Fibrosis

28.4.3 Ovaries

A single dose of 1–2 Gray will cause temporary damage and suppress menstruation for periods up to 3 years; a dose of 4 Gray will cause permanent sterility.[9]

28.4.4 Testicles

A dose of 0.1 Gray will cause low sperm counts for up to a year; 2.5 Gray will cause sterility for 2 to 3 years or more. 4 Gray will cause permanent sterility.[9]

28.5 Long-term effects

28.5.1 Cataract induction

The timespan for developing this symptom ranges from 6 months to 30 years to develop but the median time for developing them is 2–3 years.[9]

- 2 Gray of gamma rays cause opacities in a few percent

- 6-7 Gray can seriously impair vision and cause cataracts

28.5.2 Cancer induction

Cancer induction is the most significant long-term risk of exposure to a nuclear bomb. Approximately 1 out of every 80 people exposed to 1 Gray will die from cancer, in addition to the normal rate of 20 out of 80. About 1 in 40 people will get cancer, in addition to the typical rates of 16-20 out of 40. Different types of cancer take different times for them to appear:[9]

- 2 years for leukemia to appear

- 20 or more years for skin cancer or lung cancer

28.5.3 In utero (within the womb) effects on human development

A 1 Gray (unit) dose of radiation will cause between 0 and 20 extra cases of perinatal mortality, per 1,000 births and 0-20 cases per 1000 births of severe mental sub normality. A 0.05 Gray dose will increase death due to cancer 10 fold, from the normal 0.5 per 1000 birth rate to a rate of 5 per 1,000. An antenatal dose of 1 Gray in the first trimester causes the lifetime risk of fatal cancer sometime in the child's life to increase from, ~25% in non exposed humans, to 100% in the first trimester 1 Gray exposed.[9]

28.5.4 Transgenerational genetic damage

Exposure to even relatively low doses of radiation generates genetic damage in the progeny of irradiated rodents. This damage can accumulate over several generations.[10] No statistically demonstrable increase of congenital malformations was found among the *later conceived children born* to survivors of the Nuclear weapons at Hiroshima and Nagasaki.[11][12][13] The surviving women of Hiroshima and Nagasaki, that could conceive, who were exposed to substantial amounts of radiation, went on and had children with no higher incidence of abnormalities than the Japanese average.[14][15]

28.5.5 Infectious diseases resulting from nuclear attack

It was assumed in the 1983 book "Medical Consequences of Radiation Following a Global Nuclear War." that although not caused by radiation, one of the long-term effects of a nuclear war would be a massive increase in infectious diseases caused by fecal matter contaminated water from untreated sewage, crowded living conditions, poor standard of living, and lack of vaccines in the aftermath of a nuclear war, with the following list of diseases being cited:[7]

- Dysentery

- Typhoid

- Infectious hepatitis

- Salmonellosis

- Cholera

- Meningococcal meningitis

- Tuberculosis

- Diphtheria

- Whooping cough

- Polio

- Pneumonia

However although what the authors describe are conditions already prevalent in many of the world's city slums, it is inconceivable why people would try to remain living in crowded conditions by reverting to slum lifestyles, during or after a nuclear war. As many cities would already be destroyed, with urban life, slum or otherwise, therefore serving no benefit to inhabitants.

There would be billions of disease carrying vectors, in the form of city residents,[16] lying deceased in cities caused by the direct nuclear weapons effects alone, with the surviving few billion people spread out in rural communities living agrarian lifestyles, with the survivors therefore posing a way of living far less prone to creating the crowded slum living conditions required for infectious diseases to spread. Moreover, as reported in a paper published in the journal Public Health Reports, it is also one of a number of prevalent myths that infectious diseases always occur after a disaster in cities.[17][18]

Epidemics seldom occur after a disaster, and dead bodies do not lead to catastrophic outbreaks of infectious diseases. Intuitively, epidemic diseases, illnesses, and injuries might be expected following major disasters. However, as noted by de Goyet, epidemics seldom occur after disasters, and unless deaths are caused by one of a small number of infectious diseases such as smallpox, typhus, or plague, exposure to dead bodies does not cause disease...Cholera and typhoid seldom pose a major health threat after disasters unless they are already endemic.

28.6 See also

- Blast shelter

- Fallout shelter

28.7 Notes

[1] http://www.remm.nlm.gov/RemmMockup_files/ radiationlethality.jpg

[2] page 3. see *negligible*. Meaning that if you are close enough to get a harmful dose of radiation from a 1 megaton weapons, you are going to die from blast effects alone.

[3] 7 hour rule: At 7 hours after detonation the fission product activity will have decreased to about 1/10 (10%) of its amount at 1 hour. At about 2 days (49 hours-7X7) the activity will have decreased to 1% of the 1-hour value! http://www.falloutradiation.com/johnwayne7

[4] Nuclear Warfare chapter 9 see page 22

[5] Philip Yam. Nuclear Exchange, *Scientific American*, June 2010, p. 24.

[6] Alan Robock and Owen Brian Toon. Local Nuclear War, Global Suffering, *Scientific American*, January 2010, p. 74-81.

[7] Middleton, Hugh. "Medical Consequences of Radiation Following a Global Nuclear War." The Aftermath (1983): 50-56.

[8] Royle, Stephen A. (1999). "Conservation and Heritage in the Face of Insular Urbanization: The Marshall Islands and Kiribati". *Built Environment (1978-)*. **25** (3).

[9] Coggle, J.E., Lindop, Patricia J. "Medical Consequences of Radiation Following a Global Nuclear War." The Aftermath (1983): 60-71.

[10] "Long-term development of the radionuc... [Radiat Environ Biophys. 2005] - PubMed - NCBI". Ncbi.nlm.nih.gov. 25 March 2013. Retrieved 1 September 2013.

[11] Teratology in the Twentieth Century Plus Ten

[12] "JAMA Network | JAMA | The Children of Atomic Bomb Survivors: A Genetic Study". Jama.ama-assn.org. 9 January 2013. Retrieved 1 September 2013.

[13] British Journal of Cancer. "British Journal of Cancer - Sex ratio among offspring of childhood cancer survivors treated with radiotherapy". Nature.com. Retrieved 1 September 2013.

[14] "Birth defects among the children of atomic-bomb survivors (1948-1954) - Radiation Effects Research Foundation". Rerf.jp. Retrieved 1 September 2013.

[15] "Nuclear Crisis: Hiroshima and Nagasaki cast long shadows over radiation science - Monday, April 11, 2011". www.eenews.net. 11 April 2011. Retrieved 1 September 2013.

[16] "Cities and Emerging or Re-emergging Diseases". Apps.who.int. Retrieved 1 September 2013.

[17] Disaster Mythology and Fact: Hurricane Katrina and Social Attachment

[18] http://www.publichealthreports.org/issueopen.cfm?articleID=2091

Chapter 29

Peaceful nuclear explosion

Peaceful nuclear explosions (**PNEs**) are nuclear explosions conducted for non-military purposes, such as activities related to economic development including the creation of canals. During the 1960s and 1970s, both the United States and the Soviet Union conducted a number of PNEs.

Six of the explosions by the Soviet Union are considered to have been of an applied nature, not just tests.

Subsequently the United States and the Soviet Union halted their programs. Definitions and limits are covered in the Peaceful Nuclear Explosions Treaty of 1976.[1][2] The Comprehensive Nuclear-Test-Ban Treaty of 1996 prohibits all nuclear explosions, regardless of whether they are for peaceful purposes or not.

29.1 Peaceful Nuclear Explosions Treaty

In the PNE Treaty, the signatories agreed: not to carry out any individual nuclear explosions having a yield exceeding 150 kilotons; not to carry out any group explosion (consisting of a number of individual explosions) having an aggregate yield exceeding 1,500 kilotons; and not to carry out any group explosion having an aggregate yield exceeding 150 kilotons unless the individual explosions in the group could be identified and measured by agreed verification procedures. The parties also reaffirmed their obligations to comply fully with the Limited Test Ban Treaty of 1963.

The parties reserve the right to carry out nuclear explosions for peaceful purposes in the territory of another country if requested to do so, but only in full compliance with the yield limitations and other provisions of the PNE Treaty and in accord with the Non-Proliferation Treaty.

Articles IV and V of the PNE Treaty set forth the agreed verification arrangements. In addition to the use of national technical means, the treaty states that information and access to sites of explosions will be provided by each side, and includes a commitment not to interfere with verifica-

tion means and procedures.

The protocol to the PNE Treaty sets forth the specific agreed arrangements for ensuring that no weapon-related benefits precluded by the Threshold Test Ban Treaty are derived by carrying out a nuclear explosion used for peaceful purposes, including provisions for use of the hydrodynamic yield measurement method, seismic monitoring and on-site inspection.

The agreed statement that accompanies the treaty specifies that a "peaceful application" of an underground nuclear explosion would not include the developmental testing of any nuclear explosive.[3]

29.2 United States: Operation Plowshare

Operation Plowshare was the name of the U.S. program for the development of techniques to use nuclear explosives for peaceful purposes. The name was coined in 1961, taken from Micah 4:3 ("And he shall judge among the nations, and shall rebuke many people: and they shall beat their swords into plowshares, and their spears into pruning hooks: nation shall not lift up sword against nation, neither shall they learn war any more"). Twenty-eight nuclear blasts were detonated between 1961 and 1973.

One of the first U.S. proposals for peaceful nuclear explosions that came close to being carried out was Project Chariot, which would have used several hydrogen bombs to create an artificial harbor at Cape Thompson, Alaska. It was never carried out due to concerns for the native populations and the fact that there was little potential use for the harbor to justify its risk and expense. There was also talk of using nuclear explosions to excavate a second Panama Canal.[4]

The largest excavation experiment took place in 1962 at the Department of Energy's Nevada Test Site. The Sedan nuclear test carried out as part of Operation Storax displaced 12 million tons of earth, creating the largest human-made

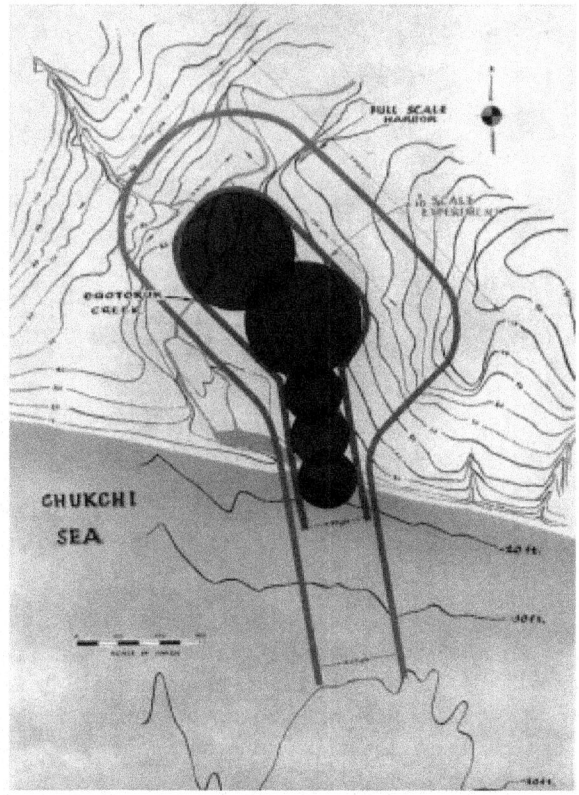

One of the Chariot *schemes involved chaining five thermonuclear devices to create the artificial harbor.*

crater in the world, generating a large nuclear fallout over Nevada and Utah. Three tests were conducted in order to stimulate natural gas production, but the effort was abandoned as impractical because of cost and radioactive contamination of the gas.[5][6]

There were many negative impacts from Project Plowshare's 27 nuclear explosions. For example, the Gasbuggy site,[6] located 55 miles east of Farmington, New Mexico, still contains nuclear contamination from a single subsurface blast in 1967.[7] Other consequences included blighted land, relocated communities, tritium-contaminated water, radioactivity, and fallout from debris being hurled high into the atmosphere. These were ignored and downplayed until the program was terminated in 1977, due in large part to public opposition, after $770 million had been spent on the project.[8]

29.3 Soviet Union: Nuclear Explosions for the National Economy

The Soviet Union conducted a much more vigorous program of 239 nuclear tests, some with multiple devices, be-

tween 1965 and 1988 under the auspices of Program No. 6 and Program No. 7 – Nuclear Explosions for the National Economy. Its aims and results were similar to those of the American effort, with the exception that many of the blasts were considered applications, not tests.[9] The best known of these in the West was the Chagan test in January 1965 as radioactivity from the Chagan test was detected over Japan by both the U.S. and Japan. The United States complained to the Soviets, but the matter was dropped.

In the 1970s, the Soviet Union started the "Deep Seismic Sounding" Program, that included the use of peaceful nuclear explosions to create seismic deep profiles. Compared to the usage of conventional explosives or mechanical methods, nuclear explosions allow the collection of longer seismic profiles (up to several thousand kilometers).[10]

There are proponents for continuing the PNE programs in modern Russia. They (e.g. A. Koldobsky) state that the program already paid for itself and saved the USSR billions of rubles and can save even more if continued. They also allege that the PNE is the only feasible way to put out large fountains and fires on natural gas deposits and the safest and most economically viable way to destroy chemical weapons.

Their opponents, including Alexey Yablokov,[11] state that all PNE technologies have non-nuclear alternatives and that many PNEs actually caused nuclear disasters.

Reports on the successful Soviet use of nuclear explosions in extinguishing out-of-control gas well fires were widely cited in United States policy discussions of options for stopping the 2010 Gulf of Mexico Deepwater Horizon oil spill.[12][13]

29.4 Other nations

Germany at one time considered manufacturing nuclear explosives for civil engineering purposes. In the early 1970s a feasibility study was conducted for a project to build a canal from the Mediterranean Sea to the Qattara Depression in the Western Desert of Egypt using nuclear demolition. This project proposed to use 213 devices, with yields of 1 to 1.5 megatons detonated at depths of 100 to 500 meters, to build this canal for the purpose of producing hydroelectric power.[14][15][16]

The Smiling Buddha, India's first explosive nuclear device was described by the Indian Government as a peaceful nuclear explosion.[17]

In Australia, nuclear blasting was proposed as a way of mining iron ore in the Pilbara.[18]

29.5 Civil engineering and energy production

See also: Athabasca oil sands § Project oilsand, and Project Gnome

Apart from their use as weapons, nuclear explosives have

The 1962 Sedan nuclear test formed a crater 100 m (330 ft) deep with a diameter of about 390 m (1,300 ft), as a means of investigating the possibilities of using peaceful nuclear explosions for large-scale earth moving. If this test was conducted in 1965+, when improvements in device design were realized, a "100-fold" reduction in radiation release was considered feasible.[19] The 140 kiloton Soviet Chagan (nuclear test), comparable in yield to the Sedan test of 104 kt, formed Lake Chagan, reportedly used as a watering hole for cattle and human swimming.[20][21][22]

been tested and used, in a similar manner to chemical high explosives, for various non-military uses. These have included large-scale earth moving, isotope production and the stimulation and the closing-off of the flow of natural gas.

At the peak of the Atomic Age, the United States initiated Operation Plowshare, involving "peaceful nuclear explosions". The United States Atomic Energy Commission chairman announced that the Plowshares project was intended to "highlight the peaceful applications of nuclear explosive devices and thereby create a climate of world opinion that is more favorable to weapons development and tests".[23] The Operation Plowshare program included 27 nuclear tests designed towards investigating these non-weapons uses from 1961 through 1973. Due to the inability of the U.S. physicists to reduce the fission fraction of small, approximately 1 kiloton, yield nuclear devices that would have been required for many civil engineering projects, when long term health and clean-up costs from fission products were included in the cost, there was virtually no economic advantage over conventional explosives, except for potentially the very largest of projects.[24][25]

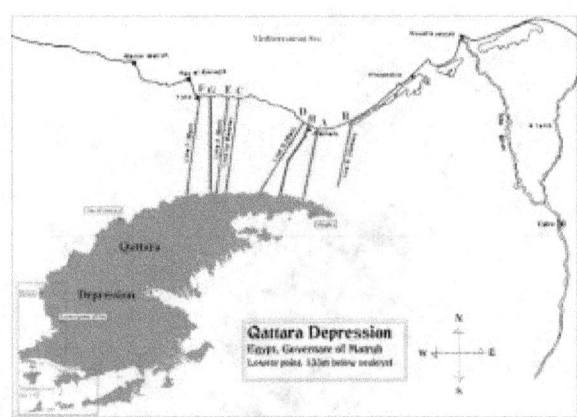

Map of all proposed routes for a tunnel and/or canal route from the Mediterranean Sea to the Qattara Depression.
No route was shorter than 55 kilometers in length. Canal-cutting investigations began with the buggy salvo shot of Operation Crosstie in 1967.

The Qattara Depression Project, as developed by Professor Friedrich Bassler, who during his appointment to the West German ministry of economics in 1968, put forth a plan to create a Saharan lake and hydroelectric power station by blasting a tunnel between the Mediterranean sea and the Qattara Depression in Egypt, an area that lies below sea level. The core problem of the entire project was the water supply to the depression. Calculations by Bassler showed that digging a canal or tunnel would be too expensive, therefore Bassler determined that the use of nuclear explosive devices, to excavate the canal or tunnel, would be the most economical. The Egyptian government declined to pursue the idea.[26]

The Soviet Union conducted a much more exhaustive program than Plowshare, with 239 nuclear tests, between 1965 and 1988. Furthermore, many of the "tests" were considered economic applications, not tests, in the Nuclear Explosions for the National Economy program.[27]

These included one 30 kiloton explosion being used to close the Uzbekistani *Urtabulak* gas well in 1966 that had been blowing since 1963, and a few months later a 47 kiloton explosive was used to seal a higher pressure blowout at the nearby *Pamuk* gas field.[28]

The public records for devices that produced the highest proportion of their yield via fusion-only reactions are possibly the Taiga Soviet peaceful nuclear explosions of the 1970s, with 98% of their 15 kiloton explosive yield being derived from fusion reactions, a total fission fraction of 0.3 kilotons in a 15 kt device.[29][30]

The repeated detonation of nuclear devices underground in salt domes, in a somewhat analogous manner to the explosions that power a car internal combustion engine (in that it would be a heat engine) has also been proposed as a means of fusion power, in what is termed PACER.[31] Other investigated uses for peaceful nuclear explosions were underground detonations to stimulate, by a process analogous to fracking, the flow of petroleum and natural gas in tight formations, this was most developed in the Soviet Union, with an increase in the production of many well heads being reported.[28]

29.6   Physics

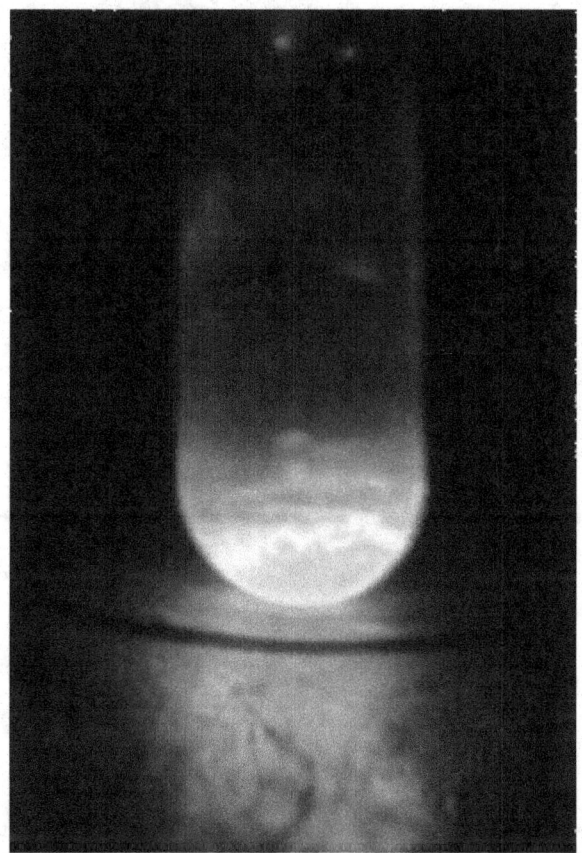

The element einsteinium was first discovered, in minute quantities, following the analysis of the fallout from the first thermonuclear atmospheric test.[32]

The discovery and synthesis of new chemical elements by nuclear transmutation, and their production in the necessary quantities to allow the studying of their properties, was carried out in nuclear explosive device testing. For example, the discovery of the short lived einsteinium and fermium, both created under the intense neutron flux environment within thermonuclear explosions, followed the first Teller-Ulam thermonuclear device test – Ivy Mike. The rapid capture of so many neutrons required in the synthesis of einsteinium would provide the needed direct experimental confirmation of the so-called r-process,

the multiple neutron absorptions needed to explain the cosmic nucleosynthesis (production) of all heavy chemical elements heavier than nickel on the periodic table, in supernova explosions, before beta decay, with the r-process explaining the existence of many stable elements in the universe.[33]

The worldwide presence of new isotopes from atmospheric testing beginning in the 1950s led to the 2008 development of a reliable way to detect art forgeries. Paintings created after that period may contain traces of caesium-137 and strontium-90, isotopes that did not exist in nature before 1945.[34][35] (Fission products were produced in the natural nuclear fission reactor at Oklo about 1.7 billion years ago, but these decayed away before the earliest known human painting.)[36]

Both climatology and particularly aerosol science, a subfield of atmospheric science, were largely created to answer the question of how far and wide fallout would travel. Similar to radioactive tracers used in hydrology and materials testing, fallout and the neutron activation of nitrogen gas served as a radioactive tracer that was used to measure and then help model global circulations in the atmosphere by following the movements of fallout aerosols.[37][38]

After the Van Allen Belts surrounding Earth were published about in 1958, James Van Allen suggested that a nuclear detonation would be one way of probing the magnetic phenomenon, data obtained from the August 1958 Project Argus test shots, a high altitude nuclear explosion investigation, were vital to the early understanding of Earth's magnetosphere.[39][40]

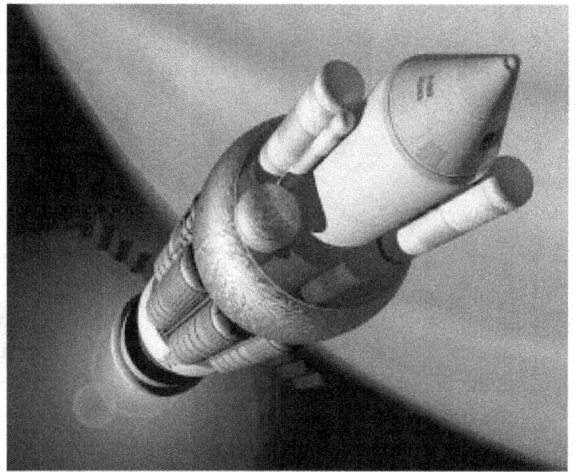

An artist's conception of the NASA reference design for the Project Orion spacecraft powered by nuclear pulse propulsion.

Soviet nuclear physicist and Nobel peace prize recipient Andrei Sakharov also proposed the idea that earthquakes could be mitigated and particle accelerators could be made by utilizing nuclear explosions,[41][42] with the latter created by connecting a nuclear explosive device with another of his inventions, the explosively pumped flux compression generator,[43] to accelerate protons to collide with each other to probe their inner workings, an endeavor that is now done at much lower energy levels with non-explosive superconducting magnets in CERN. Sakharov suggested to replace the copper coil in his MK generators by a big superconductor solenoid to magnetically compress and focus underground nuclear explosions into a shaped charge effect. He theorized this could focus 10^{23} positively charged protons per second on a 1 mm^2 surface, then envisaged making two such beams collide in the form of a supercollider.[44]

Underground nuclear explosive data from peaceful nuclear explosion test shots have been used to investigate the composition of Earth's mantle, analogous to the exploration geophysics practice of mineral prospecting with chemical explosives in "deep seismic sounding" reflection seismology.[45][46][47]

Project A119, proposed in the 1960s, which as Apollo scientist Gary Latham explained, would have been the detonating of a "smallish" nuclear device on the Moon in order to facilitate research into its geologic make-up.[48] Analogous in concept to the comparatively low yield explosion created by the water prospecting (LCROSS) Lunar Crater Observation and Sensing Satellite mission, which launched in 2009 and released the "Centaur" kinetic energy impactor, an impactor with a mass of 2,305 kg (5,081 lb), and an impact velocity of about 9,000 km/h (5,600 mph),[49] releasing the kinetic energy equivalent of detonating approximately 2 tons of TNT (8.86 GJ).

29.7   Propulsion use

Main article: Nuclear pulse propulsion

Although likely never achieving orbit due to aerodynamic drag, the first macroscopic object to obtain Earth orbital velocity was a "manhole cover" propelled by the detonation of test shot Pascal-B, before sputnik obtained orbital velocity, and also successfully became the first satellite, in October 1957. The use of a subterranean shaft and nuclear device to propel an object to escape velocity has since been termed a "thunder well".[50]

The direct use of nuclear explosives, by using the impact of propellant plasma from a nuclear shaped charge acting on a pusher plate, has also been seriously studied as a potential propulsion mechanism for space travel (see Project Orion).

Edward Teller, in the United States, proposed the use of a nuclear detonation to power an explosively pumped *soft* X-

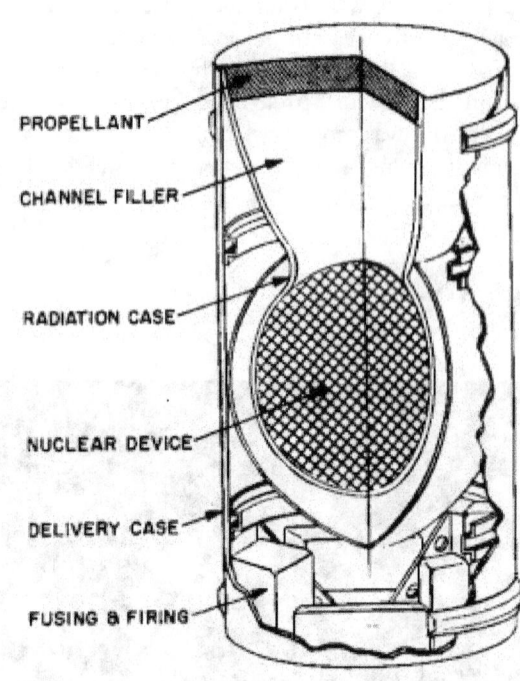

A nuclear shaped charge design that was to provide nuclear pulse propulsion to the Project Orion vehicle.

ray laser as a component of a ballistic missile defense shield, this would destroy missile components by transferring momentum to the vehicles surface by laser ablation. This ablation process is one of the damage mechanisms of a laser weapon, but it is also the basis of pulsed laser propulsion for spacecraft.

Ground flight testing by Professor Leik Myrabo, using a non-nuclear, conventionally powered pulsed laser test-bed, successfully lifted a lightcraft 72 meters in altitude by a method similar to ablative laser propulsion in 2000.[51]

A powerful solar system based *soft* X-ray, to ultraviolet, laser system has been calculated to be capable of propelling an interstellar spacecraft, by the light sail principle, to 11% of the speed of light.[52] In 1972 it was also calculated that a 1 Terawatt, 1-km diameter x-ray laser with 1 angstrom wavelength impinging on a 1-km diameter sail, could propel a spacecraft to Alpha Centauri in 10 years.[53]

29.8   Asteroid impact avoidance

Main article: Asteroid impact avoidance
See also: B83 nuclear bomb

A proposed means of averting an asteroid impacting with Earth, assuming low lead times between detection and Earth

Artist's impression of the impact event that resulted in the Cretaceous–Paleogene extinction event, which killed the Dinosaurs some 65 million years ago. A natural impact with an explosive yield of 100 teratons of TNT $(4.2 \times 10^{23}$ J).[54] The most powerful man-made explosion, the Tsar Bomba, by comparison had a yield almost 2 million times smaller – 57 megatons of TNT $(2.4 \times 10^{17}$ J).[55] The 1994 Comet Shoemaker–Levy 9 impacts on planet Jupiter, the Tunguska and Chelyabinsk asteroid–Earth collisions of 1908 and 2013 respectively, have served as an impetus for the analysis of technologies that could prevent the destruction of human life by impact events.

impact, is to detonate one, or a series, of nuclear explosive devices, on, in, or in a stand-off proximity orientation with the asteroid,[56] with the latter method occurring far enough away from the incoming threat to prevent the potential fracturing of the near-Earth object, but still close enough to generate a high thrust laser ablation effect.[57]

A 2007 NASA analysis of impact avoidance strategies using various technologies stated:[58]

> Nuclear stand-off explosions are assessed to be 10–100 times more effective than the non-nuclear alternatives analyzed in this study. Other techniques involving the surface or subsurface use of nuclear explosives may be more efficient, but they run an increased risk of fracturing the target near-Earth object. They also carry higher development and operations risks.

29.9 See also

- Operation Plowshare
- Operation Chariot
- Project Gnome

29.10 Books

- IAEA review of the 1968 book: The constructive uses of nuclear explosions by Edward Teller.

- "The Containment of Underground Nuclear Explosions", Project Director Gregory E van der Vink, U.S. Congress, Office of Technology Assessment, OTA-ISC-414, (Oct 1989).

29.11 References

[1] "Announcement of Treaty on Underground Nuclear Explosions Peaceful Purposes (PNE Treaty)" (PDF). Gerald R. Ford Museum and Library. May 28, 1976.

[2] Peters, Gerhard; Woolley, John T. "Gerald R. Ford: "Message to the Senate Transmitting United States-Soviet Treaty and Protocol on the Limitation of Underground Nuclear Explosions," July 29, 1976". *The American Presidency Project*. University of California - Santa Barbara.

[3] "Peaceful Nuclear Explosions Treaty". Arms Control Association.

[4] "Report of the Atlantic–Pacific transoceanic canal study commission", *Congressional Record: proceedings & debates of the 90th Congress, second session* (message from the President), Senate, Washington, DC, US: Congress, p. 25747, 1968-09-05, retrieved 2012-01-22, The collection of data was substantially completed on Route 17 in Panama one of the routes considered for nuclear excavations… The Atomic Energy Commission has recently conducted the first two of the planned series of nuclear excavation experiments designed to determine the feasibility of nuclear excavation of a sea level canal.

[5] "Site" (PDF), *Rulison, CO* (PDF) (fact sheet), US: Department of Energy, Office of Legacy Management.

[6] Metzger, Peter (February 22, 1970), "Project Gasbuggy and Catch-85*: *That's krypton-85, one of the radioactive by-products of nuclear explosions that release natural gas", *The New York Times*, p. SM14, It's 95 per cent safe? We worry about the other 5

[7] "Gas Buggy". Environmental Management (EM). DOE. Retrieved 2010-09-19.

[8] Sovacool, Benjamin K (2011), "Contesting the Future of Nuclear Power: A Critical Global Assessment of Atomic Energy", *World Scientific*, pp. 171–2.

[9] Nordyke, MD (2000-09-01). *The Soviet Program for Peaceful Uses of Nuclear Explosions* (PDF). Lawrence Livermore National Laboratory. pp. 34–35. doi:10.2172/793554. Report no.: UCRL-ID-124410 Rev 2; US Department of Energy contract no.: W-7405-Eng48.

[10] "DSS", *Seismic*, University of Wyoming.

[11] Яблоков, AB (1995-02-01). "Ядерная Мифология Конца XX Века (Nuclear Mythology of the 20th Century)" (in Russian). Russia: Новый Мир (New World). Retrieved 2013-09-18.

[12] Broad, William J (2010-06-02). "Nuclear Option on Gulf Oil Spill? No Way, US Says". The New York Times. Retrieved 2010-06-18.

[13] Astrasheuskaya, Nastassia; Judah, Ben; Selyukh, Alina (2010-07-02). "Special Report: Should BP nuke its leaking well?". *Reuters*. Retrieved 2010-07-08.

[14] "Managing water for peace in the Middle East". *archive.unu.edu*. Retrieved 2016-08-18.

[15] "Flooding the Qattara Depression | The Basement Geographer". *basementgeographer.com*. Retrieved 2016-08-18.

[16] Ministry of Electric Power and Energy (1978). "Development of the Qattara Project, Egypt". *Technical committee on the peaceful uses of nuclear explosions; Vienna, Austria; 22 - 24 Nov 1976*. IAEA-TC-—81-5/6.

[17] "India's Nuclear Weapons Program - Smiling Buddha: 1974". *nuclearweaponarchive.org*. Retrieved 2016-08-18.

[18] "Nuclear blasting proposed for Pilbara Iron Ore Project", *Industrial Reviews and Mining Year Book*, 1970, pp. 255–9.

[19] Declassified U.S. Nuclear Test Film #35 c. 29:30 minutes

[20] Guinness World Records. "Largest crater from an underground nuclear explosion". Retrieved October 24, 2014.

[21] "The Soviet Nuclear Weapons Program". Retrieved October 24, 2014.

[22] Russia Today documentary that visits the lake at around the 1 minute mark on YouTube

[23] Perrow, Charles (September–October 2013). "Nuclear denial: From Hiroshima to Fukushima". *Bulletin of the Atomic Scientists*. **69** (5).

[24] "Q&A with Scott Kirsch: Digging with bombs". Usnews.com. Archived from the original on January 31, 2010. Retrieved November 25, 2010.

[25] Declassified U.S. Nuclear Test Film #35 c. 12:00 minutes

[26] "Global Hyper Saline Power Generation Qattara Depression Potentials" (PDF). *MIK Technology*. December 19, 2010. Retrieved November 20, 2015.

[27] Nordyke, MD (September 1, 2000). *The Soviet Program for Peaceful Uses of Nuclear Explosions* (PDF). Lawrence Livermore National Laboratory. pp. 34–35. doi:10.2172/793554. Report no.: UCRL-ID-124410 Rev 2; US Department of Energy contract no.: W-7405-Eng48.

[28] Nordyke, M. D. (September 1, 2000). "Extinguishing Runaway Gas Well Fires". *The Soviet Program for Peaceful Uses of Nuclear Explosions* (PDF). Lawrence Livermore National Laboratory. pp. 34–35. doi:10.2172/793554. Report no.: UCRL-ID-124410 Rev 2. U. S. Department of Energy contract no.: W-7405-Eng48.

[29] Disturbing the Universe – Freeman Dyson

[30] The Soviet Program for Peaceful Uses of Nuclear Explosions by Milo D. Nordyke. *Science & Global Security*, 1998, Volume 7, pp. 1–117. See test shot "Taiga".

[31] John Nuckolls, "Early Steps Toward Inertial Fusion Energy (IFE)", LLNL, June 12, 1998

[32] EINSTEINIUM AND FERMIUM, ALBERT GHIORSO, LAWRENCE BERKELEY NATIONAL LABORATORY

[33] Byrne, J. *Neutrons, Nuclei, and Matter*, Dover Publications, Mineola, NY, 2011, ISBN 978-0-486-48238-5 (pbk.) pp. 267.

[34] Cartlidge, Edwin (July 4, 2008). "Nuclear fallout used to spot fake art". *Physics World – the member magazine of the Institute of Physics*. IOP Group. Retrieved December 7, 2014.

[35] "Can past nuclear explosions help detect forgeries?". Theartnewspaper.com. Archived from the original on November 13, 2010. Retrieved November 25, 2010.

[36] Emsley, John (2011). *Nature's Building Blocks: An A-Z Guide to the Elements* (New ed.). New York, NY: Oxford University Press. ISBN 978-0-19-960563-7.

[37] Entangled histories: Climate science and nuclear weapons research

[38] "Nuclear weapons' surprising contribution to climate science". Phys.org. July 13, 2012. Retrieved May 30, 2013.

[39] "Going Nuclear Over the Pacific | Past Imperfect". Blogs.smithsonianmag.com. Retrieved May 30, 2013.

[40] "Section VI Nuclear Weapons Effects Technology II-6-28" (PDF). *Federation of American Scientists*. Retrieved November 20, 2015.

[41] "Tsar Bomba – The Legacy". Archived from the original on November 4, 2012.

[42] Viktor Adamsky and Yuri Smirnov. 1994. "Moscow's Biggest Bomb: the 50-Megaton Test of October 1961" Cold War International History Project Bulletin, Issue 4, Fall 1994

[43] Sakharov, A. D. (1966). "Magnetoimplosive Generators". *Soviet Physics Uspekhi*. **9** (2): 294. doi:10.1070/PU1966v009n02ABEH002876.

[44] "Guide to the Victor Frederick Weisskopf Papers MC.0572". *Massachusetts Institute of Technology*. Retrieved November 20, 2015.

[45] "Travel time analysis of P waves arising from six underground nuclear explosion at Novaya Zemlya". Annalsofgeophysics.eu. Retrieved May 30, 2013.

[46] "Upper Mantle Heterogeneities from Active and Passive Seismology". Springer.com. April 16, 1997. Retrieved May 30, 2013.

[47] "A Database Of Deep Seismic Sounding Peaceful Nuclear Explosion Recordings For Seismic Monitoring Of Northern Eurasia" (PDF). Retrieved May 30, 2013.

[48] "Moon madness". *The Sydney Morning Herald*. December 21, 1969. p. 19. Retrieved September 9, 2011.

[49] "NASA's LCROSS Mission Changes Impact Crater". NASA. September 29, 2009. Retrieved November 21, 2009.

[50] "Operation Plumbbob". Retrieved October 24, 2014.

[51] "Laser Propulsion Thrusters for Space Transportation". *springer.com*. Retrieved April 1, 2015.

[52] "Roundtrip Interstellar Travel Using Laser-Pushed Lightsails. VOL. 21, NO. 2, MARCH-APRIL 1984 J. SPACECRAFT. Robert Forward et. al" (PDF). Retrieved May 30, 2013.

[53] W.E Mockel, Propulsion by impinging Laser beams. Journal of Spacecraft and rockets. 9, no 12 p 942 (1972).

[54] Covey *et al.*

[55] Adamsky and Smirnov, 19.

[56] Solem, J. C. (2000). "Deflection and disruption of asteroids on collision course with Earth". *Journal of the British Interplanetary Society*. **53**: 180–196.

[57] Dillow, Clay (April 9, 2012). "How it Would Work: Destroying an Incoming Killer Asteroid With a Nuclear Blast". *Popular Science*. Bonnier. Retrieved January 6, 2013.

[58] "Near-Earth Object Survey and Deflection Analysis of Alternatives Report to Congress March 2007". *National Aeronautics and Space Administration*. Retrieved November 20, 2015.

29.12 External links

- *Peaceful Nuclear Explosions*, Comprehensive Nuclear-Test-Ban Treaty Organization Preparatory Commission.

- *The 104Kt Sedan PNE as part of Operation Plowshare* (video), Sonic bomb.

- *The Soviet Chagan PNE* (video), Sonic bomb.

- *The Soviet Taiga PNE* (video), Sonic bomb.

- *Russia*, Nuclear weapon archive. On the Soviet nuclear program.

- *On the Soviet program for peaceful uses of nuclear weapons*, American Office of Scientific and Technical Information.

- *United States Nuclear Tests, July 1945 through September 1992*, FAS, DOE/NV-209 [Rev.14].

- *Arms control agreements*, Federation of American Scientists.

- "World Reaction", *The Indian Nuclear Tests*, Center for Nonproliferation Studies.

- *Treaty between the USA and USSR on underground nuclear explosions for peaceful purposes*, Nuclear Files, 1990-12-11.

- Kuran, Peter, *Atomic Journeys* (film) (documentary), Atom central includes tests of peaceful nuclear explosions.

Chapter 30

Nuclear weapons and the United States

The United States was the first country to manufacture nuclear weapons, and is the only country to have used them in combat, with the separate bombings of Hiroshima and Nagasaki in World War II. Before and during the Cold War, it conducted over a thousand nuclear tests and tested many long-range weapon delivery systems.[Note 1]

Between 1940 and 1996, the U.S. government spent at least $8.78 trillion in present-day terms[6] on nuclear weapons, including platforms development (aircraft, rockets and facilities), command and control, maintenance, waste management and administrative costs.[7] It is estimated that, since 1945, the United States produced more than 70,000 nuclear warheads, which is more than all other nuclear weapon states combined. The Soviet Union/Russia has produced approximately 55,000 nuclear warheads since 1949, France built 1110 warheads since 1960, the United Kingdom built 835 warheads since 1952, China built about 600 warheads since 1964, and other nuclear powers built fewer than 500 warheads all together since they developed their first nuclear weapons.[8] Until November 1962, the vast majority of U.S. nuclear tests were aboveground. After the acceptance of the Partial Test Ban Treaty, all testing was relegated underground, in order to prevent the dispersion of nuclear fallout.

By February 2006 over $1.2 billion in compensation had been paid to U.S. citizens exposed to nuclear hazards as a result of the U.S. nuclear weapons program, and by 1998 at least $759 million had been paid to the Marshall Islanders in compensation for their exposure to U.S. nuclear testing.[9][10]

In 2016, the United States maintained an arsenal of 4,500 warheads[11] and facilities for their construction and design, though many of the Cold War facilities have since been deactivated and are sites for environmental remediation.[Note 2]

30.1 Development history

30.1.1 Manhattan Project

The United States first began developing nuclear weapons

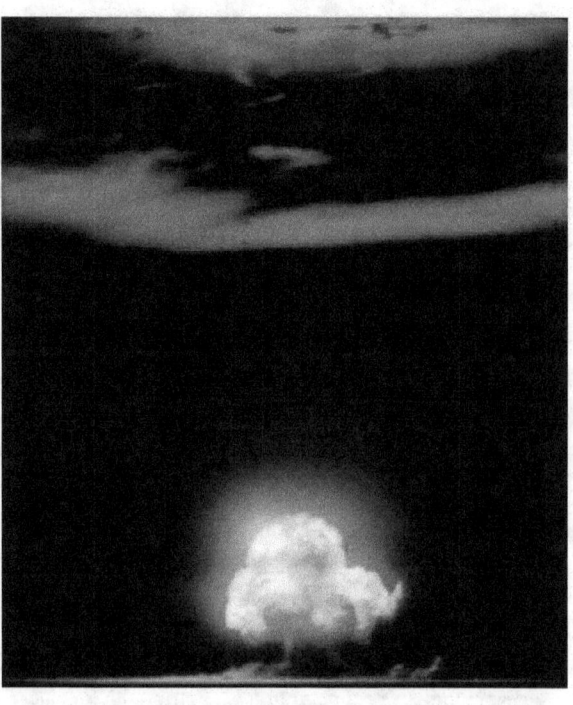

The "Trinity" explosion was the world's first nuclear weapon test

during World War II under the order of President Franklin Roosevelt in 1939, motivated by the fear that they were engaged in a race with Nazi Germany to develop such a weapon. After a slow start under the direction of the National Bureau of Standards, at the urging of British scientists and American administrators, the program was put under the Office of Scientific Research and Development, where in 1942 it was officially transferred under the auspices of the United States Army and became known as the Manhattan Project, an American, British and Canadian joint venture. Under the direction of General Leslie Groves, over thirty different sites were constructed for the research, production, and testing of components related to bomb making. These included the Los Alamos National Laboratory at Los Alamos, New Mexico, under the direction of physicist Robert Oppenheimer, the Hanford plutonium production facility in Washington, and the Y-12 National Security Complex in Tennessee.

By investing heavily in breeding plutonium in early nuclear reactors and in the electromagnetic and gaseous diffusion enrichment processes for the production of uranium-235, the United States was able to develop three usable weapons by mid-1945. The Trinity test was a plutonium implosion-design weapon tested on 16 July 1945, with around a 20 kiloton yield. Faced with an imminent invasion of the Japanese home islands and with Japan not surrendering, President Harry S. Truman ordered the atomic raids on enemy cities. On 6 August, a uranium-gun design bomb, Little Boy, was detonated over the Japanese city of Hiroshima with an energy of about 15 kilotons of TNT, destroying nearly 50,000 buildings (including the 2nd General Army and Fifth Division headquarters) and killing approximately 70,000 people, among them 20,000 Japanese combatants and 20,000 Korean slave laborers. On 9 August, a plutonium implosion-design bomb, Fat Man, was used against the Japanese city of Nagasaki with the explosion equivalent to about 20 kilotons of TNT, destroying 60% of the city and killing approximately 35,000 people, among them 23,200-28,200 Japanese munitions workers, 2,000 Korean slave laborers, and 150 Japanese combatants.[13]

30.1.2 During the Cold War

Between 1945 and 1990, more than 70,000 total warheads were developed, in over 65 different varieties, ranging in yield from around .01 kilotons (such as the man-portable Davy Crockett shell) to the 25 megaton B41 bomb.[9] Between 1940 and 1996, the U.S. spent at least $8.78 trillion in present-day terms[6] on nuclear weapons development. Over half was spent on building delivery mechanisms for the weapon. $551 billion in present-day terms was spent on nuclear waste management and environmental remediation.[7]

Richland, Washington was the first city established to support plutonium production at the nearby Hanford nuclear site, to power the American nuclear weapons arsenals. It produced plutonium for use in cold war atomic bombs.[14]

30.1.3 Post–Cold War

After the end of the Cold War following the dissolution of the Soviet Union in 1991, the U.S. nuclear program was heavily curtailed, halting its program of nuclear testing, ceasing production of new nuclear weapons, and reduc-

ing its stockpile by half by the mid-1990s under President Bill Clinton. Many of its former nuclear facilities were shut down, and their sites became targets of extensive environmental remediation. Much of the former efforts towards the production of weapons became involved in the program of stockpile stewardship, attempting to predict the behavior of aging weapons without using full-scale nuclear testing. Increased funding was also put into anti-nuclear proliferation programs, such as helping the states of the former Soviet Union eliminate their former nuclear sites, and assist Russia in their efforts to inventory and secure their inherited nuclear stockpile. By February 2006, over $1.2 billion had been paid under the Radiation Exposure Compensation Act of 1990 to U.S. citizens exposed to nuclear hazards as a result of the U.S. nuclear weapons program, and by 1998 at least $759 million had been paid to the Marshall Islanders in compensation for their exposure to U.S. nuclear testing, and over $15 million was paid to the Japanese government following the exposure of its citizens and food supply to nuclear fallout from the 1954 "Bravo" test.[9][10] In 1998, the country spent an estimated total of $35.1 billion on its nuclear weapons and weapons-related programs.[9]

In the 2013 book *Plutopia: Nuclear Families, Atomic Cities, and the Great Soviet and American Plutonium Disasters* (Oxford), Kate Brown explores the health of affected citizens in the United States, and the "slow-motion disasters" that still threaten the environments where the plants are located. According to Brown, the plants at Hanford, over a period of four decades, released millions of curies of radioactive isotopes into the surrounding environment.[14] Brown says that most of this radioactive contamination over the years at Hanford were part of normal operations, but unforeseen accidents did occur and plant management kept this secret, as the pollution continued unabated. Even today, as pollution threats to health and the environment persist, the government keeps knowledge about the associated risks from the public.[14]

During the presidency of George W. Bush, and especially after the 11 September terrorist attacks of 2001, rumors have circulated in major news sources that the U.S. has been considering design of new nuclear weapons ("bunker-busting nukes"), and the resumption of nuclear testing for reasons of stockpile stewardship. Statements by the U.S. government in 2004, however, indicated that the arsenal will drop to around 5,500 total warheads by 2012.[15] According to recent reports, much of that reduction was already accomplished by January 2008.[16]

30.2 Nuclear weapons testing

Main article: Nuclear weapons testing
See also: List of nuclear weapons tests of the United States

Between 16 July 1945 and 23 September 1992, the United

The U.S. conducted hundreds of nuclear tests at the Nevada Test Site.

Members of Nevada Desert Experience hold a prayer vigil during the Easter period of 1982 at the entrance to the Nevada Test Site.

States maintained a program of vigorous nuclear testing, with the exception of a moratorium between November 1958 and September 1961. By official count, a total of 1,054 nuclear tests and two nuclear attacks were conducted, with over 100 of them taking place at sites in the Pacific Ocean, over 900 of them at the Nevada Test Site, and ten on

Shot "Baker" of Operation Crossroads (1946) was the first underwater nuclear explosion.

miscellaneous sites in the United States (Alaska, Colorado, Mississippi, and New Mexico).[5] Until November 1962, the vast majority of the U.S. tests were atmospheric (that is, above-ground); after the acceptance of the Partial Test Ban Treaty all testing was regulated underground, in order to prevent the dispersion of nuclear fallout.

The U.S. program of atmospheric nuclear testing exposed a number of the population to the hazards of fallout. Estimating exact numbers, and the exact consequences, of people exposed has been medically very difficult, with the exception of the high exposures of Marshall Islanders and Japanese fishers in the case of the Castle Bravo incident in 1954. A number of groups of U.S. citizens—especially farmers and inhabitants of cities downwind of the Nevada Test Site and U.S. military workers at various tests—have sued for compensation and recognition of their exposure, many successfully. The passage of the Radiation Exposure Compensation Act of 1990 allowed for a systematic filing of compensation claims in relation to testing as well as those employed at nuclear weapons facilities. By June 2009 over $1.4 billion total has been given in compensation, with over $660 million going to "downwinders".[10]

A few notable U.S. nuclear tests include:

- Trinity test on 16 July 1945, was the world's first test of a nuclear weapon (yield of around 20 kt).

- Operation Crossroads series in July 1946, was the first postwar test series and one of the largest military operations in U.S. history.

- Operation Greenhouse shots of May 1951 included the first boosted fission weapon test ("Item") and a scientific test that proved the feasibility of thermonuclear weapons ("George").

- Ivy Mike shot of 1 November 1952, was the first full test of a Teller-Ulam design "staged" hydrogen bomb, with a yield of 10 megatons. It was not a deployable weapon, however—with its full cryogenic equipment it weighed some 82 tons.

- Castle Bravo shot of 1 March 1954, was the first test of a deployable (solid fuel) thermonuclear weapon, and also (accidentally) the largest weapon ever tested by the United States (15 megatons). It was also the single largest U.S. radiological accident in connection with nuclear testing. The unanticipated yield, and a change in the weather, resulted in nuclear fallout spreading eastward onto the inhabited Rongelap and Rongerik atolls, which were soon evacuated. Many of the Marshall Islanders have since suffered from birth defects and have received some compensation from the federal government. A Japanese fishing boat, *Daigo Fukuryū Maru*, also came into contact with the fallout, which caused many of the crew to grow ill; one eventually died.

- Shot Argus I of Operation Argus, on 27 August 1958, was the first detonation of a nuclear weapon in outer space when a 1.7-kiloton warhead was detonated at an altitude of 200 kilometres (120 mi) during a series of high altitude nuclear explosions.

- Shot Frigate Bird of Operation Dominic I on 6 May 1962, was the only U.S. test of an operational submarine-launched ballistic missile (SLBM) with a live nuclear warhead (yield of 600 kilotons), at Christmas Island. In general, missile systems were tested without live warheads and warheads were tested separately for safety concerns. In the early 1960s, however, there mounted technical questions about how the systems would behave under combat conditions (when they were "mated", in military parlance), and this test was meant to dispel these concerns. However, the warhead had to be somewhat modified before its use, and the missile was a SLBM (and not an ICBM), so by itself it did not satisfy all concerns.[17]

- Shot Sedan of Operation Storax on 6 July 1962 (yield of 104 kilotons), was an attempt to show the feasibility of using nuclear weapons for "civilian" and "peaceful" purposes as part of Operation Plowshare. In this instance, a 1,280-foot (390 m) diameter 320-foot (98 m) deep crater was created at the Nevada Test Site.

A summary table of each of the American operational series may be found at United States' nuclear test series.

30.3 Delivery systems

Main article: Nuclear weapons delivery
The original weapons Little Boy and Fat Man developed

Early weapons models, such as the "Fat Man" bomb, were extremely large and difficult to use.

by the United States during the Manhattan Project were relatively large (the latter had a diameter of 5 feet (1.5 m)) and heavy (around 5 tons each) weapons which required specially modified bomber planes to be adapted for their bombing missions against Japan, each of which could only carry one such weapon and only within a limited range. After these initial weapons, a considerable amount of money and research was conducted towards the goal of standardizing nuclear warheads so that they did not require highly specialized experts to assemble them before use, as in the case with the idiosyncratic wartime devices, and miniaturization of the warheads for use in more variable delivery systems.

Through the aid of brainpower acquired through Operation Paperclip at the tail end of the European theater of World War II, the United States was able to embark on an ambitious program in rocketry. One of the first products of this was the development of rockets capable of holding nuclear warheads. The MGR-1 Honest John was the first of such weapons, developed in 1953 as a surface-to-surface missile with a 15-mile (24 km) maximum range. Because of their limited range, their potential use was heavily constrained (they could not, for example, threaten Moscow with an immediate strike).

Development of long-range bombers, such as the B-29 Superfortress, during World War II was continued during the Cold War period. In 1946, the Convair B-36 Peacemaker was the first purpose-built nuclear bomber, it served with the USAF until 1959. The B-52 Stratofortress in particular was able by the mid-1950s to carry a wide arsenal of nuclear bombs, each with different capabilities and potential use situations. Starting in 1946, the U.S. based its initial de-

The MGR-1 Honest John was the first nuclear-tipped rocket developed by the U.S.

The B-36 Peacemaker in flight

terrence force on the Strategic Air Command, which, by the late 1950s maintained a number of nuclear-armed bombers in the sky at all times, prepared to receive orders to attack the USSR whenever needed. This system was, however, tremendously expensive, both in terms of natural and human resources, and raised the possibility of an accidental nuclear war.

During the 1950s and 1960s, elaborate computerized early warning systems such as Defense Support Program were developed to detect incoming Soviet attacks and to coordinate response strategies. During this same period, intercontinental ballistic missile (ICBM) systems were developed that could deliver a nuclear payload across vast distances, allowing the U.S. to house nuclear forces in the American Midwest capable of hitting the Soviet Union. Shorter-range weapons, including small tactical weapons, were fielded in Europe as well, including nuclear artillery and man-portable Special Atomic Demolition Munition. The development of submarine-launched ballistic missile systems allowed for hidden nuclear submarines to covertly launch missiles at distant targets as well, making it virtu-

ally impossible for the Soviet Union to successfully launch a first strike attack against the United States which would not guarantee a deadly response.

Improvements in warhead miniaturization in the 1970s and 1980s allowed for the development of MIRVs—missiles which could carry multiple warheads, each of which could be separately targetable. The question of whether these missiles should be based on constantly rotating train tracks (so as to avoid being easily targeted by opposing Soviet missiles) or based in heavily fortified silos (to possibly withstand a Soviet attack) was a major political controversy in the 1980s (eventually the silos won out). MIRVed systems allowed the U.S. to make the Soviet missile defense economically unfeasible, as each offensive missile would require between three and ten defensive missiles to counter.

Additional developments in weapons delivery included cruise missile systems, which allowed a plane to fire a long-distance, low-flying nuclear-tipped missile towards a target from a relatively comfortable distance.

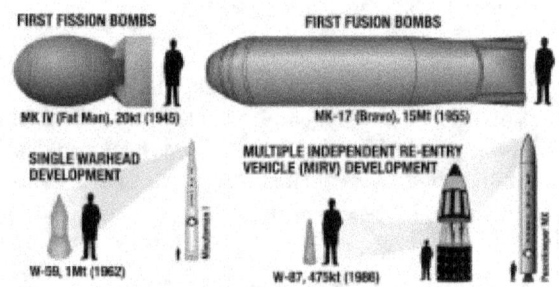

Comparing the size of U.S. nuclear weapons over time.

The current delivery systems of the U.S. make virtually any part of the Earth's surface within the reach of its nuclear arsenal. Though its land-based missile systems have a maximum range of 10,000 kilometres (6,200 mi) (less than worldwide), its submarine-based forces extend its reach from a coastline 12,000 kilometres (7,500 mi) inland. Additionally, in-flight refueling of long-range bombers and the use of aircraft carriers extends the possible range virtually indefinitely.

30.4 Accidents

Main article: Nuclear and radiation accidents

The United States nuclear program since its inception has experienced accidents of varying forms, ranging from single-casualty research experiments (such as that of Louis Slotin during the Manhattan Project), to the nuclear fallout dispersion of the Castle Bravo shot in 1954, to the accidental dropping of nuclear weapons from aircraft (broken arrows). How close any of these accidents came to being

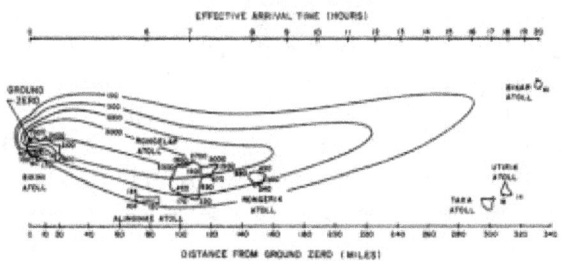

The Castle Bravo fallout plume spread dangerous levels of radioactive material over an area over 100 miles (160 km) long, including inhabited islands, in the largest single U.S. nuclear accident.

major nuclear disasters is a matter of technical and scholarly debate and interpretation.

Weapons accidentally dropped by the United States include incidents near Atlantic City, New Jersey (1957), Savannah, Georgia (1958) (see Tybee Bomb), Goldsboro, North Carolina (1961) (see 1961 Goldsboro B-52 crash), off the coast of Okinawa (1965), in the sea near Palomares, Spain (1966, see 1966 Palomares B-52 crash), and near Thule Air Base, Greenland (1968) (see 1968 Thule Air Base B-52 crash). In some of these cases (such as Palomares), the explosive system of the fission weapon discharged, but did not trigger a nuclear chain reaction (safety features prevent this from easily happening), but did disperse hazardous nuclear materials across wide areas, necessitating expensive cleanup endeavors. Eleven American nuclear warheads are thought[9] to be lost and unrecovered, primarily in submarine accidents.

The nuclear testing program resulted in a number of cases of fallout dispersion onto populated areas. The most significant of these was the Castle Bravo test, which spread radioactive ash over an area of over 100 square miles (260 km²), including a number of populated islands.[18] The populations of the islands were evacuated but not before suffering radiation burns.[18] They would later suffer long-term effects, such as birth defects and increased cancer risk. There were also instances during the nuclear testing program in which soldiers were exposed to overly high levels of radiation, which grew into a major scandal in the 1970s and 1980s, as many soldiers later suffered from what were claimed to be diseases caused by their exposures.

Many of the former nuclear facilities (see next section) produced significant environmental damages during their years of activity, and since the 1990s have been Superfund sites of cleanup and environmental remediation. The Radiation Exposure Compensation Act of 1990 allows for U.S. citizens exposed to radiation or other health risks through the U.S. nuclear program to file for compensation and damages.

30.5 Deliberate attacks on weapons facilities

Main article: Vulnerability of nuclear plants to attack

In 1972 three hijackers took control of a domestic passenger flight along the east coast of the U.S. and threatened to crash the plane into a U.S. nuclear weapons plant in Oak Ridge, Tennessee. The plane got as close as 8,000 feet above the site before the hijackers' demands were met.[19][20]

Various acts of civil disobedience since 1980 by the peace group Plowshares have shown how nuclear weapons facilities can be penetrated, and the group's actions represent extraordinary breaches of security at nuclear weapons plants in the United States. The National Nuclear Security Administration has acknowledged the seriousness of the 2012 Plowshares action. Non-proliferation policy experts have questioned "the use of private contractors to provide security at facilities that manufacture and store the government's most dangerous military material".[21] Nuclear weapons materials on the black market are a global concern,[22][23] and there is concern about the possible detonation of a small, crude nuclear weapon by a militant group in a major city, with significant loss of life and property.[24][25]

Stuxnet is a computer worm discovered in June 2010 that is believed to have been created by the United States and Israel to attack Iran's nuclear facilities.[26]

30.6 Development agencies

The United States Atomic Energy Commission (1946–1974) managed the U.S. nuclear program after the Manhattan Project.

The initial U.S. nuclear program was run by the National Bureau of Standards starting in 1939 under the edict of

President Franklin Delano Roosevelt. Its primary purpose was to delegate research and dispense funds. In 1940 the National Defense Research Committee (NDRC) was established, coordinating work under the Committee on Uranium among its other wartime efforts. In June 1941, the Office of Scientific Research and Development (OSRD) was established, with the NDRC as one of its subordinate agencies, which enlarged and renamed the Uranium Committee as the Section on Uranium. In 1941, NDRC research was placed under direct control of Vannevar Bush as the OSRD S-1 Section, which attempted to increase the pace of weapons research. In June 1942, the U.S. Army Corps of Engineers took over the project to develop atomic weapons, while the OSRD retained responsibility for scientific research.[27]

This was the beginning of the Manhattan Project, run as the Manhattan Engineering District (MED), an agency under military control that was in charge of developing the first atomic weapons. After World War II, the MED maintained control over the U.S. arsenal and production facilities and coordinated the Operation Crossroads tests. In 1946 after a long and protracted debate, the Atomic Energy Act of 1946 was passed, creating the Atomic Energy Commission (AEC) as a civilian agency that would be in charge of the production of nuclear weapons and research facilities, funded through Congress, with oversight provided by the Joint Committee on Atomic Energy. The AEC was given vast powers of control over secrecy, research, and money, and could seize lands with suspected uranium deposits. Along with its duties towards the production and regulation of nuclear weapons, it was also in charge of stimulating development and regulating civilian nuclear power. The full transference of activities was finalized in January 1947.[28]

In 1975, following the "energy crisis" of the early 1970s and public and congressional discontent with the AEC (in part because of the impossibility to be both a producer and a regulator), it was disassembled into component parts as the Energy Research and Development Administration (ERDA), which assumed most of the AEC's former production, coordination, and research roles, and the Nuclear Regulatory Commission, which assumed its civilian regulation activities.[29]

ERDA was short-lived, however, and in 1977 the U.S. nuclear weapons activities were reorganized under the Department of Energy,[30] which maintains such responsibilities through the semi-autonomous National Nuclear Security Administration. Some functions were taken over or shared by the Department of Homeland Security in 2002. The already-built weapons themselves are in the control of the Strategic Command, which is part of the Department of Defense.

In general, these agencies served to coordinate research and build sites. They generally operated their sites through contractors, however, both private and public (for example, Union Carbide, a private company, ran Oak Ridge National Laboratory for many decades; the University of California, a public educational institution, has run the Los Alamos and Lawrence Livermore laboratories since their inception, and will jointly manage Los Alamos with the private company Bechtel as of its next contract). Funding was received both through these agencies directly, but also from additional outside agencies, such as the Department of Defense. Each branch of the military also maintained its own nuclear-related research agencies (generally related to delivery systems).

30.7 Weapons production complex

This table is not comprehensive, as numerous facilities throughout the United States have contributed to its nuclear weapons program. It includes the major sites related to the U.S. weapons program (past and present), their basic site functions, and their current status of activity. Not listed are the many bases and facilities at which nuclear weapons have been deployed. In addition to deploying weapons on its own soil, during the Cold War, the United States also stationed nuclear weapons in 27 foreign countries and territories, including Okinawa, Japan (during the occupation immediately following World War II), Greenland, Germany, Taiwan, and French Morocco then independent Morocco.[31]

30.8 Proliferation

Main article: Nuclear proliferation

Early on in the development of its nuclear weapons, the United States relied in part on information-sharing with both the United Kingdom and Canada, as codified in the Quebec Agreement of 1943. These three parties agreed not to share nuclear weapons information with other countries without the consent of the others, an early attempt at nonproliferation. After the development of the first nuclear weapons during World War II, though, there was much debate within the political circles and public sphere of the United States about whether or not the country should attempt to maintain a monopoly on nuclear technology, or whether it should undertake a program of information sharing with other nations (especially its former ally and likely competitor, the Soviet Union), or submit control of its weapons to some sort of international organization (such as the United Nations) who would use them to attempt to maintain world peace. Though fear of a nuclear arms race

A sign pointing to an old fallout shelter in New York City.

The Atoms for Peace program distributed nuclear technology, materials, and know-how to many less technologically advanced countries.

spurred many politicians and scientists to advocate some degree of international control or sharing of nuclear weapons and information, many politicians and members of the military believed that it was better in the short term to maintain high standards of nuclear secrecy and to forestall a Soviet bomb as long as possible (and they did not believe the USSR would actually submit to international controls in good faith).

Since this path was chosen, the United States was, in its early days, essentially an advocate for the prevention of nuclear proliferation, though primarily for the reason of self-preservation. A few years after the USSR detonated its first weapon in 1949, though, the U.S. under President Dwight D. Eisenhower sought to encourage a program

of sharing nuclear information related to civilian nuclear power and nuclear physics in general. The Atoms for Peace program, begun in 1953, was also in part political: the U.S. was better poised to commit various scarce resources, such as enriched uranium, towards this peaceful effort, and to request a similar contribution from the Soviet Union, who had far fewer resources along these lines; thus the program had a strategic justification as well, as was later revealed by internal memos. This overall goal of promoting civilian use of nuclear energy in other countries, while also preventing weapons dissemination, has been labeled by many critics as contradictory and having led to lax standards for a number of decades which allowed a number of other nations, such as China and India, to profit from dual-use technology (purchased from nations other than the U.S.).

The United States is one of the five nuclear weapons states permitted to maintain a nuclear arsenal under the Nuclear Non-Proliferation Treaty, of which it was an original drafter and signatory on 1 July 1968 (ratified 5 March 1970).

The Cooperative Threat Reduction program of the Defense Threat Reduction Agency was established after the breakup of the Soviet Union in 1991 to aid former Soviet bloc countries in the inventory and destruction of their sites for developing nuclear, chemical, and biological weapons, and their methods of delivering them (ICBM silos, long-range bombers, etc.). Over $4.4 billion has been spent on this endeavor to prevent purposeful or accidental proliferation of weapons from the former Soviet arsenal.[32]

After India and Pakistan tested nuclear weapons in 1998, President Bill Clinton imposed economic sanctions on the countries. In 1999, however, the sanctions against India were lifted; those against Pakistan were kept in place as a result of the military government that had taken over. Shortly after the September 11 attacks in 2001, President George W. Bush lifted the sanctions against Pakistan as well, in order to get the Pakistani government's help as a conduit for US and NATO forces for operations in Afghanistan.

The U.S. government has officially taken a silent policy towards the nuclear weapons ambitions of the state of Israel, while being exceedingly vocal against proliferation of such weapons in the countries of Iran and North Korea. Until 2005 when the program was cancelled, it was violating its own non-proliferation treaties in the pursuit of so-called nuclear bunker busters. The 2003 invasion of Iraq by the U.S. was done, in part, on indications that Weapons of Mass Destruction were being stockpiled (lately, stockpiles of previously undeclared nerve agent and mustard gas shells have been located in Iraq),[33] and the Bush administration said that its policies on proliferation were responsible for the Libyan government's agreement to abandon its nuclear ambitions.[34]

The International Atomic Energy Agency (IAEA) proposed

a 2005 ban on fissile material that would greatly limit the production of weapons of mass destruction. 147 countries voted for this proposal but the United States voted against.

30.9   International relations and nuclear weapons

In 1958, the United States Air Force had considered a plan to drop nuclear bombs on China during a confrontation over Taiwan but it was overruled, previously secret documents showed after they were declassified due to the Freedom of Information Act in April 2008. The plan included an initial plan to drop 10-15 kiloton bombs on airfields in Amoy (now called Xiamen) in the event of a Chinese blockade against Taiwan's Offshore Islands.[35][36]

30.10   Occupational illness

The Energy Employees Occupational Illness Compensation Program (EEOICP) began on 31 July 2001. The program provides compensation and health benefits to Department of Energy nuclear weapons workers (employees, former employees, contractors and subcontractors) as well as compensation to certain survivors if the worker is already deceased.[37] By 14 August 2010, the program had already identified 45,799 civilians who lost their health (including 18,942 who developed cancer) due to exposure to radiation and toxic substances while producing nuclear weapons for the United States.[38]

30.11   Current status

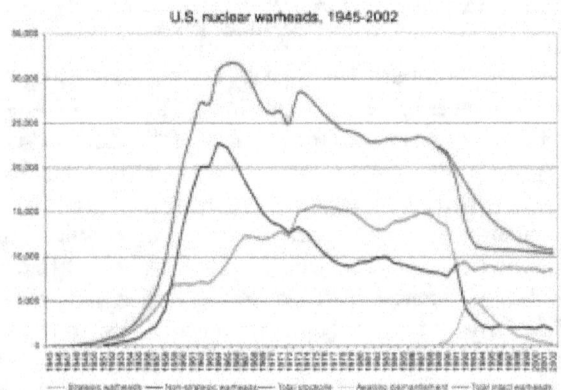

U.S. nuclear warhead stockpile, 1945–2002.

The United States is one of the five recognized nuclear powers by the signatories of the Nuclear Non-Proliferation

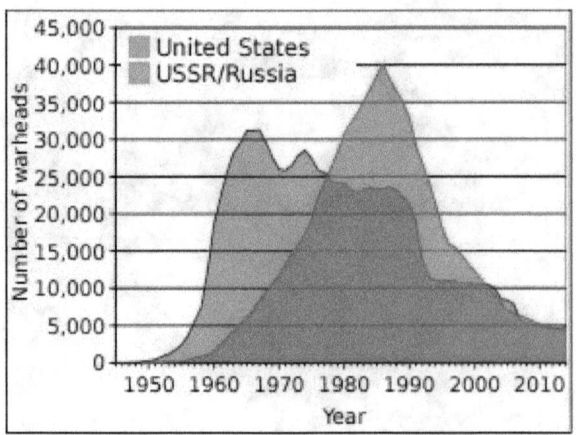

A graph showing the amount of nuclear weapons stockpiled by either country during the nuclear race.

Treaty (NPT). At the current time of writing, the US has an estimated 4,760 nuclear weapons in either deployment or storage. This figure compares to a peak of 31,225 total warheads in 1967 and 22,217 in 1989, and does not include "several thousand" warheads that have been retired and scheduled for dismantlement.

In 2009 and 2010, the administration of President Barack Obama declared policies that would invalidate the Bush-era policy for use of nuclear weapons and its motions to develop new ones. First, in a prominent 2009 speech, U.S. President Barack Obama outlined a goal of "a world without nuclear weapons".[39] To that goal, U.S. President Barack Obama and Russian Prime Minister Dmitry Medvedev signed a new START treaty on April 8, 2010, to reduce the number of active nuclear weapons from 2,200 to 1,550.[40][41] That same week Obama also revised U.S. policy on the use of nuclear weapons in a Nuclear Posture Review required of all presidents, declaring for the first time that the U.S. would not use nuclear weapons against nonnuclear, NPT-compliant states. The policy also renounces development of any new nuclear weapons.[42]

The Obama Administration, in its release of the 2012 defense budget, included plans to modernize, as well as maintain, the nation's nuclear weapons arsenal.[43]

30.12   Nuclear disarmament movement

See also: Nuclear disarmament and Anti-nuclear movement in the United States

In the early 1980s, the revival of the nuclear arms race triggered large protests about nuclear weapons.[44] On June

April 2011 OREPA rally at the Y-12 nuclear weapons plant entrance

12, 1982, one million people demonstrated in New York City's Central Park against nuclear weapons and for an end to the cold war arms race. It was the largest anti-nuclear protest and the largest political demonstration in American history.[45][46] International Day of Nuclear Disarmament protests were held on June 20, 1983 at 50 sites across the United States.[47][48] There were many Nevada Desert Experience protests and peace camps at the Nevada Test Site during the 1980s and 1990s.[49][50]

There have also been protests by anti-nuclear groups at the Y-12 Nuclear Weapons Plant,[51] the Idaho National Laboratory,[52] Yucca Mountain nuclear waste repository proposal,[53] the Hanford Site, the Nevada Test Site,[54] Lawrence Livermore National Laboratory,[55] and transportation of nuclear waste from the Los Alamos National Laboratory.[56]

On May 1, 2005, 40,000 anti-nuclear/anti-war protesters marched past the United Nations in New York, 60 years after the atomic bombings of Hiroshima and Nagasaki.[57][58] This was the largest anti-nuclear rally in the U.S. for several decades.[44] In May 2010, some 25,000 people, including members of peace organizations and 1945 atomic bomb survivors, marched from downtown New York to the United Nations headquarters, calling for the elimination of nuclear weapons.[59]

Some scientists and engineers have opposed nuclear weapons, including Paul M. Doty, Hermann Joseph Muller, Linus Pauling, Eugene Rabinowitch, M.V. Ramana and Frank N. von Hippel. In recent years, many elder statesmen have also advocated nuclear disarmament. Sam Nunn, William Perry, Henry Kissinger, and George Shultz—have called upon governments to embrace the vision of a world free of nuclear weapons, and in various op-ed columns have proposed an ambitious program of urgent steps to that end. The four have created the Nuclear Security Project to advance this agenda. Organisations such as Global Zero, an international non-partisan group of 300 world leaders dedicated to achieving nuclear disarmament, have also been established.

30.13 United States strategic nuclear weapons arsenal

Operational American strategic nuclear forces, July 1, 2016 [60]

- * Each heavy bomber is counted as one warhead (**The New START Treaty**)[61]

30.14 See also

- United States and weapons of mass destruction
- History of nuclear weapons
- National Security Strategy of the United States
- United States Strategic Command
- List of nuclear tests
- Nuclear-free zone
- Global Security Institute
- Anti-nuclear movement in the United States
- Nuclear 9/11
- International Day against Nuclear Tests
- Psychic numbing#Nuclear denial disorder

30.15 Notes

[1] According to Carey Sublette's Nuclear Weapon Archive, the United States "tested (by official count) 1054 nuclear tests" between 1945 and 1992.[5]

[2] On December 5th, 2012, the 27th subcritical underground nuclear test was conducted by the National Nuclear Security Administration (NNSA) since the U.S. signed the Comprehensive Test Ban Treaty, as listed on the NNSA website and at the Global Research organization.[12]

30.16   References

[1] "23 September 1992 - Last U.S. Nuclear Test". Infamous Anniversaries. *Preparatory Commission for the Comprehensive Nuclear-Test-Ban Treaty Organization*. Retrieved 2013-08-01.

[2] "Increasing Transparency in the U.S. Nuclear Weapons Stockpile" (pdf). *Nuclear Posture Review* (Fact Sheet). United States Department of Defense. 3 May 2010.

[3] http://fas.org/issues/nuclear-weapons/status-world-nuclear-forces/

[4] "Bulletin of the Atomic Scientists" (PDF). Retrieved 5 March 2015.

[5] "Gallery of U.S. Nuclear Tests". *The Nuclear Weapon Archive*. 6 August 2001.

[6] Federal Reserve Bank of Minneapolis Community Development Project. "Consumer Price Index (estimate) 1800–". Federal Reserve Bank of Minneapolis. Retrieved November 10, 2015.

[7] "Estimated Minimum Incurred Costs of U.S. Nuclear Weapons Programs, 1940-1996". *Brookings Institution*. Retrieved 2013-08-18.

[8] Paine, Christopher E.; Cochran, Thomas B.; Norris, Robert S. (4 January 1996). "The Arsenals of the Nuclear Weapons Powers: An Overview" (pdf). *Natural Resources Defense Council*.

[9] "50 Facts About U.S. Nuclear Weapons". *Brookings Institution*. 1998.

[10] "Radiation Exposure Compensation System Claims to Date Summary of Claims Received by 08/15/2013 All Claims" (pdf). *United States Department of Justice*. 16 August 2013. – updated regularly

[11] http://www.sipri.org/research/armaments/nuclear-forces/the-united-states

[12] Kishner, Andrew (7 December 2012). "Pollux: Subcritical Underground US Nuclear Explosion Test". *Global Research*.

[13] *Nuke-Rebuke: Writers & Artists Against Nuclear Energy & Weapons (The Contemporary anthology series)*. The Spirit That Moves Us Press. May 1, 1984. pp. 22–29.

[14] Robert Lindley (2013). "Kate Brown: Nuclear "Plutopias" the Largest Welfare Program in American History". *History News Network*.

[15] Norris, Robert S.; Kristensen, Hans M. (September–October 2007). "The U.S. stockpile, today and tomorrow" (pdf). Nuclear Notebook. *Bulletin of the Atomic Scientists*. **63** (5): 60–63. doi:10.2968/063005013.

[16] Norris, Robert S.; Kristensen, Hans M. (March–April 2008). "U.S. nuclear forces, 2008" (pdf). Nuclear Notebook. *Bulletin of the Atomic Scientists*. **64** (1): 50–53, 58. doi:10.2968/064001012.

[17] MacKenzie, Donald A. (1993). *Inventing Accuracy: A Historical Sociology of Nuclear Missile Guidance*. Cambridge, Massachusetts: MIT Press. pp. 343–344. ISBN 978-0-262-63147-1.

[18] "Race for the Superbomb". People and Events. *American Experience*. Public Broadcasting Service. 1999. |chapter= ignored (help)

[19] Threat Assessment: U.S. Nuclear Plants Near Airports May Be at Risk of Airplane Attack. *Global Security Newswire*. June 11, 2003.

[20] Newtan, Samuel Upton (2007). *Nuclear War 1 and Other Major Nuclear Disasters of the 20th Century*. AuthorHouse. p.146.

[21] Kennette Benedict (9 August 2012). "Civil disobedience". *Bulletin of the Atomic Scientists*.

[22] Jay Davis. After A Nuclear 9/11 *The Washington Post*, March 25, 2008.

[23] Brian Michael Jenkins. A Nuclear 9/11? *CNN.com*, September 11, 2008.

[24] Orde Kittrie. Averting Catastrophe: Why the Nuclear Nonproliferation Treaty is Losing its Deterrence Capacity and How to Restore It May 22, 2007, p. 338.

[25] Nicholas D. Kristof. A Nuclear 9/11 *The New York Times*, March 10, 2004.

[26] "Legal Experts: Stuxnet Attack on Iran Was Illegal 'Act of Force'". Wired. 25 March 2013.

[27] "Henry DeWolf Smyth Papers Collection Overview 1885-1987". *American Philosophical Society*. Archived from the original on 2008-05-11. Retrieved 2009-06-21.

[28] "The Atomic Energy Commissions (AEC), 1947". *Office of Science – Chicago Office*. United States Department of Energy. Retrieved 2009-06-21.

[29] "The Energy Research and Development Administration (ERDA)". *Office of Science – Chicago Office*. United States Department of Energy. Retrieved 2009-06-21.

[30] "The Department of Energy (DOE)". *Office of Science – Chicago Office*. United States Department of Energy. Retrieved 2009-06-21.

[31] "United States Secretly Deployed Nuclear Bombs In 27 Countries and Territories During Cold War". *National Security Archive* (Press release). Natural Resources Defense Council. 20 October 1999. Retrieved 2006-08-06.

[32] Larsen, Jeffrey Arthur; Smith, Dr. James M. (2005). *Historical Dictionary of Arms Control and Disarmament*. Scarecrow Press. p. 65. ISBN 9780810850606.

[33] Chivers, CJ. "The Secret Casualties of Iraq". New York Times Publishing Company. Retrieved 14 December 2014.

[34] "President Bush: Libya Pledges to Dismantle WMD Programs". *National Archives and Records Administration* (Press release). 19 December 2003. Retrieved 2009-06-21.

[35] "Air Force Histories Released through Archive Lawsuit Show Cautious Presidents Overruling Air Force Plans for Early Use of Nuclear Weapons". Washington, D.C.: National Security Archive. April 30, 2008. Retrieved 2016-03-15.

[36] "US Air Force planned nuclear strike on China over Taiwan: report". *Agence France-Presse*. 30 April 2008. Archived from the original on June 21, 2008. Retrieved 2009-06-21.

[37] "Division of Energy Employees Occupational Illness Compensation (DEEOIC)". *United States Department of Labor*.

[38] "Office of Workers' Compensation Programs (OWCP) EEOICP Program Statistics". *United States Department of Labor*. – Updated weekly

[39] "Obama sets goal of world without nuclear weapons". *The Independent*. London. 3 April 2009. Retrieved 2009-06-21.

[40] "U.S., Russia Agree To Pursue Nuclear Reduction".

[41] Michael D. Shear (8 April 2010). "Obama, Medvedev sign treaty to reduce nuclear weapons". *The Washington Post*. Retrieved 9 April 2010.

[42] David E. Sanger; Peter Baker (5 April 2010). "Obama Limits When U.S. Would Use Nuclear Arms". *New York Times*. Retrieved 8 April 2010.

[43] Hans Kristensen (17 February 2011). "The Nuclear Weapons Modernization Budget". *FAS Strategic Security Blog*. Federation of American Scientists. Retrieved 23 February 2011.

[44] Lawrence S. Wittner. "Disarmament movement lessons from yesteryear". Archived from the original on 2016-01-17. *Bulletin of the Atomic Scientists*, 27 July 2009.

[45] Jonathan Schell. The Spirit of June 12 *The Nation*, July 2, 2007.

[46] 1982 - a million people march in New York City

[47] Harvey Klehr. Far Left of Center: The American Radical Left Today Transaction Publishers, 1988, p. 150.

[48] 1,400 Anti-nuclear protesters arrested *Miami Herald*, June 21, 1983.

[49] Robert Lindsey. 438 Protesters are Arrested at Nevada Nuclear Test Site *New York Times*, February 6, 1987.

[50] 493 Arrested at Nevada Nuclear Test Site *New York Times*, April 20, 1992.

[51] Stop the Bombs! April 2010 Action Event at Y-12 Nuclear Weapons Complex.

[52] Keep Yellowstone Nuclear Free (2003). Keep Yellowstone Nuclear Free

[53] Sierra Club. (undated). Deadly Nuclear Waste Transport

[54] 22 Arrested in Nuclear Protest *New York Times*, August 10, 1989.

[55] Hundreds Protest at Livermore Lab *The TriValley Herald*, August 11, 2003.

[56] Concerned Citizens for Nuclear Safety (undated). About CCNS

[57] Lance Murdoch. Pictures: New York MayDay anti-nuke/war march *IndyMedia*, 2 May 2005.

[58] Anti-Nuke Protests in New York *Fox News*, May 2, 2005.

[59] A-bomb survivors join 25,000-strong anti-nuclear march through New York *Mainichi Daily News*, May 4, 2010.

[60] New START Treaty Aggregate Numbers of Strategic Offensive Arms. July 1, 2016

[61] New START

Sources

• Hacker, Barton C. *Elements of Controversy: The Atomic Energy Commission and Radiation Safety in Nuclear Weapons Testing, 1947-1974*. Berkeley, CA: University of California Press, 1994. ISBN 978-0-520-08323-3

• Hansen, Chuck. *U.S. Nuclear Weapons: The Secret History*. Arlington, TX: Aerofax, 1988. ISBN 978-0-517-56740-1

• Schwartz, Stephen I. *Atomic Audit: The Costs and Consequences of U.S. Nuclear Weapons*. Washington, D.C.: Brookings Institution Press, 1998. http://www.brookings.edu/about/projects/archive/nucweapons/50 ISBN 978-0-8157-7773-1

• Weart, Spencer R. *Nuclear Fear: A History of Images*. Cambridge, MA: Harvard University Press, 1985. ISBN 978-0-674-62835-9

• Biello, David. "A Need for New Warheads?" *Scientific American*, November 2007

30.17   Further reading

- "Presidency in the Nuclear Age", conference and forum at the JFK Library, Boston, October 12, 2009. Four panels: "The Race to Build the Bomb and the Decision to Use It", "Cuban Missile Crisis and the First Nuclear Test Ban Treaty", "The Cold War and the Nuclear Arms Race", and "Nuclear Weapons, Terrorism, and the Presidency".

30.18   External links

- Video archive of US Nuclear Testing at sonicbomb.com

- Nuclear Threat Initiative: United States

- NDRC's data on the US Nuclear Stockpile, 1945-2002

- Snapshot of the U.S. Nuclear Weapons Complex, April 2004 by the Los Alamos Study Group

- New nuclear warhead design for US

- Annotated bibliography of U. S. nuclear weapons programs from the Alsos Digital Library for Nuclear Issues

Chapter 31

Nuclear weapons and the United Kingdom

The United Kingdom was the third country to test an independently developed nuclear weapon, in October 1952. It is one of the five nuclear-weapon states under the NPT and a permanent member of the UN Security Council. Since the 1958 US-UK Mutual Defence Agreement, the United States and the United Kingdom have cooperated extensively on nuclear security matters. The special relationship between the two countries has involved the exchange of classified scientific data and materials such as plutonium.

The UK is thought to retain a stockpile of around 225 thermonuclear warheads, of which 160 are operational, but has refused to declare the exact size of its arsenal.[3] Since 1998, the Trident nuclear programme has been the only operational nuclear weapons system in British service. The delivery system consists of four *Vanguard* class submarines based at HMNB Clyde in Scotland. Each submarine is armed with up to 16 Trident II missiles, each carrying warheads in up to eight MIRV re-entry vehicles. With at least one submarine always on patrol, the *Vanguards* perform a strategic deterrence role and are also believed to have a sub-strategic capability. In contrast with the other permanent members of the United Nations Security Council, the United Kingdom operates only a submarine-based delivery system, having decommissioned its tactical WE.177 free-fall bombs in 1998.

The UK has not run a programme to develop an independent delivery system since the cancellation of the Blue Streak (missile) in 1960. Instead it has purchased US delivery systems for UK use, fitting them with warheads designed and manufactured by the UK's Atomic Weapons Establishment and its predecessor. In 1974, a US proliferation report discussing British nuclear and missile development noted that "In many cases, it is based on technology received from the US and could not legitimately be passed on without US permission."[4]

The Atomic Weapons Establishment is undertaking research which is largely dedicated to providing new warheads[5] and on 4 December 2006 the then Prime Minister Tony Blair announced plans for a new class of nuclear

missile submarines.[6]

31.1 Number of warheads

31.1.1 Current

Faslane Naval Base, HMNB Clyde, Scotland. Home of the Vanguard class submarines which carry the UK's current nuclear arsenal.

Since 1969 the United Kingdom has always had at least one ballistic-missile submarine on patrol, giving it a nuclear deterrent that is what the Defence Council described in 1980 as "effectively invulnerable to pre-emptive attack".[7] In the Strategic Defence Review published in July 1998, the government stated that once the Vanguard submarines became fully operational (the fourth and final one, *Vengeance*, entered service on 27 November 1999), it would "maintain a stockpile of fewer than 200 operationally available warheads".[8] The Stockholm International Peace Research Institute has estimated the figure as about 170, consisting of 144 deployed weapons plus an extra 15 percent as spares.[9]

At the same time, the UK government indicated that warheads "required to provide a necessary processing margin and for technical surveillance purposes" were not included in the "fewer than 200" figure.[10] As recently declassified archived documents on Chevaline make clear, the 15% excess (referred to by SIPRI as for spares) is normally intended to provide the 'necessary processing margin', and 'surveillance rounds do not contain any nuclear material, being completely inert. These surveillance rounds are used to monitor deterioration in the many non-nuclear components of the warhead, and are best compared with inert training rounds.' The SIPRI figures correspond accurately with the official announcements and are likely to be the most accurate. The Natural Resources Defense Council speculates that a figure of 200 is accurate to within a few tens.[11] In 2008 the National Audit Office stated that the UK stockpile was of fewer than 160 operationally available nuclear warheads.[12] During a debate on the Queen's Speech on 26 May 2010 Foreign Secretary William Hague reiterated that the UK has no more than 160 operationally available warheads, and announced that the total number will not exceed 225.[13]

31.1.2   Historical

Until the 1990s the UK deployed a wide variety of nuclear weapons around the world, such as V bombers in Singapore in the 1960s, aircraft on Cyprus and on Royal Navy carriers in the 1960s and 1970s.[14] Until August 1998, the UK retained the WE.177 nuclear weapon manufactured in the mid-1960s to late 1970s, in air-dropped free-fall bomb and depth charge versions.[15] Its withdrawal left the four *Vanguard* class submarines, which replaced the Polaris ones in the early 1990s, as Britain's only nuclear weapons platform. It has been estimated by the *Bulletin of the Atomic Scientists* that the United Kingdom has built around 1,200 warheads since the first Hurricane device of 1952.[16] In terms of number of warheads, the UK arsenal was at its maximum size of about 520 in the 1970s, but this figure does not include the large numbers of US-owned warheads, bombs, nuclear depth bombs supplied from US stocks in Europe for use by NATO allies. At its peak, these numbered 327 for the British Army of the Rhine in Germany alone.

31.2   Weapons tests

See also: Nuclear weapons tests in Australia

The United Kingdom tested its first nuclear weapons in Australia during the 1950s, on the Montebello Islands (Western Australia) and at the Woomera Prohibited Area (South Australia).

The first detonation, codenamed Operation Hurricane, occurred on 3 October 1952, in a shallow bay on Trimouille Island. Two further tests were held on the Montebello Islands during 1956. The second of these, codenamed "G2", included the largest nuclear explosion in Australia, with a yield of 98 kilotons.

Seven further nuclear tests were conducted on the Australian mainland between 1955 and 1963, within the Woomera Prohibited Area, at Emu Field and Maralinga in South Australia.

The first British hydrogen bombs were tested during Operation Grapple at Malden Island and Christmas Island in the Central Pacific Ocean. The operation consisted of nine tests in the period 1957–1959, ultimately proving that the UK had developed thermonuclear weapons.

Different sources give the total number of test explosions that the UK has conducted as either 44[17][18] or 45.[19][20] The 24 tests from December 1962 onwards were in conjunction with the United States at the Nevada Test Site[21][22] with the final test being the Julin Bristol shot which took place on 26 November 1991.[23]

Because Britain did not test as often as the United States for financial and political reasons, and did not have the Americans' state-of-the-art computer facilities, British weapons design depended more on theoretical understanding, with potential for both greater advances and greater risks between tests.[24] The low number of UK tests is misleading when compared to the large number of tests carried out by the US, the Soviet Union, China, and especially France, because the UK has had extensive access to US test data, obviating the need for UK tests. An added factor is that many tests were for 'weapon effects tests': tests not of the nuclear device itself, but of the nuclear effects on hardened components designed to resist ABM attack. Numerous such 'effects' tests were done in support of the Chevaline programme especially, and there is some evidence that some were permitted for the French programme to harden their RVs and warheads; because most French tests were under the ocean floor, access to measure 'weapon effects' was nearly impossible.[25] An independent test programme would have seen the UK numbers soar to French levels.

The UK government signed the Partial Test Ban Treaty on 5

August 1963[26] along with the United States and the Soviet Union which effectively restricted it to underground nuclear tests by outlawing testing in the atmosphere, underwater, or in outer space. The UK signed the Comprehensive Test Ban Treaty on 24 September 1996[27] and ratified it on 6 April 1998,[28] having passed the necessary legislation on 18 March 1998 as the Nuclear Explosions (Prohibition and Inspections) Act 1998.

A series summary of British testing is shown here: United Kingdom's nuclear testing series. Note that the *Vixen* safety tests are not usually listed along with the "real" nuclear testing in lists of British tests. However, they are included in totals for US, USSR, Chinese and French testing.[29] Their outcomes are not known, though they are for most other safety tests.

31.3 Nuclear defence

31.3.1 Warning systems

This solid-state phased array radar at RAF Fylingdales in North Yorkshire is a UK-controlled early warning station and part of the American-controlled Ballistic Missile Early Warning System.

Further information: Four-minute warning, RAF Fylingdales, Ballistic Missile Early Warning System, and National Missile Defense

The UK has relied on the Ballistic Missile Early Warning System (BMEWS) and, in later years, Defense Support Program (DSP) satellites for warning of a nuclear attack. Both of these systems are owned and controlled by the United States, although the UK has joint control over UK-based systems. One of the four component radars for the BMEWS is based at RAF Fylingdales in North Yorkshire.

In 2003 the UK government stated that it will consent to a request from the US to upgrade the radar at Fylingdales for use in the US National Missile Defense system.[30]

Nevertheless, missile defence is not currently a significant political issue within the UK. The ballistic missile threat is perceived to be less severe, and consequently less of a priority, than other threats to its security.[31]

31.3.2 Attack scenarios

During the Cold War, a significant effort by government and academia was made to assess the effects of a nuclear attack on the UK. There were four major exercises:

- **Exercise *Inside Right*** took place on 6–26 October 1975.

- **Exercise *Scrum Half*** was conducted in 1978.

- **Exercise *Square Leg*** was conducted in 1980. The scenario involved around 130 warheads with a total yield of 205 megatons (69 ground burst, 62 air burst) with an average of 1.5 megatons per bomb. The exercise was criticised as unrealistic as an actual exchange would be much larger, with one academic describing a 200-megaton attack as an "extremely low figure and one which we find very difficult to take seriously",[32] and did not include targets in Inner London such as Whitehall.[33] Even so, the effect of the limited attack in Square Leg was estimated to be 29 million dead (53 percent of the population) and 6.4 million seriously injured.[34]

- **Exercise *Hard Rock*** was a combined communications and civil defence exercise planned for September and October 1982. It assumed a conventional war in Europe lasting two to three days, during which the UK would be attacked with conventional weapons, then a limited nuclear exchange, with 54 nuclear warheads used against military targets in the UK. 250,000 people protested against the exercise and 24 councils refused to participate.[34] The limited scenario still assumed casualties of 7.9 million dead and 5 million injured.[34] The scenario was ridiculed by the Campaign for Nuclear Disarmament and the exercise was postponed indefinitely.[35] The *New Statesman* later claimed the Ministry of Defence insisted on having a veto over proposed targets in the exercise and several were removed to make them politically more acceptable; for example, the nuclear submarine base HMNB Clyde was removed from the target list.[36]

In the early 1980s it was thought an attack causing almost complete loss of life could be achieved with the use of less than 15 percent of the total nuclear yield available to the Soviets.[32]

31.3.3   Civil defence

Main articles: Civil defence and Protect and Survive

During the cold war, various governments developed civil defence programmes aimed to prepare civilian and local government infrastructure for a nuclear strike on the UK. A series of seven Civil Defence Bulletin films were produced in 1964, and in the 1980s the most famous such programme was probably the series of booklets and public information films entitled *Protect and Survive*.

> If the country was ever faced with an immediate threat of nuclear threat or complete annihilation, a copy of this booklet would be distributed to every household as part of a public information campaign which would include announcements on television and radio and in the press. The booklet has been designed for free and general distribution in that event. It is being placed on sale now for those who wish to know what they would be advised to do at such a time.[37]

The booklet contained information on building a nuclear refuge within a so-called "fall-out room" at home, sanitation, limiting fire hazards, and descriptions of the audio signals for attack warning, fall-out warning and all clear. It was anticipated that families might need to stay in their fall-out room for up to 14 days after an attack almost without leaving it at all.

The government also prepared a recorded announcement which was to have been broadcast by the BBC if a nuclear attack ever did occur.[38]

Sirens left over from the London Blitz during World War II were also to be used to warn the public. The system was mostly dismantled in 1993.

31.4   Historical   weapons   programmes

Further information: History of nuclear weapons

31.4.1   Tube Alloys and Manhattan Project

Main articles: Tube Alloys and British contribution to the Manhattan Project

The United Kingdom's nuclear weapons had their genesis in the Second World War when two recently exiled atomic

The United Kingdom worked in partnership with the United States and Canada on the Manhattan Project, resulting in the development of the first nuclear weapons, and the first-ever nuclear detonation at the Trinity test of 16 July 1945.

scientists, Otto Frisch and Rudolf Peierls, wrote a memorandum on the construction of "a radioactive super-bomb". Forwarded to the Ministry of Aircraft Production (MAP), the secret MAUD Committee to evaluate the possibilities was soon set up.[39] British scientists worked initially alone on the atomic bomb under the cover name of *Tube Alloys*, later becoming a partner in the tri-national Manhattan Project under the Quebec Agreement.   The Manhattan Project resulted in the two nuclear weapons dropped over Japan.

31.4.2   Post-war development programme

End of American cooperation

After Hiroshima and Nagasaki revealed the existence of the atomic bomb to the world, Labour Prime Minister Clement Attlee published a detailed account, prepared by his Conservative predecessor Winston Churchill, of the United Kingdom's participation in developing the bomb. On 8 August 1945 Attlee sent a message to President Harry Truman in which he referred to themselves as "heads of the Governments which have control of this great force". For the next year he attempted to persuade Truman to grant access to information which the British believed they deserved given their involvement.[40]

The Americans disagreed. Manhattan Project head Leslie Groves had excluded British scientists from participating in the manufacturing of the Hiroshima and Nagasaki bombs, contrary to the intentions of his government for close co-operation, for security reasons. Postwar spy scandals in both countries increased American concerns over atomic secrecy. More importantly, Truman hoped to establish international control over atomic weapons, and sharing information with even a close ally like Britain might have made such controls impossible. Nonetheless, the Americans' refusal to share information, formalised by the McMahon Act of 1946 restricting foreign access to US nuclear technology, shocked and disappointed the British.[40]

Resumption of independent UK efforts

The United Kingdom started independently developing nuclear weapons again shortly after the war. Attlee set up a cabinet sub-committee, the Gen 75 Committee (GEN.75) (known informally as the "Atomic Bomb Committee"), to examine the feasibility as early as 29 August 1945.[40] A September 1945 study forecast that an enemy could build 500 bombs during "ten years of 'peace'", and warned that if 10 percent of the arsenal was used on the United Kingdom, "over night the main base of the British Empire could be rendered ineffective", with enough left for other British forces around the world.[41]:391 The Chief of the Air Staff Arthur Tedder officially requested an atomic weapon in August 1946,[24] but work on a British equivalent to the vast American facilities at Hanford, Washington and Oak Ridge, Tennessee began in February 1946. American refusal to continue nuclear cooperation (except in certain non-weapons-related areas in exchange for uranium from the British-controlled supply in the Belgian Congo) only affected the amount of cooperation the British expected to receive,[40] for the government had decided that atomic weapons were vital to the nation regardless of cost:

> In October 1946, Attlee called a small cabinet sub-committee meeting to discuss building a gaseous diffusion plant to enrich uranium. The meeting was about to decide against it on grounds of cost, when [Ernest] Bevin arrived late and said "We've got to have this thing. I don't mind it for myself, but I don't want any other Foreign Secretary of this country to be talked at or to by the Secretary of State of the US as I have just been... We've got to have this thing over here, whatever it costs ... We've got to have the bloody Union Jack on top of it."[42]

The committee, under pressure from Hugh Dalton and Sir Stafford Cripps to opt out of building the bomb due to its cost, eventually decided to go ahead because of the likely industrial importance of atomic energy and to increase Britain's prestige and influence.[43] The nation's leaders wished for close cooperation with the Americans but were unsure whether it would continue. Bevin believed that Britain and Europe could, with help from the Commonwealth, become an independent "Third Force" equal to the United States and the Soviet Union. Military leaders disagreed, seeing an American partnership as the only way for Britain to resist the Soviets. It is likely that covert communications between the two nations' militaries on nuclear issues, unknown to and sometimes contradicting civilian leaders' wishes, began at this time.[41]:68–69,72–74

A nuclear programme started in 1946 under the control of the Atomic Energy Research Establishment (incorporated into the United Kingdom Atomic Energy Authority (UKAEA) in 1954), that was civilian in character, but was also tasked with the job of producing the fissile material, initially only plutonium-239, that was expected to be required for a military programme. It was based in a former airfield, Harwell, Berkshire, and a former Royal Ordnance Factory, Risley in Cheshire. Risley became the headquarters of the Industrial Division of UKAEA, and there were other sites under its control, notably the Calder Hall reactors at Windscale (later Sellafield) used to produce weapons grade Pu-239. The first nuclear pile in the UK, GLEEP, went critical at Harwell on 15 August 1947. The first plutonium metal was ready at Windscale in March 1952. AWRE was established at Aldermaston by the Ministry of Supply, later becoming the Weapons Division of the (civilian) UKAEA, before being subsumed into the Ministry of Defence in the 1970s.

William Penney, a physicist specialising in hydrodynamics, was asked in October 1946 to prepare a report on the viability of building a UK weapon. Joining the Manhattan Project in 1944, he had been in the observation plane *Big Stink* over Nagasaki, and had also done damage assessment on the ground following Japan's surrender. He had subsequently participated in the American Operation Crossroads test at Bikini Atoll. As a result of his report, the decision to proceed was formally made on 8 January 1947 at a meeting of the GEN.163 committee of six cabinet members including Attlee, with Penney appointed to take charge of the programme.[24]

The project was hidden under the code name *High Explosive Research* or *HER* and was based initially at the Ministry of Supply's Armament Research and Development Establishment (ARDE) at Fort Halstead in Kent,[44] and also at the Royal Arsenal, Woolwich. In 1951 it moved to a new site at AWRE Aldermaston in Berkshire. The Attlee ministry revealed the existence of a British atomic program in Parliament on 12 May 1948;[24] the announcement was viewed by Parliament, the press, and the people as uncontroversial.[40] However, although British scientists knew well the areas of the Manhattan Project in which they had worked they knew little of the other areas. Building a full-scale plant for production of weapons-grade U-235 would be very expensive,[24] and the McMahon Act prevented American technical aid. The government refused to provide public details on its progress beyond stating that atomic weapons research was of the highest priority, but it was assumed that the project was following the Americans' precedents. While Leader of the Opposition, Churchill criticised the government in February 1951 for not having completed an atomic weapon.[40][45]

Unsuccessful attempt to renew American partnership

By 1949, international control of atomic weapons seemed almost impossible to achieve, and Truman proposed to the Joint Committee on Atomic Energy in July a "full partnership" with Britain in exchange for uranium;[40] negotiations between the two countries began that month. While the first Soviet atomic bomb test in August 1949 was embarrassing to the British (who had not expected a Soviet atomic weapon until 1954) for having been beaten, it was for the Americans another reason for cooperation. Although they would soon have their own nuclear capability, the British proposed that instead of building their own uranium-enrichment plant they would send most of their scientists to work in the US, as well as plutonium from Windscale. While Britain would not formally give up building or researching its own weapons, the United States would manufacture all bombs and allocate some to Britain.[46][41]:75–76[47]

By agreeing to subsume its own weapons program within the Americans', the plan would have given Britain nuclear weapons much sooner than its own target date of late 1952. Although a majority of Americans including Truman supported the proposal, several key officials, including the Atomic Energy Commission's Lewis Strauss and Senator Arthur Vandenberg, did not. Their opposition, and security concerns caused by the arrest in early 1950 of Klaus Fuchs, a Soviet spy working at Harwell, ended the negotiations in January 1950.[46] After Britain developed atomic weapons through its own efforts, the scientist Sir Leonard Owen nonetheless stated that "the McMahon Act was probably one of the best things that happened ... as it made us work and think for ourselves along independent lines."[40]

31.4.3   First test and early systems

Main articles: Operation Hurricane, Blue Danube (nuclear weapon), and Blue Peacock

Churchill, now again prime minister, announced on 17

HMS Plym *in 1943.*

The UK's first nuclear test, Operation Hurricane, in 1952.

A Blue Danube bomb. *The first Blue Danube weapons issued to the RAF were of 10–12-kiloton-of-TNT (42–50 TJ) yield, approximately the same yield as the Hiroshima bomb, although Blue Danube was of the implosion type similar to the Nagasaki bomb. This airframe design was used for all the devices detonated at Christmas Island in the Operation Grapple tests.*

February 1952 that the first British weapon test would occur before the end of the year. Operation Hurricane was detonated below the frigate HMS *Plym* anchored in the Monte Bello Islands, Western Australia on 3 October 1952.[40] This led to the first deployed weapon, the Blue Danube free-fall bomb, in November 1953. It was very similar to the American Mark 4 weapon in having a 60-inch (1,500 mm) diameter, 32 lens implosion system with a levitated core suspended within a natural uranium tamper. The warhead was contained within a bomb casing measuring 62 inches (1.6 m) diameter and 24 feet (7.3 m) long, and being so large, could only be carried by the V bomber fleet.

A nuclear landmine dubbed Brown Bunny, later Blue Bunny, and finally Blue Peacock that used the Blue Danube warhead was developed from 1954 with the goal of deployment in the Rhine area of Germany. The system would have been set to an eight-day timer in the case of invasion of Western Europe by the Soviets but was cancelled in February 1958 with only two built. It was judged that the risks posed by the nuclear fallout and the political aspects of preparing for destruction and contamination of allied territory were too high to justify. Another reason for cancellation revealed by numerous archived declassified documents was that the Army felt it was too unwieldy and diverted their efforts into a successor, Violet Vision, based on the smaller successor to Blue Danube, Red Beard. None were ever built, the Army instead receiving US ADMs or Atomic Demolition Munitions under the established procedures for supply of NATO allies from US stocks held in US custody in Europe. A sea mine based on the Blue Danube warhead and codenamed Cudgel was also envisaged for delivery by midget submarines, referred to by naval sources as "sneak craft"; perhaps reflecting a belief that these craft were really rather ungentlemanly methods of waging war. None were built.

A gaseous diffusion plant was built at Capenhurst, near Chester and started production in 1953 producing low enriched uranium (LEU). By 1957 it was capable of annually producing 125 kg of highly enriched uranium (HEU). The capacity was further increased and by 1959 it may have been producing as much as 1600 kg per year.[48] At the end of 1961, having produced between 3.8 and 4.9 tonnes of HEU it was switched over to LEU production for civil use. Additional plutonium production was provided by eight electricity generating Magnox reactors at Calder Hall and Chapelcross which started operating in 1956 and 1959 respectively.

A Blue Danube bomb released from a Valiant bomber. The fins are not yet extended to quickly stabilise the bomb into a predictable ballistic trajectory. Fuzing was by means of a barometric 'gate' to switch on the radar altimeter controlled firing circuit. These bomb casings were used for all the air-drop tests at Christmas Island and Maralinga, Australia. Detonation was approximately 52 seconds after release from the aircraft.

31.4.4   Thermonuclear weaponry

Debate

A month after Britain's first atomic weapons test, America tested the first thermonuclear (hydrogen) bomb. The Soviets tested their first in 1953.[45] Penney believed that Britain could not afford to develop a hydrogen bomb.[24] Henry Tizard believed that the nation should focus on conventional forces instead of duplicating the nuclear capabilities of the American Strategic Air Command, which already defended Britain and Europe:[41]:86–87 "We are a great nation, but if we continue to behave like a Great Power we shall soon cease to be a great nation. Let us take warning from the fate of the Great Powers of the past and not burst ourselves with pride."[41]:86–87

First Sea Lord Lord Mountbatten and Chief of the Imperial General Staff Gerald Templer supported the development of a hydrogen bomb, but preferred more support for conventional forces. They believed that the large American and Soviet nuclear forces acted as mutual deterrents for nuclear war, making conventional war more likely.[49]:145–147 Others proposed that, instead of repeated unsuccessful attempts to increase cooperation with the Americans, Britain work with Australia, Canada, and other Commonwealth countries. (Britain could not disclose atomic information to Australia despite testing weapons there because of restrictions in existing agreements with the United States.)[41]:162–163

The Chiefs of Staff Committee[50][41]:87[49]:145 and the Churchill ministry, however, believed that

> If we did not develop megaton weapons we would sacrifice immediately and in perpetuity

our position as a firstclass power. We would have to rely on the whim of the United States for the effectiveness of the whole basis of our strategy.[49]:145

The government decided on 27 July 1954 to begin development of a thermonuclear bomb and announced its plans in February 1955.[45][24][41]:160–163,179–185

An independent deterrent

Believing that the United Kingdom was extremely vulnerable to a nuclear attack to which defence was impossible, the Chiefs of Staff and the RAF first advocated a British nuclear deterrence—not just nuclear weapons—in 1945: "It is our opinion that our only chance of securing a quick decision is by launching a devastating attack upon [enemy cities] with absolute weapons." In 1947 the Chiefs of Staff stated that even with American help the United Kingdom could not prevent the "vastly superior" Soviet forces from overrunning Western Europe, from which Russia could destroy Britain with missiles without using atomic weapons. Only "the threat of large-scale damage from similar weapons" could prevent the Soviet Union from using atomic weapons in a war.[41]:48,397–398

John Slessor, who became Chief of the Air Staff in 1950, wrote that year that the Soviet superiority in European forces was so great that even "an ultimatum by Russia within the next two to three years" might cause Western Europe to surrender without a war. He feared that the United Kingdom might also do so "unless we can make ourselves far less defenceless than we are now." By 1952 the Air Ministry had abandoned the concept of a conventional defense of Western Europe.[49]:71,78–79 The hydrogen bomb increased the threat to Britain. In 1957, a government study stated that although RAF fighters would "unquestionably be able to take a heavy toll of enemy bombers, a proportion would inevitably get through. Even if it were only a dozen, they could with megaton bombs inflict widespread devastation." Although disarmament remained a British goal, "the only existing safeguard against major aggression is the power to threaten retaliation with nuclear weapons."[41]:429–430

Churchill stated in a 1955 speech that deterrence would be "the parents of disarmament" and that, unless Britain contributed to Western deterrence with its own weapons, during a war the targets that threatened it the most might not be prioritized. Harold Macmillan stated that nuclear weapons would give Britain influence over targeting and American policy, and would affect strategy in the Middle East and Far East. Duncan Sandys stated that nuclear weapons reduced Britain's dependence on the United States.[45] The Suez Crisis increased the value to Britain of a deterrent that would give it greater influence with the US and USSR.[51]

Independent targeting was also vital. The Chiefs of Staff believed that—contrary to Tizard's view—once the USSR became able to attack the United States itself with nuclear weapons in the late 1950s, America might not risk its own cities to defend Europe, or not emphasize targets that endangered the United Kingdom more than the United States:[41]:185–187[50]

> When New York is vulnerable to attack the United States will not use her strategic weapon in defence of London. The United Kingdom must, therefore, have its own retaliatory defence. Similarly, however, we will not be prepared to sacrifice the United Kingdom in the defence of say Darwin, and eventually each political unit must have its own means of retaliation.[41]:416

Britain thus needed the ability to convince the USSR that attacking Europe would be too costly regardless of American participation. Part of the perceived effectiveness of an independent deterrent was the willingness to target enemy cities. Slessor saw atomic weapons as a way to avoid a third devastating world war given that the two previous ones had begun without them. While he sought to deemphasize city targeting in British plans as Air Chief,[41]:110–112,114 Slessor wrote in 1954 after retirement:[50]

> And if [war] is forced upon us, we must be able to instantly deliver a crushing counter attack upon aggression at its source—not merely at its airfields, its launching sites and submarine bases, at its armies in the field but at the heart of the aggressor country. There will be the battlefield if battlefield there must be.[50]

When Nigel Mills became head of RAF Bomber Command in 1955 he similarly insisted on targeting Soviet cities, writing "Whoever would be afraid of launching a sudden attack if he thought the greater part of our retaliation would come back to his airfields?"[52] The belief in the importance of retaining an independent capability has continued over several decades and changes in government. As the Defence Council stated in 1980,[7]

> our force has to be visibly capable of making a massive strike on its own ... We need to convince Soviet leaders that even if they thought ... the US would hold back, the British force could still inflict a blow so destructive that the penalty for aggression would have proved too high.[7]

There was little dissent in the House of Commons; nuclear weapons had almost bipartisan support until 1960, with only the Liberals dissenting in 1958. Despite opposition from its left wing the Labour party supported British nuclear weapons but opposed tests, and Labour Opposition Leader Hugh Gaitskell and shadow foreign secretary Aneurin Bevan agreed with Sandys on the importance of reducing dependence on the American deterrent. The left-wing Bevan told his colleagues that their demand for unilateral nuclear disarmament would send a future Labour government "naked into the conference chamber" during international negotiations.[45][51] The *Manchester Guardian* and other newspapers critical of the Conservative government supported the British deterrent, although the *Guardian* did criticise the government for relying on developing bombers rather than missiles to carry the weapons.[53] In 1962 it stated that the forthcoming Chinese nuclear weapon was a reason for having more than one Western nuclear nation.[45] From 1955 the government chose to emphasize the nuclear deterrent and de-emphasize conventional forces.[51] *The Economist*, the *New Statesman*, and many left-wing newspapers supported the reliance on nuclear deterrence and nuclear weapons, but in their view considered that of the United States would suffice, and that of the costs of the "nuclear umbrella" was best left to be borne by the United States alone.[54]

Renewed American partnership

A Yellow Sun thermonuclear bomb

The first prototype, Short Granite, was detonated on 15 May 1957 in Operation Grapple, with disappointing results at 300 kilotons of TNT (1.3 PJ), when the target requirement was 1 Mt (4.2 PJ). A further test of Purple Granite yielded less at 200 kt (0.84 PJ). An interim weapon was deployed in the V-bomber fleet until a true thermonuclear weapon could be devised from the Christmas Island tests. This interim weapon was never tested; it was a very large unboosted pure fission weapon estimated to yield 400 kt (1.7 PJ). It was derived from the Orange Herald warhead tested on 31 May 1957 yielding 720 kt (3.0 PJ)[55] known as Green Grass.

After British scientists demonstrated to the United States in

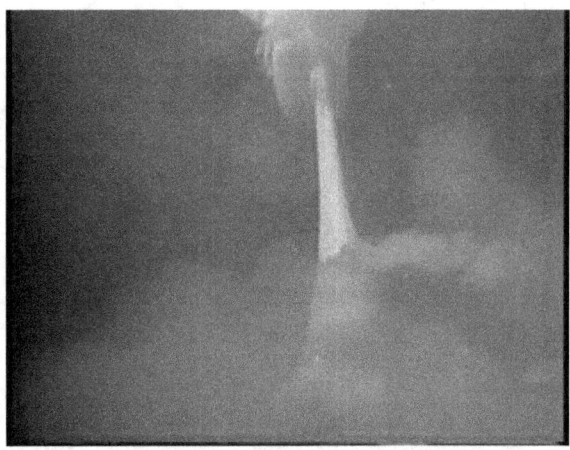

The Orange Herald test on 31 May 1957, claimed to be Britain's first H-bomb test at the time, as reported by Universal International Newsreel a few days later. In fact it was a large fusion boosted fission weapon test, but the fusion boosting worked very poorly.

late 1957 that they had developed a Teller-Ulam design different from American methods and thus understood how to build a hydrogen bomb, the 1958 US-UK Mutual Defence Agreement made fully developed and tested American designs available more quickly and more cheaply. The first of these was the US B28 nuclear bomb, which was anglicised and manufactured in the UK as Red Snow and quickly deployed as Yellow Sun Mk.2 in the V-bomber fleet. Red Snow became the warhead of choice for the Blue Steel stand-off missile and some of the Skybolt missiles intended for carriage by the V-bombers. (The American B28 design had reliability issues. The British soon withdrew their warheads from deployment, and never again simply copied an American design.)[24] Under the Mutual Defence Agreement 5.4 tonnes of UK produced plutonium was sent to the US in return for 6.7 kg of tritium and 7.5 tonnes of HEU over the period 1960–1979, replacing Capenhurst production, although much of the HEU was used not for weapons, but as fuel for the growing UK fleet of nuclear submarines, both of the Polaris variety and others numbering approx twelve.

Fifty-eight Blue Danube bombs were produced, although archived declassified files indicate that only a small proportion of these were ever serviceable at any one time. It remained in service until 1963, when it was replaced by Red Beard, a smaller tactical boosted fission weapon that used the same fissile core as Blue Danube and was deployed on many smaller aircraft than the V-bombers, both ashore and at sea aboard five carriers. Stocks of Red Beard were maintained in Cyprus, Singapore, and a smaller number in the UK.

It was the largest pure fission weapon ever deployed by any nuclear state. Green Grass was deployed first in a modified Blue Danube casing and known as Violet Club. A later variant was deployed in a Yellow Sun Mk.1 casing.

In 1960 the government cancelled the Blue Streak missile based on the Chiefs of Staff's conclusion that it was too vulnerable to attack and thus was only useful for a first strike and decided to purchase the American Skybolt missile instead.[41]:286–288 In 1962 it cancelled the Blue Steel extended range upgrade (Blue Steel Mk2) for Skybolt. Similarly, reassessments of Soviet capabilities changed military perceptions and led to the removal of Thor IRBM missiles in the UK; and Jupiter IRBMs in Italy and Turkey; although the Turkish sites were implicated in an alleged deal following the Cuban Missile Crisis. To consternation, and considerable protests, the incoming Kennedy administration cancelled Skybolt at the end of 1962 because it was believed by the US Secretary of State for Defense, Robert McNamara, that other delivery systems were progressing better than expected, and a further expensive system was surplus to US requirements.

End of cross-party support

Gaitskell's Labour party ceased supporting an independent deterrent in 1960 via its new "Policy for Peace", after the cancellation of Blue Streak made nuclear independence less likely. Labour also adopted a resolution favoring unilateral disarmament. Although Gaitskell opposed the resolution and it was reversed in 1961 in favor of continuing support of a general Western nuclear deterrent, the party's opposition to a British deterrent remained and became more prominent. This became a campaign issue during the 1964 general election. Alec Douglas-Home's incumbent Conservatives stated that the British deterrent was both necessary for independence from the Americans and maintaining British world influence, and was "working for peace" in such cases as the passage of the Nuclear Test Ban Treaty. Led by Gaitskell's successor Harold Wilson, Labour emphasized domestic economic issues but called deterrence the "Tory Nuclear Pretense" which would be neither independent nor effective. The populace's greater interest in domestic over foreign policy likely contributed to Labour's victory.[51]

31.4.5 Polaris

Main article: UK Polaris programme

After the cancellation of Skybolt, the UK purchased Polaris missiles for use in UK-built ballistic missile submarines. The agreement between US President John F. Kennedy and Prime Minister of the United Kingdom Harold Macmillan, the Polaris Sales Agreement, was announced on 21 December 1962 and HMS *Resolution* made her first Polaris-armed

The Polaris A1 or A2 missile, seen here on a launch pad in Cape Canaveral, was a submarine-launched ballistic missile purchased from the US. The UK purchased the A3T variant, the final production model, that incorporated hardened missile electronic components to resist ABM attack in the boost phase, although neither the three re-entry vehicles or UK-manufactured warheads were hardened, leading to the Chevaline programme.

operational patrol on 15 June 1968.[56] In the 1970s the UK Polaris RVs and warheads were vulnerable to the Soviet ABM screen concentrated around Moscow, and the UK developed a Polaris improved-front-end (IFE) codenamed Chevaline, designed to counter this ABM defence which threatened to completely nullify an independent UK deterrent posture. When Chevaline became public knowledge in 1980, it generated huge controversy as it had been kept secret by the four governments of Wilson, Heath, Wilson (again) and Callaghan, whilst costs rocketed during a period of high inflation, until disclosed by the Thatcher government. By the time it entered service in 1982 it had cost approx £1bn. The final Polaris/Chevaline patrol took place in 1996, two years after the first Trident-carrying submarine sailed on its first patrol.

As well as the establishment at Aldermaston, the UK nuclear weapons programme also has a factory at Burghfield nearby which assembled the weapons and is responsible for their maintenance, and had another in Cardiff which fabricated non-fissile components and a 2000-acre (8 km²) test range at Foulness. Since 1993 the sites have been managed by private consortia. The Foulness and Cardiff facilities closed in October 1996 and February 1997 respectively.

31.5 Current weapons programmes

31.5.1 Trident

Main article: Trident nuclear programme

The UK currently has four *Vanguard*-class submarines

A Trident missile launched from a ballistic missile submarine.

HMS Vanguard, *one of four* Vanguard *class ballistic missile submarines of the Royal Navy, which serve as the UK's nuclear delivery system.*

based at HMNB Clyde in Scotland, armed with nuclear-tipped Trident missiles. The principle of operation is based on maintaining deterrent effect by always having at least one submarine at sea, and was designed for the Cold War period. One submarine is normally undergoing maintenance and the remaining two in port or on training exercises.

Rare WE.177A sectioned instructional example of an operational round, one of only two in existence at the Boscombe Down Aviation Collection

Each submarine carries 16 Trident II D-5 missiles, which can each carry up to 12 warheads. However, the UK government announced in 1998 that each submarine would carry only 48 warheads, an increase of 50% over the 32 warheads carried by Trident's predecessor, Chevaline, (halving the limit specified by the previous government), which is an average of three per missile. However one or two missiles per submarine are probably armed with fewer warheads for "sub-strategic" use causing others to be armed with more; but this is speculative.

The UK-designed warheads are thought to be selectable between 0.3, 5–10 and 100 kt (1.3, 21–42 and 420 TJ); the yields obtained using either the unboosted primary, the boosted primary, or the entire "physics package"; these yields and similar data are speculative. Although the UK designed, manufactured and owns the warheads, there is evidence that the warhead design is similar to, or even based on, the US W76 warhead fitted in some US Navy Trident missiles, with design data being supplied by the United States through the 1958 US-UK Mutual Defence Agreement.[57][58] The United Kingdom owns 58 missiles which are shared in a joint pool with the United States government and these are exchanged when requiring maintenance with missiles from the United States Navy's own pool and vice versa.

31.5.2 Trident renewal

Further information: Trident nuclear programme § Trident renewal

A decision on the renewal of Trident was made on 4 December 2006. Prime Minister Tony Blair told MPs it would be "unwise and dangerous" for the UK to give up its nuclear weapons. He outlined plans to spend up to £20bn on a new generation of submarines for Trident missiles. He said submarine numbers may be cut from four to three, while the number of nuclear warheads would be cut by 20% to 160. Blair said although the Cold War had ended, the UK needed nuclear weapons, as no-one could be sure another nuclear threat would not emerge in the future.

The 2010 coalition government agreed "that the renewal of Trident should be scrutinised to ensure value for money. Liberal Democrats will continue to make the case for alternatives." Research and development work continued with an 'Initial Gate' procurement decision, but the 'Main Gate' decision to manufacture a replacement was re-scheduled for 2016, after the next election.[59] A vote in the House of Commons on whether to replace the existing four Vanguard-class submarines was scheduled for 18 July 2016.[60]

That vote on the so-called Trident renewal programme motion was held on that date[61] and it passed with a significant majority with 472 MPs voting in favour and 117 against. (As expected, Jeremy Corbyn and 47 Labour MPs had voted against it; 41 did not vote but 140 Labour votes were cast in favour of the motion.[62]) The new Successor-class submarines were expected to come into operation by 2028[63] according to some estimates and certainly by the 2030s,[64] extending the programme's life until at least the 2060s.[65] At that time, there was already some urgency to move ahead because some experts predicted it could take 17 years to develop the replacement for the Vanguard-class submarines [66][67] The new class would continue to carry the Trident D-5 missiles.[68]

31.6 Deployment of US tactical nuclear weapons

Until 1992 UK forces also deployed US tactical nuclear weapons as part of a US-UK dual-key NATO nuclear sharing role.[69][70] This arrangement commenced in 1958 as Project E to provide nuclear weapons to the RAF prior to a sufficient number of Britain's own nuclear weapons becoming available.

The weapons deployed included nuclear artillery, nuclear demolition mines and warheads for Corporal and Lance missiles in Germany; theatre nuclear weapons on RAF aircraft;[71] Mark 101 nuclear depth bombs on RAF Shackleton maritime patrol aircraft, later replaced by a modern successor, the B-57 deployed on RAF Nimrod aircraft.

The Lance missiles were purchased in 1975,[72] to replace Honest John missiles which had been bought in 1960;[73][74]

and were themselves a replacement for the US Corporal missiles deployed in Germany by the Royal Artillery. Not generally recognised is the fact that the Royal Artillery deployed a numerically greater quantity of US nuclear weapons than the RAF and Royal Navy combined, peaking at 277 in 1976–78; with a further 50 ADMs deployed with another British Army unit, the Royal Engineers, peaking in 1971–81.[75] The dual-key agreement for controlling US tactical nuclear weapons, known as the Heidelberg Agreement, was made on 30 August 1961. The UK sponsored access for the Canadian Army Honest John missile deployments to the US/UK nuclear warhead storage sites.[76]

During the 1980s nuclear armed USAF Ground Launched Cruise Missiles were deployed at RAF Greenham Common and RAF Molesworth. Until about 2006 the US continued to store nuclear weapons in the UK, when approximately 110 tactical B61 nuclear bombs stored at RAF Lakenheath for deployment by USAF F-15E Strike Eagle aircraft were removed.[77][78]

The UK continues to permit the US to deploy nuclear weapons from its territory, the first having arrived in 1954.[79]

31.7 Research and development facilities

31.7.1 Atomic Weapons Establishment, Aldermaston

Main article: Atomic Weapons Establishment

The Atomic Weapons Establishment (AWE), Aldermaston (formerly the Atomic Weapons Research Establishment, Aldermaston) is situated just 7 miles (11 km) north of Basingstoke and approximately 14 miles (23 km) south-west of Reading, Berkshire, near a village called Aldermaston, bordering with Tadley. It was built in 1949 on the site of a former World War II Royal Air Force base and converted to nuclear weapons research, design and development in the 1950s. Although some early test devices were probably assembled on this site, final assembly of Service-engineered weapons takes place at the nearby site of Burghfield.

31.7.2 Royal Ordnance Factories, Cardiff and Burghfield

Main articles: Royal Ordnance Factory, ROF Burghfield, and ROF Cardiff

Other nuclear weapons sites could be found in Cardiff and Burghfield near Reading, Berkshire. These were the only two Royal Ordnance Factories (ROFs) not privatised in the 1980s.

ROF Cardiff, which closed in 1997, was involved in nuclear weapons programmes since 1961. The site was used for the task of recycling old nuclear weapons and precisely shaping uranium 235 (U235) and metallic beryllium components for the boosted fission devices used as primaries or 'triggers' in modern thermonuclear weapons.[80] ROF Burghfield was a former Filling Factory, opened in 1942, and run as an *Agency Factory*, by Imperial Tobacco, to fill Oerlikon 20 mm ammunition.[81]

31.8 Politics, decision making and nuclear posture

31.8.1 Anti-nuclear movement

Main article: Anti-nuclear movement in the United Kingdom

The anti-nuclear movement in the United Kingdom consists

The now-familiar peace symbol was originally the Campaign for Nuclear Disarmament logo.

of groups who oppose nuclear technologies such as nuclear power and nuclear weapons. Many different groups and individuals have been involved in anti-nuclear demonstrations and protests over the years.

One of the most prominent anti-nuclear groups in the UK is the Campaign for Nuclear Disarmament (CND). CND's

Aldermaston Marches began in 1958 and continued into the late 1960s when tens of thousands of people took part in the four-day marches. One significant anti-nuclear mobilisation in the 1980s was the Greenham Common Women's Peace Camp. In London, in October 1983, more than 300,000 people assembled in Hyde Park as part of the largest protest against nuclear weapons in British history. In 2005 in Britain, there were many protests about the government's proposal to replace the ageing Trident weapons system with a newer model.

31.8.2 Nuclear posture

UK nuclear posture during the cold war was informed by dependence on the United States. Operational control of the UK Polaris force was assigned to SACLANT, while targeting policy for its missiles was determined, as for the V-bomber force before it, by NATO's SACEUR, while maintaining an independent wholly UK targeting policy for circumstances when a critical national emergency required it to be used alone, without the UK's NATO allies.[82][83] In these circumstances, the Moscow criterion referred to the ability of the UK to strike back at the highly centralised Soviet decision-making apparatus concentrated in the Moscow area, intended to destroy the ability of the Soviet leadership to remain in control of a Soviet Union otherwise untouched. The early beginnings of studies to increase the likelihood of successful penetration of the Polaris warheads to Moscow can be traced back to 1964,[84] before the Polaris system was deployed, in order to preserve this capability in the face of anti-ballistic missile batteries around Moscow. These studies later materialised as Chevaline.[85][86]

The UK has relaxed its nuclear posture since the collapse of the Soviet Union. The Labour government's 1998 Strategic Defence Review made reductions from the plans announced by the previous Conservative government:[87]

- The stockpile of *"operationally available warheads"* was reduced from 300 to *'less than 200"*

- The final batch of missile bodies would not be purchased, limiting the fleet to 58.

- A submarine's load of warheads were reduced from 96 to 48. This reduced the explosive power of the warheads on a Vanguard class Trident submarine to *"one third less than a Polaris submarine armed with Chevaline."* However 48 warheads per Trident submarine represents a 50% increase on the 32 warheads per submarine of Chevaline. Total explosive power has been in decline for decades as the accuracy of missiles has improved, therefore requiring less power to destroy each target. Trident can destroy 48 targets per submarine, as opposed to 32 targets that could be destroyed by Chevaline.

- Submarines missiles would not be targeted, but rather at several days "notice to fire".

- Although one submarine would always be on patrol it will operate on a "reduced day-to-day alert state". A major factor in maintaining a constant patrol is to avoid *"misunderstanding or escalation if a Trident submarine were to sail during a period of crisis."*

Current UK posture as outlined in the Strategic Defence Review of 1998[88] is as it has been for many years; Trident SLBMs still provide the long-range strategic element. Until 1998 the aircraft-delivered, free-fall WE.177A, WE.177B and WE.177C bombs provided an sub-strategic option in addition to their designed function of tactical battlefield weapons. With the retirement of WE.177, a sub-strategic warhead is to be used with some (but not all) deployed Trident missiles. The exact mix of warheads is unknown, as are their number and yield. The 2010 Strategic Defence and Security Review further pledged to reduce its requirement for operationally available warheads from fewer than 160 to no more than 120.[89] In a January 2015 written statement, Defence Secretary Michael Fallon reported that " All Vanguard Class SSBNs on continuous at-sea deterrent patrol now carry 40 nuclear warheads and no more than eight operational missiles".[90]

31.8.3 Nuclear weapons control

The precise details of how a British Prime Minister would authorise a nuclear strike remain secret, although the principles of the Trident missile control system is believed to be based on the plan set up for Polaris in 1968, which has now been declassified. A closed-circuit television system was set up between 10 Downing Street and the SSBN Control Officer at the Northwood Headquarters of the Royal Navy. Both the Prime Minister and the SSBN Control Officer would be able to see each other on their monitors when the command was given. If the link failed – for instance during a nuclear attack or when the PM was away from Downing Street – the Prime Minister would send an authentication code which could be verified at Northwood. The PM would then broadcast a firing order to the SSBN submarines via the Very Low Frequency radio station at Rugby. The UK has not deployed control equipment requiring codes to be sent before weapons can be used, such as the U.S. Permissive Action Link, which if installed would preclude the possibility that military officers could launch British nuclear weapons without authorisation.

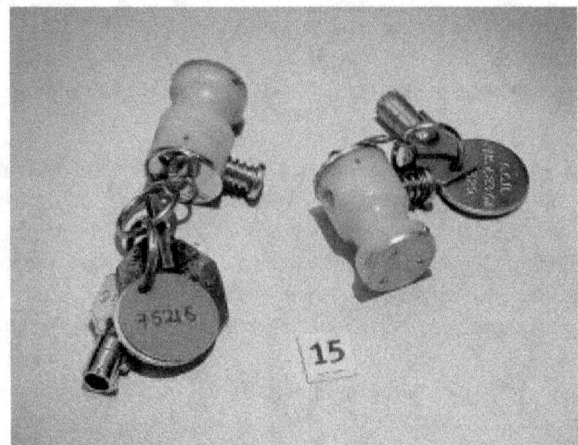

WE.177 safety and arming keys

Until 1998, when it was withdrawn from service, the WE.177 bomb was armed with a standard tubular pin tumbler lock (as used on bicycle locks) and a standard allen key was used to set yield and burst height. Currently, British Trident missile commanders are able to launch their missiles without authorisation, whereas their American colleagues cannot. At the end of the Cold War the U.S. Fail Safe Commission recommended installing devices to prevent rogue commanders persuading their crews to launch unauthorised nuclear attacks. This was endorsed by the Nuclear Posture Review and Trident missile Coded Control Devices were fitted to all U.S. SSBNs by 1997. These devices prevented an attack until a launch code had been sent by the Chiefs of Staff on behalf of the President. The UK took a decision not to install Trident CCDs or their equivalent on the grounds that an aggressor might be able to wipe out the British chain of command before a launch order had been sent.[91][92][93]

In December 2008 BBC Radio 4 made a programme titled *The Human Button*, providing new information on the manner in which the United Kingdom could launch its nuclear weapons, particularly relating to safeguards against a rogue launch. Former Chief of the Defence Staff (most senior officer of all British armed forces) and Chief of the General Staff (most senior officer in the British Army), General Lord Guthrie of Craigiebank, explained that the highest level of safeguard was against a prime minister ordering a launch without due cause: Lord Guthrie stated that the constitutional structure of the United Kingdom provided some protection against such an occurrence, as while the Prime Minister is the chief executive and so practically commands the armed services, the formal commander-in-chief is the Monarch, to whom the chief of the defence staff could appeal: *"the chief of the defence staff, if he really did think the prime minister had gone mad, would make quite sure that that order was not obeyed... You have to remember that ac-*

tually prime ministers give direction, they tell the chief of the defence staff what they want, but it's not prime ministers who actually tell a sailor to press a button in the middle of the Atlantic. The armed forces are loyal, and we live in a democracy, but actually their ultimate authority is the Queen."[94]

Weapons Engineer Officers Tactical Trigger used to launch a Trident Missile. Taken in 2012 aboard HMS Vigilant during a test launch of an unarmed Trident ballistic missile at sea.

The same interview pointed out that while the Prime Minister would have the constitutional authority to fire the Chief of the Defence Staff, he could not appoint a replacement as the position is appointed by the monarch. During the Cold War the Prime Minister was also required to name a senior member of the cabinet as his/her designated survivor, who would have the authority to order a nuclear response in the event of an attack incapacitating the Prime Minister, and this system was re-adopted after the September 11 attacks.

The programme also addressed the workings of the system; detailing that two persons are required to authenticate each stage of the process before launching, with the submarine captain only able to access the firing trigger after two safes have been opened with keys held by the ship's executive and weapons engineering officers. It was explained that all Prime Ministers issue hand-written orders, termed the letters of last resort,[95] seen by their eyes only, sealed and stored within the safes of each of the four Royal Navy Vanguard class submarines. These notes instruct the submarine commander of what action to take in the event of the United Kingdom being attacked with nuclear weapons that destroy Her Majesty's Government in the United Kingdom and/or the chain of command.

Although the final orders of the Prime Minister are at his or her discretion, and no fixed options exist, according to the December 2008 BBC Radio 4 documentary *The Human Button* there were four known options: retaliating with nuclear weapons, not retaliating with nuclear weapons, the submarine commander uses his own judgement, or the submarine commander places himself under United States or Australian command if possible. This system of issuing notes containing orders in the event of the head of government's death is said to be unique to the United Kingdom (although the concept of written last orders, particularly of a ship's captain, is a naval tradition), with other nuclear powers using different procedures. The letters are destroyed unopened whenever a Prime Minister leaves office.

All relevant former prime ministers have supported an "independent nuclear deterrent", including David Cameron[96] and the incumbent Prime Minister Theresa May.[97] Only one former Prime Minister, Lord Callaghan, has given any insight on his orders: Callaghan stated that, although in a situation where nuclear weapon use was required – and thus the whole purpose and value of the weapon as a deterrent had failed – he would have ordered use of nuclear weapons, if needed: *...if we had got to that point, where it was, I felt it was necessary to do it, then I would have done it (used the weapon)...but if I had lived after pressing that button, I could have never forgiven myself.*[98] Lord Healey, Secretary of State for Defence and "alternate decision-taker" under Prime Minister Harold Wilson, said that in the event of Soviet nuclear weapons attacking the United Kingdom and the Prime Minister had been killed or incapacitated, he would not have ordered a retaliation.[98]

The process by which a Trident missile-armed submarine would determine if the British government continued to function included, among other checks, establishing whether BBC Radio 4 continued broadcasting.[99]

31.8.4 The special relationship

Main article: Special relationship

The 1958 "Agreement For Cooperation on the Uses of Atomic Energy for Mutual Defence Purposes", also known as the "Mutual Defence Agreement", was renewed in 1994 and again in 2005.[100]

31.8.5   Cost

The current Trident system cost £12.6bn (at 1996 prices) and costs £280m a year to maintain. Options for replacing Trident range from £5bn for the missiles alone to £20-30bn for missiles, submarines and research facilities. At minimum, for the system to continue after around 2020, the missiles will need to be replaced.[101] The price of replacement of submarine has risen to £31bn and it's estimated by Ministry of defence that the cost of trident replacement program for 30 years to be at £167bn and the Campaign for Nuclear Disarmament puts the cost as high as £205bn.[102]

31.9   Legality

Further information: International Court of Justice advisory opinion on the Legality of the Threat or Use of Nuclear Weapons and Nuclear Non-Proliferation Treaty

After the British government announced its plans to refurbish its Trident SLBM missiles and build new submarines to carry them,[103] it published a white paper *The Future of the United Kingdom's Nuclear Deterrent* in which it stated that the renewal is fully compatible with the United Kingdom's treaty commitments and international law.[104] These arguments are summarised in a question and answer briefing published by UK Permanent Representative to the Conference on Disarmament[105]

> • Is Trident replacement legal under the Non Proliferation Treaty (NPT)? Renewal of the Trident system is fully consistent with our international obligations, including those on disarmament. ...
>
> • Is retaining the deterrent incompatible with NPT Article VI? The NPT does not establish any timetable for nuclear disarmament. Nor does it prohibit maintenance or renewal of existing capabilities. Renewing the current Trident system is fully consistent with the NPT and with all our international legal obligations. ...

At the start of the House of Commons debate to authorise the replacement of Trident,[106] Margaret Beckett stated:

> Article VI of the NPT imposes an obligation on all states: "to pursue negotiations in good faith on effective measures relating to cessation of the nuclear arms race at an early date and to

nuclear disarmament, and on a Treaty on general and complete disarmament". The NPT Review Conference held in 2000 agreed, by consensus, 13 practical steps towards nuclear disarmament. The UK remains committed to these steps and is making progress on them. We have been disarming. Since the Cold War ended, we have withdrawn and dismantled our tactical maritime and airborne nuclear capabilities. We have terminated our nuclear capable Lance missiles and artillery. We have the smallest nuclear capability of any recognised nuclear weapon state accounting for less than one per cent of the global inventory. And we are the only nuclear weapon state that relies on a single nuclear system.

The subsequent vote was won overwhelmingly, including unanimous support from the opposition Conservative Party.[107]

The Government position remains that it is abiding by the NPT legally in renewing Trident and Britain has the right to possess nuclear weapons, a position reiterated by Tony Blair in PMQs on 21 February 2007.[108]

In contrast, reports by Philippe Sands QC, and by Rabinder Singh QC and Professor Christine Chinkin, used in the case against, were commissioned by the activist groups Greenpeace and Peace Rights respectively.[109] Both groups are opposed to the renewal, use, or proliferation of nuclear weapons.

Furthermore, the British Government and NATO do not recognise advisory opinion of the ICJ,[110] as interpreter of IHL and referred to by Sands et al., (see Advisory Opinion) with regard to use of nuclear weaponry as legally binding.[111]

This position is held in common with all five nuclear states as defined in the NPT. However, only the United Kingdom has expressed its opposition to the establishment of a new legally binding treaty to prevent the threat or use of nuclear weapons against non-nuclear states[112] by its vote in the United Nations General Assembly in 1998.[113]

One view is that the white paper *The Future of the United Kingdom's Nuclear Deterrent* stands in contrast to two counsel's opinions. The first, commissioned by Peacerights,[114] was given on 19 December 2005 by Rabinder Singh QC and Professor Christine Chinkin of Matrix Chambers. It addressed '...whether Trident or a likely replacement to Trident breaches customary international law'[115]

Drawing on the International Court of Justice (ICJ) opinion, Singh and Chinkin advised that:

> The use of the Trident system would breach

customary international law, in particular because it would infringe the "intransgressible" [principles of international customary law] requirement that a distinction must be drawn between combatants and non-combatants.[115]

The second opinion was commissioned by Greenpeace[116] and given by Philippe Sands QC and Helen Law, also of Matrix Chambers, on 13 November 2006.[117] The opinion addressed

> The compatibility with international law, in particular the jus ad bellum, international humanitarian law ('IHL') and Article VI of the Treaty on the Non-Proliferation of Nuclear Weapons ('NPT'), of the current UK strategy on the use of Trident...The compatibility with IHL of deploying the current Trident system...[and] the compatibility with IHL and Article VI NPT of the following options for replacing or upgrading Trident: (a) Enhanced targeting capability; (b) Increased yield flexibility; (c) Renewal of the current capability over a longer period.[118]

With regards to the *jus ad bellum*, Sands and Law advised that

> Given the devastating consequences inherent in the use of the UK's current nuclear weapons, we are of the view that the proportionality test is unlikely to be met except where there is a threat to the very survival of the state. In our view, the 'vital interests' of the UK as defined in the Strategic Defence Review are considerably broader than those whose destruction threaten the survival of the state. The use of nuclear weapons to protect such interests is likely to be disproportionate and therefore unlawful under Article 2(4) of the UN Charter.[119]

The phrase "very survival of the state" is a direct quote from paragraph 97 of the ICJ ruling. With regards to international humanitarian law, they advised that

> it [is] hard to envisage any scenario in which the use of Trident, as currently constituted, could be consistent with the IHL prohibitions on indiscriminate attacks and unnecessary suffering. Further, such use would be highly likely to result in a violation of the principle of neutrality.[120]

Finally, with reference to the NPT, Sands and Law advised that

> A broadening of the deterrence policy to incorporate prevention of nonnuclear attacks so as to justify replacing or upgrading Trident would appear to be inconsistent with Article VI; b) Attempts to justify Trident upgrade or replacement as an insurance against unascertainable future threats would appear to be inconsistent with Article VI; c) Enhancing the targeting capability or yield flexibility of the Trident system is likely to be inconsistent with Article VI; d) Renewal or replacement of Trident at the same capability is likely to be inconsistent with Article VI; and e) In each case such inconsistency could give rise to a material breach of the NPT.[121]

31.10 See also

- Anti-nuclear movement in the United Kingdom
- C. P. Snow - writer and scientist who wrote several novels based on his experience in the British nuclear programme
- Global Security Institute
- Letters of last resort
- Nuclear disarmament
- Nuclear testing
- *The War Game* (An earlier, documentary-style TV film, also dealing with the effects of nuclear attack on the United Kingdom, that was made in 1965 and banned from being broadcast until the mid-1980s)
- *Threads* (A fictional film about the effects of nuclear war on the United Kingdom)
- United Kingdom and weapons of mass destruction

31.11 References

[1] "UK to be "more open" about nuclear warhead levels". BBC News. 26 May 2010. Retrieved 30 July 2010.

[2] "Trident missile factfile". BBC News. 23 September 2009. Retrieved 11 May 2012.

[3] "Federation of American Scientists :: Status of World Nuclear Forces". Fas.org. 26 May 2010. Retrieved 30 July 2010.

[4] *Prospects for Further Proliferation of Nuclear Weapons* (PDF). Special National Intelligence Estimate. CIA. 23 August 1974. p. 40. SNIE 4-1-74. Retrieved 20 January 2008.

[5] Woolf, Marie (29 October 2006). "So, minister, are we developing new nuclear weapons or not?; Scientists say they are designing a new warhead design, despite government denials". *The Independent on Sunday*. Newspaper Publishing plc. p. 6.

[6] "Blair's Trident statement in full". BBC News. 4 December 2006. Retrieved 4 December 2006.

[7] "The Future United Kingdom Strategic Deterrent Force" (PDF). The Defence Council. July 1980. Retrieved 17 May 2012.

[8] Point 64, Strategic Defence Review, Presented to Parliament by the Secretary of State for Defence, George Robertson, July 1998

[9] SIPRI project on nuclear technology and arms Archived 3 May 2006 at the Wayback Machine.

[10] House of Commons Written Answers, Hansard, 14 July 1998 : Column:171

[11] Table of Global Nuclear Weapons Stockpiles, 1945–2002, National Resources Defense Council, 25 November 2002

[12] *Ministry of Defence: The United Kingdom's Future Nuclear Deterrent Capability*. National Audit Office. 5 November 2008. ISBN 978-0-10-295436-4. Retrieved 9 November 2008.

[13] "UK to be "more open" about nuclear warhead levels". BBC News. 26 May 2010. Retrieved 26 May 2010.

[14] Rogers, P. (2006). "Big Boats and Bigger Skimmers: Determining Britain's Role in the Long War". *International Affairs (Royal Institute of International Affairs 1944-)*. **82** (4): 651–665. JSTOR 3874150.

[15] "RAF nuclear frontline Order-of-Battle 1966–94". Retrieved 30 July 2010.

[16] Norris, Robert S.; Kristensen, Hans M. (2013). "The British nuclear stockpile, 1953-2013". *Bulletin of the Atomic Scientists*. **69** (4): 69–75. doi:10.1177/0096340213493260.

[17] Press Release, Verification Technology Information Centre, 17 August 1995

[18] Weapons around the world, Jon Wolfsthal, physicsweb, August 2005

[19] Nuclear Weapons Milestons (Part 1-B), compiled by Wm. Robert Johnston, 3 June 2005

[20] History of Nuclear Weapons Testing, Greenpeace, April 1996

[21] Database of nuclear tests, United Kingdom, compiled by Wm. Robert Johnston, last modified 19 June 2005

[22] Office of the Deputy Administrator for Defense Programs (January 2001). *Highly Enriched Uranium: Striking A Balance – A Historical Report On The United States Highly Enriched Uranium Production, Acquisition, And Utilization Activities From 1945 Through September 30, 1996* (Revision 1 (Redacted For Public Release) ed.). U.S. Department of Energy, National Nuclear Security Administration. Retrieved 13 June 2009.

[23] History of the British Nuclear Arsenal, Last changed 30 April 2002

[24] Spinardi, Graham (August 1997). "Aldermaston and British Nuclear Weapons Development: Testing the 'Zuckerman Thesis'". *Social Studies of Science*. **27** (4): 547–582. doi:10.1177/030631297027004001.

[25] Public Record Office, London, DEFE 19/180, E66. Declassified Jan 2006 using the FOI Act.

[26] Inventory of International Nonproliferation Organizations and Regimes, Center for Nonproliferation Studies.

[27] House of Commons Debate, Nuclear Explosions (Prohibition and Inspections) Bill, Hansard, 6 November 1997 : Column 455

[28] Status of CTBT Ratification, British American Security Information Council, last updated on 14 June 2001

[29] See *Project-56, −57, −58*, and *Hardtack II* (among others) for the US; *1961, 1962* and *1978* series for the USSR; *1971-1974* and *1975-1978* for the French; and tests *39, 41-43* for the Chinese.

[30] Statement by the Secretary of State for Defence, Hansard 15 January 2003 : Column 697

[31] Royal United Services Institute – Ballistic Missile Defence and the UK April 2005

[32] Possible Nuclear Attack Scenarios on Britain, Paul Rogers, Proceedings of Conference on Nuclear Deterrence: Implications and Policy Options for the 1980s, September 1981

[33] Stan Openshaw, Philip Steadman and Owen Greene (1983). *Doomsday, Britain after Nuclear Attack*. Basil Blackwell. p. 105. ISBN 0-631-13394-1.

[34] Duncan Campbell (1983). *War Plan UK*. Paladin. p. 32. ISBN 0-586-08479-7.

[35] Lawrence S. Wittner (2003). *Toward Nuclear Abolition: A History of the World Nuclear Disarmament Movement, 1971–Present*. Stanford University Press. p. 294. ISBN 0-8047-4862-4.

[36] "Hard Rock is cancelled again". New Statesman. 11 March 1983. p. 6.

[37] Protect and Survive, prepared for the Home Office by the Central Office of Information, May 1980

[38] Britain planned taped messages after nuclear war, by Gregory Katz, Associated Press, 3 October 2008.

[39] Peter Hennessy, *Cabinets and the Bomb*, The British Academy/Oxford University Press, 2007, p. 31.

[40] Goldberg, Alfred (July 1964). "The Atomic Origins of the British Nuclear Deterrent". *International Affairs*. **40** (3): 409–429. doi:10.2307/2610825.

[41] Baylis, John (1995). *Ambiguity and Deterrence: British Nuclear Strategy 1945–1964*. Oxford: Clarendon Press. ISBN 0-19-828012-2.

[42] Sir M. Perrin, who was present, *The Listener*, 7 October 1982. Also quoted in *How Nuclear Weapons Decisions are Made*, p.137, 1986, Oxford Research Group.

[43] Peter Hennessy, "Cabinets and the Bomb", The British Academy/Oxford University Press, 2007, p. 48.

[44] Arnold, Lorna (2001).*Britain and the H-Bomb: the official history*. Published by Palgrave. ISBN 0-333-94742-8 in North America, ISBN 0-312-23518-6 elsewhere. p71

[45] Gott, Richard (April 1963). "The Evolution of the Independent British Deterrent". *International Affairs*. **39** (2): 238–252. doi:10.2307/2611300.

[46] Dawson, R.; Rosecrance, R. (1966). "Theory and Reality in the Anglo-American Alliance". *World Politics*. **19** (1): 21–51. doi:10.2307/2009841. JSTOR 2009841.

[47] Aldrich, Richard J. (July 1998). "British Intelligence and the Anglo-American 'Special Relationship' during the Cold War". *Review of International Studies*. **24** (3): 331–351. doi:10.1017/s0260210598003313. JSTOR 20097530.

[48] "Britain's Nuclear Weapons – British Nuclear Testing". Nuclearweaponarchive.org. Retrieved 30 July 2010.

[49] Ball, S. J. *The Bomber in British Strategy: Britain's World Role, 1945–1960*. Boulder, Colorado: Westview Press. ISBN 0-8133-8934-8.

[50] Wheeler, N. J. (1985–1986). "British Nuclear Weapons and Anglo-American Relations 1945-54". *International Affairs (Royal Institute of International Affairs 1944-)*. **62** (1): 71–86. doi:10.2307/2618068. JSTOR 2618068.

[51] Epstein, L. D. (1966). "The Nuclear Deterrent and the British Election of 1964". *Journal of British Studies*. **5** (2): 139–163. doi:10.1086/385523. JSTOR 175321.

[52] Young, Ken (Spring 2007). "A Most Special Relationship: The Origins of Anglo-American Nuclear Strike Planning". *Journal of Cold War Studies*. **9** (2): 5–31. doi:10.1162/jcws.2007.9.2.5.

[53] Lorna Arnold, Britain and the H-Bomb, the official history. Published 2001. Palgrave. ISBN 0-333-94742-8 and ISBN 0-312-23518-6 in North America, p65.

[54] A.J.R.Groom, *"British thinking about nuclear weapons"*, pps 131–154. Published Frances Pinter 1974. ISBN 0-903804-01-8. (Note that this is an old 10 digit ISBN, as used at publication in 1974)

[55] Lorna Arnold p147, ibid

[56] Amazon.co.uk review, The Impact of Polaris: The Origins of Britain's Seaborne Nuclear Deterrent, J.E. Moore.

[57] "Britain's Next Nuclear Era". Federation of American Scientists. 7 December 2006. Archived from the original on 6 February 2007. Retrieved 15 March 2007.

[58] "Stockpile Stewardship Plan: Second Annual Update (FY 1999)" (PDF). US Department of Energy. April 1998. Retrieved 15 March 2007.

[59] "Trident: The Initial Gate Decision" (PDF). *Briefings on Nuclear Security*. British Pugwash. July 2011. Retrieved 31 December 2012.

[60] "Defence Secretary welcomes vote on CASD". UK Ministry of Defence. 8 July 2016. Retrieved 16 July 2016.

[61] "UK's Nuclear Deterrent". *UK Hansard*. 18 July 2016. Retrieved 20 July 2016.

[62] Kuenssberg, Laura (19 July 2016). "MPs vote to renew Trident weapons system". *BBC News*. BBC. Retrieved 19 July 2016. Jeremy Corbyn has been heckled and accused of lying by his own MPs and told he was "defending the countries' enemies" as he announced he would vote against renewing Trident.

[63] "Everything you need to know about Trident - Britain's nuclear deterrent". *ITV News*. ITV plc. 18 July 2016. Retrieved 20 July 2016. The £40 billion construction of a new fleet, Successor, could begin this year and be operational by 2028 while the current fleet will be phased out by 2032.

[64] "MPs approve Trident renewal". BBC News. 18 July 2016. Retrieved 18 July 2016. MPs approve Trident renewal

[65] Tom Peck (18 July 2016). "Theresa May warns threat of nuclear attack has increased ahead of Trident vote". *Independent*. Retrieved 18 July 2016.

[66] "A guide to Trident and the debate about replacement". *BBC News*. BBC. 18 July 2016. Retrieved 20 July 2016.

[67] Norton-Taylor, Richard; Scruton, Paul (16 July 2016). "Trident: what you need to know before the parliamentary vote". *The Guardian*. The Guardian. Retrieved 20 July 2016. Parliament will decide on Monday if the UK's nuclear submarine fleet will be replaced at an estimated cost of £41bn

[68] Allison, George (18 July 2016). "British parliament votes to renew Trident". *UK Defence Journal*. UK Defence Journal. Retrieved 20 July 2016. The Successor class is the proposed replacement for the Vanguard class ballistic missile submarines. They will carry Trident D-5 missiles, the vehicle for delivering the UK's nuclear weapons.

[69] Nuclear Weapons – Mr. Hanley, Hansard, 22 June 1993 : Column 154

[70] Operational Selection Policy OSP 11, Nuclear Weapons Policy 1967–1998, The National Archives, November 2005

[71] Lord Garden, Lords Hansard, 24 January 2007 : Column 1179

[72] Lance Chronology, Redstone Arsenal Historical Information

[73] P G Barry (October 2001). *Deployment of Lance, 50th Missile Regiment Royal Artillery*. Royal Artillery. Retrieved 10 November 2008.

[74] *Honest John*. Windscreen, Military Vehicle Trust. Summer 2006. Retrieved 10 November 2008.

[75] Norris, Burrows, Fieldhouse. Nuclear Weapons Databook Vol 5. British, French and Chinese Nuclear Weapons, p63. Published Westview Press, Oxford, 1994. ISBN 0-8133-1611-1

[76] John Clearwater (1998). *Canadian Nuclear Weapons: The Untold Story of Canada's Cold War Arsenal*. Dundurn Press Ltd. p. 284. ISBN 1-55002-299-7. Retrieved 10 November 2008.

[77] Hans M. Kristensen (February 2005). *U.S. Nuclear Weapons in Europe* (PDF). Natural Resources Defense Council. Retrieved 23 May 2006.

[78] Hans M. Kristensen (4 August 2010). "NATO Nuclear Weapons Policy: Mission and Forces at the Crossroads" (PDF). Federation of American Scientists. Retrieved 23 April 2012.

[79] Hans M. Kristensen (February 1978). *History of the Custody and Deployment of Nuclear Weapons: July 1945 through September 1977*. US Department of Defence. Retrieved 2006-05-23.

[80] How Nuclear Weapons Decisions are Made, Scilla McClean (ed), Oxford Research Group, 1984, [ISBN 0-333-40583-8], p. 120

[81] Cocroft, Wayne D. (2000). *Dangerous Energy: The archaeology of gunpowder and military explosives manufacture*. Swindon; English Heritage. ISBN 1-85074-718-0.

[82] PRO, London, DEFE 25/335, E93 classified until 2010, obtained Jan 2006 using the FOI Act.

[83] PRO, London, T225/3280, E32.

[84] Kate Pyne, The AWRE Contribution to Chevaline. Proceedings of the Royal Aeronautical Society Symposium on Chevaline, 2004. Published 2005 as ISBN 1-85768-109-6

[85] PRO, London, DEFE 25/335, E44 Annex A, 'Unacceptable Damage', plus maps. Classified until 2010, and obtained Jan 2006 using the FOI Act.

[86] British Nuclear Doctrine: The 'Moscow Criterion' and the Polaris Improvement Programme, John Baylis, Contemporary British History, Vol. 19, No. 1, Spring 2005, pp.53–65

[87] http://www.mod.uk/NR/rdonlyes/ 65F3D7AC-4340-4119-93A2-20825848E50E/0/ sdr1998_complete.pdf

[88] http://www.mod.uk/NR/rdonlyre/ 65F3D7AC-4340-4119-93A2-20825848E/0?sdr1998_ complete.pdf

[89] https://www.gov.uk/government/uploads/ system/uploads/attachment_data/file/62482/ strategic-defence-security-review.pdf page 38

[90] "Nuclear Deterrent:Written statement - HCWS210". Retrieved 1 August 2016.

[91] Ministry of Defence (UK) (15 November 2007). "Nuclear weapons security – MoD statement". BBC Newsnight. Retrieved 19 May 2008.

[92] British nuclear weapon control (streaming video), Susan Watts, BBC Newsnight, November 2007

[93] "BBC press release November 2007". Bbc.co.uk. 15 November 2007. Retrieved 30 July 2010.

[94] Knight, Richard (2 December 2008). "Whose hand is on the button?". *BBC News*. Retrieved 20 May 2010.

[95] Rosenbaum, Ron (January 2009). "The Letter of Last Resort". *Slate Magazine*. Retrieved 18 May 2009. In the control room of the sub, the Daily Mail reports, "there is a safe attached to a control room floor. Inside that, there is an inner safe. And inside that sits a letter. It is addressed to the submarine commander and it is from the Prime Minister.

[96] "Brown backs Trident replacement". *BBC News*. 21 June 2006. Retrieved 20 May 2010.

[97] "Theresa May calls for urgent go-ahead on Trident replacement". *The Telegraph*. Retrieved 16 July 2016.

[98] Knight, Richard (2 December 2008). "Finger on the nuclear button". *BBC News*. Retrieved 20 May 2010.

[99] Peter Hennessy. The Secret State: Whitehall and the Cold War, 1945–1970. Allen Lane, The Penguin Press. 256 pages. ISBN 0-7139-9626-9

[100] US-UK nuclear weapons collaboration under the Mutual Defence Agreement, Nigel Chamberlain, Nicola Butler and Dave Andrews, British American Security Council, June 2004

[101] "Trident: the done deal". New Statesman. Retrieved 30 July 2010.

[102] MacAskill, Ewen (2016-07-17). "Trident renewal: would £205bn be a price worth paying?". *the Guardian*. Retrieved 2016-07-30.

[103] Memoranda on the Future of the UK's Strategic Nuclear Deterrent: the White Paper to the House of Commons Defence Committee

[104] The Future of the United Kingdom's Nuclear Deterrent(pdf) December 2006:

[105] "Britain's Nuclear Deterrent by Conference on Disarmament-UK Permanent Representative to the Conference on Disarmament". Retrieved 1 August 2016.

[106] "Trident". Hansard. 14 March 2007.

[107] "Trident plan wins Commons support". BBC NEWS. 15 March 2007.

[108] "Blair wins Trident vote after telling UK Parliament that the NPT gives Britain the Right to have nuclear weapons". Disarmament Diplomacy. Spring 2007.

[109] Norton-Taylor, Richard (20 December 2005). "Use of Trident 'would be illegal'". London: The Guardian.

[110] "Legality of the threat or use of nuclear weapons" (PDF). International Court of Justice. 8 July 1996.

[111] "Q&A: Nuclear disarmament". BBC NEWS. 11 December 2006.

[112] United Nations General Assembly Session 53 *Resolution 77*.*Operative Paragraph 17, Resolution Y* **A/RES/53/77** page 42. Retrieved 11 March 2008.

[113] United Nations General Assembly Session 53 *Verbotim Report 79*.**A/53/PV.79** page 27. 4 December 1998. Retrieved 11 March 2008.

[114] "Account Suspended". Retrieved 1 August 2016.

[115] Singh, Rabinder; and Chinkin, Christine; The Maintenance and Possible Replacement of the Trident Nuclear Missile System Introduction and Summary of Advice for Peacerights (paragraph 1 and 2)

[116] Greenpeace Trident replacement may be illegal under international law

[117] Sands, Philippe; and Law, Helen; The United Kingdom's nuclear deterrent:Current and future issies of legality (see References)

[118] Sands, Philippe; and Law, Helen; References, paragraph 1

[119] Sands, Philippe; and Law, Helen; References, paragraph 4(i)

[120] Sands, Philippe; and Law, Helen; References, paragraph 4(iii)

[121] Sands, Philippe; and Law, Helen; References, paragraph 4(iv)

31.12   Further reading

- Arnold, Lorna (2001). *Britain and the H-Bomb. The official history up to the 1958 US-UK Mutual Defence Agreement.* Copyright MoD. Published Palgrave. ISBN 0-312-23518-6 in North America, ISBN 0-333-94742-8 outside North America.

- Blakeway, Denys, and Sue Lloyd-Roberts (1985). *Fields of Thunder: Testing Britain's Bomb* London, Unwin.

- Gill, David James. *Britain and the Bomb: Nuclear Diplomacy, 1964-1970* (Stanford University Press, 2014) 304pp online review

- Gowing, Margaret and Arnold, Lorna (1974). *Independence and Deterrence: Britain and Atomic Energy, 1945–1952. Volume 1: Policy Making* London: The Macmillan Press. ISBN 0-333-15781-8.

- Gowing, Margaret and Arnold, Lorna (1974). *Independence and Deterrence: Britain and Atomic Energy, 1945–1952. Volume 2: Policy Execution* London: The Macmillan Press. ISBN 0-333-16695-7.

- Hicks, George and Roy Dommet, "History of the RAE [Farnborough] and Nuclear Weapons". *Prospero*, refereed journal of the BROHP, Spring 2005.

- Rogers, Paul. "Possible Nuclear Attack Scenarios on Britain", *Proceedings of the Conference on Nuclear Deterrence, Implications and Policy Options for the 1980s*, International Standing Conference on Conflict and Peace Studies, London, 1982.

- Roy Dommett, "The Blue Streak Weapon". *Prospero*, refereed journal of the BROHP, Spring 2005.

- Proceedings of the Royal Aeronautical Society, Symposium on Chevaline 2004, ISBN 1-85768-109-6. See note on sources at Talk:Nuclear weapons and the United Kingdom

- Dr Peter Jones, Director AWE (Ret), "The Chevaline Technical Programme". *Prospero*, the refereed journal of the BROHP, Spring 2005.

- Peter Nailor, *The Nassau Connection: the organisation and management of the British POLARIS project*, London: H.M.S.O, (1988).

- Wynne, Humphrey (1997). *RAF Strategic Nuclear Deterrent Forces, their origins, roles and deployment, 1946–69. The documentary history.* Copyright MoD. Published by The Stationery Office. ISBN 0-11-772833-0.

- Dr Frank Panton, "The Unveiling of Chevaline". *Prospero*, the refereed journal of the BROHP, Spring 2005.

- Dr Frank Panton, "Polaris Improvements and the Chevaline Programme". *Prospero*, the refereed journal of the BROHP, Spring 2004.

31.13   External links

- British Nuclear Weapons Stockpile, 1953-2013 at History in Pieces

- Video archive of the UK's Nuclear Testing at sonicbomb.com

- British Nuclear Policy, BASIC

- Table of UK Nuclear Weapons models

- Trident: the done deal, Robert Fox, *New Statesman*, 13 June 2005

- Text of the Nuclear Explosions (Prohibition and Inspections) Act 1998

- Nuclear Notebook: British nuclear forces, 2001, *Bulletin of the Atomic Scientists*, Nov/Dec 2001.

- The United Kingdom's Defence Nuclear Programme, UK Ministry of Defence, 4 September 2001

- British Nuclear Forces, 2005, by Robert S. Norris and Hans M. Kristensen. *Bulletin of the Atomic Scientists*, November/December 2005.

- Britain's secret nuclear blueprint *The Sunday Times*, 12 March 2006

- Revealed: UK develops secret nuclear warhead by Michael Smith, *The Sunday Times*, 12 March 2006

- Government White Paper Cm 6994 *The Future of the United Kingdom's Nuclear Deterrent* (December 2006)

- Richard Moore (March 2004). *The Real Meaning of the Words: a Pedantic Glossary of British Nuclear Weapons* (PDF). Mountbatten Centre for International Studies, University of Southampton.

- http://www.nuclear-weapons.info/

- Annotated bibliography for the British nuclear weapons program from the Alsos Digital Library for Nuclear Issues

Chapter 32

Russia and weapons of mass destruction

According to the Federation of American Scientists, an organization that assesses nuclear weapon stockpiles, as of 2016, Russia possesses 7,300 total nuclear warheads, of which 1,790 are strategically operational.[2] This is in large part due to the special bomber counting rules allowed by the treaty which counts each strategic nuclear bomber as one warhead irrespective of the number of warheads—gravity bombs and/or cruise missiles carried by the aircraft. The figures are, by necessity, only estimates because "the exact number of nuclear weapons in each country's possession is a closely held national secret."[4] In addition to nuclear weapons, Russia declared an arsenal of 39,967 tons of chemical weapons in 1997,[5] of which 57% have been destroyed.[6][7] The Soviet Union ratified the Geneva Protocol on April 5, 1928 with reservations. The reservations were later dropped on January 18, 2001. Russia is also party to the Biological Weapons Convention and the Chemical Weapons Convention. The Soviet Union had a peak stockpile of 45,000 nuclear warheads in 1988.[8] It is estimated that from 1949 to 1991 the Soviet Union produced approximately 55,000 nuclear warheads.[9]

32.1 Nuclear weapons

32.1.1 History

Soviet era

Main article: Soviet atomic bomb project

Post-Soviet era

At the dissolution of the Soviet Union in 1991, Soviet nuclear weapons were deployed in four of the new republics: Russia, Ukraine, Belarus and Kazakhstan. In May 1992, these four states signed the Lisbon Protocol, agreeing to join the Treaty on the Non-Proliferation of Nuclear Weapons,

with Russia the successor to the Soviet Union as a nuclear state, and the other three states joining as non-nuclear states.

Ukraine agreed to give up its weapons to Russia, in exchange for guarantees of Ukrainian territory from Russia, the UK and the USA, known as the Budapest Memorandum on Security Assurances. China and France also made statements in support of the memorandum.[10]

32.1.2 Nuclear arsenal of Russia

The exact number of nuclear warheads is a state secret and is therefore a matter of guesswork. The Federation of American Scientists estimates that Russia possesses 4,490 nuclear warheads, while the U.S. has 4,500; Russia has 1,790 active strategic nuclear warheads, compared with 1,750.[2] According to 2016 data from the New START Treaty Aggregate Numbers of Strategic Offensive Arms facts sheet, the United States has fewer operationally deployed strategic warheads than Russia.[11] On the other hand, the Russia is estimated to have roughly 1,500 tactical nuclear weapons, all of which are declared to be in central storage.[12]

32.1.3 Nuclear weapons in Russian military doctrine

Main article: Military doctrine of Russia

According to a Russian military doctrine stated in 2010, nuclear weapons could be used by Russia "in response to the use of nuclear and other types of weapons of mass destruction against it or its allies, and also in case of aggression against Russia with the use of conventional weapons when the very existence of the state is threatened".[13]

32.1.4   Nuclear proliferation

After the Korean War, the Soviet Union transferred nuclear technology and weapons to the People's Republic of China as an adversary of the United States and NATO. According to Ion Mihai Pacepa, "Khrushchev's nuclear-proliferation process started with Communist China in April 1955, when the new ruler in the Kremlin consented to supply Beijing a sample atomic bomb and to help with its mass production. Subsequently, the Soviet Union built all the essentials of China's new military nuclear industry."[14]

Russia is one of the five "Nuclear Weapons States" (NWS) under the Nuclear Non-Proliferation Treaty (NPT), which Russia ratified (as the Soviet Union) in 1968.

Following the dissolution of the Soviet Union in 1991, a number of Soviet-era nuclear warheads remained on the territories of Belarus, Ukraine, and Kazakhstan. Under the terms of the Lisbon Protocol to the NPT, and following the 1995 Trilateral Agreement between Russia, Belarus, and the USA, these were transferred to Russia, leaving Russia as the sole inheritor of the Soviet nuclear arsenal. It is estimated that the Soviet Union had approximately 45,000 nuclear weapons stockpiled at the time of its collapse.

The collapse of the Soviet Union allowed for a warming of relations with NATO. Fears of a nuclear holocaust lessened. In September 1997, the former secretary of the Russian Security Council Alexander Lebed claimed 100 "suitcase sized" nuclear weapons were unaccounted for. He said he was attempting to inventory the weapons when he was fired by President Boris Yeltsin in October 1996.[15] In 2005, Sergey Sinchenko, a legislator from the Yulia Tymoshenko Bloc, said 250 nuclear weapons were unaccounted for. When comparing documents of nuclear weapons transferred from Ukraine to weapons received by Russia, there was a 250-weapon discrepancy.[16] Indeed, several US politicians have expressed worries and promised legislation addressing the threat.[17]

In 2002, the United States and Russia agreed to reduce their stockpiles to not more than 2,200 warheads each in the SORT treaty. In 2003, the US rejected Russian proposals to further reduce each nation's nuclear stockpiles to 1,500. Russia, in turn, refused to discuss reduction of tactical nuclear weapons.[18]

Russia is actively producing and developing new nuclear weapons. Since 1997 it manufactures Topol-M (SS-27) ICBMs.

There were allegations that Russia contributed to North Korean nuclear program, selling it the equipment for the safe storage and transportation of nuclear materials.[19] Nevertheless, Russia condemned Korean nuclear tests since then.[20]

According to high-ranking Russian SVR defector Sergei Tretyakov, a businessman told him that he keeps his own nuclear bomb at his dacha outside Moscow.[21]

32.1.5   Nuclear sabotage allegations from Russia

The highest-ranking GRU defector Stanislav Lunev described alleged Soviet plans for using tactical nuclear weapons for sabotage against the United States in the event of war. He described Soviet-made suitcase nukes identified as RA-115s (or RA-115-01s for submersible weapons) which weigh from fifty to sixty pounds. These portable bombs can last for many years if wired to an electric source. "In case there is a loss of power, there is a battery backup. If the battery runs low, the weapon has a transmitter that sends a coded message – either by satellite or directly to a GRU post at a Russian embassy or consulate.".[22]

Lunev was personally looking for hiding places for weapons caches in the Shenandoah Valley area.[22] He said that "it is surprisingly easy to smuggle nuclear weapons into the US" either across the Mexican border or using a small transport missile that can slip though undetected when launched from a Russian airplane.[22] US Congressman Curt Weldon supported claims by Lunev, but "Weldon said later the FBI discredited Lunev, saying that he exaggerated things." [23] Searches of the areas identified by Lunev – who admits he never planted any weapons in the US – have been conducted, "but law-enforcement officials have never found such weapons caches, with or without portable nuclear weapons" in the US.[24]

32.2   Biological weapons

Main article: Soviet biological weapons program

Soviet program of biological weapons was initially developed by the Ministry of Defense of the Soviet Union (between 1945 and 1973).[25]

The Soviet Union signed the Biological Weapons Convention on April 10, 1972 and ratified the treaty on March 26, 1975. However, it subsequently augmented its biowarfare programs. After 1975, the program of Biological weapons was run primarily by the "civilian" Biopreparat agency, although it also included numerous facilities run by the Soviet Ministry of Defense, Ministry of Agriculture, Ministry of Chemical Industry, Ministry of Health, and Soviet Academy of Sciences.[25]

According to Ken Alibek, who was deputy-director of Biopreparat, the Soviet biological weapons agency, and who defected to the USA in 1992, weapons were developed in labs in isolated areas of the Soviet Union including mobilization facilities at Omutininsk, Penza and Pokrov and research facilities at Moscow, Stirzhi and Vladimir. These weapons were tested at several facilities most often at "Rebirth Island" (Vozrozhdeniya) in the Aral Sea by firing the weapons into the air above monkeys tied to posts, the monkeys would then be monitored to determine the effects. According to Alibek, although Soviet offensive program was officially ended in 1992, Russia may be still involved in the activities prohibited by BWC.[25]

In 1993, the story about the Sverdlovsk anthrax leak was published in Russia. The incident occurred when spores of anthrax were accidentally released from a military facility in the city of Sverdlovsk (formerly, and now again, Yekaterinburg) 1,500 km (930 mi) east of Moscow on April 2, 1979. The ensuing outbreak of the disease resulted in 94 people becoming infected, 64 of whom died over a period of six weeks.[25]

32.3 Chemical weapons

Russia signed the Chemical Weapons Convention on January 13, 1993, and ratified it on November 5, 1997. Russia declared an arsenal of 39,967 tons of chemical weapons in 1997 consisting of:

- blister agents: Lewisite, mustard, Lewisite-mustard-mix (HL)

- nerve agents: Sarin, Soman, VX

Ratification was followed by three years of inaction on chemical weapons destruction because of the August 1998 Russian financial crisis.

Russia met its treaty obligations by destroying 1% of its chemical agents by the Chemical Weapons Convention's 2002 deadline,[26] but requested technical and financial assistance and extensions on the deadlines of 2004 and 2007 due to the environmental challenges of chemical disposal. This extension procedure spelled out in the treaty has been utilized by other countries, including the United States. The extended deadline for complete destruction (April 2012) was not met.[6] As of October 2011, Russia has destroyed 57% of its stockpile. Russia also destroyed all of its declared Category 2 (10,616 MTs) and Category 3 chemicals.[7]

Russia has stored its chemical weapons (or the required chemicals) which it declared within the CWC at 8 locations: in Gorny (Saratov Oblast) (2.9% of the declared stockpile by mass) and Kambarka (Udmurt Republic) (15.9%) stockpiles already have been destroyed. In Shchuchye (Kurgan

Oblast) (13.6%), Maradykovsky (Kirov Oblast) (17.4%) and Leonidovka (Penza Oblast) (17.2%) destruction takes place, while installations are under construction in Pochep (Bryansk Oblast) (18.8%) and Kizner (Udmurt Republic) (14.2%).[5]

32.3.1 Novichok agents

Main article: Novichok agent

In addition to the chemical weapons declared under the convention, Russia is expected to be in possession of a series of nerve agents developed in the 1970s and 1980s, some of which are one order of magnitude more lethal (based on LD50 exposure testing) than VX (the agent with the lowest LD50 in the US arsenal).[27] The agents are termed Novichok (*newcomer*) agents.

32.3.2 Disposal facilities

Russia has a number of factories for destruction of its chemical weapons arsenal: Gorny in Saratov Oblast, Kambarka in Udmurtia, Leonidovka Penza Oblast, Maradykovsky in Kirov Oblast, Shchuchye in Kurgan Oblast and the latest one Pochep in the Bryansk Oblast 70 km from the border with Ukraine, built with funds from Italy in accordance with the agreement signed between the two countries.[28][29] The last Russian chemical disposal facility in Kizner, Udmurtia, was opened on December 2013.[30]

32.4 See also

- Father of all bombs

- United States and weapons of mass destruction

- Nuclear weapons and the United States

- List of Russian weaponry makers

- Defence industry of Russia

- Military doctrine of Russia

32.5 References

[1] http://www.tandfonline.com/doi/full/10.1080/00963402.2016.1170359#aHR0cDozNOUBAQDA=/

[2] http://fas.org/issues/nuclear-weapons/ status-world-nuclear-forces/

[3] http://www.tandfonline.com/doi/full/ 10.1080/00963402.2016.1170359# aHR0cDovL3d3dy50YW5kZm9ubGluZS5jb20vZG9p

[4] Federation of American Scientists :: Status of World Nuclear Forces

[5] "Russia profile". NTI.org. 2009. Retrieved 2010-09-17.

[6] Global Campaign to Destroy Chemical Weapons Passes 60 Percent Mark. OPCW. 8 July 2010 (Accessed 19 August 2010)

[7] "Opening Statement by the Director-General to the Conference of the States Parties at its Sixteenth Session". OPCW. 28 November 2011. Retrieved 1 May 2012.

[8] Robert S. Norris and Hans M. Kristensen, "Global nuclear stockpiles, 1945-2006," Bulletin of the Atomic Scientists 62, no. 4 (July/August 2006), 64-66.

[9] "Bulletin of the Atomic Scientists". Retrieved October 24, 2014.

[10] "The Budapest Memorandum and Crimea". *VOA*. Retrieved October 24, 2014.

[11] http://www.state.gov/t/avc/rls/2016/255377.htm

[12] http://www.tandfonline.com/doi/pdf/10.1080/00963402. 2016.1170359

[13] Russian military doctrine (in Russian)

[14] Tyrants and the Bomb - by Ion Mihai Pacepa, *National Review*, October 17, 2006

[15] "Russian Officials Deny Claims Of Missing Nuclear Weapons". Retrieved October 24, 2014.

[16] Russian and Ukrainian Officials Deny New Allegations That Nuclear Warheads Were Lost in the 1990s

[17] "Nuclear Dangers: Fear Increases of Terrorists Getting Hands on 'Loose' Warheads as Security Slips". October 19, 1997. Retrieved October 24, 2014.

[18] Russia's Nuclear Policy in the 21st Century Environment - analysis by Dmitri Trenin, IFRI Proliferation Papers n°13, 2005

[19] Russia secretly offered North Korea nuclear technology - by a Special Correspondent in Pyongyang and Michael Hirst, *Telegraph*, September 7, 2006.

[20] "Russia expresses serious concern over DPRK nuke issue". Retrieved October 24, 2014.

[21] Pete Earley, "Comrade J: The Untold Secrets of Russia's Master Spy in America After the End of the Cold War", Penguin Books, 2007. ISBN 978-0-399-15439-3, pages 114-121.

[22] Stanislav Lunev. *Through the Eyes of the Enemy: The Autobiography of Stanislav Lunev*, Regnery Publishing, Inc., 1998. ISBN 0-89526-390-4.

[23] Nicholas Horrock, "FBI focusing on portable nuke threat", *UPI* (20 December 2001).

[24] Steve Goldstein and Chris Mondics, "Some Weldon-backed allegations unconfirmed; Among them: A plot to crash planes into a reactor, and missing suitcase-size Soviet atomic weapons." *Philadelphia Inquirer* (15 March 2006) A7.

[25] Alibek, K. and S. Handelman. Biohazard: The Chilling True Story of the Largest Covert Biological Weapons Program in the World– Told from Inside by the Man Who Ran it. Delta (2000) ISBN 0-385-33496-6

[26] News Archived April 6, 2004, at the Wayback Machine.

[27] Tucker, J. B.; War of Nerves; Anchor Books; New York; 2006; pp 232-233.

[28] ""Russia opens new chemical weapons destruction plant", RIA Novosti. November 2010". *RIA Novosti*. Retrieved October 24, 2014.

[29] "Italy to help Russia destroy chemical weapons". *RIA Novosti*. Retrieved October 24, 2014.

[30] New Chemical Weapons Destruction Facility Opens at Kizner in the Russian Federation

32.6   External links

- Video archive of the Soviet Union's Nuclear Testing at sonicbomb.com

- New Video: A World Without Nuclear Weapons

- Abolishing Weapons of Mass Destruction: Addressing Cold War and Other Wartime Legacies in the Twenty-First Century By Mikhail S. Gorbachev

- Russia's Nuclear Policy in the 21st Century Environment - analysis by Dmitri Trenin, IFRI Proliferation Papers n°13, 2005

- Nuclear Threat Initiative on Russia by National Journal

- UK statement on the chemical weapons convention - Link is not available now

- 1999 Nuclear stockpile estimate

- Nuclear Notebook: Russian nuclear forces, 2006, Bulletin of the Atomic Scientists, March/April 2006.

- Nuclear Files.org Current information on nuclear stockpiles in Russia

- Chemical Weapons in Russia: History, Ecology, Politics by Lev Fedorov, Moscow, Center of Ecological Policy of Russia, 27 July 1994

- History of the Russian Nuclear Weapons Program

- The Arsenals of Nuclear Weapons Powers

- Nuclear pursuits, 2012

Chapter 33

China and weapons of mass destruction

The **People's Republic of China** has developed and possesses **weapons of mass destruction**, including chemical and nuclear weapons. The first of China's nuclear weapons tests took place in 1964, and its first hydrogen bomb test occurred in 1967. Tests continued until 1996, when China signed the Comprehensive Test Ban Treaty (CTBT). China has acceded to the Biological and Toxin Weapons Convention (BWC) in 1984 and ratified the Chemical Weapons Convention (CWC) in 1997.

The number of nuclear warheads in China's arsenal is a state secret and is therefore unknown. There are varying estimates of the size of China's arsenal. China is estimated by the Federation of American Scientists to have an arsenal of about 260 total warheads as of 2015, which would make it the second smallest nuclear arsenal amongst the five nuclear weapon states acknowledged by the Treaty on the Non-Proliferation of Nuclear Weapons; in terms of warheads, they are ranked 3rd in megatonnage. According to some estimates, the country could "more than double" the "number of warheads on missiles that could threaten the United States by the mid-2020s".[5]

Early in 2011, China published a defense white paper, which repeated its nuclear policies of maintaining a minimum deterrent with a no-first-use pledge. Yet China has yet to define what it means by a "minimum deterrent posture". This, together with the fact that "it is deploying four new nuclear-capable ballistic missiles, invites concern as to the scale and intention of China's nuclear upgrade".[5]

33.1   Chemical weapons

China signed the Chemical Weapons Convention (CWC) on January 13, 1993. China ratified the CWC on April 25, 1997.[6] In the official declaration submitted to the OPCW, the Chinese government declared that it had possessed a small arsenal of chemical weapons in the past but that it had destroyed it before ratifying the Convention. It has declared only three former chemical production facilities that may

have produced mustard gas, phosgene and Lewisite.[7]

China was found to have supplied Albania with a small stockpile of chemical weapons in the 1970s during the Cold War.[8]

33.2   Biological weapons

China is currently a signatory of the Biological and Toxin Weapons Convention and Chinese officials have stated that China has never engaged in biological activities with offensive military applications. However, China was reported to have had an active biological weapons program in the 1980s.[9]

Kanatjan Alibekov, former director of one of the Soviet germ-warfare programs, said that China suffered a serious accident at one of its biological weapons plants in the late 1980s. Alibekov asserted that Soviet reconnaissance satellites identified a biological weapons laboratory and plant near a site for testing nuclear warheads. The Soviets suspected that two separate epidemics of hemorrhagic fever that swept the region in the late 1980s were caused by an accident in a lab where Chinese scientists were weaponizing viral diseases.[10]

US Secretary of State Madeleine Albright expressed her concerns over possible Chinese biological weapon transfers to Iran and other nations in a letter to Senator Robert E. Bennett (R-Utah) in January 1997.[11] Albright stated that she had received reports regarding transfers of dual-use items from Chinese entities to the Iranian government which concerned her and that the United States had to encourage China to adopt comprehensive export controls to prevent assistance to Iran's alleged biological weapons program. The United States acted upon the allegations on January 16, 2002, when it imposed sanctions on three Chinese firms accused of supplying Iran with materials used in the manufacture of chemical and biological weapons. In response to this, China issued export control protocols on dual use biological technology in late 2002.[12]

33.3 Nuclear weapons

33.3.1 History

Mao Zedong decided to begin a Chinese nuclear-weapons program during the First Taiwan Strait Crisis of 1954–1955 over the Quemoy and Matsu Islands. While he did not expect to be able to match the large American nuclear arsenal, Mao believed that even a few bombs would increase China's diplomatic credibility. Construction of uranium-enrichment plants in Baotou and Lanzhou began in 1958, and a plutonium facility in Jiuquan and the Lop Nur nuclear test site by 1960. The Soviet Union provided assistance in the early Chinese program by sending advisers to help in the facilities devoted to fissile material production,[13] and in October 1957 agreed to provide a prototype bomb, missiles, and related technology. The Chinese, who preferred to import technology and components to developing them within China, exported uranium to the Soviet Union, and the Soviets sent two R-2 missiles in 1958.[14]

That year, however, Soviet leader Nikita Khruschev told Mao that he planned to discuss arms control with the United States and Britain. China was already opposed to Khruschev's post-Stalin policy of "peaceful coexistence". Although Soviet officials assured China that it was under the Soviet nuclear umbrella, the disagreements widened the emerging Sino-Soviet split. In June 1959 the two nations formally ended their agreement on military and technology cooperation,[14] and in July 1960 all Soviet assistance with the Chinese nuclear program was abruptly terminated and all Soviet technicians were withdrawn from the program.[15]

The American government under John F. Kennedy and Lyndon B. Johnson was concerned about the program and studied ways to sabotage or attack it, perhaps with the aid of Taiwan or the Soviet Union, but Khruschev was not interested. The Chinese conducted their first nuclear test, code-named 596, on 16 October 1964,[13] and acknowledged that their program would have been impossible to complete without the Soviet help.[14] China's last nuclear test was on July 29, 1996. According to the Australian Geological Survey Organization in Canberra, the yield of the 1996 test was 1–5 kilotons. This was China's 22nd underground test and 45th test overall.[16]

33.3.2 Size

China has made significant improvements in its miniaturization techniques since the 1980s. There have been accusations, notably by the Cox Commission, that this was done primarily by covertly acquiring the U.S.'s W88 nuclear warhead design as well as guided ballistic missile technology. Chinese scientists have stated that they

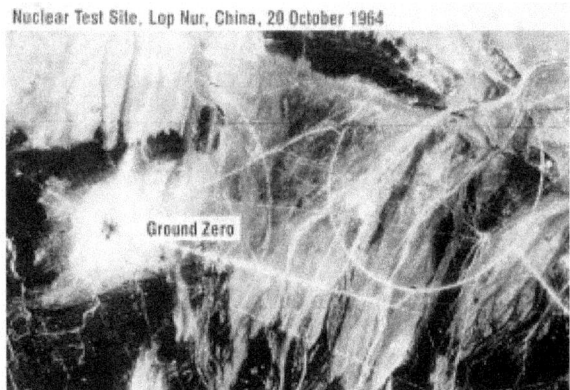

Nuclear Test Site, Lop Nur, China, 20 October 1964

Ground Zero

Satellite image of the testing site 4 days after China's first atomic bomb test

have made advances in these areas, but insist that these advances were made without espionage.

The international community has debated the size of the Chinese nuclear force since the nation first acquired such technology. Because of strict secrecy it is very difficult to determine the exact size and composition of China's nuclear forces. Estimates vary over time. Several declassified U.S. government reports give historical estimates. The 1984 Defense Intelligence Agency's Defense Estimative Brief estimates the Chinese nuclear stockpile as consisting of between 150 and 160 warheads.[17] A 1993 United States National Security Council report estimated that China's nuclear deterrent force relied on 60 to 70 nuclear armed ballistic missiles.[18] The Defense Intelligence Agency's *The Decades Ahead: 1999 - 2020* report estimates the 1999 Nuclear Weapons' Inventory as between 140 and 157.[19] In 2004 the U.S. Department of Defense assessed that China had about 20 intercontinental ballistic missiles capable of targeting the United States.[20] In 2006 a U.S. Defense Intelligence Agency estimate presented to the Senate Armed Services Committee was that "China currently has more than 100 nuclear warheads." [21]

A variety of estimates abound regarding China's current stockpile. Although the total number of nuclear weapons in the Chinese arsenal is unknown, as of 2005 estimates vary from as low as 80 to as high as 2,000. The 2,000-warhead estimate has largely been rejected by diplomats in the field. It appears to have been derived from a 1990s-era Usenet post, in which a Singaporean college student made unsubstantiated statements concerning a supposed 2,000 warhead stockpile.[22][23]

In 2004, China stated that "among the nuclear-weapon states, China... possesses the smallest nuclear arsenal," implying China has fewer than the United Kingdom's 200 nuclear weapons.[24] Several non-official sources estimate that China has around 400 nuclear warheads. However, U.S.

A mock-up of China's first nuclear bomb.

intelligence estimates suggest a much smaller nuclear force than many non-governmental organizations.[25]

In 2011, high estimates of the Chinese nuclear arsenal again emerged. One three-year study by Georgetown University raised the possibility that China had 3,000 nuclear weapons, hidden in a sophisticated tunnel network.[26] The study was based on state media footage showing tunnel entrances, and estimated a 4,800 km (3,000 mile) network. The tunnel network was revealed after the 2008 Sichuan earthquake collapsed tunnels in the hills. China has confirmed the existence of the tunnel network.[27][28] In response, the US military was ordered by law to study the possibility of this tunnel network concealing a nuclear arsenal.[29] However, the tunnel theory has come under substantial attack due to several apparent flaws in its reasoning. From a production standpoint, China probably does not have enough fissile material to produce 3,000 nuclear weapons. Such an arsenal would require 9-12 tons of Plutonium as well as 45-75 tons of enriched uranium and a substantial amount of Tritium.[30][31] The Chinese are estimated to have only 2 tons of weapons grade plutonium, which limits their arsenal to 450-600 weapons, despite a 16-ton disposable supply of uranium, theoretically enough for 1,000 warheads. Additionally, the PRC's supply of Tritium limits its stockpile to around 300 weapons.[30]

In 2012, a retired Russian officer, Viktor Yesin, stated that the Chinese arsenal was at 1,800 nuclear weapons.[32] Yesin's statements, however, have incited backlash. His

claims may have originated from the same Usenet post that previous dubious assertions of 2,000 or more nuclear warheads stemmed from.[33]

As of 2011, the Chinese nuclear arsenal was estimated to contain 55-65 ICBM's.[34]

In 2012, STRATCOM commander C. Robert Kehler said that the best estimates where "in the range of several hundred" warheads and FAS estimated the current total to be "approximately 240 warheads".[35]

The U.S. Department of Defense 2013 report to Congress on China's military developments stated that the Chinese nuclear arsenal consists of 50-75 ICBM's, located in both land-based silos and Ballistic missile submarine platforms. In addition to the ICBM's, the report stated that China has approximately 1,100 Short-range ballistic missiles, although it does not have the warhead capacity to equip them all with nuclear weapons.[36]

33.3.3 Nuclear policy

China is one of the five nuclear weapons states (NWS) recognized by the Nuclear Non-Proliferation Treaty, which China ratified in 1992. China is the only NWS[37] to give an unqualified security assurance to non-nuclear-weapon states:

> "China undertakes not to use or threaten to use nuclear weapons against non-nuclear-weapon States or nuclear-weapon-free zones at any time or under any circumstances."[38]

Chinese public policy has always been one of the "no first use rule" while maintaining a deterrent retaliatory force targeted for countervalue targets.[1]

In 2005, the Chinese Foreign Ministry released a white paper stating that the government "would not be the first to use [nuclear] weapons at any time and in any circumstance". In addition, the paper went on to state that this "no first use" policy would remain unchanged in the future and that China would not use or threaten to use nuclear weapons against any non-nuclear-weapon states or nuclear-weapon-free zones.[39]

China normally stores nuclear warheads separately from their launching systems, unless there is a heightened threat level.[40]

Historically, China has been implicated in the development of the Pakistani nuclear program. In the early 1980s, China is believed to have given Pakistan a "package" including uranium enrichment technology, high-enriched uranium, and the design for a compact nuclear weapon.[41]

Amitai Etzioni of the Institute for Communitarian Policy Studies has suggested that the prevention of nuclear proliferation could be a fruitful area of cooperation between China and the United States, by which each country could "trust but verify" the other's intentions and "help them move away from the current distrust both sides exhibit in their dealings with each other."[42]

33.3.4 Delivery systems estimates

2010 IISS Military Balance

The following are estimates of China's strategic missile forces from the International Institute of Strategic Studies *Military Balance 2010*.[43] According to these estimates, China has up to 90 inter-continental range ballistic missiles (66 land-based ICBMs and 24 submarine-based JL-2 SLBMs), not counting MIRV warheads.

2010 DoD annual PRC military report

The following are estimates from the United States Department of Defense 2010 report to Congress concerning the *Military Power of the People's Republic of China*[44]

2006 FAS & NRDC report

The following table is an overview of PRC nuclear forces taken from a November 2006 report by Hans M. Kristensen, Robert S. Norris, and Matthew G. McKinzie of the Federation of American Scientists and the Natural Resources Defense Council titled *Chinese Nuclear Forces and U.S. Nuclear War Planning*.[45]:202

Situation in 2013–14

After increasing under Bush, the number of Chinese nuclear armed missiles capable of reaching North America leveled off under Obama with delays in bringing forth new capabilities such as MIRV and operational sub launched missiles.[46] The U.S. DOD 2013 report to Congress continued to state that China had 50–75 ICBM's.[36] However the United States-China Economic and Security Review Commission stated that it was possible for China to finally have an operational Submarine-launched ballistic missile capability by the end of the year.[47] The US–China Economic and Security Review Commission stated in November 2014 that patrols with nuclear-armed submarines would take place before the end of the year, "giving China its first credible sea-based nuclear deterrent".[48]

33.3.5 Land-based intercontinental ballistic missiles

Main articles: People's Liberation Army Rocket Force, DF-4, DF-5, DF-31, and DF-41

The Dongfeng 5A is a single-warhead, three-stage, liquid-fueled missile with a range of 13,000+ km. In 2000, General Eugene Habiger of the U.S. Air Force, then-commander of the U.S. Strategic Command, testified before Congress that China has 18 silo-based DF-5s.[49] Since the early 21st century, the Second Artillery Corps have also deployed up to 10 Solid-fueled mobile DF-31 ICBMs, with a range of 7,200+ km and possibly up to 3 MIRVs.[50] China has also developed the DF-31A, an intercontinental ballistic missile with a range of 11,200+ km with possibly 3-6 multiple independently targetable reentry vehicle (MIRV) capability.

China stores many of its missiles in huge underground tunnel complexes; US Representative Michael Turner[51] referring to 2009 Chinese media reports said "This network of tunnels could be in excess of 5,000 kilometers (3,110 miles), and is used to transport nuclear weapons and forces.",[52] the Chinese Army newsletter calls this tunnel system an underground Great Wall of China.[53]

33.3.6 Medium range ballistic missiles

Approximately 55% of China's missiles are in the medium range category, targeted at regional theater targets.[45]:61

DF-3A/CSS-2

Main article: DF-3A

DF-21/CSS-5

Main article: DF-21

33.3.7 Tactical cruise missiles

The CJ-10 long-range cruise missile made its first public appearance during the military parade on the 60th Anniversary of the People's Republic of China as a part of the Second Artillery Corps' long range conventional missile forces; the CJ-10 represents the next generation in rocket weapons technology in the PLA. A similar naval cruise missile, the YJ-62, was also revealed during the parade; the

YJ-62 serves as the People's Liberation Army Navy's latest development into naval rocketry.

33.3.8 Long range ballistic missiles

The Chinese categorize long-range ballistic missiles as ones with a range between 3000 and 8000 km.[45]:103

DF-4/CSS-3

Main article: DF-4

The Dong Feng 4 or DF-4 (also known as the CSS-3) is a long-range two-stage Chinese intermediate-range ballistic missile with liquid fuel (nitric acid/UDMH). It was thought to be deployed in limited numbers in underground silos beginning in 1980.[45]:67 The DF-4 has a takeoff thrust of 1,224.00 kN, a takeoff weight of 82,000 kg, a diameter of 2.25 m, a length of 28.05 m, and a fin span of 2.74 m. It is equipped with a 2190 kg nuclear warhead with 3300 kt explosive yield, and its range is 5,500 km.[45]:68 The missile uses inertial guidance, resulting in a relatively poor CEP of 1,500 meters.

33.3.9 Intercontinental ballistic missiles (ICBMs)

DF-5A/CSS-4 Mod 2

Main article: DF-5

The Dongfeng 5 or DF-5 is a 3-stage Chinese ICBM. It has a length 32.6 m and a diameter of 3.35 m. It weighs 183 tonnes and has an estimated range of 12,000–15,000 kilometers.[45]:71-72 The DF-5 had its first flight in 1971 and was in operational service 10 years later. One of the downsides of the missile was that it took between 30 and 60 minutes to fuel.

DF-31/CSS-10

Main article: DF-31

The Dong Feng 31 (or CSS-10) is a medium-range, three stage, solid propellant intercontinental ballistic missile developed by the People's Republic of China. It is a land-based variant of the submarine-launched JL-2.

DF-41/CSS-X-10

Main article: DF-41

The DF-41 (or CSS-X-10) is an intercontinental ballistic missile believed to be under development by China. It may be designed to carry Multiple independently targetable reentry vehicles (MIRV), delivering multiple nuclear warheads.

33.3.10 Nuclear cruise missiles

The US DoD estimated in 2006 that the PRC was developing ground- and air-launched cruise missiles that could easily be converted to carry nuclear warheads once developed.[54]

DH-10

Main article: DH-10

The DongHai 10 (DH-10) is a cruise missile developed in the People's Republic of China. According to Jane's Defense Weekly, the DH-10 is a second-generation land-attack cruise missile (LACM), with over 4,000 km range, integrated inertial navigation system, GPS, terrain contour mapping system, and digital scene-matching terminal-homing system. The missile is estimated to have a circular error probable (CEP) of 10 meters.

CJ-10

Main article: CJ-10

The ChangJian-10 (Long Sword 10) is a cruise missile developed by China, based on the Hongniao missile family. It has a range of 2,200 km. Although not confirmed, it is suspected that the CJ-10 could carry nuclear warheads. An air-launched variant (named CJ-20) has also been developed.[55][56]

HongNiao missile family

Main article: Hongniao missile

There are three missiles in this family: the HN-1, HN-2, and HN-3. Reportedly based on the Kh-SD/65 missiles, the Hongniao (or Red Bird) missiles are some of the first nuclear-capable cruise missiles in China. The HN-1 has a range of 600 km, the HN-2 has a range of 1,800 km, and the HN-3 has a range of 3,000 km.[57][58][59]

ChangFeng missile family

Main article: Changfeng missile

There are 2 missiles in the Chang Feng (or Long Wind) family: CF-1 and CF-2. These are the first domestically developed long-range cruise missiles for China. The CF-1 has a range of 400 km while the CF-2 has a range of 800 km. Both variants can carry a 10 kt nuclear warhead.[57][58]

33.3.11   Sea-based weapons

Main articles: People's Liberation Army Navy, JL-1, and JL-2

The submarine-launched ballistic missile (SLBM) stockpile of the People's Liberation Army Navy (PLAN) is thought to be relatively new. China launched its first second-generation nuclear submarine in April 1981. The navy currently has a 1 Type 092 *Xia* class SSBN at roughly 8000 tons displacement. A second Type 092 was reportedly lost in an accident in 1985. The Type 092 is equipped with 12 JL-1 SLBMs with a range of 2150–2500 km. The JL-1 is a modified DF-21 missile. It is suspected that the Type 092 is being converted into a cruise missile submarine.

The Chinese navy has developed Type 094 ballistic missile submarine, open source satellite imagery has shown that at least 2 of these have been completed. This submarine will be capable of carrying 12 of the longer ranged, more modern JL-2s with a range of approximately 14000 km.

China is also developing the Type 096 submarine, claimed to be able to carry up to 24 JL-2 ballistic missiles each. Some Chinese sources states that the submarine is already undergoing trials.[60]

33.3.12   Nuclear Bomber Group

Main article: People's Liberation Army Air Force

China's bomber force consists mostly of Chinese-made versions of Soviet aircraft. The People's Liberation Army Air Force has 120 H-6s (a variant of the Tupolev Tu-16). These bombers are outfitted to carry nuclear as well as conventional weapons. While the H-6 fleet is aging, it is not as old as the American B-52 Stratofortress.[45]:93–98 The Chinese have also produced the Xian JH-7 *Flying Leopard* fighter-bomber with a range and payload exceeding the F-111 (cur-

rently about 80 are in service) capable of delivering a nuclear strike. China has also bought the advanced Sukhoi Su-30 from Russia; currently, about 100 Su-30s (MKK and MK2 variants) have been purchased by China. The Su-30 is capable of carrying tactical nuclear weapons.[45]:102

China is alleged to be testing rumored new H-8 and H-9 strategic bombers which are either described as an upgraded H-6 or an aircraft in the same class as the US B-2, able to carry nuclear weapons.[61][62][63]

33.4   Missile ranges

- Maximum Ranges for China's Conventional SRBM Force (2006). Note: China currently is capable of deploying ballistic missile forces to support a variety of regional contingencies.

- Medium and Intercontinental Range Ballistic Missiles (2007). Note: China currently is capable of targeting its nuclear forces throughout the region and most of the world, including the continental United States. Newer systems, such as the DF-31, DF-31A, and JL-2, will give China a more survivable nuclear force.

- Surface-to-Air Missile Coverage over the Taiwan Strait (2006). Note: This map depicts notional coverage provided by China's SA-10, SA-20 SAM systems, as well as the soon-to-be acquired S-300PMU2. Actual coverage would be non-contiguous and dependent upon precise deployment sites.

33.5   See also

- People's Liberation Army

- Program 863

- China space program

- List of states with nuclear weapons

- Republic of China and weapons of mass destruction

- China's nuclear test series

- Underground Great Wall of China

33.6   References

[1] "Fact Sheet:China: Nuclear Disarmament and Reduction of". Ministry of Foreign Affairs. 27 April 2004. Retrieved 2010-04-06.

[2] "China's nuclear arsenal". *BBC News*. 1999-07-15.

[3] http://bos.sagepub.com/content/71/4/77.full.pdf

[4] http://www.idsa.in/idsacomments/
DF-41ChinasanswertotheUSBMDefforts_
ArjunSubramaniam_121112

[5] Hans M. Kristensen and Robert S. Norris (November/December 2011 vol. 67 no. 6). "Chinese nuclear forces, 2011". *Bulletin of the Atomic Scientists*. pp. 81–87. Check date values in: |date= (help)

[6] States Parties to the Chemical Weapons Convention

[7] NTI Research Library: country profile: China

[8] "Albania's Chemical Cache Raises Fears About Others", *Washington Post*, 10 January 2005, page A01.

[9] Roland Everett Langford, Introduction to Weapons of Mass Destruction: Radiological, Chemical, and Biological, Wiley-IEEE, 2004

[10] William J Broad, Soviet Defector Says China Had Accident at a Germ Plant, New York Times, April 5, 1999

[11] Leonard Spector, Chinese Assistance to Iran's Weapons of Mass Destruction and Missile Programs, Carnegie Endowment for International Peace, September 12, 1996

[12] Nuclear Threat Initiative, Country Profile: China

[13] Burr, W.; Richelson, J. T. (2000–2001). "Whether to "Strangle the Baby in the Cradle": The United States and the Chinese Nuclear Program, 1960-64". *International Security*. **25** (3): 54–99. doi:10.2307/2626706 (inactive 2015-01-09). JSTOR 2626706.

[14] Jersild, Austin. "Sharing the Bomb among Friends: The Dilemmas of Sino-Soviet Strategic Cooperation". Cold War International History Project, Wilson Center. Retrieved 28 October 2013.

[15] John Lewis and Litai Xue, *China Builds the Bomb* (Stanford University Press, 1991), 53, 61, 121.

[16] "Chinese Nuclear Tests Allegedly Cause 750,000 Deaths" *Epoch Times*. March 30, 2009.

[17] http://www.gwu.edu/~{}nsarchiv/news/19990527/01-01.htm

[18] "Report to Congress on Status of China, India and Pakistan Nuclear and Ballistic Missile Programs". Fas.org. Retrieved 2010-04-06.

[19] Archived July 10, 2007, at the Wayback Machine.

[20] "1.doc" (PDF). Retrieved 2010-04-06.

[21] http://www.dia.mil/publicaffairs/Testimonies/statement24.html

[22] https://groups.google.com/forum/?fromgroups#!topic/soc.culture.taiwan/tOzuUZc1C_c

[23] http://lewis.armscontrolwonk.com/archive/4799/collected-thoughts-on-phil-karber

[24] name=MOFA-factsheet-2004>"Fact Sheet:China: Nuclear Disarmament and Reduction of". Ministry of Foreign Affairs. 27 April 2004. Retrieved 2010-04-06.

[25] "The ambiguous arsenal | thebulletin.org". Web.archive.org. Archived from the original on 2006-09-28. Retrieved 2010-04-06.

[26] https://fas.org/nuke/guide/china/Karber_UndergroundFacilities-Full_2011_reduced.pdf

[27] Fernandez, Yusuf. "Obama against Chinese Nuclear Great Wall". PressTV. Retrieved 25 March 2013.

[28] "China 'has up to 3,000 nuclear weapons hidden in tunnels', three-year study of secret documents reveals". *Daily Mail*. London. 2011-11-30. Retrieved 25 March 2013.

[29] MINNICK, WENDELL (Jan 5, 2013). "New U.S. Law Seeks Answers On Chinese Nuke Tunnels". *Defense News*. Retrieved 25 March 2013.

[30] https://fas.org/blogs/security/2011/12/chinanukes/

[31] http://www.armed-services.senate.gov/statemnt/2009/March/Maples%2003-10-09.pdf

[32] Robertson, Matthew (June 28, 2012). "Nuclear Arsenal in China Much Bigger Than Believed, Says Expert Strategists and arms control experts disagree over recent report". *Epoch Times*. Retrieved 25 March 2013.

[33] http://lewis.armscontrolwonk.com/archive/5460/yesin-on-chinas-nukes

[34] http://www.defense.gov/pubs/pdfs/2011_cmpr_final.pdf

[35] Kristensen, Hans. "STRATCOM Commander Rejects High Estimates for Chinese Nuclear Arsenal." *FAS*, 22 August 2012.

[36] Annual Report to Congress: Military and Security Developments Involving the People.s Republic of China 2013 (PDF) (Report). Office of the Secretary of Defense. 2013. Retrieved 23 October 2013.

[37] Kaegan McGrath and Vasileios Savvidis (1 February 2009). "UNSC Resolution 1887: Packaging Nonproliferation and Disarmament at the United Nations". Nuclear Threat Initiative. Retrieved 17 September 2012.

[38] "Statement on security assurances issued on 5 April 1995 by the People's Republic of China" (PDF). United Nations. 6 April 1995. S/1995/265. Retrieved 20 September 2012.

[39] "China Publishes White Paper on Arms Control". *China.org.cn*. 1 September 2005. Retrieved 15 October 2013.

[40] Hugh Chalmers (January 2014). A Disturbance in the Force (PDF) (Report). Royal United Services Institute. p. 4. Retrieved 4 February 2014.

[41] Matthew Kroenig, *Exporting the Bomb: Technology Transfer and the Spread of Nuclear Weapons* (Cornell University Press, 2010), 1.

[42] Etzioni, Amitai, "MAR: A Model for US-China Relations," The Diplomat, September 20, 2013, .

[43] IISS Military Balance 2010

[44] Office of the Secretary of Defense - Annual Report to Congress: Military Power of the People's Republic of China 2010 (PDF)

[45] Kristensen, Hans M; Robert S. Norris; Matthew G. McKinzie. *Chinese Nuclear Forces and U.S. Nuclear War Planning*. Federation of American Scientists and Natural Resources Defense Council, November 2006.

[46] Kristensen, Hans M. (19 April 2013). "Chinese ICBM Force Leveling Out?". *Strategic Security Blog*. Federation of American Scientists. Retrieved 19 April 2013.

[47] MINNICK, WENDELL (11 November 2013). "US Report: 1st Sub-launched Nuke Missile Among China's Recent Strides". *defensenews.com*. Gannett Government Media Corporation. Retrieved 11 November 2013.

[48] Tweed, David (9 December 2014). "China Takes Nuclear Weapons Underwater Where Prying Eyes Can't See". *bloomberg.com*. Retrieved 9 December 2014.

[49] Archived May 25, 2005, at the Wayback Machine.

[50] "DongFeng 31A (CSS-9) Intercontinental Ballistic Missile". SinoDefence.com. Retrieved 2010-04-06.

[51] http://chinadigitaltimes.net/2011/10/u-s-lawmaker-warns-of-chinas-nuclear-strategy

[52] straitstimes.com

[53] China Builds Underground 'Great Wall' Against Nuke Attack The Chosun Ilbo, Dec. 14, 2009.

[54] U.S. Department of Defense, Office of the Secretary of Defense, Military Power of the People's Republic of China, 2006, May 23, 2006, pp. 26, 27.

[55] "Sword —20 cruise missiles loaded on to H-6M bombers". Global Military. 2009-12-10. Archived from the original on December 11, 2010. Retrieved 2010-04-06.

[56] "□□□28□□:CJ-10 □□□□□□(□□)_□□□". Mil.huanqiu.com. Retrieved 2010-04-06.

[57] John Pike. "Land-Attack Cruise Missiles (LACM)". Globalsecurity.org. Retrieved 2010-04-06.

[58] "Land-Attack Cruise Missile (LACM)". SinoDefence.com. 2007-05-07. Retrieved 2010-04-06.

[59] "HN-2". MissileThreat. Retrieved 2010-04-06.

[60] "Global Security Newswire". NTI. Retrieved 2010-04-06.

[61] "□□□□□□□□□□□ —8□□□□□□□□□(□)_□□□□_□□□". Mil.news.sina.com.cn. Retrieved 2010-04-06.

[62] "□□□□□□□□□□□□□ —8□□□□□□_□□□□_□□□". News.xinhuanet.com. Retrieved 2010-04-06.

[63] "Google Translate". Translate.google.com. 2008-11-11. Retrieved 2010-04-06.

33.7 Further reading

- Federation of American Scientists et al. (2006). Chinese Nuclear Forces and U.S. Nuclear War Planning

- China Nuclear Forces Guide Federation of American Scientists

33.8 External links

- Archival Documents on the Chinese Nuclear Program at The Wilson Center Digital Archive

- Chinese Nuclear Weapon Testing Video at sonicbomb.com

- First nuclear test Video - 596 test

- Conference on U.S.-China Strategic Nuclear Dynamics, June 20–21, 2006

- Fact Sheet: China: Nuclear Disarmament and Reduction, Ministry of Foreign Affairs, People's Republic of China, 2004/04/27

- FY04 Report to Congress on PRC Military Power, U.S. Department of Defense

- Status of Nuclear Powers and Their Nuclear Capabilities, Federation of American Scientists

- Nuclear Threat Initiative on China

- PLA Strategic Missile Force - Chinese Defence Today

- Jeffrey Lewis, "The ambiguous arsenal", *Bulletin of the Atomic Scientists*, May/June 2005.

- Nuclear Notebook: Chinese nuclear forces, 2003, *Bulletin of the Atomic Scientists*, Nov/Dec 2003.

- Defense Estimative Brief, Nuclear Weapons Systems in China, Defense Intelligence Agency, 24 April 1984

- Report to Congress on Status of China, India and Pakistan Nuclear and Ballistic Missile Programs, United States National Security Council, July 28, 1993

- Nuclear Files.org Information on the background of nuclear weapons in China

- Nuclear Files.org Current information on nuclear stockpiles in China

- Parallel History Project On Cooperative Security, Account of Soviet-China nuclear technology transfer, October 2002

- Chinese nuclear forces, 2008, Robert S. Norris and Hans M. Kristensen, Bulletin of the Atomic Scientists

- Annotated bibliography for the Chinese nuclear weapons program from the Alsos Digital Library for Nuclear Issues

Chapter 34

India and weapons of mass destruction

India possesses weapons of mass destruction in the form of nuclear weapons and, in the past, chemical weapons. Though India has not made any official statements about the size of its nuclear arsenal, recent estimates suggest that India has 110 nuclear weapons[3][7] consistent with earlier estimates that it had produced enough weapons-grade plutonium for up to 75–110 nuclear weapons.[8] In 1999 India was estimated to have 800 kg of separated reactor-grade plutonium, with a total amount of 8300 kg of civilian plutonium, enough for approximately 1,000 nuclear weapons.[9][10] India is not a signatory to the 1968 Nuclear Non-Proliferation Treaty (NPT), which it argues entrenches the status quo of the existing nuclear weapons states whilst preventing general nuclear disarmament.[11]

India has signed and ratified the Biological Weapons Convention and the Chemical Weapons Convention. India is also a member of the Missile Technology Control Regime and a subscribing state to the Hague Code of Conduct.

34.1 Biological weapons

Further information: History of biological warfare

India has a well-developed biotechnology infrastructure that includes numerous pharmaceutical production facilities bio-containment laboratories (including BSL-3 and BSL-4) for working with lethal pathogens. It also has highly qualified scientists with expertise in infectious diseases. Some of India's facilities are being used to support research and development for biological weapons (BW) defence purposes. India has ratified the Biological Weapons Convention (BWC) and pledges to abide by its obligations. There is no clear evidence, circumstantial or otherwise, that directly points toward an offensive BW program. India does possess the scientific capability and infrastructure to launch an offensive BW program, but has chosen not to do so. In terms of delivery, India also possesses the capability to produce aerosols and has numerous potential delivery systems ranging from crop dusters to sophisticated ballistic missiles.[12]

No information exists in the public domain suggesting interest by the Indian government in delivery of biological agents by these or any other means. To reiterate the latter point, in October 2002, the then President Dr. A. P. J. Abdul Kalam asserted that "India will not make biological weapons. It is cruel to human beings".[12]

34.2 Chemical weapons

Further information: Chemical weapon

In 1992, India signed the Chemical Weapons Convention (CWC), stating that it did not have chemical weapons and the capacity or capability to manufacture chemical weapons. By doing this India became one of the original signatories of the Chemical Weapons Convention [CWC] in 1993,[13] and ratified it on 2 September 1996. According to India's ex-Army Chief General Sunderji, a country having the capability of making nuclear weapons does not need to have chemical weapons, since the dread of chemical weapons could be created only in those countries that do not have nuclear weapons. Others suggested that the fact that India has found chemical weapons dispensable highlighted its confidence in the conventional weapons system at its command.

In June 1997, India declared its stock of chemical weapons (1,045 tonnes of sulphur mustard).[14][15] By the end of 2006, India had destroyed more than 75 percent of its chemical weapons/material stockpile and was granted extension for destroying (the remaining stocks by April 2009) and was expected to achieve 100 percent destruction within that time frame.[14] India informed the United Nations in May 2009 that it had destroyed its stockpile of chemical weapons in compliance with the international Chemical Weapons Convention. With this India has become third country after South Korea and Albania to do so.[16][17] This was cross-checked by inspectors of the United Nations.

India has an advanced commercial chemical industry, and produces the bulk of its own chemicals for domestic consumption. It is also widely acknowledged that India has an extensive civilian chemical and pharmaceutical industry and annually exports considerable quantities of chemicals to countries such as the United Kingdom, United States and Taiwan.[18]

34.3 Nuclear weapons

Further information: Strategic Forces Command
As early as 26 June 1946, Jawaharlal Nehru, soon to be

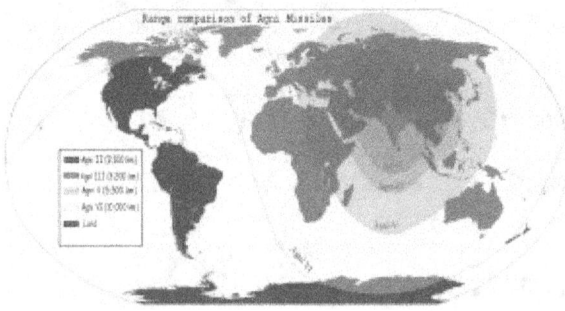

Range of Indian missiles

India's first Prime Minister, announced:

India's nuclear programme started on March 1944 and its three-stage efforts in technology were established by Dr. Homi Bhabha when he founded the nuclear research centre, the Institute of Fundamental Research.[20][21] India's loss of territory to China in a brief Himalayan border war in October 1962, provided the New Delhi government impetus for developing nuclear weapons as a means of deterring potential Chinese aggression.[22] India first tested a nuclear device in 1974 (code-named "Smiling Buddha"), which it called a "peaceful nuclear explosion." The test used plutonium produced in the Canadian-supplied CIRUS reactor, and raised concerns that nuclear technology supplied for peaceful purposes could be diverted to weapons purposes. This also stimulated the early work of the Nuclear Suppliers Group.[23] India performed further nuclear tests in 1998 (code-named "Operation Shakti"). In 1998, as a response to the continuing tests, the United States and Japan imposed sanctions on India, which have since been lifted.[24]

34.3.1 India's no-first-use policy

India has a declared nuclear no-first-use policy and is in the process of developing a nuclear doctrine based on "credible minimum deterrence." In August 1999, the Indian government released a draft of the doctrine[25] which asserts that nuclear weapons are solely for deterrence and that India will pursue a policy of "retaliation only". The document also maintains that India "will not be the first to initiate a nuclear first strike, but will respond with punitive retaliation should deterrence fail" and that decisions to authorise the use of nuclear weapons would be made by the Prime Minister or his 'designated successor(s)'.[25] According to the NRDC, despite the escalation of tensions between India and Pakistan in 2001–2002, India remained committed to its nuclear no-first-use policy.

India's Strategic Nuclear Command was formally established in 2003, with an Air Force officer, Air Marshal Asthana, as the Commander-in-Chief. The joint services SNC is the custodian of all of India's nuclear weapons, missiles and assets. It is also responsible for executing all aspects of India's nuclear policy. However, the civil leadership, in the form of the CCS (Cabinet Committee on Security) is the only body authorised to order a nuclear strike against another offending strike. The National Security Advisor Shivshankar Menon signalled a significant shift from "No first use" to "no first use against non-nuclear weapon states" in a speech on the occasion of Golden Jubilee celebrations of National Defence College in New Delhi on 21 October 2010, a doctrine Menon said reflected India's "strategic culture, with its emphasis on minimal deterrence."[26][27] In April 2013 Shyam Saran, convener of the National Security Advisory Board, affirmed that regardless of the size of a nuclear attack against India, be it a miniaturised version or a "big" missile, India will retaliate massively to inflict unacceptable damage.[28]

34.3.2 Land-based ballistic missiles

The Indian Army's Agni II missile on parade.

The land-based nuclear weapons of India are under the control of and deployed by the Indian Army, using a variety of both vehicles and launching silos. They currently consist of three different types of ballistic missiles, the Agni-I, the Agni-II, Agni-III and the Army's variant of the Prithvi missile family – the Prithvi-I. Additional variants of the Agni missile series are currently under-development, including the most recent, the Agni-IV and Agni-V, which are due to enter full operational service in the near future. Agni-VI is also under development, with an envisioned range of 8000-12000 km and features such as Multiple independently targetable reentry vehicles (MIRVs) or Maneuverable reentry vehicles (MARVs).[29][30]

34.3.3 Strategic bombing

Conceptual drawing of the INS Arihant.

The Indian Air Force's Jaguar attack aircraft are believed to have a secondary nuclear-strike role.

Surface warships such as the Shivalik class frigates (shown) may in future be equipped with the nuclear armed Dhanush ballistic missiles.

The current status of India's air-based nuclear weapons is unclear. In addition to their ground-attack role, however, it is believed that the Dassault Mirage 2000s and SEPECAT Jaguars of the Indian Air Force are able to provide a secondary nuclear-strike role.[31] The SEPECAT Jaguar was designed to be able to carry and deploy nuclear weapons and the Indian Air Force has identified the jet as being capable of delivering Indian nuclear weapons.[32] The most likely delivery method would be the use of bombs that were free-falling and unguided.[33]

34.3.4 Sea-based ballistic missiles

The Indian Navy has developed two sea-based delivery systems for nuclear weapons, completing Indian ambitions for a nuclear triad, which may have been deployed in 2015.[34][35]

The first is a submarine-launched system consisting of at least four 6,000 tonne (nuclear-powered) ballistic missile submarines of the Arihant class. The first vessel, INS *Arihant*, has been launched and will complete extensive sea-trials before being commissioned and declared operational. She is the first nuclear-powered submarine to be built by India.[36][37] A CIA report claimed that Russia provided technological aid to the naval nuclear propulsion program.[38][39] The submarines will be armed with up to 12 Sagarika (K-15) missiles armed with nuclear warheads. Sagarika is a submarine-launched ballistic missile with a range of 700 km. This missile has a length of 8.5 meters, weighs seven tonnes and can carry a pay load of up to 500 kg.[40] Sagarika has already been test-fired from an underwater pontoon, but now DRDO is planning a full-fledged test of the missile from a submarine and for this purpose may use the services of the Russian Navy.[41] India's DRDO is also working on a submarine-launched ballistic missile version of the Agni-III missile, known as the Agni-III SL. According to Indian defence sources, the Agni-III SL will have a range of 3,500 kilometres (2,200 mi).[42] The new missile will complement the older and less capable Sagarika submarine-launched ballistic missiles. However, the Arihant class ballistic missile submarines will be only capable of carrying a maximum of four Agni-III SL.

The second is a ship-launched system based around the short range ship-launched Dhanush ballistic missile (a variant of the Prithvi missile). It has a range of around 300 km. In the year 2000 the missile was test-fired from INS *Subhadra* (a Sukanya class patrol craft). INS Subhadra was modified for the test and the missile was launched from the reinforced helicopter deck. The results were considered partially successful.[43] In 2004, the missile was again tested from INS *Subhadra* and this time the results were reported successful.[44] In December 2005 the missile was tested again, but this time from the destroyer INS *Rajput*. The test was a success with the missile hitting the land based target.[45]

34.4   International response

India is not a signatory to either the Nuclear Non-Proliferation Treaty (NPT) or the Comprehensive Test Ban Treaty (CTBT), but did accede to the Partial Test Ban Treaty in October 1963. India is a member of the International Atomic Energy Agency (IAEA), and four of its 17 nuclear reactors are subject to IAEA safeguards. India announced its lack of intention to accede to the NPT as late as 1997 by voting against the paragraph of a General Assembly Resolution[48] which urged all non-signatories of the treaty to accede to it at the earliest possible date.[49] India voted against the UN General Assembly resolution endorsing the CTBT, which was adopted on 10 September 1996. India objected to the lack of provision for universal nuclear disarmament "within a time-bound framework." India also demanded that the treaty ban laboratory simulations. In addition, India opposed the provision in Article XIV of the CTBT that requires India's ratification for the treaty to enter into force, which India argued was a violation of its sovereign right to choose whether it would sign the treaty. In early February 1997, Foreign Minister I. K. Gujral reiterated India's opposition to the treaty, saying that "India favors any step aimed at destroying nuclear weapons, but considers that the treaty in its current form is not comprehensive and bans only certain types of tests."

In August 2008, the International Atomic Energy Agency (IAEA) approved safeguards agreement with India under which the former will gradually gain access to India's civilian nuclear reactors.[50] In September 2008, the Nuclear Suppliers Group granted India a waiver allowing it to access civilian nuclear technology and fuel from other countries.[51] The implementation of this waiver makes India the only known country with nuclear weapons which is not a party to the NPT but is still allowed to carry out nuclear commerce with the rest of the world.[52]

Since the implementation of the NSG waiver, India has signed nuclear deals with several countries including France,[53] United States,[54] Mongolia, Namibia,[55] Kazakhstan[56] and Australia[57] while the framework for similar deals with Canada and United Kingdom are also being prepared.[58][59]

34.5   See also

- India–United States Civil Nuclear Agreement

- Indian Armed Forces

- Weapons of mass destruction

- Nuclear Command Authority (India)

34.6   References

[1] Sachin Parashar, TNN, 28 August 2009, 12.55am IST (28 August 2009). "Kalam certifies Pokharan II, Santhanam stands his ground – India". *The Times of India*. Retrieved 31 August 2010.

[2] Carey Sublette. "What Are the Real Yields of India's Test?". *Carey Sublette*. Retrieved 12 January 2013.

[3] "Federation of American Scientists: Status of World Nuclear Forces". Fas.org. Retrieved 4 June 2013.

[4] Saran, Shyam (25 April 2013). "Is India's Nuclear Deterrent Credible? (Statement given by Shyam Saran, Chairman of India's National Security Advisory Board)". *irgamag.com*. [...] These include a modest arsenal, nuclear-capable aircraft and missiles, both in fixed underground silos as well as [...] mounted on mobile rail and road-based platforms. These land-based missiles include both Agni-II (1,500 km) as well as Agni-III (2,500 km) missiles. The range and accuracy of further versions – for example, Agni V (5,000 km), which was tested successfully only recently – will improve with the acquisition of further technological capability and experience

[5] "New chief of India's military research complex reveals brave new mandate". *India Today*. 4 July 2013. Retrieved 4 July 2013.

[6] "Strategic Forces Command fires AGNI-3 successfully". *Business Standard*. 23 December 2013. Retrieved 23 December 2013. (Second operational test firing by the Strategic Forces Command).

[7] "Pakistan has 10 more nuclear weapons than India, finds study".

[8] "Weapons around the world". physicsworld.com. Retrieved 31 August 2010.

[9] "India's Nuclear Weapons Program". *nuclearweaponarchive.org*. Retrieved 26 June 2012.

[10] "India's and Pakistan's Fissile Material and Nuclear Weapons Inventories, end of 1999". Institute for Science and International Security. Retrieved 26 June 2012.

[11] US wants India to sign NPT *Business Standard*, 7 May 2009.

[12] "Research Library: Country Profiles: India Biological Chronology". NTI. Archived from the original on 4 June 2011. Retrieved 16 July 2010.

[13] [pointer]=49

[14] "India to destroy chemical weapons stockpile by 2009". Dominican Today. Archived from the original on 7 September 2013. Retrieved 30 April 2013.

[15] Smithson, Amy Gaffney, Frank, Jr.; 700+ words. "India declares its stock of chemical weapons". Retrieved 30 April 2013.

[16] "Zee News – India destroys its chemical weapons stockpile". Zeenews.india.com. 14 May 2009. Retrieved 30 April 2013.

[17] Archived 21 May 2009 at the Wayback Machine.

[18] "Research Library: Country Profiles: India Biological Chronology". NTI. Retrieved 16 July 2010.

[19] B. M. Udgaonkar, *India's nuclear capability, her security concerns and the recent tests*, Indian Academy of Sciences, January 1999.

[20] Chengappa, Raj (2000). *Weapons of peace : the secret story of India's quest to be a nuclear power*. New Delhi: Harper Collins Publishers, India. ISBN 81-7223-330-2.

[21] et. al (30 March 2001). "The Beginning: 1944–1960". Nuclear weapon archive. Retrieved 15 January 2013.

[22] Bruce Riedel (28 June 2012). "JFK's Overshadowed Crisis". The National Interest. Retrieved 7 July 2012.

[23] Archived 1 October 2008 at the Wayback Machine.

[24] "Bush Waives Nuclear-Related Sanctions on India, Pakistan - Arms Control Association".

[25] "Draft Report of National Security Advisory Board on Indian Nuclear Doctrine". Indianembassy.org. Archived from the original on 5 December 2009. Retrieved 30 April 2013.

[26] "Speech by NSA Shri Shivshankar Menon at NDC on "The Role of Force in Strategic Affairs"". Retrieved 27 July 2015.

[27] NSA Shivshankar Menon at NDC (Speech) : india Blooms Archived 10 August 2011 at the Wayback Machine.

[28] Bagchi, Indrani. "Even a midget nuke strike will lead to massive retaliation, India warns Pak – The Economic Times". *The Economic Times*. Retrieved 30 April 2013.

[29] "Advanced Agni-6 missile with multiple warheads likely by 2017". Retrieved 1 October 2013.

[30] Subramanian, T.S. "Agni-VI all set to take shape". Retrieved 1 October 2013.

[31] Indian Nuclear Forces, 14 July 2012.

[32] India plans to impart power punch to Jaguar fighters, October 2012.

[33] "CDI Nuclear Issues Area – Nuclear Weapons Database: French Nuclear Delivery Systems". cdi.org. Retrieved 16 July 2010.

[34] Peri, Dinakar (12 June 2014). "India's Nuclear Triad Finally Coming of Age". The Diplomat. Retrieved 10 March 2015.

[35] "Nuclear triad weapons ready for deployment: DRDO".

[36] Unnithan, Sandeep (28 January 2008). "The secret undersea weapon". *India Today*. Retrieved 11 November 2012.

[37] "Indian nuclear submarine", India Today, August 2007 edition

[38] "Russia helped India's nuke programme: CIA". Press Trust of India. 9 January 2003. Retrieved 2 January 2013.

[39] "Russia helped Indian nuclear programme, says CIA". *The Dawn*. 9 January 2009. Retrieved 2 January 2013.

[40] "Sagarika missile test-fired successfully". *The Hindu*. Chennai, India. 27 February 2008. Retrieved 31 August 2010.

[41] "Coming from India's defense unit: ASTRA missile". Rediff.com. 31 December 2004. Retrieved 31 August 2010.

[42] "Agni-III test-fired successfully". Hinduonnet.com. 7 May 2008. Retrieved 31 August 2010.

[43] "Nuclear Data – Table of Indian Nuclear Forces, 2002". NRDC. Retrieved 16 July 2010.

[44] Archived 18 September 2009 at the Wayback Machine.

[45] "Dhanush, naval surface-to-surface missile, test fired successfully". domain-b.com. 31 March 2007. Retrieved 31 August 2010.

[46] "India s Dhanush Undergoes 1st Night Test - SP's Naval Forces". Retrieved 27 July 2015.

[47] Press Trust of India (25 March 2014). "India test fires long range N-missile launched from under sea". Retrieved 27 July 2015.

[48] United Nations General Assembly Session 52 *Verbatim* 67.**A/52/PV.67** 9 December 1997. Retrieved 22 August 2007.

[49] United Nations General Assembly Session 52 *Resolution* **A/RES/52/38** page 16. Retrieved 22 August 2007.

[50] "IAEA approves India nuclear inspection deal — IAEA". iaea.org. Retrieved 2 October 2008.

[51] "Nuclear Suppliers Group Grants India Historic Waiver — MarketWatch". Marketwatch.com. 6 October 2008. Archived from the original on 20 October 2008. Retrieved 2 October 2008.

[52] "AFP: India energised by nuclear pacts". Google News. Agence France-Presse. Archived from the original on 20 May 2011. Retrieved 2 October 2008.

[53] "India, France agree on civil nuclear cooperation". Rediff.com. Retrieved 16 July 2010.

[54] "Bush signs India-US nuclear deal into law – Home". livemint.com. 9 October 2008. Retrieved 16 July 2010.

[55] TNN, 15 September 2009, 02.41am IST (15 September 2009). "India, Mongolia sign civil, nuclear cooperation pact – India". The Times of India. Retrieved 16 July 2010.

[56] Sanjay Dutta, TNN, 23 January 2009, 01.35am IST (23 January 2009). "Kazakh nuclear, oil deals hang in balance – International Business – Business". The Times of India. Retrieved 16 July 2010.

[57] SUHASINI HAIDARSeptember 6, 2014 05:35 IST. "India, Australia seal civil nuclear deal - The Hindu".

[58] UK, Canada eye India's nuclear business (18 January 2009). "UK, Canada eye India's nuclear business". NDTV.com. Retrieved 16 July 2010.

[59] Sitakanta Mishra, THE PAPER (12 June 2016). "India - From 'Nuclear Apartheid' to Nuclear Multi-Alignment". IndraStra.

34.7 Further reading

Abraham, Itty (1998). *The Making of the Indian Atomic Bomb. Science, Secrecy, and the Postcolonial State*. London and New York: Zed Books. ISBN 9788125016151.

Pahuja, Om Parkash (2001). *India: A Nuclear Weapon State*. New Delhi: Ocean Books. ISBN 978-81-87100-69-0.

Perkovich, George (1999). *India's Nuclear Bomb: The Impact on Global Proliferation*. Berkeley, Los Angeles, and London: University of California Press. ISBN 978-0-520-23210-5.

Szalontai, Balázs (2011). The Elephant in the Room: The Soviet Union and India's Nuclear Program, 1967-1989. Nuclear Proliferation International History Project Working Paper #1. Washington, D.C.: Woodrow Wilson Center Press.

34.8 External links

- Indian nuclear weapons program at nuclearweaponarchive.org.

- At nuclearfiles.org:

Nuclear India's nuclear confrontation with Pakistan
Nuclear weapon stockpiles

- CIA on India's nuclear program

- India's missile testing ranges

- Video interviews taken at the 2008 NPT PrepCom on the United States-India Peaceful Atomic Energy Cooperation Act

- Annotated bibliography for India's nuclear weapons program at the Alsos Digital Library for Nuclear Issues.

- Woodrow Wilson Center's Nuclear Proliferation International History Project, including a collection of primary-source documents on Indian nuclear development.

- The National Security Archive's "Nuclear Vault" features a number of compilations of declassified US government documents related to India's nuclear program.

Chapter 35

Pakistan and weapons of mass destruction

Pakistan is one of nine states to possess nuclear weapons, and the only Muslim majority country to do so. Pakistan began development of nuclear weapons in January 1972 under Prime Minister Zulfikar Ali Bhutto, who delegated the program to the Chairman of PAEC Munir Ahmad Khan with a commitment to have the bomb ready by the end of 1976.[10][11][12] Since PAEC, consisting of over twenty laboratories and projects under nuclear engineer, Munir Ahmad Khan[13] was falling behind schedule and having considerable difficulty producing fissile material, Abdul Qadeer Khan was brought from Europe by Zulfikar Ali Bhutto at the end of 1974. As pointed out by Houston Wood, Professor of Mechanical & Aerospace Engineering, University of Virginia, Charlottesville, USA in his article on gas centrifuges, *"The most difficult step in building a nuclear weapon is the production of fissile material"*,[14][15] so this work in producing fissile material as head of the *Kahuta Project* was pivotal to Pakistan developing the capability to detonate a nuclear bomb by the end of 1984.[16][17]

The Kahuta Project started under the supervision of a coordination board that oversaw the activities of KRL and the Pakistan Atomic Energy Commission (PAEC). The Board consisted of Mr A G N Kazi (secretary general, finance), Mr Ghulam Ishaq Khan (secretary general, defence)[18] and Mr Agha Shahi (secretary general, foreign affairs) and reported directly to Prime Minister Zulfikar Ali Bhutto. Mr. Ghulam Ishaq Khan and General Tikka Khan [19] appointed military engineer, Major General Ali Nawab to the program. Eventually, the supervison passed to Lt General Zahid Ali Akbar in President General Muhammad Zia-ul-Haq's Administration. Moderate uranium enrichment for the production of fissile material was achieved at KRL by April 1978.[20]

Pakistan's nuclear weapons development was in response to the loss of East Pakistan in 1971's Bangladesh Liberation War. Bhutto called a meeting of senior scientists and engineers on 20 January 1972, in Multan, which came to known as "*Multan meeting*".[21][22] Bhutto was the main architect of this programme, and it was here that Bhutto orches-

trated nuclear weapons programme and rallied Pakistan's academic scientists to build the atomic bomb in three years for national survival.[23]

At the Multan meeting, Bhutto also appointed Munir Ahmad Khan as chairman of Pakistan Atomic Energy Commission (PAEC), who, until then, had been working as director at the nuclear power and Reactor Division of the International Atomic Energy Agency (IAEA), in Vienna, Austria. In December 1972, Abdus Salam led the establishment of Theoretical Physics Group (TPG) as he called scientists working at ICTP to report to Munir Ahmad Khan. This marked the beginning of Pakistan's pursuit of nuclear deterrence capability. Following India's surprise nuclear test, codenamed *Smiling Buddha* in 1974, the first confirmed nuclear test by a nation outside the permanent five members of the United Nations Security Council, the goal to develop nuclear weapons received considerable impetus.[24]

Finally, on 28 May 1998, a few weeks after India's second nuclear test (*Operation Shakti*), Pakistan detonated five nuclear devices in the Ras Koh Hills in the Chagai district, Balochistan. This operation was named *Chagai-I* by Pakistan, the underground iron-steel tunnel having been long-constructed by provincial martial law administrator General Rahimuddin Khan during the 1980s. The last test of Pakistan was conducted at the sandy Kharan Desert under the codename *Chagai-II*, also in Balochistan, on 30 May 1998. Pakistan's fissile material production takes place at Nilore, Kahuta, and Khushab/Jauharabad, where weapons-grade plutonium is refined. Pakistan thus became the seventh country in the world to successfully develop and test nuclear weapons.[25] Although, according to a letter sent by A.Q. Khan to General Zia, the capability to detonate a nuclear bomb using highly enriched uranium as fissile material produced at KRL had been achieved by KRL in 1984.[16][17]

35.1   History

See also: Project-706

After the Partition of India in 1947, India and Pakistan have been in conflict over several issues, including the disputed territory of Azad Jammu and Kashmir.[26] The uneasy relationships with India, Afghanistan, the former Soviet Union, and the energy shortage explains its motivation to become a nuclear power as part of its defence and energy strategies.[27]

35.1.1   Initial non-weapon policy

Main article: Nuclear energy in Pakistan

On 8 December 1953, Pakistan media welcomed the

In 1948, Mark Oliphant sent a letter to Muhammad Ali Jinnah recommending that Pakistan should start a nuclear programme.[28]

U.S. Atoms for Peace initiatives, followed by the establishment of Pakistan Atomic Energy Commission (PAEC) in 1956.[29] In 1953, Foreign minister Muhammad Zafarullah Khan publicly stated that "Pakistan does not have a policy towards the atom bombs".[30] Following the announcement, on 11 August 1955, the United States and Pakistan reached an understanding concerning the peaceful and industrial use of nuclear energy which also includes a $350,000 worth

pool-type reactor.[30] Before 1971, Pakistan's nuclear development was peaceful but an effective deterrent against India, as Benazir Bhutto maintained in 1995.[27] Pakistan's nuclear energy programme was established and started in 1956 following the establishment of PAEC. Pakistan became a participant in U.S. President Eisenhower's Atoms for Peace program. PAEC's first chairman was Dr. Nazir Ahmad. Although proposals to develop nuclear weapons were made in the 1960s by several officials and senior scientists, Pakistan followed a strict non-nuclear weapon policy from 1956 until 1971, as PAEC under its chairman Ishrat Hussain Usmani made no efforts to acquire nuclear fuel cycle for the purposes of an active nuclear weapons programme.[30]

In 1961, the PAEC set up a Mineral Center at Lahore and a similar multidisciplinary Center was set up in Dhaka, in the then East Pakistan. With these two centres, the basic research work started.

The first thing that was to be undertaken was the search for uranium. This continued for about three years from 1960 to 1963. Uranium deposits were discovered in the Dera Ghazi Khan district, and the first-ever national award was given to the PAEC. Mining of uranium began in the same year. Dr. Abdus Salam and Dr. Ishrat Hussain Usmani also sent a large number of scientists to pursue doctorate degrees in the field of nuclear technology and nuclear reactor technology. In December 1965, then-foreign minister Zulfikar Ali Bhutto visited Vienna where he met IAEA nuclear engineer, Munir Ahmad Khan. At a Vienna meeting on December, Khan informed Bhutto about the status of Indian nuclear program.

The next landmark under Dr. Abdus Salam was the establishment of PINSTECH – Pakistan Institute of Nuclear Science and Technology, at Nilore near Islamabad. The principal facility there was a 5MW research reactor, commissioned in 1965 and consisting of the PARR-I, which was upgraded to 10 MWe by Nuclear Engineering Division under Munir Ahmad Khan in 1990.[31] A second Atomic Research Reactor, known as PARR-II, was a Pool-type, light-water, 27–30 kWe, training reactor that went critical in 1989 under Munir Ahmad Khan.[32] The PARR-II reactor was built and provided by PAEC under the IAEA safeguards as IAEA had funded this mega project.[32] The PARR-I reactor was, under the agreement signed by PAEC and ANL, provided by the U.S. Government in 1965, and scientists from PAEC and ANL had led the construction.[31] Canada build Pakistan's first civil-purpose nuclear power plant. The Ayub Khan Military Government made then-science advisors to the Government Abdus Salam as the head of the IAEA delegation. Abdus Salam began lobbying for commercial nuclear power plants, and tirelessly advocated for nuclear power in Pakistan.[33] In 1965, Salam's efforts finally paid off, and a Canadian firm signed a deal to provide 137MWe CANDU reactor in Paradise Point, Karachi. The construction began in 1966 as PAEC its general contractor as GE Canada provided nuclear materials and financial assistance. Its project director was Parvez Butt, a nuclear engineer, and its construction completed in 1972. Known as *KANUPP-I*, it was inaugurated by Zulfikar Ali Bhutto as President, and began its operations in November 1972. Currently, Pakistan Government is planning to build another 400MWe commercial nuclear power plant. Having known as *KANUPP-II*, the PAEC completed its feasibility studies in 2009. However, the work is put on hold since 2009.

In 1965,[34] amidst skirmishes that led up to the Indo-Pakistani War of 1965, Zulfikar Ali Bhutto announced:

> If India builds the bomb, we will eat grass and leaves for a thousand years, even go hungry, but we will get one of our own. The Christians have the bomb, the Jews have the bomb and now the Hindus have the bomb. Why not the Muslims too have the bomb?[35][36]

In the Indo-Pakistani War of 1965, which was the second of four openly declared Indo-Pakistani wars and conflicts, Pakistan solicited Central Treaty Organization (CENTO) assistance,[37] but came under arms supply embargo in United Nations Security Council Resolution 211.[38] Foreign minister (later Prime minister) Zulfikar Ali Bhutto aggressively began the advocating the option of "nuclear weapons programmes" but such attempts were dismissed by Finance minister Muhammad Shoaib and chairman Ishrat Hussain Usmani.[30] Pakistani scientists and engineers' working at IAEA became aware of advancing Indian nuclear program towards making the bombs. Therefore, In October 1965, Munir Khan, director at the Nuclear Power and Reactor Division of the International Atomic Energy Agency (IAEA), met with Bhutto on emergency basis in Vienna, revealing the facts about the Indian nuclear programme and Bhabha Atomic Research Centre in Trombay. At this meeting Munir Khan concluded: "a (nuclear) India would further undermine and threaten Pakistan's security, and for her survival, Pakistan needed a nuclear deterrent...".

Understanding the sensitivity of the issue, Bhutto arranged a meeting with President Ayub Khan 11 December 1965 at Dorchester Hotel in London. Munir Khan pointed out to the President that Pakistan must acquire the necessary facilities that would give the country a nuclear weapon capability, which were available free of safeguards and at an affordable cost, and there were no restrictions on nuclear technology, that it was freely available, and that India was moving forward in deploying it, as Munir Khan maintained. When asked about the economics of such programme, Mu-

nir Ahmad Khan estimated the cost of nuclear technology at that time. Because things were less expensive, the then costs were not more than US$150 million. After hearing the proposal President Ayub Khan swiftly denied the proposal, saying that Pakistan was too poor to spend that much money and that, if Pakistan ever needed the atomic bomb, it could somehow acquire it off the shelf.

Pakistan's weaker conventional weapon military in comparison to India and the Indian nuclear programme that started in 1967 promped Pakistan's clandestine development of nuclear weapons.[39] Although Pakistan began the development of nuclear weapons in 1972, Pakistan responded to India's 1974 nuclear test (see *Smiling Buddha*) with a number of proposals for a nuclear-weapon-free zone to prevent a nuclear arms race in South Asia.[40] On many different occasions, India rejected the offer.[40]

In 1969, after a long negotiation, the United Kingdom Atomic Energy Authority (UKAEA) signed a formal agreement to supply Pakistan with a nuclear fuel reprocessing plant capable of extracting 360 grams (13 oz) of weapons-grade plutonium annually.[29] PAEC selected a team five senior scientists, including geophysicist Dr. Ahsan Mubarak,[29] who were sent to Sellafield to receive technical training.[29] Later Mubarak's team advised the government not to acquire the whole reprocessing plant, only key parts important to building the weapons, while the plant would be built indigenously.[29]

The PAEC in 1970 began work on a pilot-scale plant at Dera Ghazi Khan for the concentration of uranium ores. The plant had a capacity of 10,000 pounds a day.[41] In 1989, Munir Ahmad Khan signed a nuclear cooperation deal and, since 2000, Pakistan has been developing two more nuclear power plants with an agreement signed with China. Both these plants are of 300 MW capacity and are being built at Chashma city of Punjab province. The first of these, *CHASNUPP-I*, began producing electricity in 2000, and 'CHASNUPP-II', began its operation in fall of 2011. In 2011, the board of governors of International Atomic Energy Agency gave approval of Sino-Pak Nuclear Deal, allowing Pakistan legally to build the 300-MW 'CHASNUPP-III' and 'CHASNUPP-VI' reactors.[42]

35.1.2 Development of nuclear weapons

Main articles: Bangladesh liberation war, Indo-Pakistani War of 1971, and Project-706

The Indo-Pakistani War of 1971 was a crushing defeat for Pakistan, which led to it losing roughly 56,000 square miles (150,000 km^2) of territory as well as losing millions of its citizens to the newly created state of Bangladesh.[43] In addition to the psychological setback for Pakistan,[43] it had failed to gather any significant material support or assistance from its key allies, the United States and the People's Republic of China.[44][45] Pakistan seemed to be isolated internationally, and in great danger; it felt that it could rely on no one but itself.[44] Prime Minister Zulfiqar Ali Bhutto was "obsessed" with India's nuclear program.[46][47] At a United Nations Security Council meeting, Bhutto drew comparisons between the Instrument of Surrender that ended the 1971 war, and the Treaty of Versailles, which Germany was forced to sign in 1919. There, Bhutto vowed never to allow a repeat.

At the Multan meeting on 20 January 1972, Bhutto stated, "What Raziuddin Siddiqui, a Pakistani, contributed for the United States during the Manhattan Project, could also be done by scientists in Pakistan, for their own people."[48] Siddiqui was a Pakistani theoretical physicist who, in the early 1940s, worked on both the British nuclear program and the Manhattan Project.[49]

In December 1972, Dr. Abdus Salam directed a secretly coded memo to Pakistani scientists working at the International Centre for Theoretical Physics (ICTP) in Italy to report to the Chairman of the Pakistan Atomic Energy Commission (PAEC), Munir Ahmad Khan, informing them about the program what was to be equivalent of the U.S. "Manhattan Project."[50] In an effort to instill a sense of pride, Salam noted that the heads of the Manhattan Engineer District were theoreticians, and informed the scientists at ICTP that a similar division was being established at PAEC; this marked the beginning of the "Theoretical Physics Group" (TPG).[51][52] Other theoreticians at Quaid-e-Azam University would also join the TPG, then led by Salam who had done ground-breaking work for TPG.[53] Among them was Riazuddin, Fayyazuddin, Masud Ahmad, and Faheem Hussain who were the cornerstone of the TPG.[54][55]

Tedious mathematical work on fast neutron calculations, relativity, complex hydrodynamics and quantum mechanics were conducted by the TPG led by Salam until 1974 when he left Pakistan in protest, though he kept close contact with TPG.[56] No such endeavours of the kind had taken place in the country and computerized numerical control (CNC) and basic computing facilities were non-existent at that time (though later acquired).[57] For this purpose, the calculations on the high-performance computing and numerical analysis were performed by Dr. Tufail Naseem, a PhD graduate in mathematics from Cambridge University, assisted by other members of Mathematics Division– the division of pure mathematics at PAEC under Dr. Raziuddin Siddiqui and Asghar Qadir.[58] About the lack of CNC facilities, Munir Ahmad Khan famously marked: "If the Americans could do it without CNC machines in the 1940s, why can't we do the same now.".[59] With Abdus Salam de-

parting, Munir Ahmad eventually led the TPG and assisted in the calculations.[60] Two types of weapon design were analyzed: the Gun-type fission weapon and the implosion nuclear weapon.[61] The program turned to the more technically difficult implosion-type weapon design, contrary to the relatively simple 'gun-type' weapon.[62]

In 1974, Abdul Qadeer Khan a metallurgist, joined the program and pushed for the feasibility of highly enriched uranium (HEU) fissile material and collaborated under Bashiruddin Mahmood at the PAEC– a moved that irked Khan.[63] Preliminary studies on gaseous centrifuge were already studied by PAEC in 1967 but yielded few results.[64] Khan advanced uranium enrichment from the expertise he had from the Urenco Group in the Netherlands. Under Khan's supervision, the Kahuta Research Laboratories (KRL) was set-up and engaged in clandestine efforts to obtain the necessary materials technology and electronic components for its developing uranium enrichment capabilities.[65]

The TPG succeeded in the earlier implosion-type weapon design in 1977–78, with the first cold test conducted in 1983 by Ishfaq Ahmad.[67] The program evolved towards the boosted fission weapon designs that were eventually used in the Chagai-I tests in 1998.[68] Enormous production was undertaken by the Pakistan Atomic Energy Commission for feasibility of weapons grade plutonium but parallel efforts were mounted toward weapons-grade uranium after India's test, the Smiling Buddha, in 1974.[69]

In 1983, Khan was convicted in absentia by the Court of Amsterdam for stealing centrifuge blueprints, though the conviction was overturned on a legal technicality.[70] A nuclear proliferation ring was established by Khan through Dubai to smuggle URENCO nuclear technology to KRL after founding the Zippe method for the gas centrifuge [70][71][72][73][74]

On 11 March 1983, PAEC, led by Munir Ahmad Khan, carried out its first subcritical testing of a working nuclear device. This is also called a cold test, and was codenamed *Kirana-I*. There were 24 more cold tests from 1983–94.[75]

Coordination between each site was overseen by the Directorate of Technical Development (DTD) under Dr. Zaman Sheikh (a chemical engineer) and Hafeez Qureshi, a mechanical engineer.[76] The DTD was established by Munir Ahmad Khan in 1974 at the Metallurgical Laboratory and was tasked with development of tampers, reflective and explosive lenses, optics, and triggering mechanisms that are crucial in atomic weapons.[76] First implosion design was built by TPG in 1977 and the DTD eventually conducted the cold-test on 11 March 1983, codename *Kirana-I*.[76] Between 1983 and 1990, PAEC carried out 24 more cold tests of various nuclear weapon designs and shifted its focused towards tactical designs in 1987 that could be delivered by all Pakistan Air Force fighter aircraft.[77]

Dr. Ishrat Hussain Usmani's contribution to the nuclear energy programme is also fundamental to the development of atomic energy for civilian purposes as he, with efforts led by Salam, established PINSTECH, that subsequently developed into Pakistan's premier nuclear research institution.[78] In addition to sending hundreds of young Pakistanis abroad for training, he laid the foundations of the Muslim world's first nuclear power reactor KANUPP, which was inaugurated by Munir Ahmad Khan in 1972.[79] Scientists and engineers under Khan developed the nuclear capability for Pakistan within the late 1970s, and under his leadership PAEC had carried out a cold test of nuclear devices at Kirana Hills, evidently made from non-weaponized plutonium. The former chairman of PAEC, Munir Khan, was credited as one of the pioneers of Pakistan's atomic bomb by a study from the London International Institute for Strategic Studies (IISS), on Pakistan's atomic bomb program.[23]

35.1.3 Policy

Main article: Minimum Credible Deterrence (Pakistan)

Pakistan acceded to the Geneva Protocol on 15 April 1960. As for its Biological warfare capability, Pakistan is not widely suspected of either producing biological weapons or having an offensive biological programme.[80] However, the country is reported to have well developed biotechnological facilities and laboratories, devoted entirely to the medical research and applied healthcare science.[80] In 1972, Pakistan signed and ratified the Biological and Toxin Weapons Convention (BTWC) in 1974.[80] Since then Pakistan has been a vocal and staunch supporter for the success of the BTWC. During the various BTWC Review Conferences, Pakistan's representatives have urged more robust participation from state signatories, invited new states to join the treaty, and, as part of the non-aligned group of countries, have made the case for guarantees for states' rights to engage in peaceful exchanges of biological and toxin materials for purposes of scientific research.[80]

Pakistan is not known to have an offensive chemical weapons programme, and in 1993 Pakistan signed and ratified the Chemical Weapons Convention (CWC), and has committed itself to refrain from developing, manufacturing, stockpiling, or using chemical weapons.[81]

Pakistan is not a party to the Non-Proliferation Treaty (NPT) and is not bound by any of its provisions. In 1999, Prime Ministers Nawaz Sharif of Pakistan and Atal Bihari Vajpayee of India signed the *Lahore Declaration*, agreeing to a bilateral moratorium on further nuclear testing. This

initiative was taken a year after both countries had publicly tested nuclear weapons. (See Pokhran-II, Chagai-I and II)

Since the early 1980s, Pakistan's nuclear proliferation activities have not been without controversy. However, since the arrest of Abdul Qadeer Khan, the government has taken concrete steps to ensure that Nuclear proliferation is not repeated and have assured the IAEA about the transparency of Pakistan's upcoming Chashma Nuclear Power Complex series of Nuclear Power Plants. In November 2006, The International Atomic Energy Agency Board of Governors approved an agreement with the Pakistan Atomic Energy Commission to apply safeguards to new nuclear power plants to be built in the country with Chinese assistance.[82]

35.1.4 Protections

In May 1999, during the anniversary of Pakistan's first nuclear weapons test, former Prime Minister of Pakistan Nawaz Sharif claimed that Pakistan's nuclear security is the strongest in the world.[83] According to Dr. Abdul Qadeer Khan, Pakistan's nuclear safety program and nuclear security program is the strongest program in the world and there is no such capability in any other country for radical elements to steal or possess nuclear weapons.[84]

35.1.5 Modernisation and expansion

Pakistan is increasing its capacity to produce plutonium at its Khushab nuclear facility, a Washington-based science think tank has reported.[85] The sixth nuclear test (codename: Chagai-II) on 30 May 1998, at Kharan was a quiet successful test of a sophisticated, compact, but "powerful plutonium bomb" designed to be carried by aircraft, vessels, and missiles. The Pakistanis are believed to be spiking their plutonium based nuclear weapons with tritium. Only a few grams of tritium can result in an increase of the explosive yield by 300% to 400%."[86] Citing new satellite images of the facility, the Institute for Science and International Security (ISIS) said the imagery suggests construction of the second Khushab reactor is "likely finished and that the roof beams are being placed on top of the third Khushab reactor hall".[87] A third and a fourth[88] reactor and ancillary buildings are observed to be under construction at the Khushab site.

In an opinion published in *The Hindu*, former Indian Foreign Secretary Shyam Saran wrote that Pakistan's expanding nuclear capability is "no longer driven solely by its oft-cited fears of India" but by the "paranoia about U.S. attacks on its strategic assets."[89][90] Noting recent changes in Pakistan's nuclear doctrine, Saran said "the Pakistan Military and civilian elite is convinced that the United States has also become a dangerous adversary, which seeks to disable, disarm or take forcible possession of Pakistan's nuclear arsenals and its status as nuclear power."[90]

As of 2014, Pakistan has been reportedly developing smaller, more tactical nuclear weapons for potential use on the battlefield exclusively. This is consistent with earlier statements from a meeting of the National Command Authority (which directs nuclear policy and development) saying Pakistan is developing "a full-spectrum deterrence capability to deter all forms of aggression." [91]

35.1.6 Arms control proposals

Pakistan has over the years proposed a number of bilateral or regional non-proliferation steps and confidence building measures to India, including :[92]

- A joint Indo-Pakistan declaration renouncing the acquisition or manufacture of nuclear weapons, in 1978.[93]

- South Asian Nuclear Weapons Free Zone, in 1978.[94]

- Mutual inspections by India and Pakistan of each other's nuclear facilities, in 1979.[95]

- Simultaneous adherence to the NPT by India and Pakistan, in 1979.[96]

- A bilateral or regional nuclear test-ban treaty, in 1987.[97]

- A South Asia Zero-Missile Zone, in 1994.[98]

India rejected all six proposals.[99][100]

However, India and Pakistan reached three bilateral agreements on nuclear issues. In 1989, they agreed not to attack each other's nuclear facilities.[101] Since then they have been regularly exchanging lists of nuclear facilities on 1 January of each year.[102] Another bilateral agreement was signed in March 2005 where both nations would alert the other on ballistic missile tests.[103] In June 2004, the two countries signed an agreement to set up and maintain a hotline to warn each other of any accident that could be mistaken for a nuclear attack. These were deemed essential risk reduction measures in view of the seemingly unending state of misgiving and tension between the two countries, and the extremely short response time available to them to any perceived attack. None of these agreements limits the nuclear weapons programs of either country in any way.[104]

35.1.7 Disarmament policy

Pakistan has blocked negotiation of a Fissile Material Cutoff Treaty as it continues to produce fissile material for weapons.[105][106]

In a recent statement at the Conference on Disarmament, Pakistan laid out its nuclear disarmament policy and what it sees as the proper goals and requirements for meaningful negotiations:

- A commitment by all states to complete verifiable nuclear disarmament;

- Eliminate the discrimination in the current non-proliferation regime;

- Normalize the relationship of the three ex-NPT nuclear weapon states with those who are NPT signatories;

- Address new issues like access to weapons of mass destruction by non-state actors;

- Non-discriminatory rules ensuring every state's right to peaceful uses of nuclear energy;

- Universal, non-discriminatory and legally binding negative security assurances to non-nuclear weapon states;

- A need to address the issue of missiles, including development and deployment of Anti-ballistic missile systems;

- Strengthen existing international instruments to prevent the militarisation of outer space, including development of ASATs;

- Tackle the growth in armed forces and the accumulation and sophistication of conventional tactical weapons.

- Revitalise the UN disarmament machinery to address international security, disarmament and proliferation challenges.

Pakistan has repeatedly stressed at international fora like the Conference on Disarmament that it will give up its nuclear weapons only when other nuclear armed states do so, and when disarmament is universal and verifiable. It rejects any unilateral disarmament on its part.[107]

35.2 Infrastructure

35.2.1 Uranium

Pakistan's uranium infrastructure is based on the use of gas centrifuges to produce highly enriched uranium (HEU) at the Khan Research Laboratories (KRL) at Kahuta.[4] Responding to India's nuclear test in 1974, Munir Khan launched the uranium program, codename *Project-706* under the aegis of the PAEC.[108] Physical chemist, Dr. Khalil Qureshi, did most of the calculations as a member of the uranium division at PAEC, which undertook research on several methods of enrichment, including gaseous diffusion, jet nozzle and molecular laser isotope separation techniques, as well as centrifuges.[109] Abdul Qadeer Khan officially joined this program in 1976, bringing with him centrifuge designs he mastered at URENCO, the Dutch firm where he had worked as a senior scientist. Later that year, the government separated the program from PAEC and moved the program to the Engineering Research Laboratories (ERL), with A.Q. Khan as its senior scientist.[110] To acquire the necessary equipment and material for this program, Khan developed a procurement ring. Electronic materials were imported from the United Kingdom by two liaison officers posted to the High Commission of Pakistan in London and Bonn Germany.[111] The army engineer and ex-technical liaison officer, Major-General Syed Ali Nawab discreetly oversaw KRL operations in the 1970s including procuring the electronics that were marked as "common items."[111][112] This ring was also illicitly used decades later, in the late 1980s and 90s to provide technology to Libya (under Muammar Gaddafi), North Korea, and Iran.[113] Despite these efforts, it is claimed Khan Research Laboratories suffered setbacks until PAEC provided technical assistance.[114] Although, A.Q. Khan disputes it and counter claims that PAEC is merely trying to take credit for KRL's success and that PAEC hindered progress at KRL after the two programs had been separated by Bhutto in 1976.[20] In any case, KRL achieved modest enrichment of Uranium by 1978 and was ready to detonate an HEU uranium bomb by 1984. In contrast PAEC was unable to enrich any Uranium or produce weapons grade fissile material until 1998.

The uranium program proved to be a difficult, challenging and most enduring approach to scale up to industrial levels to military-grade.[115] Producing HEU as a fissile material is even more difficult and challenging than extracting plutonium and Pakistan experimented with HEU as an implosion design as contrary to other nuclear states.[116] Little and rudimentary knowledge was available of gas centrifuges at that time, and HEU fissile material was only known to the world for nuclear power usage; its military applications for HEU were non-existent.[117] Commenting on the difficulty, mathematician Tasneem Shah; who worked with A.Q. Khan, was quoted in the book *Eating Grass* that "hydrodynamical problem in centrifuge was simply stated, but extremely difficult to evaluate, not only in order of magnitude but in detailing also."[115] Many of Khan's fellow theorists were unsure about the feasibility of the enriched uranium on time despite Khan's strong advocacy.[115] One scientist recalled his memories in *Eating Grass*: "No one

in the world has used the [gas] centrifuge method to produce weapon grade material.... [T]his was not going to work, he [A.Q. Khan] is simply wasting time."[115] Despite A.Q. Khan having difficulty getting his peers to listen to him, he aggressively continued his research and the program was made feasible in the shortest time possible.[115] His efforts won him praise from Pakistan's politicians and military science circles, and he was now debuted as the "father of the uranium" bomb.[115] On 28 May 1998, it was the KRL's HEU that ultimately created the nuclear chain reaction which led the successful detonation of boosted fission devices in a scientific experiment codenamed Chagai-I.[115]

35.2.2 Plutonium

The televised screen-shot of Chagai-I *on 28 May 1998.*

As for Pakistan's plutonium capability, it has always been there, from the early 1970s onwards. However, there were only two logistic problems faced by PAEC. One was that Pakistan did not want to be an irresponsible state and the PAEC under Munir Ahmed Khan did not divert spent fuel from the safeguarded KANUPP for reprocessing at the New Labs. The second one was allocation of resources.

Alternatively, it is hard to believe that Pakistan had the engineering capability to produce plutonium for nuclear bombs in the early 70s, when it couldn't even complete the less difficult step of subcritical, cold testing until 1983 in Kirana Hills. After all, according to Houston Wood, Professor of Mechanical & Aerospace Engineering, University of Virginia, Charlottesville, USA in his article on gas centrifuges, "The most difficult step in building a nuclear weapon is the production of fissile material".[14][15] There would have been no need to bring A.Q. Khan if Pakistan had the capability to produce fissile plutonium in the early 70s. Pakistan had to bring Dr. A.Q. Khan from Europe to develop fissile uranium after PAEC failed to produce fissile plutonium by the 1976 deadline.[12]

As opposed to uranium, the parallel plutonium programme is indigenous, locally developed and culminated under the scientific directorship of PAEC chairman Munir Ahmad Khan.[24] Since 1972, earlier efforts were directed towards plutonium and necessary infrastructure was built by Bhutto as early as the 1970s.[24] Contrary to popular perception, Pakistan did not forego or abandon the plutonium program and pursued it along with the uranium route.[24] Despite many setbacks and international embargo, PAEC continued its research on plutonium and created a separated electromagnetic isotope separation program alongside the enrichment program, under Dr. G D Allam, a theoretical physicist.[24]

Towards the end of the 1970s, the PAEC began to pursue plutonium production capabilities. Consequently, Pakistan built the 40–50 MW (megawatt, thermal) Khushab Reactor Complex at Joharabad, and in April 1998, Pakistan announced that the nuclear reactor was operational. The Khushab reactor project was initiated in 1986 by Munir Khan, who informed the world that the reactor was totally indigenous, i.e. that it was designed and built by Pakistani scientists and engineers. Various Pakistani industries contributed in 82% of the reactor's construction. The Project-Director for this project was Sultan Bashiruddin Mahmood. According to public statements made by the U.S. Government officials, this heavy-water reactor can produce up to 8 to 10 kg of plutonium per year with increase in the production by the development of newer facilities,[118] sufficient for at least one nuclear weapon.[119] The reactor could also produce H^3 if it were loaded with Li^6, although this is unnecessary for the purposes of nuclear weapons, because modern nuclear weapon designs use ^6Li directly. According to J. Cirincione of Carnegie Endowment for International Peace, Khushab's Plutonium production capacity has allowed Pakistan to develop lighter nuclear warheads that would be easier to deliver to any place in the range of the ballistic missiles.

The Plutonium electromagnetic separation takes place at the *New Laboratories*, a reprocessing plant, which was completed by 1981 by PAEC and is next to the Pakistan Institute of Nuclear Science and Technology (PINSTECH) near Islamabad, which is not subject to IAEA inspections and safeguards.

In late 2006, the Institute for Science and International Security released intelligence reports and imagery showing the construction of a new plutonium reactor at the Khushab nuclear site. The reactor is deemed to be large enough to produce enough plutonium to facilitate the creation of as many as "40 to 50 nuclear weapons a year."[120][121][122] The *New York Times* carried the story with the insight that this would be Pakistan's third plutonium reactor,[123] signalling a shift

to dual-stream development, with Plutonium-based devices supplementing the nation's existing HEU stream to atomic warheads. On 30 May 1998, Pakistan proved its plutonium capability in a scientific experiment and sixth nuclear test: codename Chagai-II.[115]

35.2.3 Stockpile

Pakistani Missiles on display at the IDEAS 2008 defence exhibition in Karachi, Pakistan.

Estimates of Pakistan's stockpile of nuclear warheads vary. The most recent analysis, published in the Bulletin of the Atomic Scientists in 2010, estimates that Pakistan has 70–90 nuclear warheads.[124] In 2001, the U.S.-based Natural Resources Defense Council (NRDC) estimated that Pakistan had built 24–48 HEU-based nuclear warheads with HEU reserves for 30–52 additional warheads.[125][126] In 2003, the U.S. Navy Center for Contemporary Conflict estimated that Pakistan possessed between 35 and 95 nuclear warheads, with a median of 60.[127] In 2003, the Carnegie Endowment for International Peace estimated a stockpile of approximately 50 weapons. By contrast, in 2000, U.S. military and intelligence sources estimated that Pakistan's nuclear arsenal may be as large as 100 warheads.[128]

The actual size of Pakistan's nuclear stockpile is hard for

A truck-mounted launch system (TEL) armed with 4 Babur cruise missiles on display at the IDEAS 2008 defence exhibition in Karachi, Pakistan.

Truck-mounted Missiles on display at the IDEAS 2008 defence exhibition in Karachi, Pakistan.

experts to gauge owing to the extreme secrecy which surrounds the program in Pakistan. However, in 2007, retired Pakistan Army's Brigadier-General Feroz Khan, previously second in command at the Strategic Arms Division of Pakistans' Military told a Pakistani newspaper that Pakistan had "about 80 to 120 genuine warheads."[129][130]

Pakistan tested plutonium capability in the sixth nuclear test, codename Chagai-II, on 30 May 1998 at Kharan Desert.

The critical mass of a bare mass sphere of 90% enriched uranium-235 is 52 kg. Correspondingly, the critical mass of a bare mass sphere of plutonium-239 is 8–10 kg. The bomb that destroyed Hiroshima used 60 kg of U-235 while the Nagasaki Pu bomb used only 6 kg of Pu-239. Since all Pakistani bomb designs are implosion-type weapons, they will typically use between 15–25 kg of U-235 for their cores. Reducing the amount of U-235 in cores from 60 kg in gun-

type devices to 25 kg in implosion devices is only possible by using good neutron reflector/tamper material such as beryllium metal, which increases the weight of the bomb. And the uranium, like plutonium, is only usable in the core of a bomb in metallic form.

However, only 2–4 kg of plutonium is needed for the same device that would need 20–25 kg of U-235. Additionally, a few grams of tritium (a by-product of plutonium production reactors and thermonuclear fuel) can increase the overall yield of the bombs by a factor of three to four. "The sixth Pakistan nuclear test, codename Chagai-II, (30 May 1998) at Kharan Desert was a successful test of a sophisticated, compact, but powerful bomb designed to be carried by missiles.

Ultra-centrifugation for obtaining U-235 cannot be done simply by putting natural uranium through the centrifuges. It requires the complete mastery over the front end of the nuclear fuel cycle, beginning at uranium mining and refining, production of uranium ore or yellow cake, conversion of ore into uranium dioxide (UO$_2$) (which is used to make nuclear fuel for natural uranium reactors like Khushab and KANUPP), conversion of UO$_2$ into uranium tetrafluoride (UF$_4$) and then into the feedstock for enrichment (UF$_6$).

The complete mastery of fluorine chemistry and production of highly toxic and corrosive hydrofluoric acid and other fluorine compounds is required. The UF$_6$ is pumped into the centrifuges for enrichment. The process is then repeated in reverse until UF$_4$ is produced, leading to the production of uranium metal, the form in which U-235 is used in a bomb.

It is estimated that there are approximately 10,000–20,000 centrifuges in Kahuta. This means that with P2 machines, they would be producing between 75–100 kg of HEU since 1986, when full production of weapons-grade HEU began. Also the production of HEU was voluntarily capped by Pakistan between 1991 and 1997, and the five nuclear tests of 28 May 1998 also consumed HEU. So it is safe to assume that between 1986 and 2005 (prior to the 2005 earthquake), KRL produced 1500 kg of HEU. Accounting for losses in the production of weapons, it can be assumed that each weapon would need 20 kg of HEU; sufficient for 75 bombs as in 2005.

Pakistan's first nuclear tests were made in May 1998, when six warheads were tested under codename Chagai-I and Chagai-II. It is reported that the yields from these tests were 12 kt, 30 to 36 kt and four low-yield (below 1 kt) tests. From these tests Pakistan can be estimated to have developed operational warheads of 20 to 25 kt and 150 kt in the shape of low weight compact designs and may have 300–500 kt[131] large-size warheads. The low-yield weapons are probably in nuclear bombs carried on fighter-bombers such as the Dassault Mirage III and fitted to Pakistan's short-range ballistic missiles, while the higher-yield warheads are probably fitted to the Shaheen series and Ghauri series ballistic missiles.[131]

35.2.4 Second strike capability

According to a U.S. congressional report, Pakistan has addressed issues of survivability in a possible nuclear conflict through second strike capability. Pakistan has been dealing with efforts to develop new weapons and at the same time, have a strategy for surviving a nuclear war. Pakistan has built hard and deeply buried storage and launch facilities to retain a second strike capability in a nuclear war.[132] In January 2000, two years past after the atomic tests, U.S. intelligence officials stated that previous intelligence estimates "overstated the capabilities of India's homegrown arsenal and understate those of Pakistan".[133] The United States Central Command commander, General Anthony Zinni, a friend of Musharraf,[133] told the NBC that longtime assumptions, that "India had an edge in the South Asian strategic balance of power, were questionable at best. Don't assume that the Pakistan's nuclear capability is inferior to the Indians", General Zinni quoted to NBC.[133]

It was confirmed that Pakistan has built Soviet-style road-mobile missiles, state-of-the-art air defences around strategic sites, and other concealment measures. In 1998, Pakistan had 'at least six secret locations' and since then it is believed Pakistan may have many more such secret sites. In 2008, the United States admitted that it did not know where all of Pakistan's nuclear sites are located. Pakistani defence officials have continued to rebuff and deflect American requests for more details about the location and security of the country's nuclear sites.[134]

35.2.5 Personnel

In 2010, Russian foreign ministry official Yuriy Korolev stated that there are somewhere between 120,000 and 130,000 people directly involved in Pakistan's nuclear and missile programs, a figure considered extremely large for a developing country.[135]

35.2.6 Alleged foreign co-operation

Historically, the People's Republic of China (PRC) has been repeatedly charged with allegedly transferring missile and related materials to Pakistan.[136] Despite China strongly dismissing the charges and accusations, the United States alleged China to have played a major role in the

establishment of Pakistan's atomic bomb development infrastructure.[136] There are also unofficial reports in Western media that the nuclear weapon technology and the weapon-grade enriched uranium was transferred to Pakistan by China.[137][138] China has consistently maintained that it has not sold any weapon parts or components to Pakistan or anyone else.[136] On August 2001, it was reported that U.S. officials confronted China numerous times over this issue and pointed out "rather bluntly"[136] to Chinese officials that the evidences from intelligence sources was "powerful."[136] But they had been rebuffed by the Chinese, who have retorted by referring to the U.S. support for Taiwan's military build-up which Beijing says is directed against it.[136]

The former U.S. officials have also disclosed that China had allegedly transferred technology to Pakistan and conducting putative test for it in 1980.[139] However, senior scientists and officials strongly dismissed the U.S. disclosure, and in 1998 interview given to Kamran Khan, Abdul Qadeer Khan maintained to the fact that, "due to its sensitivity, no country allows another country to use their tests site to explode the devices," although the UK conducted such tests in Australia and the United States.[3] His statement was also traced by Samar Mubarakmand who acknowledged that cold tests were carried out, under codename Kirana-I, in a test site which was built by the Corps of Engineers under the guidance of the PAEC.[3][140] According to a 2001 Department of Defense report, China has supplied Pakistan with nuclear materials and has provided critical technical assistance in the construction of Pakistan's nuclear weapons development facilities, in violation of the Nuclear Non-Proliferation Treaty, of which China is a signatory.[141][142] In 2001 visit to India, the Chairman of the Standing Committee of the National People's Congress Li Peng rejected all the accusations against China to Indian media and strongly maintained on the ground that "his country was not giving any nuclear arms to Pakistan nor transferring related-technology to it."[143] Talking to a media correspondents and Indian parliamentarians, Li Peng frankly quoted: "We do not help Pakistan in its atomic bomb projects. Pakistan is a friendly country with whom we have good economic and political relations."[143]

In 1986, it was reported that both countries have signed a mutual treaty of peaceful use of civil nuclear technology agreement in which China would supply Pakistan a civil-purpose nuclear power plant. A grand ceremony was held in Beijing where Pakistan's then-Foreign Minister Yakub Khan signed on behalf of Pakistan in the presence of Munir Khan and Chinese Prime Minister. Therefore, in 1989, Pakistan reached agreement with China for the supply of the 300-MW commercial CHASHNUPP-1 nuclear power plant.

In February 1990, President François Mitterrand of France

visited Pakistan and announced that France had agreed to supply a 900 MWe commercial nuclear power plant to Pakistan. However, after the Prime Minister Benazir Bhutto was dismissed in August 1990, the French nuclear power plant deal went into cold storage and the agreement could not be implemented due to financial constraints and the Pakistani government's apathy. Also in February 1990, Soviet Ambassador to Pakistan, V.P. Yakunin, said that the USSR was considering a request from Pakistan for the supply of a nuclear power plant. The Soviet and French civilian nuclear power plant was on its way during the 1990s. However, Bob Oakley, the U.S. Ambassador to Pakistan, expressed U.S. displeasure at the recent agreement made between France and Pakistan for the sale of a nuclear power plant.[144] After the U.S. concerns the civilian-nuclear technology agreements were cancelled by France and Soviet Union.

Declassified documents from 1982, released in 2012 under the U.S. Freedom of Information Act, said that U.S. intelligence detected that Pakistan was seeking suspicious procurements from Belgium, Finland, Japan, Sweden and Turkey.[145]

According to more recent reports, it has been alleged that North Korea had been secretly supplying Pakistan with ballistic missile technology in exchange for nuclear weapons technology.[146]

35.2.7 Doctrine

See also: Nuclear doctrine of Pakistan

Pakistan refuses to adopt a "no-first-use" doctrine, indicating that it would strike India with nuclear weapons even if India did not use such weapons first. Pakistan's asymmetric nuclear posture has significant influence on India's decision and ability to retaliate, as shown in 2001 and 2008 crises, when non-state actors carried out deadly attacks on Indian soil, only to be met with a relatively subdued response from India. A military spokesperson stated that "Pakistan's threat of nuclear first-use deterred India from seriously considering conventional military strikes."[147] India is Pakistan's primary geographic neighbour and primary strategic competitor, helping drive Pakistan's conventional warfare capability and nuclear weapons development: The two countries share an 1800-mile border and have suffered a violent history—four wars in less than seven decades. The past three decades have seen India's economy eclipse that of Pakistan's, allowing the former to outpace the latter in defence expenditure at a decreasing share of GDP. In comparison to population, India is more powerful than Pakistan by almost every metric of military, economic, and political power—and the gap continues to grow," a Belfer Center for Science and International Affairs report claims.[148]

35.2.8 Theory of deterrence

Main articles: N-deterrence and Nuclear deterrence

The theory of "N-deterrence" has been frequently being interpreted by the various government-in-time of effect of Pakistan. Although the nuclear deterrence theory was officially adopted in 1998 as part of Pakistan's defence theory,[149] on the other hand, the theory has had been interpreted by the government since in 1972. The relative weakness in defence warfare is highlighted in Pakistan's nuclear posture, which Pakistan considers its primary deterrent from Indian conventional offensives or nuclear attack. Nuclear theorist Brigadier-General Feroz Hassan Khan adds: "The Pakistani situation is akin to NATO's position in the Cold War. There are geographic gaps and corridors similar to those that existed in Europe ... that are vulnerable to exploitation by mechanized Indian forces ... With its relatively smaller conventional force, and lacking adequate technical means, especially in early warning and surveillance, Pakistan relies on a more proactive nuclear defensive policy."[150]

Indian political scientist Vipin Narang, however, argues that Pakistan's asymmetric escalation posture, or the rapid first use of nuclear weapons against conventional attacks to deter their outbreak, increases instability in South Asia. Narang supports his arguments by noting to the fact that since India's assured retaliation nuclear posture has not deterred these provocations, Pakistan's passive nuclear posture has neutralised India's conventional options for now; limited retaliation would be militarily futile, and more significant conventional retaliation is simply off the table."[147]

The strategists in Pakistan Armed Forces has ceded nuclear assets and a degree of nuclear launch code authority to lower-level officers to ensure weapon usability in a "fog of war" scenario, making credible its deterrence doctrine.[147] On further military perspective, the Pakistan Air Force (PAF), has retrospectively contended that "theory of defense is not view to enter into a "nuclear race", but to follow a policy of "peaceful co-existence" in the region, it cannot remain oblivious to the developments in South Asia."[151] The Pakistan Government officials and strategists have consistently emphasised that nuclear deterrence is intended by maintaining a balance to safeguard its sovereignty and ensure peace in the region.[152]

Pakistan's motive for pursuing a nuclear weapons development program is never to allow another invasion of Pakistan.[153] President Muhammad Zia-ul-Haq allegedly told the Indian Prime Minister Rajiv Gandhi in 1987 that,

"If your forces cross our borders by an inch, we are going to annihilate your cities."[154]

Pakistan has not signed the Non-Proliferation Treaty (NPT) or the Comprehensive Test Ban Treaty (CTBT). According to the United States Department of Defense report cited above, "Pakistan remains steadfast in its refusal to sign the NPT, stating that it would do so only after India joined the Treaty. Pakistan has responded to the report by stating that the United States itself has not ratified the CTBT. Consequently, not all of Pakistan's nuclear facilities are under IAEA safeguards. Pakistani officials have stated that signature of the CTBT is in Pakistan's best interest, but that Pakistan will do so only after developing a domestic consensus on the issue, and have disavowed any connection with India's decision."

The Congressional Research Service, in a report published on 23 July 2012, said that in addition to expanding its nuclear arsenal, Pakistan could broaden the circumstances under which it would be willing to use nuclear weapons.[155]

35.2.9   Nuclear Command and Control

Main articles: Pakistan National Command Authority, Strategic Plans Division Force, Joint Chiefs of Staff Committee, National Security Council of Pakistan, Strategic Plans Division, Defence Committee of the Cabinet (Pakistan), Nuclear command and control, and Threat Matrix (database)

The government institutional organisation authorised to make critical decisions about Pakistan's nuclear posturing is the NCA.[156] The NCA has its genesis since the 1970s[156] and has been constitutionally established in February 2000.[156] The NCA is composed of two civic-military committees that advises and console both Prime minister and the President of Pakistan, on the development and deployment of nuclear weapons; it is also responsible for war-time command and control. In 2001, Pakistan further consolidated its nuclear weapons infrastructure by placing the Khan Research Laboratories and the Pakistan Atomic Energy Commission under the control of one Nuclear Defense Complex. In November 2009, Pakistan President Asif Ali Zardari announced that he will be replaced by Prime Minister Yusuf Raza Gilani as the chairman of NCA.[157] The NCA consists of the Employment Control Committee (ECC) and the Development Control Committee (DCC), both now chaired by the Prime Minister.[158] The Foreign minister and Economic Minister serves as a deputy chairmen of the ECC, the body which defines nuclear strategy, including the deployment and employment of strategic forces, and would advise the prime minister on nuclear use. The committee includes

key senior cabinet ministers as well as the respective military chiefs of staff.[158] The ECC reviews presentations on strategic threat perceptions, monitors the progress of weapons development, and decides on responses to emerging threats.[158] It also establishes guidelines for effective command-and-control practices to safeguard against the accidental or unauthorised use of nuclear weapons.[158]

The chairman of the Joint Chiefs of Staff Committee is the deputy chairman of the Development Control Committee (DCC), the body responsible for weapons development and oversight which includes the nation's military and scientific, but not its political, leadership.[158] Through DCC, the senior civilian scientists maintains a tight control of scientific and ethical research; the DCC exercises technical, financial and administrative control over all strategic organisations, including national laboratories and scientific research and development organisations associated with the development and modernisation of nuclear weapons and their delivery systems.[158] Functioning through the SPD, the DCC oversees the systematic progress of weapon systems to fulfil the force goals set by the committee.[158]

Under the Nuclear Command Authority, its secretariat, Strategic Plans Division (SPD), is responsible for the physical protection and to ensure security of all aspects of country's nuclear arsenals and maintains dedicated force for this purpose.[159] The SPD functions under the Joint Chiefs of Staff Committee at the Joint Headquarters (JS HQ) and reports directly to the Prime Minister.[159] The comprehensive nuclear force planning is integrated with conventional war planning at the National Security Council (NSC).[159] According to the officials of Pakistan's military science circles, it is the high-profile civic-military committee consisting the Cabinet ministers, President, Prime minister and the four services chiefs, all of whom who reserves the right to order the deployment and the operational use of the nuclear weapons.[159] The final and executive political decisions on nuclear arsenals deployments, operational use, and nuclear weapons politics are made during the sessions of the Defence Committee of the Cabinet, which is chaired by the Prime minister.[160] It is this DCC Council where the final political guideles, discussions and the nuclear arsenals operational deployments are approved by the Prime minister.[160] The DCC reaffirmed its policies on development of nuclear energy and arsenals through the country's media.[160]

35.2.10   U.S. security assistance

From the end of 2001 the United States has provided material assistance to aid Pakistan in guarding its nuclear material, warheads and laboratories. The cost of the program has been almost $100 million. Specifically the USA has provided helicopters, night-vision goggles and nuclear detection equipment.[161]

During this period Pakistan also began to develop a modern export control regulatory regime with U.S. assistance. It supplements the U.S. National Nuclear Security Administration Megaports program at Port Qasim, Karachi, which deployed radiation monitors and imaging equipment monitored by a Pakistani central alarm station.[162]

Pakistan turned down the offer of Permissive Action Link (PAL) technology, a sophisticated "weapon release" program which initiates use via specific checks and balances, possibly because it feared the secret implanting of "dead switches". But Pakistan is since believed to have developed and implemented its own version of PAL and U.S. military officials have stated they believe Pakistan's nuclear arsenals to be well secured.[163][164]

Security concerns of the United States

Since 2004 the U.S. government has reportedly been concerned about the safety of Pakistani nuclear facilities and weapons. Press reports have suggested that the United States has contingency plans to send in special forces to help "secure the Pakistani nuclear arsenal".[165][166] In 2007, Lisa Curtis of The Heritage Foundation, while giving testimony before the United States House Foreign Affairs Subcommittee on Terrorism, Nonproliferation, and Trade, concluded that "preventing Pakistan's nuclear weapons and technology from falling into the hands of terrorists should be a top priority for the U.S."[167] However Pakistan's government has ridiculed claims that the weapons are not secure.[165]

Diplomatic reports published in the United States diplomatic cables leak revealed American and British worries over a potential threat posed by Islamists. In February 2009 cable from Islamabad, former US Ambassador to Pakistan Anne W. Patterson said "Our major concern is not having an Islamic militant steal an entire weapon but rather the chance someone working in [Pakistani government] facilities could gradually smuggle enough material out to eventually make a weapon."[168]

A report published by The Times in early 2010 states that the United States is training an elite unit to recover Pakistani nuclear weapons or materials should they be seized by militants, possibly from within the Pakistani nuclear security organisation. This was done in the context of growing Anti-Americanism in the Pakistani Armed Forces, multiple attacks on sensitive installations over the previous 2 years and rising tensions. According to former U.S. intelligence official Rolf Mowatt-Larssen, U.S. concerns are justified because militants have struck at several Pakistani military facilities and bases since 2007. According to this report,

the United States does not know the locations of all Pakistani nuclear sites and has been denied access to most of them.[169] However, during a visit to Pakistan in January 2010, the U.S. Secretary of Defense Robert M. Gates denied that the United States had plans to take over Pakistan's nuclear weapons.[170]

A study by Belfer Center for Science and International Affairs at Harvard University titled 'Securing the Bomb 2010', found that Pakistan's stockpile "faces a greater threat from Islamic extremists seeking nuclear weapons than any other nuclear stockpile on earth".[171]

According to Rolf Mowatt-Larssen, a former investigator with the CIA and the U.S. Department of Energy there is "a greater possibility of a nuclear meltdown in Pakistan than anywhere else in the world. The region has more violent extremists than any other, the country is unstable, and its arsenal of nuclear weapons is expanding."[172]

Nuclear weapons expert David Albright author of 'Peddling Peril' has also expressed concerns that Pakistan's stockpile may not be secure despite assurances by both Pakistan and U.S. government. He stated Pakistan "has had many leaks from its program of classified information and sensitive nuclear equipment, and so you have to worry that it could be acquired in Pakistan."[173]

A 2010 study by the Congressional Research Service titled 'Pakistan's Nuclear Weapons: Proliferation and Security Issues' noted that even though Pakistan had taken several steps to enhance Nuclear security in recent years 'Instability in Pakistan has called the extent and durability of these reforms into question.'[174]

In April 2011, IAEA's deputy director general Denis Flory declared Pakistan's nuclear programme safe and secure.[175][176] According to the IAEA, Pakistan is currently contributing more than $1.16 million in IAEA's Nuclear Security Fund, making Pakistan as 10th largest contributor.[177]

In response to a November 2011 article in The Atlantic written by Jeffrey Goldberg highlighting concerns about the safety of Pakistan's nuclear weapons program, the Pakistani Government announced that it would train an additional 8,000 people to protect the country's nuclear arsenal. At the same time, the Pakistani Government also denounced the article. Training will be completed no later than 2013.[178]

Pakistan consistently maintains that it has tightened the security over the several years.[179] In 2010, the Chairman Joint Chiefs General Tariq Majid exhorted to the world delegation at the National Defence University that, "World must accept Pakistan as nuclear power."[179] While dismissing all the concerns on the safety of country's nuclear arsenal, General Majid maintains to the fact: "We are shouldering our responsibility with utmost vigilance and confidence.

We have put in place a very robust regime that includes "multilayered mechanisms" and processes to secure our strategic assets, and have provided maximum transparency on our practices. We have reassured the international community on this issue over and over again and our track record since the time our atomic bomb programme was made overt has been unblemished".[179]

On 7 September 2013, the U.S. State Department said "Pakistan has a professional and dedicated security force that fully understands the importance of nuclear security." Pakistan had earlier rejected claims in U.S. media that the Obama Administration was worried about the safety of Pakistani nuclear weapons, saying the country has a professional and robust system to monitor its nukes.[180]

35.2.11   National Security Council

- Economic Coordination Committee (ECC)
- Development Control Committee (DCC)
- Employment Control Committee (ECC)
- Financial Monitoring Unit (FMU)

Strategic combat commands

- Air Force Strategic Command (AFSC)
- Army Strategic Forces Command (ASFC)
- Naval Strategic Forces Command (NSFC)

35.3   Weapons development agencies

35.3.1   National Engineering & Scientific Commission (NESCOM)

- National Development Complex (NDC), Islamabad
- Project Management Organization (PMO), Khanpur
- Air Weapon Complex (AWC), Hasanabdal
- National Centre for Physics (NCP), Islamabad
- Maritime Technologies Complex (MTC), Karachi

35.3.2   Ministry of Defense Production

- Pakistan Ordnance Factories (POF), Wah
- Pakistan Aeronautical Complex (PAC), Kamra
- Defense Science and Technology Organization (DESTO), Chattar

35.3.3   Pakistan Atomic Energy Commission (PAEC)

- Directorate of Technical Development
- Directorate of Technical Equipment
- Directorate of Technical Procurement
- Directorate of Science & Engineering Services
- Institute of Nuclear Power, Islamabad
- Pakistan Institute of Nuclear Science & Technology (PINSTECH)
- New Laboratories, Rawalpindi
- Pilot Reprocessing Plant
- PARR-1 and PARR-2 Nuclear Research Reactors
- Center for Nuclear Studies (CNS), Islamabad
- Computer Training Center (CTC), Islamabad
- Nuclear Track Detection Center (Solid State Nuclear Track Detection Center)
- Khushab Reactor, Khushab
- Atomic Energy Minerals Centre, Lahore
- Hard Rock Division, Peshawar
- Mineral Sands Program, Karachi
- Baghalchur Uranium Mine, Baghalchur
- Dera Ghazi Khan Uranium Mine, Dera Ghazi Khan
- Issa Khel/Kubul Kel Uranium Mines and Mills, Mianwali
- Multan Heavy Water Production Facility, Multan, Punjab
- Uranium Conversion Facility, Islamabad
- Golra Ultracentrifuge Plant, Golra
- Sihala Ultracentrifuge Plant, Sihala
- Directorate of Quality Assurance, Islamabad
- New Labs Nilore, Islamabad

35.3.4   Ministry of Industries & Production

- State Engineering Corporation (SEC)
- Heavy Mechanical Complex Ltd. (HMC)
- Pakistan Steel Mills Limited, Karachi.

35.4 Delivery systems

35.4.1 Land

As of 2011, Pakistan possesses a wide variety of nuclear capable medium range ballistic missiles with ranges up to 2500 km.[181] Pakistan also possesses nuclear tipped Babur cruise missiles with ranges up to 700 km. In April 2012, Pakistan launched a Hatf-4 Shaheen-1A, said to be capable of carrying a nuclear warhead designed to evade missile-defense systems.[182] The Babur cruise missile range can also be extended to 1000 km or more. These land-based missiles are controlled by Army Strategic Forces Command of Pakistan Army.

Pakistan is also believed to be developing tactical nuclear weapons for use on the battlefield with ranges up to 60 km such as the Nasr missile. According to Jeffrey Lewis, director of the East Asia Non-proliferation Program at the Monterey Institute of International Studies, citing a Pakistani news article,[183] Pakistan is developing its own equivalent to the Davy Crockett launcher with miniaturised warhead that may be similar to the W54.[184]

35.4.2 Air

The Pakistan Air Force (PAF) is believed to have practised "toss-bombing" in the 1980s and 1990s, a method of launching weapons from fighter-bombers which can also be used to deliver nuclear warheads. The PAF has two dedicated units (No. 16 *Black Panthers* and No. 26 *Black Spiders*) operating 18 aircraft in each squadron (36 aircraft total) of the JF-17 Thunder, believed to be the preferred vehicle for delivery of nuclear weapons.[185] These units are major part of the Air Force Strategic Command, a command responsible for nuclear response. The PAF also operates a fleet of F-16 fighters, of which 18 were delivered in 2012 and confirmed by General Ashfaq Parvez Kayani, are capable of carrying nuclear weapons. With a third squadron being raised, this would bring the total number of dedicated nuclear capable aircraft to a total of 54.[186] The PAF also possesses the Ra'ad air-launched cruise missile which has a range of 350 km and can carry a nuclear warhead with a yield of between 10kt to 35kt.[187]

It has also been reported that an air-launched cruise missile (ALCM) with a range of 350 km has been developed by Pakistan, designated Hatf 8 and named Ra'ad ALCM, which may theoretically be armed with a nuclear warhead. It was reported to have been test-fired by a Mirage III fighter and, according to one Western official, is believed to be capable of penetrating some air defence/missile defence systems.[188]

35.4.3 Sea

The Pakistan Navy was first publicly reported to be considering deployment of nuclear weapons on submarines in February 2001. Later in 2003 it was stated by Admiral Shahid Karimullah, then Chief of Naval Staff, that there were no plans for deploying nuclear weapons on submarines but if "*forced to*" they would be. In 2004, Pakistan Navy established the Naval Strategic Forces Command and made it responsible for countering and battling naval-based weapons of mass destruction. It is believed by most experts that Pakistan is developing a sea-based variant of the Hatf VII Babur, which is a nuclear-capable ground-launched cruise missile.[189] With a stockpile of plutonium, Pakistan would be able to produce a variety of miniature nuclear warheads which would allow it to nuclear-tip the C-802 and C-803 anti-ship missiles as well as being able to develop nuclear torpedoes, nuclear depth bombs and nuclear naval mines.

Nuclear submarine

In response to INS *Arihant*, India's first nuclear submarine, the Pakistan Navy pushed forward a proposal to build its own nuclear submarine as a direct response to the Indian nuclear submarine program.[190][191] Many military experts believe that Pakistan has the capability of building a nuclear submarine and is ready to build such a fleet.[190] Finally in February 2012, the Navy announced it would start work on the construction of a nuclear submarine to better meet the Indian Navy's nuclear threat.[192] According to the Navy, the nuclear submarine is an ambitious project, and will be designed and built indigenously. However, the Navy stressed that "the project completion and trials would take anywhere from between 5 to 8 years to build the nuclear submarine after which Pakistan would join the list of countries that has a nuclear submarine."[190][197]

35.5 See also

- Chronology of Pakistan's rocket tests

- Nuclear power in Pakistan

- Pakistan Army

- Pakistan Navy

- List of countries with nuclear weapons

35.6   References

[1] "Pakistan Nuclear Weapons – A Chronology". *Federation of American Scientists (FAS)*. FAS (Pakistan Nuclear Weapons – A Chronology). Retrieved 5 May 2012.

[2] Samdani, Zafar (25 March 2000). "Pakistan can build hydrogen bomb: Scientist". *Dawn*. Retrieved 23 December 2012.

[3] Khan, Kamran (30 May 1998). "Interview with Abdul Qadeer Khan". *Kamran Khan, director of the News Intelligence Unit of "The News International"*. Jang Media Group, Co. Retrieved 30 May 2011.

[4] "Pakistan Nuclear Weapons". Fas.org. Retrieved 21 August 2010.

[5] Sublette, Carey (10 September 2001). "1998 Year of Testing". Nuclear Weapon Archives. Retrieved 12 January 2013.

[6] Approximating and calculating the exact, accurate and precise yields are difficult to calculate. Even under very controlled conditions, precise yields can be very hard to determine, and for less controlled conditions the margins of error can be quite large. There are number of different ways that the yields can be determined, including calculations based on blast size, blast brightness, seismographic data, and the strength of the shock wave. The Pakistan Government authorities puts up the yield range from 20-~40kt (as noted by Carey Sublette of the Nuclear Weapon Archives in her report. The explosion measured 5.54 degrees on the Richter Scale, the PAEC provided the data as public domain in the KNET sources.

[7] "Pakistan has 10 more nuclear weapons than India, finds study".

[8] Wall, Robert (3 June 2013). "China Nuclear Stockpile Grows as India Matches Pakistan Rise". *Bloomberg*. Reuters.

[9] "Test launch of Pakistan's 'Shaheen-III' surface-to-surface ballistic missile successful". *mid-day*. 9 March 2015. Retrieved 26 April 2016.

[10] Weissman, Steve R. and Herbert Krosney, The Islamic Bomb. New York: Times Book). 1981: page 45.

[11] Chakma, Bhumitra, pg 42, The Politics of Nuclear Weapons in South Asia, Ashgate Publishing Company, Burlington, VT, USA, 2011

[12] "An indomitable man". Retrieved 26 April 2016.

[13] "Owl's Tree: Pakistani Nuclear Program 2–5". Owlstree.blogspot.com. 10 June 2006. Retrieved 21 August 2010.

[14] Wood, Houston; Glasser, Alexander; Kemp, Scott (2008). "The gas centrifuge and nuclear weapons proliferation". *Physics Today*. September: 40–45. doi:10.1063/1.2982121.

[15] Houston G. Wood-Alexander Glaser-R. Scott Kemp. "The gas centrifuge and nuclear weapons proliferation". Retrieved 26 April 2016.

[16] Levy, Adrian and Catherine Scott-Clark, Deception: Pakistan, the United States, and the Secret Trade in Nuclear Weapons. New York. Walker Publishing Company. 1977: page 112. Print.

[17] "The News International: Latest News Breaking, Pakistan News". Retrieved 26 April 2016.

[18] "An indomitable man". Retrieved 26 April 2016.

[19] "The News International: Latest News Breaking, Pakistan News". Retrieved 26 April 2016.

[20] Khan, A. Quadeer (4 August 2014). "Unsung heroes". *The News*. Archived from the original on 4 August 2014.

[21] Khan (2012, pp. 174–178)

[22] Rehman (1999, pp. 16–17)

[23] "Bhutto was father of Pakistan's nuclear weapons programme". Archived from the original on 14 March 2012. Retrieved 11 April 2011.

[24] Ahmad, Mansoor; Usman Shabbir, Syed Ahmad H, Khan (2006). "Multan Conference January 1972: The Birth of Pakistan's Nuclear Weapons Program." (PDF). *Pakistan Military Consortium*. Islamabad, Pakistan: Pakistan Military Consortium. **1** (1): 16. Retrieved 2010. Check date values in: |access-date= (help)

[25] "Pakistan Nuclear Weapons". Retrieved 22 February 2007.

[26] Ganguly S, Kapur SP. India, Pakistan, and the Bomb: Debating Nuclear Stability in South Asia (Contemporary Asia in the World) 2010 Columbia University Press. ISBN 9780231512824

[27] Siddiqi, Muhammad Ali (20 April 1995). "N-deterrent vital to security, says PM Benazir Bhutto". *Dawn Newspapers, 1995*. Dawn Media Group. pp. 3–6. Retrieved 13 May 2012. Pakistanis are "security conscious" because of the 1971 trauma and the three wars with India. Pakistan's programme was peaceful but was "a deterrent to India" because New Delhi had detonated a nuclear device. Pakistan, thus, had to take every step to ensure its territorial integrity and sovereignty

[28] "Eating Grass". Retrieved 26 April 2016.

[29] Kapur, Ashok; U.N. Media release (8 December 1953). "Origins and Early history of Pakistan's cover nuclear development" (PDF). *Atom for Peace: Eisenhowever's UN Speech*. New York, United States: New York: Croom Helm, 1987). Retrieved 21 March 2014.

[30] NTI. "1950s Nuclear Policy of Pakistan". *Resources obtained from Shahid-ur-Rehman, "Z.A. Bhutto, A Man in Hurry for the Bomb," Long Road to Chagai, (Islamabad: 1999, Print Wise Publication), p. 22*. NTI Pakistan Overview (1950s activities).

[31] Pervez., S.; M. Latif, I.H. Bokhari and S.Bakhtyar (2004). "Performance of PARR-I reactor with LEU fuel". *Nuclear Engineering Division of Pakistan Institute of Nuclear Science and Technology (PINSTECH) and Reduced Enrichment for Research and Test Reactors, Nuclear Engineering Division at Argonne National Laboratory (U.S. Department of Energy)*. Argonne National Laboratory. Retrieved 2011. Check date values in: |access-date= (help)

[32] PAEC and IAEA (3 December 2008). "Research Reactor Details – PARR-2". *Nuclear Engineering Division of Pakistan Institute of Nuclear Science and Technology*. Pakistan Institute of Nuclear Science and Technology. Retrieved 2011. Check date values in: |access-date= (help)

[33] Duff, Michael (2007). Salam + 50: proceedings of the conference, §*Abdus Salam and Pakistan*. London, United Kingdom: Imperial College Press. pp. 42.

[34] "Bhutto on Nuclear weapons". Nuclearweaponarchive.org. Retrieved 3 January 2014.

[35] Yasin, Rahil (16 January 2009). "War clouds hovering over South Asia". *Weekly Blitz*. Dhaka. Archived from the original on 16 October 2009. Retrieved 21 August 2010.

[36] Ponnatt, Sijo Joseph (1 March 2006). "The normative approach to nuclear proliferation.". *International Journal on World Peace* (Viewpoint essay). XXIII (1): 89–. Retrieved 21 August 2010 – via HighBeam Research. (subscription required (help)).

[37] CENTO nation help sought by Pakistan. Chicago Tribune. 7 September 1965

[38] The India-Pakistan War of 1965. Office of the Historian, Bureau of Public Affairs, United States Department of State

[39] Volha Charnysh (3 September 2009). "Pakistan's Nuclear Bomb Program" (PDF). Nuclear Age Peace Foundation.

[40] Ahmad, Shamshad (July–August 1999). "The Nuclear Subcontinent: Bringing Stability to South Asia". *Foreign Affairs*. It proposed a nuclear weapons-free zone in South Asia; a joint renunciation of acquisition or manufacture of nuclear weapons; mutual inspection of nuclear facilities; adherence to the Nuclear Nonproliferation Treaty and International Atomic Energy Agency safeguards on nuclear facilities; a bilateral nuclear test ban; and a missile-free zone in South Asia.

[41] "NUCLEAR AND MISSILE PROLIFERATION (Senate – May 16, 1989)". Fas.org. Retrieved 21 August 2010.

[42] "IAEA gave legal justification to Pakistan to build nuclear power plants". Jang News. 2011. Retrieved 2011. Check date values in: |access-date= (help)

[43] Haqqani, Hussain (2005). *Pakistan: Between Mosque and Military*. United Book Press. ISBN 978-0-87003-214-1., Chapter 3, p. 87.

[44] Langewiesche. William (November 2005). "The Wrath of Khan". *The Atlantic*. Retrieved August 2011. Check date values in: |access-date= (help)

[45] "India's First Bomb: 1967–1974". India's First Bomb: 1967–1974. Retrieved 14 January 2013.

[46] Stengel, Richard (3 June 1985). "Who has the Bomb?". *Time magazine*. pp. 7–13. Archived from the original on 3 June 1985. Retrieved 23 February 2011

[47] Khan (2012, pp. 404)

[48] Rehman (1999, pp. 21–23)

[49] Shahid-Ur-, Rehman (1999). "A Manhattan Project Scientist". *Long Road To Chagai*. Islamabad, Pakistan: Print Wise Publication. p. 23. Retrieved 21 March 2014.

[50] Khan (2012, pp. 177–178)

[51] Khan (2012, pp. 178)

[52] Rehman (1999, pp. 38–39)

[53] Rehman (1999, pp. 39–40)

[54] Rehman (1999, pp. 41–42)

[55] Hoodbhoy, Pervez (13 November 2013). "The Man who Designed the Bomb". *http://newsweekpakistan.com/*. Islamabad, Pakistan: Newsweek. Retrieved 23 June 2015. External link in |website= (help)

[56] Rehman (1999, pp. 42–43)

[57] Khan (2012, pp. 180–181)

[58] Rehman (1999, pp. 44–45)

[59] Khan (2012, pp. 180)

[60] Shahidur Rehman, Long Road to Chagai, *Professor Abdus Salam and Pakistan's Fission Weapons Programme*, pp51-89, Printwise publications, Islamabad, 1999

[61] Rehman (1999, pp. 53–55)

[62] Khan (2012, pp. 177–176)

[63] Rehman (1999, pp. 72–80)

[64] Khan (2012, pp. 141–142)

[65] "Pakistan Nuclear Weapons". Fas.org. Retrieved 21 August 2010.

[66] WILLIAM J. BROAD (12 December 2008). "The Hidden Travels of The Bomb". *The New York Times, 2008*. Retrieved 13 March 2012.

[67] Khan (2012, pp. 184-183)

[68] Khan (2012, pp. 281–282)

[69] Munir Ahmad Khan, "How Pakistan made its nuclear fuel cycle", *The Nation*, (Islamabad) 7 and 9 February 1998.

[70] "A.Q. Khan". globalsecurity.org. Retrieved 10 April 2009.

[71] Armstrong, David; Joseph John Trento, National Security News Service. *America and the Islamic Bomb: The Deadly Compromise*. Steerforth Press, 2007. p. 165. ISBN 9781586421373.

[72] "Eye To Eye: An Islamic Bomb". CBS News.

[73] Archived 13 January 2012 at the Wayback Machine.

[74] "On the trail of the black market bombs". BBC News. 12 February 2004.

[75] Mubarakmand, Samar (former Technical member and former director of Fast-Neutron Physics Group) (2004). "Pakistan became nuclear state in 1983". *The News International*. Karachi: Jang Group of Newspapers. pp. 1–2. Check date values in: |access-date= (help);

[76] Mubarakmand, Samar (30 November 1998). "Science Odyssey". *http://pakdef.org/*. Lahore, Punjab: Samar Mubarakmand's witness accounts. Retrieved 23 June 2015. External link in |website= (help)

[77] "When Mountains Move – The Story of Chagai". Defencejournal.com. Retrieved 21 August 2010.

[78] Rehman (1999, pp. 19–22)

[79] Khan (2012, pp. 50–70)

[80] NTI, Nuclear Threat Initiatives. "Pakistan: Biological Review". NTI Research on Countries.

[81] NTI, Nuclear Threat Initiatives. "Pakistan: Chemical Weapons Review". NTI Research on Countries with Chemical facilities and capabilities.

[82] "Pakistan gets IAEA approval for new N-plant". Payvand.com. 22 November 2006. Retrieved 21 August 2010.

[83] "Leading News Resource of Pakistan". *Daily Times*. 29 May 2009. Retrieved 21 August 2010.

[84] "Security Verification". *The News International*. Retrieved 21 August 2010.

[85] "Global Beat: Mark Hibbs' Nuclear Watch: July 17, 1998". Bu.edu. Retrieved 21 August 2010.

[86] "The dangers of India – Pakistan war". 1913 Intel. Retrieved 21 August 2010.

[87] "'Pakistan building third nuclear reactor at Khushab' | David Albright". *The New Indian Express*. 25 April 2009. Retrieved 21 August 2010.

[88] "ISIS Online Jan 15 2011". Isis-online.org. Retrieved 3 January 2014.

[89] Daily News (7 December 2012). "Pakistan 'expanding nuclear arsenal to deter US attack'". *Daily News, India*. New York. Retrieved 15 January 2013.

[90] The Hindu (7 December 2012). "Dealing with Pakistan's brinkmanship". *The Hindu, 2012*. Chennai, India. Retrieved 15 January 2013.

[91] Pierre Bienaimé (26 September 2014). "Pakistan Is Working To Create Tactical Nuclear Weapons - Business Insider". *Business Insider*. Retrieved 26 April 2016.

[92] "Print: Background Note: India". Nuclearfiles.org. Retrieved 3 January 2014.

[93] http://www.stimson.org/images/uploads/research-pdfs/PAKISTAN_ISRAEL.pdf

[94] http://www.worldlii.org/int/other/UNGARsn/1979/100.pdf

[95] *Military Capacity and the Risk of War: China, India, Pakistan, and Iran*. Retrieved 3 January 2014.

[96] "Introduction". Acronym.org.uk. 14 May 1998. Retrieved 3 January 2014.

[97] "Pakistan Prime Minister Urges India To Accept Bilateral Nuclear Test Ban Pact". Apnewsarchive.com. 24 September 1987. Retrieved 3 January 2014.

[98] *Containing Missile Proliferation: Strategic Technology, Security Regimes ... –Dinshaw Mistry*. Retrieved 3 January 2014.

[99] John Pike. "Pakistan Nuclear Program Chronology – 1965 – 1979". Globalsecurity.org. Retrieved 3 January 2014.

[100] "IPRI :: Islamabad Policy Research Institute". Ipripak.org. Retrieved 3 January 2014.

[101] "Pakistan and India resume conventional CBM talks". *Dawn*. Pakistan. 27 December 2011. Retrieved 3 January 2014.

[102] APP (1 January 2011). "Pakistan, India exchange lists of nuclear installations". *Dawn*. Retrieved 3 January 2014.

[103] "India, Pakistan Sign Missile Notification Pact". Arms Control Association. Retrieved 3 January 2014.

[104] "India, Pakistan to Set Up Hotline". *The Washington Post*. 21 June 2004. Retrieved 3 January 2014.

[105] Crail, Peter (March 2011). "Pakistan's Nuclear Buildup Vexes FMCT Talks". *Arms Control Today*. Retrieved 18 November 2012.

[106] Krepon, Michael (15 February 2012). "Pakistan and the FMCT". *Dawn*. Retrieved 18 November 2012.

[107] http://library.fes.de/pdf-files/iez/global/05652.pdf

[108] Rehman (1999, pp. 72–73)

[109] Chaudhry, MA. "Separating Myth from Reality". *M.A. Chaudhry; originally published in Defense Journal on May 2006*. MA Chaudhry. Retrieved 9 January 2013.

[110] Rehman (1999, pp. 74–76)

[111] "The News International: Latest News Breaking, Pakistan News". Retrieved 26 April 2016.

[112] "The News International: Latest News Breaking, Pakistan News". Retrieved 26 April 2016.

[113] Albright, David. "Holding Khan Accountable, An ISIS Statement Accompanying Release of Libya: A Major Sale at Last". ISIS. Retrieved 22 March 2012.

[114] Rehman (1999, pp. 75–79)

[115] Khan, Feroz Hassan (7 November 2012). "10: Mastery of Uranium enrichment". *Eating Grass: The Making of the Pakistani Bomb* (google book (free preview)). Stanford, California: Stanford University Press. pp. 148–160. ISBN 978-0804776011. Retrieved 24 June 2015.

[116] Khan (2012, pp. 177)

[117] Khan (2012, pp. 139–172)

[118] "Uranium Institute News Briefing 00.25 14–22 June 2000". Uranium Institute. 2000. Archived from the original on 23 September 2006. Retrieved 7 May 2006.

[119] "Key Issues: Nuclear Energy: Issues: IAEA: World Plutonium Inventories". Nuclearfiles.org. Retrieved 3 January 2014.

[120] Jillani, Shahzeb (7 August 2006). "World I Pakistan nuclear report disputed". BBC News. Retrieved 3 January 2014.

[121] Warrick, Joby (24 July 2006). "Pakistan Expanding Nuclear Program –". *The Washington Post*. Retrieved 3 January 2014.

[122] "World I Pakistan 'building new reactor'". BBC News. 24 July 2006. Retrieved 3 January 2014.

[123] "U.S. Group Says Pakistan Is Building New Reactor". *The New York Times*. 23 June 2007. Retrieved 26 April 2016.

[124] Global nuclear weapons inventories, 1945–2010, Robert S. Norris and Hans M. Kristensen, Bulletin of the Atomic Scientists, July/August 2010.

[125] "Federation of American Scientists". Fas.org. Retrieved 3 January 2014.

[126] Center for Defense Information

[127] "US Navy Strategic Insights. Feb 2003". U.S. Navy. 2003. Retrieved 28 October 2006.

[128] "Pakistan's Nuclear Arsenal Underestimated, Reports Say". Fas.org. Retrieved 3 January 2014.

[129] Anwar Iqbal (3 December 2007). "Impact of US wargames on Pakistan N-arms 'negative' -DAWN – Top Stories; 3 December 2007". *Dawn*. Pakistan. Retrieved 3 January 2014.

[130] Ricks, Thomas E. (2 December 2007). "Calculating the Risks in Pakistan –". *The Washington Post*. Retrieved 3 January 2014.

[131] "Pakistan's Nukes – Al-Qaeda's Next Strategic Surprise? – Defense Update News Analysis". Defense-update.com. Retrieved 21 August 2010.

[132] "World I Pakistan enhances second strike N-capability: US report". *Dawn*. Pakistan. Retrieved 21 August 2010.

[133] Haider, Moin (10 January 2000). "Pakistan has edge over India in Nuclear Capability". *Dawn Archives January 2000*. Retrieved 23 December 2012.

[134] Sanger, David E. (4 May 2009). "Strife in Pakistan Raises U.S. Doubts Over Nuclear Arms". *The New York Times*. Retrieved 27 March 2010.

[135] Rodriguez, Alex (29 November 2010). "Cables reveal doubts about Pakistani nuclear security". *Los Angeles Times*.

[136] Our Correspondents (17 August 2001). "US senators accuse China of selling arms". *Dawn archives 2001 August*. Retrieved 23 December 2012.

[137] Smith, R. Jeffrey; Warrick, Joby (13 November 2009). "Pakistani nuclear scientist's accounts tell of Chinese proliferation". *The Washington Post*. Retrieved 3 January 2014.

[138] "China, Pakistan, and the Bomb". Retrieved 26 April 2016.

[139] China tested N-weapons for Pak: US insider The Times of India 6 September 2008

[140] Mir, Hamid (3 May 2004). "Interview of Dr. Samar Mubarak – Head of Pakistan Missile Program.". *Hamid Mir, director of the Political Intelligence Directorate of "The News International"*. Geo Television Network. Retrieved 13 May 2011.

[141] "US Report: China gifted nuclear bomb and Pakistan stole the technology". TheWorldReporter.com. 18 November 2009.

[142] "Report No. 2001/10: Nuclear Weapons Proliferation". Csis-scrs.gc.ca. 15 May 2008. Retrieved 21 August 2010.

[143] Jawed Naqvi, New Delhi correspondents. (31 January 2001). "No Nuclear arms for Pakistan: Li". *Dawn Archives January 2001*. Retrieved 23 December 2012.

[144] "Research Library: Country Profiles: Pakistan". NTI. Retrieved 21 August 2010.

[145] "AFP: Reagan era envoy found Pakistan 'lied' on nukes". Google. Retrieved 3 January 2014.

[146] https://fas.org/sgp/crs/nuke/RL31900.pdf

[147] Narang, Vipin (January 2010). "Pakistan's Nuclear Posture: Implications for South Asian Stability" (PDF). *Harvard Kennedy School, Belfer Center for Science and International Affairs Policy Brief*. Retrieved 4 January 2013.

[148] Aguilar, Francisco, Randy Bell, Natalie, Black (July 2011). "An Introduction To Pakistan's Military" (PDF): 8. Retrieved 4 January 2013.

[149] Abidi, Zawar Haider. "Threat Reduction in South Asia" (PDF). *Zawar Haider Abidi*. pp. 6–15. Retrieved 21 July 2012.

[150] Khan, Feroz (Spring 2003). "Challenges to Nuclear Stability in South Asia," (PDF). *Nonproliferation Review*. No. 1. **10**: 65. doi:10.1080/10736700308436917. Retrieved 4 January 2013.

[151] ANI, ANI (22 March 2011). "Peace-loving' Pakistan to continue credible minimum nuke deterrence policy.". *The Yahoo News!*. Retrieved 21 July 2012. Pakistan does not wish to enter into a nuclear arms race, but will continue to maintain the policy of credible minimum deterrence, Pakistan's Air Force chief Marshal Rao Qamar Suleman has said

[152] Staff report (28 February 2006). "Pakistan will maintain minimum credible deterrence". *Daily Times*. Retrieved 21 July 2012.

[153] Khan, Abdul Qadeer. "I saved my country from nuclear blackmail'". Newsweek; The Tribune; The NTI; various others. Retrieved 3 December 2011. The State of [P]akistan's motivation for atomic weapons arose from a need to prevent "*nuclear blackmail*" by India. Had Iraq and Libya been nuclear powers, they wouldn't have been destroyed in the way we have seen recently.... If (Pakistan) had [atomic] capability before 1971, we [Pakistanis] would not have lost half of our country after a disgraceful defeat.Professor Dr. Abdul Qadeer Khan

[154] Hashimi, Shafik H. "The Nuclear Danger in South Asia". Pakistan Link. Retrieved 22 July 2012.

[155] "Pakistan swelling nuclear arsenal to counter India, says US Congressional report". *The Times Of India*. 9 August 2012.

[156] NTI, Nuclear Threat Initiatives (5 May 1994). "Bare All and Be Damned" (PDF). Far Eastern Economic Review, 5 May 1994, Pg. 23; in NTI Nuclear and Missile Database. p. 47. Retrieved 17 May 2012. The NCA determines the state of readiness which has to be maintained at all times...and lays down in great detail the policy of how the various components will be placed, protected and safeguarded

[157] "Pakistani PM takes charge of nuclear weapons". Reuters. 29 November 2009. Retrieved 29 November 2009.

[158] IISS Report. "The Nuclear control and Command in Pakistan". IISS. Retrieved 2 March 2013.

[159] Khan, Feroz Hassan; Feroze Hassan Khan. "Towards the Operational Deterrent". *Eating grass: the making of the Pakistani bomb* (google book). Stanford, California: Stanford University Press, (FH Khan). pp. 210–390. ISBN 978-0804776011. Retrieved 21 March 2014.

[160] Abrar, Saeed. "No diversion". The Nations, Pakistan. Retrieved 2 March 2013.

[161] Sanger, David E.; Broad, William J. (18 November 2007). "U.S. Secretly Aids Pakistan in Guarding Nuclear Arms". *The New York Times*. Retrieved 18 November 2007.

[162] "Nuclear Security Cooperation Between the United States and Pakistan". *name*. Retrieved 26 April 2016.

[163] "U.S. Secretly Aids Pakistan in Guarding Nuclear Arms". *The New York Times*. 18 November 2007. Retrieved 26 April 2016.

[164] "International Institute for Strategic Studies Pakistan's nuclear oversight reforms". Iiss.org. Retrieved 21 August 2010.

[165] Are Pakistan's nuclear weapons safe?, BBC, 23 January 2008

[166] Obama's Worst Pakistan Nightmare, The New York Times, 11 January 2009

[167] U.S. Policy and Pakistan's Nuclear Weapons: Containing Threats and Encouraging Regional Security, The Heritage Foundation, 6 July 2007

[168] Leigh, David (30 November 2010). "WikiLeaks cables expose Pakistan nuclear fears". *The Guardian*.

[169] Elite US troops ready to combat Pakistani nuclear hijacks. *The Times*, 17 January 2010

[170] Elisabeth Bumiller, "Gates Sees Fallout From Troubled Ties With Pakistan", *The New York Times*, 23 January 2010. Retrieved 24 January 2010.

[171] Pakistan nuclear weapons at risk of theft by terrorists, US study warns, *The Guardian*, 12 April 2010

[172] Could terrorists get hold of a nuclear bomb?, BBC, 12 April 2010

[173] "Official: Terrorists seek nuclear material, but lack ability to use it", CNN, 13 April 2010

[174] "Pakistan's Nuclear Weapons: Proliferation and Security Issues", Congressional Research Service, 23 February 2010

[175] "Tehran Times". *Tehran Times*. Retrieved 26 April 2016.

[176] IAEA declared Pakistan's Nuke programme safe and secure, *Tehran Times*

[177] "IAEA terms Pakistan's programme, safe and secure". *The News International*. Retrieved 3 January 2014.

[178] Lustig, Robert H. (7 November 2011). "Pakistan Announces Move to Increase Nuclear Security – Jeffrey Goldberg". *The Atlantic*. Retrieved 3 January 2014.

[179] Khan, Iftikhar A. "World must accept Pakistan as nuclear power: Gen Majid". *Dawn Archives, 2010*. Retrieved 3 December 2012.

[180] Anwar Iqbal (6 September 2013). "Pakistan understands importance of N-security: US". *Dawn*. Pakistan. Retrieved 3 January 2014.

[181] "Design Characteristics of Pakistan's Ballistic Missiles". NTI. Retrieved 4 July 2012.

[182] Salman Masood (25 April 2012). "Pakistan Says It Test-Fires Nuclear-Capable Missile". *The New York Times*. Retrieved 26 April 2012.

[183] Abbassi, Ansar (30 November 2011). "Pakistan has developed smartest nuclear tactical devices". *The News*. Retrieved 22 January 2012.

[184] Lewis, Jeffrey. "Pakistan's Nuclear Artillery?". Arms Control Wonk. Retrieved 22 January 2012.

[185] "First Squadron of JF-17 Thunder inducted in PAF". Associated Press of Pakistan. Retrieved 3 January 2014.

[186] Gishkori, Zahid. "Boosting air defence: F-16s replace Americans at Jacobabad airbase". *The Express Tribune*. Retrieved 3 January 2014.

[187] Archived 17 May 2008 at the Wayback Machine.

[188] "Pakistan Unveils Cruise Missile". Power Politics. 13 August 2005. Retrieved 21 August 2010.

[189] NTI, Nuclear Threat Initiatives (June 2011). "Pakistan's Naval capabilities: Submarine system". *Research: Submarine Proliferation by countries*. NTI: Research: Submarine Proliferation by countries. Retrieved 2011. Check date values in: |access-date= (help)

[190] Ansari, Usman (11 February 2012). "Pakistani Navy to Develop Nuclear-Powered Submarines: Reports". Defense News. Retrieved 13 August 2013.

[191] Ghosh, Palash (4 April 2012). "India Joins Nuclear Submarine Community; Pakistan Alarmed". *International Business Times*. Retrieved 13 August 2013.

[192] "Pakistan Navy to build nuclear submarine". ARY News. 10 February 2012. Retrieved 8 April 2012.

Bibliography and literature

- Rais, Rasul Baksh (25 November 2008). "Debating first use" (asp). *Daily Times*.

- Ganguly, Šumit; Kapur, S. Paul (2010). *India, Pakistan, and the bomb debating nuclear stability in South Asia* ([Online-Ausg.] ed.). New York: Columbia University Press. ISBN 0231512821.

- Haider, Taj (27 March 2000). "CTBT: Security Perspectives". *Dawn*. Pakistan.

- Lodhi, Maliha (6 November 2012). "Nuclear compulsion". *The News International*.

- Lieven, Anatol. *Pakistan a hard country* (1st ed.). New York: PublicAffairs. ISBN 1610390237.

- Luongo, Kenneth N. (December 2007). "Building Confidence in nuclear safety in Pakistan" (asp). Arms Control Associations.

- Saleem, Farukh. "Tipping point". *Daily Times*. Retrieved 11 October 2013.

- Cohen, Stephen P. (2004). *The idea of Pakistan* (1. paperback ed.). Washington, D.C.: Brookings Institution Press. ISBN 0815715021.

- Khan, Feroze Hassan (7 November 2012). *Eating Grass: The making of Pakistan atomic bomb*. Stanford, CA, United States,: Stanford Security Studies. ISBN 978-0804776011.

- Sattar, Abdul (1994). "Reducing Nuclear Dangers in South Asia" (PDF). *The Nonproliferation Review*. Retrieved 11 October 2013.

- Rehman, Shahid-ur- (1999). *Long Road to Chagai*. Islamabad, Pakistan: Printwise Publications. ISBN 9698500006.

- Bhutto, ZA (15 April 1969). *The Myth of Independence*. Berkeley, CA, US: Oxford University Press, USA. ISBN 978-0192151674.

35.7 External links

- Why He Went Nuclear by Douglas Frantz and Catherine Collins

- Nuclear Files.org Pakistan's nuclear conflict with India- background and the current situation

- Defense Export Promotion Organization – Ministry of Defense

- Pakistani & Indian Missile Forces (Tarmuk missile mentioned here)

- Annotated bibliography on Pakistan's nuclear weapons from the Alsos Digital Library

- The Woodrow Wilson Center's Nuclear Proliferation International History Project The Wilson Center's Nuclear Proliferation International History Project contains a collection of primary source documents on Pakistani nuclear development.

- THE MAN WHO DESIGNED PAKISTAN'S BOMB

Chapter 36

Nuclear weapons and Israel

See also: Israel and weapons of mass destruction

Israel is widely believed to possess nuclear weapons[9][10] and to be the sixth country in the world to have developed them, allegedly having built its first nuclear weapon in December 1966.[11][12] It is one of four nuclear-armed countries not recognized as a Nuclear Weapons State by the Nuclear Non-Proliferation Treaty (NPT), the others being India, Pakistan and North Korea.[13] Israel maintains a policy known as "nuclear ambiguity" (also known as "nuclear opacity").[14][15] Israel has never officially admitted to having nuclear weapons, instead repeating over the years that it would not be the first country to "introduce" nuclear weapons to the Middle East, leaving ambiguity as to whether it means it will not create, will not disclose, will not make first use of the weapons or possibly some other interpretation of the phrase.[16] The "not be the first" formulation goes back to the Eshkol-Comer (sic[17]) memorandum of understanding made between Israel and the United States on March 10, 1965, which contained Israel's written assurance for the first time that it would not be the first to introduce nuclear weapons in the Middle East.[18][19] Israel has refused to sign the NPT despite international pressure to do so, and has stated that signing the NPT would be contrary to its national security interests.[20]

Additionally, Israel has made extensive efforts to deny other regional actors the ability to acquire their own nuclear weapons.[11] The counter-proliferation, preventive strike Begin Doctrine added another dimension to Israel's existing nuclear policy. Israel remains the only country in the Middle East believed to possess them.[11]

Israel started investigating the nuclear field soon after its founding in 1948 and with French support secretly began building the Negev Nuclear Research Center, a facility near Dimona housing a nuclear reactor and reprocessing plant in the late 1950s. Israel is alleged to have built its first nuclear weapon in December 1966,[12] but it is not publicly confirmed.[21] In 1986,[22][23] Mordechai Vanunu, a former Israeli nuclear technician, provided explicit details

and photographs to the *Sunday Times* of a nuclear weapons program[24] in which he had been employed for nine years, "including equipment for extracting radioactive material for arms production and laboratory models of thermonuclear devices."[25] In 1987, an unclassified US DoD report [26] (released in February 2015 in response to a FOIA request) stated that "As far as nuclear technology is concerned, the Israelis are roughly where the U.S. [w]as in the fission weapon field in about 1955 to 1960. It should be noted that the Israelis are developing the kind of codes which will enable them to make hydrogen bombs."[27]

Estimates as to the size of the Israeli nuclear arsenal vary between 75 and 400 nuclear warheads. It is estimated that the Israel nuclear deterrent force has the ability to deliver them by intermediate-range ballistic missile, intercontinental ballistic missile, aircraft, and submarine-launched cruise missile.[5] The Stockholm International Peace Research Institute estimates that Israel has approximately 80 intact nuclear weapons, of which 50 are for delivery by Jericho II medium-range ballistic missiles and 30 are gravity bombs for delivery by aircraft.[5]

36.1 Development history

36.1.1 Pre-Dimona 1949–1956

Israel's first Prime Minister David Ben-Gurion was "nearly obsessed" with obtaining nuclear weapons to prevent the Holocaust from recurring. He stated, "What Einstein, Oppenheimer, and Teller, the three of them are Jews, made for the United States, could also be done by scientists in Israel, for their own people".[28] Ben-Gurion decided to recruit Jewish scientists from abroad even before the end of the 1948 Arab–Israeli War that established Israel's independence. He and others, such as head of the Weizmann Institute of Science and defense ministry scientist Ernst David Bergmann, believed and hoped that Jewish scientists such as Oppenheimer and Teller would help Israel.[29]

In 1949 a unit of the Israel Defense Forces Science Corps, known by the Hebrew acronym HEMED GIMMEL, began a two-year geological survey of the Negev. While a preliminary study was initially prompted by rumors of petroleum fields, one objective of the longer two year survey was to find sources of uranium; some small recoverable amounts were found in phosphate deposits.[5] That year Hemed Gimmel funded six Israeli physics graduate students to study overseas, including one to go to the University of Chicago and study under Enrico Fermi, who had overseen the world's first artificial and self-sustaining nuclear chain reaction.[30] In early 1952 Hemed Gimmel was moved from the IDF to the Ministry of Defense and was reorganized as the Division of Research and Infrastructure (EMET). That June, Bergmann was appointed by Ben-Gurion to be the first chairman of the Israel Atomic Energy Commission (IAEC).[31]

Hemed Gimmel was renamed Machon 4 during the transfer, and was used by Bergmann as the "chief laboratory" of the IAEC; by 1953, Machon 4, working with the Department of Isotope Research at the Weizmann Institute, developed the capability to extract uranium from the phosphate in the Negev and a new technique to produce indigenous heavy water.[5][32] The techniques were two years more advanced than American efforts.[29] Bergmann, who was interested in increasing nuclear cooperation with the French, sold both patents to the Commissariat à l'énergie atomique (CEA) for 60 million francs. Although they were never commercialized, it was a consequential step for future French-Israeli cooperation.[33] In addition, Israeli scientists probably helped construct the G-1 plutonium production reactor and UP-1 reprocessing plant at Marcoule. France and Israel had close relations in many areas. France was principal arms supplier for the young Jewish state, and as instability spread through French colonies in North Africa, Israel provided valuable intelligence obtained from contacts with Sephardi Jews in those countries.[1] At the same time Israeli scientists were also observing France's own nuclear program, and were the only foreign scientists allowed to roam "at will" at the nuclear facility at Marcoule.[34] In addition to the relationships between Israeli and French Jewish and non-Jewish researchers, the French believed that cooperation with Israel could give them access to international Jewish nuclear scientists.[29]

After U.S. President Dwight Eisenhower announced the Atoms for Peace initiative, Israel became the second country to sign on (following Turkey), and signed a peaceful nuclear cooperation agreement with the United States on July 12, 1955.[35][29] This culminated in a public signing ceremony on March 20, 1957, to construct a "small swimming-pool research reactor in Nachal Soreq", which would be used to shroud the construction of a much larger facility with the French at Dimona.[36]

In 1986 Francis Perrin, French high-commissioner for atomic energy from 1951 to 1970 stated publicly that in 1949 Israeli scientists were invited to the Saclay Nuclear Research Centre, this cooperation leading to a joint effort including sharing of knowledge between French and Israeli scientists especially those with knowledge from the Manhattan Project.[1][2][37] According to Lieutenant Colonel Warner D. Farr in a report to the USAF Counterproliferation Center while France was previously a leader in nuclear research "Israel and France were at a similar level of expertise after the war, and Israeli scientists could make significant contributions to the French effort. Progress in nuclear science and technology in France and Israel remained closely linked throughout the early fifties." Furthermore, according to Farr, "There were several Israeli observers at the French nuclear tests and the Israelis had 'unrestricted access to French nuclear test explosion data.'"[1]

36.1.2 Dimona 1956–1965

Main article: Negev Nuclear Research Center

Negotiation

The French justified their decision to provide Israel a nuclear reactor by claiming it was not without precedent. In September 1955 Canada publicly announced that it would help the Indian government build a heavy-water research reactor, the CIRUS, for "peaceful purposes".[38] When Egyptian President Gamal Abdel Nasser nationalized the Suez Canal, France proposed Israel attack Egypt and invade the Sinai as a pretext for France and Britain to invade Egypt posing as "peacekeepers" with the true intent of seizing the Suez Canal (see Suez Crisis). In exchange, France would provide the nuclear reactor as the basis for the Israeli nuclear weapons program. Shimon Peres, sensing the opportunity on the nuclear reactor, accepted. On September 17, 1956, Peres and Bergmann reached a tentative agreement in Paris for the CEA to sell Israel a small research reactor. This was reaffirmed by Peres at the Protocol of Sèvres conference in late October for the sale of a reactor to be built near Dimona and for a supply of uranium fuel.[39][29]

Israel benefited from an unusually pro-Israel French government during this time.[29] After the Suez Crisis led to the threat of Soviet intervention and the British and French were being forced to withdraw under pressure from the U.S., Ben-Gurion sent Peres and Golda Meir to France. During their discussions the groundwork was laid for France to build a larger nuclear reactor and chemical reprocessing plant, and Prime Minister Guy Mollet, ashamed at having abandoned his commitment to fellow socialists in Israel, supposedly told an aide, "I owe the bomb to them,"[40] while General Paul Ely, Chief of the Defence Staff, said that "We must give them this to guarantee their security, it is vital." Mollet's successor Maurice Bourgès-Maunoury stated "I gave you [Israelis] the bomb in order to prevent another Holocaust from befalling the Jewish people and so that Israel could face its enemies in the Middle East."[29]

The French–Israeli relationship was finalized on October 3, 1957, in two agreements whose contents remain secret:[29] One political that declared the project to be for peaceful purposes and specified other legal obligations, and one technical that described a 24 megawatt EL-102 reactor. The one to actually be built was to be two to three times as large[41] and be able to produce 22 kilograms of plutonium a year.[42] When the reactor arrived in Israel, Prime Minister Ben-Gurion declared that its purpose was to provide a pumping station to desalinate a billion cubic gallons of seawater annually and turn the desert into an "agricultural paradise". Six of seven members of the Israel Atomic Energy Commission promptly resigned, protesting that the reactor was the precursor to "political adventurism which will unite the world against us".[43]

Excavation

Before construction began it was determined that the scope of the project would be too large for the EMET and IAEC team, so Shimon Peres recruited Colonel Manes Pratt, then Israeli military attaché in Burma, to be the project leader. Building began in late 1957 or early 1958, bringing hundreds of French engineers and technicians to the Beersheba and Dimona area . In addition, thousands of newly immigrated Sephardi Jews were recruited to do digging; to circumvent strict labor laws, they were hired in increments of 59 days, separated by one day off.[44]

Creation of LEKEM

By the late 1950s Shimon Peres had established and appointed a new intelligence service assigned to search the globe and clandestinely secure technology, materials and equipment needed for the program, by any means necessary. The new service would eventually be named LEKEM (pronounced LAKAM, the Hebrew acronym for 'Science liaison Bureau'). Peres appointed IDF Internal Security Chief, Benjamin Blumberg, to the task. As head of the LEKEM, Blumberg would rise to become a key figure in Israel's intelligence community, coordinating agents worldwide and securing the crucial components for the program.[45][46][47][48]

Rift between Israel and France

When Charles de Gaulle became French President in late 1958 he wanted to end French–Israeli nuclear cooperation, and said that he would not supply Israel with uranium unless the plant was opened to international inspectors, declared peaceful, and no plutonium was reprocessed.[49] Through an extended series of negotiations, Shimon Peres finally reached a compromise with Foreign Minister Maurice Couve de Murville over two years later, in which French companies would be able to continue to fulfill their contract obligations and Israel would declare the project peaceful.[50] Due to this, French assistance did not end until 1966.[51] However the supply of uranium fuel was stopped earlier, in 1963.[52] Despite this, a French uranium company based in Gabon may have sold Israel uranium in 1965. The US government launched an investigation but was unable to determine if such a sale had taken place.[53]

British aid

Top secret British documents[54][55] obtained by BBC *Newsnight* show that Britain made hundreds of secret shipments of restricted materials to Israel in the 1950s and 1960s. These included specialist chemicals for reprocessing and samples of fissile material—uranium-235 in 1959, and plutonium in 1966, as well as highly enriched lithium-6, which is used to boost fission bombs and fuel hydrogen bombs.[56] The investigation also showed that Britain shipped 20 tons of heavy water directly to Israel in 1959 and 1960 to start up the Dimona reactor.[57] The transaction was made through a Norwegian front company called Noratom, which took a 2% commission on the transaction. Britain was challenged about the heavy water deal at the International Atomic Energy Agency after it was exposed on Newsnight in 2005. British Foreign Minister Kim Howells claimed this was a sale to Norway. But a former British intelligence officer who investigated the deal at the time confirmed that this was really a sale to Israel and the Noratom contract was just a charade.[58] The Foreign Office finally admitted in March 2006 that Britain knew the destination was Israel all along.[59] Israel admits running the Dimona reactor with Norway's heavy water since 1963. French engineers who helped build Dimona say the Israelis were expert operators, so only a relatively small portion of the water was lost during the years since the reactor was first put into operation.[60]

36.1.3   Criticality

In 1961, the Israeli Prime Minister David Ben-Gurion informed the Canadian Prime Minister John Diefenbaker that a pilot plutonium-separation plant would be built at Dimona. British intelligence concluded from this and other information that this "can only mean that Israel intends to produce nuclear weapons".[54] The nuclear reactor at Dimona went critical in 1962.[1] After Israel's rupture with France, the Israeli government reportedly reached out to Argentina. The Argentine government agreed to sell Israel yellowcake (uranium oxide).[53][61] Between 1963 and 1966, about 90 tons of yellowcake were allegedly shipped to Israel from Argentina in secret.[52] By 1965 the Israeli reprocessing plant was completed and ready to convert the reactor's fuel rods into weapons grade plutonium.[62]

Costs

The exact costs for the construction of the Israeli nuclear program are unknown, though Peres later said that the reactor cost $80 million in 1960,[63] half of which was raised by foreign Jewish donors, including many American Jews. Some of these donors were given a tour of the Dimona complex in 1968.[64]

36.1.4 Weapons production 1966–present

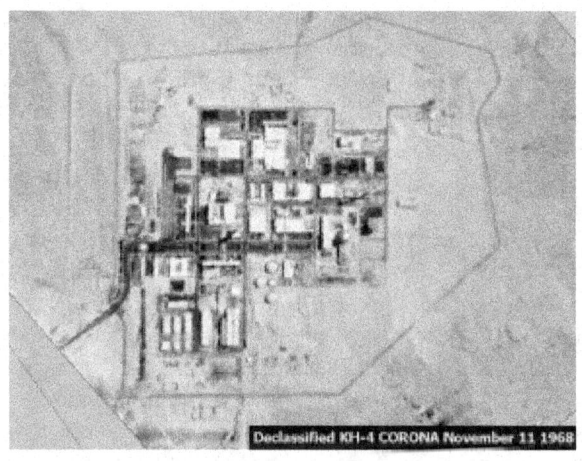

Completed Dimona complex as seen by US Corona satellite on November 11, 1968

Israel is believed to have begun full-scale production of nuclear weapons following the 1967 Six-Day War, although it had built its first operational nuclear weapon by December 1966.[12] A CIA report from early 1967 stated that Israel had the materials to construct a bomb in six to eight weeks[65] and some authors suggest that Israel had two crude bombs ready for use during the war.[1] According to US journalist Seymour Hersh, everything was ready for production at this time save an official order to do so. Another CIA report from 1968 states that "...Israel might undertake a nuclear weapons program in the next several years."[66] Moshe Dayan, then Defense Minister, believed that nuclear

weapons were cheaper and more practical than indefinitely growing Israel's conventional forces.[67] He convinced the Labor Party's economic boss Pinchas Sapir of the value of commencing the program by giving him a tour of the Dimona site in early 1968, and soon after Dayan decided that he had the authority to order the start of full production of four to five nuclear warheads a year. Hersh stated that it is widely believed that the words "Never Again" were welded, in English and Hebrew, onto the first warhead.[68]

In order to produce plutonium the Israelis needed a large supply of uranium ore. In 1968, the Mossad purchased 200 tons from Union Minière du Haut Katanga, a Belgian mining company, on the pretense of buying it for an Italian chemical company in Milan. Once the uranium was shipped from Antwerp it was transferred to an Israeli freighter at sea and brought to Israel. The orchestrated disappearance of the uranium, named Operation Plumbat, became the subject of the 1978 book *The Plumbat Affair*.[69]

Estimates as to how many warheads Israel has built since the late 1960s have varied, mainly based on the amount of fissile material that could have been produced and on the revelations of Israeli nuclear technician Mordechai Vanunu.

Mordechai Vanunu's photograph of a Negev Nuclear Research Center glove box containing nuclear materials in a model bomb assembly, one of about 60 photographs he later gave to the British press

By 1969, U.S. Defense Secretary Melvin Laird believed that Israel might have a nuclear weapon that year.[70][71] Later that year, U.S. President Richard Nixon in a meeting with Israeli Prime Minister Golda Meir pressed Israel to "make no visible introduction of nuclear weapons or undertake a nuclear test program", so maintaining a policy of nuclear ambiguity.[72] Before the Yom Kippur War Peres nonetheless wanted Israel to publicly demonstrate its nuclear capability to discourage an Arab attack, and fear of Israeli nuclear weapons may have discouraged Arab military strategy during the war from being as aggressive as it could have been.[1]

The CIA believed that Israel's first bombs may have been

made with highly enriched uranium stolen in the mid-1960s from the U.S. Navy nuclear fuel plant operated by the Nuclear Materials and Equipment Corporation, where sloppy material accounting would have masked the theft.[73][74]

By 1974, the U.S. intelligence community believed Israel had stockpiled a small number of fission weapons,[75] and by 1979 were perhaps in a position to test a more advanced small tactical nuclear weapon or thermonuclear weapon trigger design.[76]

The CIA believed that the number of Israeli nuclear weapons stayed from 10 to 20 from 1974 until the early 1980s.[5] Vanunu's information in October 1986 said that based on a reactor operating at 150 megawatts and a production of 40 kg of plutonium per year, Israel had 100 to 200 nuclear devices. Vanunu revealed that between 1980 and 1986 Israel attained the ability to build thermonuclear weapons.[77] By the mid 2000s estimates of Israel's arsenal ranged from 75 to 400 nuclear warheads.[5][6]

Several reports have surfaced claiming that Israel has some uranium enrichment capability at Dimona. Vanunu asserted that gas centrifuges were operating in Machon 8, and that a laser enrichment plant was being operated in Machon 9 (Israel holds a 1973 patent on laser isotope separation). According to Vanunu, the production-scale plant has been operating since 1979–80. The scale of a centrifuge operation would necessarily be limited due to space constraints. Laser isotope separation, however, if developed to operational status, could be quite compact. If highly enriched uranium is being produced in substantial quantities, then Israel's nuclear arsenal could be much larger than estimated solely from plutonium production.[78] Uranium enrichment could also be used to re-enrich reprocessed uranium into reactor fuel to more efficiently use Israel's uranium supply.

In 1991 alone, as the Soviet Union dissolved, nearly 20 top Jewish Soviet scientists reportedly emigrated to Israel, some of whom had been involved in operating nuclear power plants and planning for the next generation of Soviet reactors. In September 1992, German intelligence was quoted in the press as estimating that 40 top Jewish Soviet nuclear scientists had emigrated to Israel since 1989.[79]

In a 2010 interview, Uzi Eilam, former head of the Israeli Atomic Energy Commission, told the Israeli daily *Maariv* that the nuclear reactor in Dimona had been through extensive improvements and renovations and is now functioning as new, with no safety problems or hazard to the surrounding environment or the region.[80]

36.2 Nuclear testing

Main articles: Nuclear weapons testing and Vela Incident

According to Lieutenant Colonel Warner D. Farr in a report to the USAF Counterproliferation Center much lateral proliferation happened between pre-nuclear Israel and France stating "the French nuclear test in 1960 made two nuclear powers not one—such was the depth of collaboration" and "the Israelis had unrestricted access to French nuclear test explosion data." minimizing the need for early Israeli testing though this cooperation cooled following the success of the French tests.[1]

In June 1976, a West Germany army magazine, *Wehrtechnik* ("military technology"), claimed that Western intelligence reports documented Israel had conducted an underground test in the Negev in 1963. The book *Nuclear Weapons in the Middle East: Dimensions and Responsibilities* by Taysir Nashif cites other reports that on November 2, 1966, the country may have carried out a non-nuclear test, speculated to be zero yield or implosion in nature in the Israeli Negev desert.[5][1]

On September 22, 1979 Israel may have been involved in a possible nuclear bomb test also known as the Vela Incident in the southern Indian Ocean. A committee was set up under then U.S. president Jimmy Carter headed by Prof. Jack Ruina of MIT. Most of the committee's members assumed that South African navy vessels had sailed out of Simonstown port, near Cape Town, to a secret location in the Indian Ocean, where they conducted the nuclear test. The committee defined the nuclear device tested as compact and especially clean, emitting little radioactive fallout, making it very nearly impossible to pinpoint. Another committee assessment concluded a cannon had fired a nuclear artillery shell and the detected test was focused on a small tactical nuclear weapon. After renouncing their nuclear weapons program South Africa was revealed to only have six large, primitive, aircraft-deliverable atomic bombs with a seventh being built, but no sophisticated miniaturized devices of the artillery shell size.[81]

36.3 Revelations

36.3.1 Negev Nuclear Research Center (Dimona)

Main article: Negev Nuclear Research Center

The Israeli nuclear program was first revealed publicly on December 13, 1960, in a small *Time* article,[82] which said that a non-Communist non-NATO country had made an "atomic development". On December 16, the *Daily Express* revealed this country to be Israel, and on December 18, US Atomic Energy Commission chairman John McCone appeared on *Meet the Press* to officially confirm the Israeli construction of a nuclear reactor and announce his resignation.[83] The following day *The New York Times*, with the help of McCone, revealed that France was assisting Israel.[84]

The news led Ben-Gurion to make the only statement by an Israeli Prime Minister about Dimona. On December 21 he announced to the Knesset that the government was building a 24 megawatt reactor "which will serve the needs of industry, agriculture, health, and science", and that it "is designed exclusively for peaceful purposes".[85] Bergmann, who was chairman of the Israel Atomic Energy Commission from 1954 to 1966, however said that "There is no distinction between nuclear energy for peaceful purposes or warlike ones"[86] and that "We shall never again be led as lambs to the slaughter".[87]

36.3.2 Weapons production

The first public revelation of Israel's nuclear capability (as opposed to development program) came from NBC News, which reported in January 1969 that Israel decided "to embark on a crash course program to produce a nuclear weapon" two years previously, and that they possessed or would soon be in possession of such a device.[88] This was initially dismissed by Israeli and U.S. officials, as well as in an article in *The New York Times*. Just one year later on July 18, *The New York Times* made public for the first time that the U.S. government believed Israel to possess nuclear weapons or to have the "capacity to assemble atomic bombs on short notice".[89] Israel reportedly assembled 13 bombs during the Yom Kippur War as a last defense against total defeat, and kept them usable after the war.[67]

The first extensive details of the weapons program came in the London-based *Sunday Times* on October 5, 1986, which printed information provided by Mordechai Vanunu, a technician formerly employed at the Negev Nuclear Research Center near Dimona. For publication of state secrets Vanunu was kidnapped by the Mossad in Rome, brought back to Israel, and sentenced to 18 years in prison for treason and espionage. Although there had been much speculation prior to Vanunu's revelations that the Dimona site was creating nuclear weapons, Vanunu's information indicated that Israel had also built thermonuclear weapons.[77]

Theodore Taylor, a former U.S. weapon designer leading the field in small, efficient nuclear weapons, reviewed the 1986 leaks and photographs of the Israeli nuclear program by Mordechai Vanunu in detail. Taylor concluded

that Israel's thermonuclear weapon designs appeared to be "less complex than those of other nations," and as of 1986 "not capable of producing yields in the megaton or higher range." Nevertheless, "they may produce at least several times the yield of fission weapons with the same quantity of plutonium or highly enriched uranium." In other words, Israel could "boost" the yield of its nuclear fission weapons. According to Taylor, the uncertainties involved in the process of boosting required more than theoretical analysis for full confidence in the weapons' performance. Taylor therefore concluded that Israel had "unequivocally" tested a miniaturized nuclear device. The Institute for Defense Analyses (IDA) concluded after reviewing the evidence given by Vanunu that as of 1987, "the Israelis are roughly where the U.S. was in the fission weapon field in about 1955 to 1960." and would require supercomputers or parallel computing clusters to refine their hydrogen bomb designs for improved yields without testing, though noting they were already then developing the computer code base required.[2] Israel was first permitted to import US built supercomputers beginning in November 1995.[90]

According to a 2013 report by the Bulletin of the Atomic Scientists, which cited US Defense Intelligence Agency figures, Israel began the production of nuclear weapons in 1967, when it produced its first two nuclear bombs. According to the report's calculations, Israel produced nuclear weapons at an average rate of two per year, and stopped production in 2004. The report stated that Israel has 80 nuclear warheads and has enough fissile material to produce 190 more.[91][92] In 2014, former US president Jimmy Carter stated that "Israel has, what, 300 or more, nobody knows exactly how many" nuclear weapons.[93]

36.3.3 South African documents

See also: Israel–South Africa relations and South Africa and weapons of mass destruction

In 2010, *The Guardian* released South African government documents that it alleged confirmed the existence of Israel's nuclear arsenal. According to the newspaper, the documents are minutes taken by the South African side of alleged meetings between senior officials from the two countries in 1975. *The Guardian* alleged that these documents reveal that Israel had offered to sell South Africa nuclear weapons that year. The documents appeared to confirm information disclosed by a former South African naval commander Dieter Gerhardt – jailed in 1983 for spying for the Soviet Union, who said there was an agreement between Israel and South Africa involving an offer by Israel to arm eight Jericho missiles with atomic bombs.[94][95] Waldo Stumpf—who led a project to dismantle South Africa's nuclear weapons program[96]—doubted Israel or South Africa would have contemplated a deal seriously, saying that Israel could not have offered to sell nuclear warheads to his country due to the serious international complications that such a deal could entail. Shimon Peres, former Israeli President and then Defense Minister, has rejected the newspaper's claim that the negotiations took place. He also asserted that *The Guardian*'s conclusions were "based on the selective interpretation of South African documents and not on concrete facts".[97]

Avner Cohen, author of *Israel and the Bomb* and *The Worst-Kept Secret: Israel's Bargain with the Bomb*, said "Nothing in the documents suggests there was an actual offer by Israel to sell nuclear weapons to the regime in Pretoria."[98]

36.3.4 US pressure

Main article: Israel–United States relations

The United States was concerned over possible Israeli nuclear proliferation. US intelligence began to notice the Dimona reactor shortly after construction began, when American U-2 spy planes overflew the reactor,[99] leading to a diplomatic clash. In 1960, the outgoing Eisenhower administration asked the Israeli government for an explanation for the mysterious construction near Dimona. Israel's response was that the site was a future textile factory, but that no inspection would be allowed. When Ben-Gurion visited Washington in 1960, he held a series of meetings with State Department officials, and was bluntly told that for Israel to possess nuclear weapons would affect the balance of power in the region.[43] After John F. Kennedy took office as US President in 1961, he put continuous pressure on Israel to open the plant to American inspection. Reportedly, every high-level meeting and communication between the US and Israeli governments contained a demand for an inspection of Dimona. To increase pressure, Kennedy denied Ben-Gurion a meeting at the White House – when they met in May 1961, it was at the Waldorf Astoria Hotel in New York. The meeting itself was dominated by this issue. Ben-Gurion was evasive on the issue for two years, in the face of persistent US demands for an inspection. Finally, in a personal letter dated May 18, 1963, Kennedy threatened Israel with total isolation unless inspectors were allowed into Dimona. However, Ben-Gurion resigned as Prime Minister shortly afterward. His successor, Levi Eshkol, received a similar letter from Kennedy.[100]

Israel eventually accepted an inspection, and Kennedy made two concessions – the US would sell Israel Hawk anti-aircraft missiles after having refused to sell Israel any major weapon systems for years. In addition, the US government agreed to the Israeli demand that the inspections would be

carried out by an all-American team which would schedule its visits weeks in advance, rather than the IAEA.

In 1964, the US government tried to prevent Argentina's sale of yellowcake to Israel, with no success.[101]

Allegedly, because Israel knew the schedule of the inspectors' visits, it was able to disguise the true purpose of the reactor. The inspectors eventually reported that their inspections were useless, due to Israeli restrictions on what parts of the facility they could investigate. According to British writer and intelligence expert Gordon Thomas, former Mossad agent Rafi Eitan told him how the inspectors were fooled:[43]

> A bogus control center was built over the real one at Dimona, complete with fake control panels and computer-lined gauges that gave a credible impression of measuring the output of a reactor engaged in an irrigation scheme to turn the Negev into a lush pastureland. The area containing the "heavy" water smuggled from France and Norway was placed off-limits to the inspectors "for safety reasons". The sheer volume of heavy water would have been proof the reactor was being readied for a very different purpose.

In 1968, the CIA stated in a top-secret National Intelligence Estimate that Israel had nuclear weapons. This assessment was given to President Lyndon B. Johnson. The basis for this claim was the CIA's belief, although never proven, that the uranium that went missing in the Apollo Affair had been diverted to Israel, as well as evidence gathered from NSA electronic eavesdropping on Israeli communications, which proved that the Israeli Air Force had engaged in practice bombing runs that only made sense for the delivery of nuclear weapons.[99]

In 1969, the US government terminated the inspections. That same year, Richard Nixon became President. According to US government documents declassified in 2007, the Nixon administration was concerned with Israel's nuclear program, worrying that it could set off a regional nuclear arms race, with the Soviet Union possibly granting the Arab states a nuclear guarantee. In a memorandum dated July 19, 1969, National Security Adviser Henry Kissinger warned that "The Israelis, who are one of the few peoples whose survival is genuinely threatened, are probably more likely than almost any other country to actually use their nuclear weapons." However, Kissinger warned that attempting to force Israel to disarm could have consequences, writing that "Israel will not take us seriously on the nuclear issue unless they believe we are prepared to withhold something they very much need" (Kissinger was referring to a pending sale of F-4 Phantom fighter jets to Israel). Kissinger wrote that "if we withhold the Phantoms and they make this fact public in the United States, enormous political pressure will be mounted on us. We will be in an indefensible position if we cannot state why we are withholding the planes. Yet if we explain our position publicly, we will be the ones to make Israel's possession of nuclear weapons public with all the international consequences this entails." Among the suggestions Kissinger presented to Nixon was the idea of the United States adopting a policy of "nuclear ambiguity", or pretending not to know about Israel's nuclear program.[102]

According to Israeli historian Avner Cohen, author of *Israel and the Bomb*, historical evidence indicates that when Nixon met with Israeli Prime Minister Golda Meir at the White House in September 1969, they reached a secret understanding, where Israel would keep its nuclear program secret and refrain from carrying out nuclear tests, and the United States would tolerate Israel's possession of nuclear weapons and not press it to sign the Nuclear Non-Proliferation Treaty.[102] However, no one else was present, not even Henry Kissinger, and no written record on what was discussed has ever surfaced.

36.4 Stockpile

The State of Israel has never made public any details of its nuclear capability or arsenal. The following is a history of estimates by many different sources on the size and strength of Israel's nuclear arsenal. Estimates may vary due to the amount of material Israel has on store versus assembled weapons, and estimates as to how much material the weapons actually use, as well as the overall time in which the reactor was operated.

- 1967 (Six Day War)– 2 bombs;[103][104] 13 bombs[105]

- 1969– 5–6 bombs of 19 kilotons yield each[106]

- 1973 (Yom Kippur War)– 13 bombs;[67] 20 nuclear missiles plus developed a suitcase bomb[107]

- 1974– 3 capable artillery battalions each with 12 175 mm tubes and a total of 108 warheads;[108][109] 10 bombs[110]

- 1976– 10–20 nuclear weapons[lower-alpha 1]

- 1980– 100–200 bombs[112][113]

- 1984– 12–31 atomic bombs;[114] 31 plutonium bombs and 10 uranium bombs[115]

- 1985– at least 100 nuclear bombs[116][117]

- 1986– 100 to 200 fission bombs and a number of fusion bombs[118]

- 1991– 50–60 to 200–300[119]

- 1992– more than 200 bombs[117]

- 1994– 64–112 bombs (5 kg/warhead);[120] 50 nuclear tipped Jericho missiles, 200 total[121]

- 1994– 300 nuclear weapons.[122]

- 1995– 66–116 bombs (at 5 kg/warhead);[120] 70–80 bombs;[123] "A complete Repertoire" (neutron bombs, nuclear mines, suitcase bombs, submarine-borne)[124]

- 1996– 60–80 plutonium weapons, maybe more than 100 assembled, ER variants, variable yields[125]

- 1997– More than 400 deliverable thermonuclear and nuclear weapons[6]

- 2002– Between 75 and 200 weapons[126]

- 2004– 82[127]

- 2006– Federation of American Scientists believes that Israel "could have produced enough plutonium for at least 100 nuclear weapons, but probably not significantly more than 200 weapons".[5]

- 2008– 150 or more nuclear weapons.[128]

- 2008– 80 intact warheads, of which 50 are re-entry vehicles for delivery by ballistic missiles and the rest bombs for delivery by aircraft. Total military plutonium stockpile 340–560 kg[129]

- 2009– Estimates of weapon numbers differ sharply with plausible estimates varying from 60 to 400.[130]

- 2010– According to *Jane's Defense Weekly* Israel has between 100 and 300 nuclear warheads, most of them are probably being kept in unassembled mode but can become fully functional "in a matter of days".[131]

- 2010– "More than 100 weapons, mainly two-stage thermonuclear devices, capable of being delivered by missile, fighter-bomber, or submarine"[28]

- 2014– Approximately 80 nuclear warheads for delivery by two dozen missiles, a couple of squadrons of aircraft, and perhaps a small number of sea-launched cruise missiles.[132]

- 2014 – "300 or more" nuclear weapons.[93]

36.5   Delivery systems

Main articles: Nuclear weapons delivery and Nuclear triad

Israeli military forces possess land, air, and sea based methods for deploying their nuclear weapons, thus forming a nuclear triad that is mainly medium to long ranged, the backbone of which is submarine-launched cruise missiles and medium and intercontinental ballistic missiles, with Israeli Air Force long range strike aircraft on call to perform nuclear interdiction and strategic strikes.[133] During 2008 the Jericho III ICBM became operational, giving Israel extremely long range nuclear strike abilities.[134][135]

36.5.1   Missiles

Main article: Jericho missile

Israel is believed to have nuclear second-strike abilities in the form of its submarine fleet and its nuclear-capable ballistic missiles that are understood to be buried deeply enough that they would survive a pre-emptive nuclear strike.[28][136] Ernst David Bergmann was the first to seriously begin thinking about ballistic missile capability and Israel test-fired its first Shavit II sounding rocket in July 1961.[137][138] In 1963 Israel put a large-scale project into motion, to jointly develop and build 25 short-range missiles with the French aerospace company Dassault. The Israeli project, codenamed Project 700, also included the construction of a missile field at Hirbat Zacharia, a site west of Jerusalem.[139] The missiles that were first developed with France became the Jericho I system, first operational in 1971. It is possible that the Jericho I was removed from operational service during the 1990s. In the mid-1980s the Jericho II medium-range missile, which is believed to have a range of 2800–5000 km, entered service.[140][141][1] It is believed that Jericho II is capable of delivering nuclear weapons with a superior degree of accuracy.[142] The Shavit three stages solid fuel space launch vehicle produced by Israel to launch many of its satellites into low earth orbit since 1988 is a civilian version of the Jericho II.[143] The Jericho III ICBM, became operational in January 2008[144][145] and some reports speculate that the missile may be able to carry MIRVed warheads.[146] The maximum range estimation of the Jericho III is 11,500 km with a payload of 1000–1300 kg (up to six small nuclear warheads of 100 kt each or one 1 megaton nuclear warhead),[8][147] and its accuracy is considered high.[144] In January 2008 Israel carried out the successful test launch of a long-range, ballistic missile capable of carrying a nuclear warhead from the reported launch site at the Palmachim Airbase south of Tel Aviv.[148] Israeli radio identified the missile as a Jericho III and the Hebrew

YNet news Web site quoted unnamed defence officials as saying the test had been "dramatic"[149][150] and that the new missile can reach "extremely long distances", without elaborating.[151] Soon after the successful test launch, Isaac Ben-Israel, a retired army general and Tel Aviv University professor, told Israeli Channel 2 TV:

> Everybody can do the math and understand that the significance is that we can reach with a rocket engine to every point in the world

The test came two days after Ehud Olmert, then Israel's Prime Minister, warned that "all options were on the table to prevent Tehran from acquiring nuclear weapons" and a few months later Israel bombed a suspected Syrian nuclear facility built with extensive help from North Korea.[152] At the same time, regional defense experts said that by the beginning of 2008, Israel had already launched a programme to extend the range of its existing Jericho II ground attack missiles.[145] The Jericho-II B missile is capable of sending a one ton nuclear payload 5,000 kilometers.[1] The range of Israels' Jericho II B missiles is reportedly capable of being modified to carry nuclear warheads no heavier than 500 kg over 7,800 km, making it an ICBM.[153] It is estimated that Israel has between 50 and 100 Jericho II B missiles based at facilities built in the 1980s.[154] The number of Jericho III missiles that Israel possesses is unknown.

36.5.2   Aircraft

Main articles: Israeli Air Force and F-15I

Israel is believed to use fighter bombers as a means to deliver nuclear weapons. The Israeli Aerial refueling fleet of modified Boeing 707s and the use of external and conformal fuel tanks gives Israeli F-15, F-15I and F-16 fighter bombers strategic reach as demonstrated in Operation Wooden Leg. F-16 fighter aircraft have been cited as possible nuclear delivery systems.[155][156][157][158]

Present

The Israeli Air Force possesses Lockheed Martin F-16I Sufa ("Storm") Multirole combat aircraft and McDonnell Douglas/Boeing F-15I Ra'am ("Thunder") strike fighters.

36.5.3   Marine

Main article: Popeye Turbo

The Israeli Navy operates modern German-built *Dolphin*-class submarines.[159] The first three Dolphins were delivered to Israel in 1999 and replaced the aging Gal class submarines, which had served in the Israeli navy since the late 1970s.[160] Various reports[142] indicate that these submarines are equipped with Popeye Turbo cruise missiles that can deliver nuclear and conventional[161] warheads with extremely high accuracy. The proven effectiveness of cruise missiles of its own production may have been behind Israel's recent acquisition of these submarines which are equipped with torpedo tubes suitable for launching long-range (1500–2400 km) nuclear-capable cruise missiles[162][163] that would offer Israel a second strike capability.[164] Israel is reported to possess a 200 kg nuclear warhead, containing 6 kg of plutonium, that could be mounted on cruise missiles.[162] The missiles were reportedly test launched in the Indian Ocean near Sri Lanka in June 2000, and are reported to have hit their target at a range of 1500 km. In June 2002, former State Department and Pentagon officials confirmed that the US Navy observed Israeli missile tests in the Indian Ocean in 2000, and that the Dolphin-class vessels have been fitted with nuclear-capable cruise missiles of a new design. It is believed by some to be a version of Rafael Armament Development Authority's Popeye turbo cruise missile while some believe that the missile may be a version of the Gabriel 4LR that is produced by Israel Aircraft Industries. However, others claim that such a range implies an entirely new type of missile.[165][166][167] During the second half of the 1990s, Israel asked the United States to sell it 50 Tomahawk land-attack cruise missiles to enhance its deep-strike capabilities. Washington rejected Israel's request in March 1998, since such a sale would have violated the Missile Technology Control Regime guidelines, which prohibit the transfer of missiles with a range exceeding 300 km. Shortly after the rejection, an Israeli official told Defense News, "History has taught us that we cannot wait indefinitely for Washington to satisfy our military requirements. If this weapon system is denied to us, we will have little choice but to activate our own defense industry in pursuit of this needed capability." In July 1998, the Air Intelligence Center warned the US Congress that Israel was developing a new type of cruise missile.[168]

According to Israeli defense sources, in June 2009 Israeli Dolphin-class submarine sailed from the Mediterranean to the Red Sea via the Suez Canal during a drill that showed that Israel can access the Indian Ocean, and the Persian Gulf, far more easily than before.[169] IDF sources said the decision to allow navy vessels to sail through the canal was made recently and was a definite "change of policy" within the service. Israeli officials said the submarine was surfaced when it passed through the canal. In the event of a conflict with Iran, and if Israel decided to involve its Dolphin-

class submarines, the quickest route would be to send them through the Suez Canal.[170]

The Israeli fleet was expanded after Israel signed a 1.3 billion euro contract to purchase two additional submarines from ThyssenKrupp's subsidiary HDW in 2006. These two U212s are to be delivered to the Israeli navy in 2011 and are "Dolphin II" class submarines.[171] The submarines are believed to be capable of launching cruise missiles carrying nuclear warheads, despite statements by the German government in 2006, in confirming the sale of the two vessels, that they were not equipped to carry nuclear weapons.[172] The two new boats are an upgraded version of the old Dolphins, and equipped with an air-independent propulsion system, that allow them to remain submerged for longer periods of time than the three nuclear arms-capable submarines that have been in Israel's fleet since 1999.[136][173] In October 2009 it was reported that the Israeli navy sought to buy a sixth Dolphin class submarine.[164]

On June 4, 2012, *Der Spiegel* published an investigative article stating that Israel has armed its newest submarines with nuclear missiles.[174] Numerous Israeli and German officials were quoted testifying to the nuclear capabilities of the submarines and the placement of nuclear missiles aboard the ships. In response to the article, officials from both Germany and Israel refused to comment.[175] Several papers have stated the implications of Israel attaining these nuclear weapon carrying submarines are increased due to the threat of attacks upon Iran by Israel.[176]

36.5.4   Other

It has been reported that Israel has several other nuclear weapons capabilities:

- Suitcase bomb: Seymour Hersh reports that Israel developed the ability to miniaturize warheads small enough to fit in a suitcase by the year 1973.[177]

- Tactical nuclear weapon: Israel may also have 175 mm and 203 mm self-propelled artillery pieces, capable of firing nuclear shells. There are three battalions of the 175mm artillery (36 tubes), reportedly with 108 nuclear shells and more for the 203mm tubes. If true, these low yield, tactical nuclear artillery rounds could reach at least 25 miles (40 km), while by some sources it is possible that the range was extended to 45 miles (72 km) during the 1990s.[1]

- EMP strike capabilities: Israel allegedly possesses several 1 megaton bombs,[178][179] which give it a very large EMP attack ability.[180] For example, if a megaton-class weapon were to be detonated 400 kilometers above Omaha, Nebraska, US, nearly the entire continental United States would be affected with potentially damaging EMP experience from Boston to Los Angeles and from Chicago to New Orleans.[181] A similar high-altitude airburst above Iran could cause serious damage to all of the electrical systems in the Middle East, and much of Europe.[182]

- Enhanced Radiation Weapon (ERW): Israel also is reported to have an unknown number of neutron bombs.[1]

- Nuclear land mine: Israel supposedly has deployed multiple defensive nuclear land mines in the Golan Heights.[183][184]

36.6   Policy

Main article: Policy of deliberate ambiguity

Israel's deliberately ambiguous policy to confirm or deny its own possession of nuclear weapons, or to give any indication regarding their potential use, make it necessary to gather details from other sources, including diplomatic and intelligence sources and 'unauthorized' statements by its political and military leaders. Alternatively, with the Begin Doctrine, Israel is very clear and decisive regarding the country's policy on potential developments of nuclear capability by any other regional adversaries, which it will not allow.

36.6.1   Possession

Although Israel has officially acknowledged the existence of Dimona since Ben-Gurion's speech to the Knesset in December 1960, Israel has never officially acknowledged its construction or possession of nuclear weapons.[185] In addition to this policy, on May 18, 1966, Prime Minister Levi Eshkol told the Knesset that "Israel has no atomic weapons and will not be the first to introduce them into our region," a policy first articulated by Shimon Peres to U.S. President John F. Kennedy in April 1963.[186] In the late 1960s, Israeli Ambassador to the US Yitzhak Rabin informed the United States State Department that its understanding of "introducing" such weapons meant that they would be tested and publicly declared, while merely possessing the weapons did not constitute "introducing" them.[187][188] Avner Cohen defines this initial posture as "nuclear ambiguity", but he defines the stage after it became clear by 1970 that Israel possessed nuclear weapons as a policy of *amimut*,[28] or "nuclear opacity".[189]

In 1998, former Prime Minister Shimon Peres said that Israel "built a nuclear option, not in order to have a Hiroshima

but an Oslo".[190] The "nuclear option" may refer to a nuclear weapon or to the nuclear reactor near Dimona, which Israel claims is used for scientific research. Peres, in his capacity as the Director General of the Ministry of Defense in the early 1950s, was responsible for building Israel's nuclear capability.[191]

In a December 2006 interview, Israeli Prime Minister Ehud Olmert stated that Iran aspires "to have a nuclear weapon as America, France, Israel and Russia".[192] Olmert's office later said that the quote was taken out of context; in other parts of the interview, Olmert refused to confirm or deny Israel's nuclear weapon status.[193]

36.6.2   Doctrine

Main articles: Nuclear strategy, Deterrence theory, and Assured destruction

Israel's nuclear doctrine is shaped by its lack of strategic depth: a subsonic fighter jet could cross the 72 kilometres (39 nmi) from the Jordan River to the Mediterranean Sea in just 4 minutes. It additionally relies on a reservist-based military which magnifies civilian and military losses in its small population. Israel tries to compensate for these weaknesses by emphasising intelligence, maneuverability and firepower.[183]

As a result, its strategy is based on the premise that it cannot afford to lose a single war, and thus must prevent them by maintaining deterrence, including the option of preemption. If these steps are insufficient, it seeks to prevent escalation and determine a quick and decisive war outside of its borders.[183]

Strategically, Israel's long-range missiles, nuclear-capable aircraft, and possibly its submarines present an effective second strike deterrence against unconventional and conventional attack, and if Israel's defences fail and its population centers are threatened, the Samson Option, an all-out attack against an adversary, would be employed. Its nuclear arsenal can also be used tactically to destroy military units on the battlefield.[183]

Although nuclear weapons are viewed as the ultimate guarantor of Israeli security, as early as the 1960s the country has avoided building its military around them, instead pursuing absolute conventional superiority so as to forestall a last-resort nuclear engagement.[183]

According to historian Avner Cohen, Israel first articulated an official policy on the use of nuclear weapons in 1966, which revolved around four "red lines" that could lead to a nuclear response:[194]

1. A successful military penetration into populated areas

within Israel's post-1949 (pre-1967) borders.

2. The destruction of the Israeli Air Force.

3. The exposure of Israeli cities to massive and devastating air attacks or to possible chemical or biological attacks.

4. The use of nuclear weapons against Israeli territory.

36.6.3 Use

On October 8, 1973, just after the start of the Yom Kippur War, Golda Meir and her closest aides decided to put eight nuclear armed F-4s at Tel Nof Airbase on 24-hour alert and as many nuclear missile launchers at Sedot Mikha Airbase operational as possible. Seymour Hersh adds that the initial target list that night "included the Egyptian and Syrian military headquarters near Cairo and Damascus".[195] This nuclear alert was meant not only as a means of precaution, but to push the Soviets to restrain the Arab offensive and to convince the US to begin sending supplies. One later report said that a Soviet intelligence officer did warn the Egyptian chief of staff, and colleagues of US National Security Advisor Henry Kissinger said that the threat of a nuclear exchange caused him to urge for a massive Israeli resupply.[196] Hersh points out that before Israel obtained its own satellite capability, it engaged in espionage against the United States to obtain nuclear targeting information on Soviet targets.[197]

Israeli military and nuclear doctrine increasingly focused on preemptive war against any possible attack with conventional, chemical, biological or nuclear weapons, or even a potential conventional attack on Israel's weapons of mass destruction.[1][198]

Louis René Beres, who contributed to Project Daniel, urges that Israel continue and improve these policies, in concert with the increasingly preemptive nuclear policies of the United States, as revealed in the Doctrine for Joint Nuclear Operations.[199]

After Iraq attacked Israel with Scud missiles during the 1991 Gulf War, Israel went on full-scale nuclear alert and mobile nuclear missile launchers were deployed.[200] In the buildup to the United States 2003 invasion of Iraq, there were concerns that Iraq would launch an unconventional weapons attack on Israel. After discussions with President George W. Bush, the then Israeli Prime Minister Ariel Sharon warned "If our citizens are attacked seriously — by a weapon of mass destruction, chemical, biological or by some mega-terror attack act — and suffer casualties, then Israel will respond." Israeli officials interpreted President Bush's stance as allowing a nuclear Israeli retaliation on Iraq, but only if Iraq struck before the U.S. military invasion.[201]

36.6.4 Maintaining nuclear superiority

Alone or with other nations, Israel has used diplomatic and military efforts as well as covert action to prevent other Middle Eastern countries from acquiring nuclear weapons.[202]

For example, it is believed that Israel filed a false laser patent in the late 1970s to mislead Arab nuclear research. Mossad agents triggered explosions in April 1979 at a French production plant near Toulouse, damaging the two reactor cores destined for the Iraqi reactors. Mossad agents may also have been behind the assassinations of an Egyptian nuclear engineer in Paris as well as two Iraqi engineers, all working for the Iraqi nuclear program.[203]

On June 7, 1981, Israel launched a preemptive air strike against Saddam Hussein's breeder reactor in Osirak, Iraq, in Operation Opera. The Mossad — as well as any number of other intelligence agencies — are also frequently said to have assassinated professor Gerald Bull, an artillery expert, who was allegedly building a massive cannon or "super gun" for Saddam Hussein in the 1980s, which was capable of delivering a tactical nuclear payload.[204]

On September 6, 2007, Israel launched an air strike dubbed Operation Orchard against a target in the Deir ez-Zor region of Syria. While Israel refused to comment, unnamed US officials said Israel had shared intelligence with them that North Korea was cooperating with Syria on some sort of nuclear facility.[205] Both Syria and North Korea denied the allegation and Syria filed a formal complaint with the United Nations.[206][207] The International Atomic Energy Agency concluded in May 2011 that the destroyed facility was "very likely" an undeclared nuclear reactor.[208]

Journalist Seymour Hersh speculated that this air strike might have been intended as a trial run for striking alleged Iranian nuclear weapons facilities.[209] On January 7, 2007, *The Sunday Times* reported that Israel had drawn up plans to destroy three Iranian nuclear facilities with low-yield nuclear bunker-busters that would be launched by aircraft through "tunnels" created by conventional laser-guided bombs. These tactical nuclear weapons would then explode underground to reduce radioactive fallout.[210] Israel swiftly denied the specific allegation and analysts expressed doubts about its reliability.[211] However, in 2004 its then Defense minister said that it rules out no option.[212] The death of the Iranian physicist Ardeshir Hassanpour, who may have been involved in the nuclear program, has been reported by the intelligence group Stratfor to have been a Mossad assassination.[213] Iran is currently conducting atomic research that Israel fears is aimed at building a nuclear weapon. Israel has pressed for United Nations economic sanctions against Iran,[214] and has repeatedly threatened to launch a military strike on Iran if the United States does not do so first.[28][215][216]

The 2010 Stuxnet malware – which targeted Iran's nuclear program – is widely believed to have been sponsored by Israel. In 2009, a year before Stuxnet was discovered, researcher Scott Borg suggested that Israel might prefer to mount a cyber-attack rather than a military strike on Iran's nuclear facilities.[217][218] Iran uses IR-1 centrifuges at Natanz, which are based on the P-1 centrifuge, the design A. Q. Khan stole in 1976 and took to Pakistan.[219] His black market nuclear-proliferation network sold P-1s to, among other customers, Iran and Libya. Experts believe that Israel also somehow acquired P-1s and tested Stuxnet on the centrifuges, installed at the Dimona facility that is part of its own nuclear program.[220] The equipment may be from the United States, which received P-1s from Libya's former nuclear program.[221][220]

36.6.5 Nuclear Non-Proliferation Treaty and United Nations' Resolutions

Israel was originally expected to sign the 1968 Nuclear Non-Proliferation Treaty (NPT) and on June 12, 1968, Israel voted in favor of the treaty in the UN General Assembly. But when the invasion of Czechoslovakia in August by the Soviet Union delayed ratification around the world, Israel's internal division and hesitation over the treaty became public.[222] The Johnson administration attempted to use the sale of 50 F-4 Phantoms to pressure Israel to sign the treaty that fall, culminating in a personal letter from Lyndon Johnson to Israeli PM Levi Eshkol. But by November Johnson had backed away from tying the F-4 sale with the NPT after a stalemate in negotiations, and Israel would neither sign nor ratify the treaty.[223] After the series of negotiations, U.S. assistant secretary of defense for international security Paul Warnke was convinced that Israel already possessed nuclear weapons.[224] In 2007 Israel sought an exemption to non-proliferation rules in order to import atomic material legally.[225]

In 1996, the United Nations General Assembly passed a resolution[226] calling for the establishment of a nuclear-weapon-free zone in the region of the Middle East.[227] Arab nations and annual conferences of the International Atomic Energy Agency (IAEA) repeatedly have called for application of IAEA safeguards and the creation of a nuclear-free Middle East. Arab nations have accused the United States of practicing a double standard in criticizing Iran's nuclear program while ignoring Israel's possession of nuclear weapons.[228][229][230] According to a statement by

the Arab League, Arab states will withdraw from the NPT if Israel acknowledges having nuclear weapons but refuses to open its facilities to international inspection and destroy its arsenal.[231]

In a statement to the May 2009 preparatory meeting for the 2010 NPT Review Conference, the US delegation reiterated the longstanding US support for "universal adherence to the NPT", but uncharacteristically named Israel among the four countries that have not done so. An unnamed Israeli official dismissed the suggestion that it would join the NPT and questioned the effectiveness of the treaty.[232] *The Washington Times* reported that this statement threatened to derail the 40-year-old secret agreement between the U.S. and Israel to shield Israel's nuclear weapons program from international scrutiny,[233] while Avner Cohen, author of *Israel and the Bomb*, argued that acknowledging its nuclear program would allow Israel to take part constructively in efforts to control nuclear weapons.[234]

The Final Document of the 2010 NPT Review Conference called for a conference in 2012 to implement a resolution of the 1995 NPT Review Conference that called for the establishment of a Middle East Zone free of weapons of mass destruction. The United States joined the international consensus for Final Document, but criticized the section on the Middle East resolution for singling out Israel as the only state in the region that is not party to the NPT, while at the same time ignoring Iran's "longstanding violation of the NPT and UN Security Council Resolutions."[235]

36.7   Notes

[1] Data from the CIA.[111]

36.8   References

[1] Farr 1999.

[2] http://www.wisconsinproject.org/countries/israel/nuke.html

[3] "WRMEA – Mohammed Omer Wins Norwegian PEN Prize". *Washington Report on Middle East Affairs*.

[4] https://books.google.com/books?id=VAwAAAAAMBAJ&pg=PA1&hl=iw#v=onepage&q&f=false. Page 42

[5] "Nuclear weapons – Israel". Federation of American Scientists. Retrieved July 1, 2007.

[6] Brower 1997. Brower notes that he makes a high estimate of the number of weapons.

[7] "Status of World Nuclear Forces". Federation of American Scientists. Retrieved July 28, 2012.

[8] Feikert, Andrew (March 5, 2004), *Missile Survey: Ballistic and Cruise Missiles of Foreign Countries* (PDF), Congressional Research Service.

[9] Cohen 1998a, p. 349.

[10] ElBaradei, Mohamed (July 27, 2004). "Transcript of the Director General's Interview with Al-Ahram News". International Atomic Energy Agency. Retrieved June 3, 2007.

[11] "Nuclear Overview" (profile). *Israel*. NTI. Retrieved June 23, 2009.

[12] *My Promised Land*, by Ari Shavit, (London 2014), page 188

[13] "Background Information". *Review Conference of the Parties to the Treaty on the Non-Proliferation of Nuclear Weapons*. United Nations. 2005. Retrieved July 2, 2006.

[14] Bronner, Ethan (October 13, 2010). "Vague, Opaque and Ambiguous — Israel's Hush-Hush Nuclear Policy". *The New York Times*. Retrieved March 6, 2012.

[15] Korb, Lawrence (November 1, 1998). "The Quiet Bomb". *The New York Times*. Retrieved March 6, 2012.

[16] "Foreign Relations of the United States, 1964–1968" (historical documents). *Office of the Historian*. Department of State. December 12, 1968. Document 349. Retrieved July 3, 2012.

[17] Avni, Benny,"Iran and Syria Eye Israel's Nukes", *Newsweek*, October 17, 2013. The Avni source and others misspell Robert W. Komer's last name. For confirmation of correct spelling and confirmed identification, for example: Avner Cohen, *Israel and the Bomb* (Columbia University Press) p. 207; or Seymour Hersh, *The Samson option: Israel's nuclear arsenal* (NY: Random House, 1991), p. 134; both via Google Books. Retrieved 2015-04-04.

[18] Ami Gluska,*The Israeli Military and the Origins of the 1967 War: Government, Armed Forces and Defence Policy 1963–67* Routledge, 2007 p.30. The U.S.signatory, Robert Comer (*sic*; see Avni footnote), was a U.S. National Security Council advisor.

[19] "Foreign Relations of the United States, 1964–1968" (historical documents). *Office of the Historian*. Department of State. March 11, 1965. Document 185. Retrieved July 3, 2012.:"The Government of Israel has reaffirmed that Israel will not be the first to introduce nuclear weapons into the Arab-Israel area."

[20] *Israel Rejects Offer to Join UN Atomic Agency*, Shalom Life, September 21, 2010.

[21] "Israel's Quest for Yellowcake: The Secret Argentina-Israel Connection, 1963–1966". Nuclear Proliferation International History Project http://www.wilsoncenter.org/publication/israels-quest-for-yellowcake-the-secret-argentina-israel

[22] John Pike. "Nuclear Weapons". *globalsecurity.org*.

[23] "Nuclear Weapons". *fas.org*.

[24] "Revealed: the secrets of Israel's nuclear arsenal", *Sunday Times*, pp. 1, 4–5, October 5, 1986.

[25] "Vanunu: Israel's nuclear telltale". BBC. April 20, 2004. Retrieved October 17, 2012.

[26] Townsley, Edwin S. and Clarence A. Robinson (April 1987). "Critical Technological Assessment in Israel and NATO Nations" (PDF).

[27] DOD Report Details Israel's Quest for Hydrogen Bomb, Janet McMahon, February 12, 2015

[28] Goldberg, Jeffrey. "The Point of No Return" *The Atlantic*, September 2010.

[29] Pinkus, Binyamin; Tlamim, Moshe (Spring 2002). "Atomic Power to Israel's Rescue: French-Israeli Nuclear Cooperation, 1949–1957". *Israel Studies*. **7** (1): 104–38. doi:10.1353/is.2002.0006.

[30] Cohen 1998a, p. 26.

[31] Cohen 1998a, pp. 30–1.

[32] Hersh 1991, p. 19.

[33] Cohen 1998a, pp. 33–4.

[34] Hersh 1991, p. 30.

[35] Cohen 1998a, p. 44.

[36] Cohen 1998a, p. 65.

[37] "WRMEA – Mohammed Omer Wins Norwegian PEN Prize". *Washington Report on Middle East Affairs*.

[38] Hersh 1991, p. 37.

[39] Cohen 1998a, pp. 53–4.

[40] Hersh 1991, pp. 42–43.

[41] Cohen 1998a, p. 59.

[42] Hersh 1991, pp. 45–6.

[43] Thomas, Gordon (1999), *Gideon's Spies: The Secret History of the Mossad*.

[44] Hersh 1991, pp. 60–1.

[45] Hoffman, Gil (June 22, 2010). "Netanyahu: Pollard acted as Israeli agent". *Jerusalem Post*. Retrieved June 23, 2010. One of those agents, Arnon Milchan, was considered among the most successful and prolific agents who secured items such a uranium enrichment centrifuges and Krytron nuclear triggers, would latter become one of the most successful movie producers in Hollywood history.

[46] Cieply, Michael (July 17, 2011). "New Book Tells Tale of Israeli Arms Dealer in Hollywood". *The New York Times*.

[47] "Lekem" Federation of American Scientists

[48] "Report of an Investigation Commission on the Pollard Case" Jewish Virtual Library

[49] Cohen 1998a, pp. 73–4.

[50] Cohen 1998a, p. 75.

[51] Hersh 1991, p. 70.

[52] "The Israel-Argentina Yellowcake Connection". *National Security Archive*. George Washington University. June 25, 2013. Retrieved August 6, 2013.

[53] 'Argentina sold Israel yellowcake uranium in 1960s' – *The Jerusalem Post*

[54] *Atomic Activities in Israel* (PDF). *UK Cabinet Submission from Joint Intelligence Bureau*. Cabinet Office, Government of the United Kingdom. July 17, 1961. JIC/1103/61. Retrieved July 2, 2006.

[55] *Secret Atomic Activities in Israel* (PDF). *UK Cabinet Submission from Joint Intelligence Bureau*. Cabinet Office, Government of the United Kingdom. March 27, 1961. JIC/519/61. Retrieved July 2, 2006.

[56] Secret sale of UK plutonium to Israel, BBC, March 10, 2006

[57] Crick, Michael, "How Britain helped Israel get the bomb", *Newsnight*, United Kingdom: BBC.

[58] Jones, Meirion (March 13, 2006). "Britain's dirty secret". New Statesman. Retrieved July 2, 2006.

[59] "Statement from the Foreign Office". *Newsnight*. BBC. March 9, 2006. Retrieved July 2, 2006.

[60] *Norway's Heavy Water Scandals* (editorial), Wisconsin project, September 14, 1988, retrieved June 4, 2011.

[61] Report: Argentina sold yellowcake to Israel for nuclear program – *Israel Hayom*

[62] Hersh, p. 130.

[63] Cohen 1998a, p. 70.

[64] Hersh 1991, pp. 66–7.

[65] Cohen 1998, p. 298.

[66] "Special National Intelligence Estimate 11-12-68: Emplacement of Weapons of Mass Destruction on the Seabed" (PDF). Central Intelligence Agency. August 15, 1968.

[67] "Violent Week: The Politics of Death". *Time*. April 12, 1976. Retrieved March 4, 2011.

[68] Hersh 1991, p. 179–80.

[69] Hersh 1991, p. 181.

[70] Cohen, Avner; Burr, William (May–June 2006). "Israel crosses the threshold". *Bulletin of the Atomic Scientists*. pp. 22–30.

[71] Laird, Melvin (March 17, 1969). "Stopping the introduction of nuclear weapons into the Middle East" (PDF). *Memorandum to the secretary of state*. National Security Archive. Retrieved July 2, 2006.

[72] Kissinger, Henry (October 7, 1969). "Discussions with the Israelis on nuclear matters" (PDF). *Memorandum for the President*. National Security Archive. Retrieved July 2, 2006.

[73] Gilinsky, Victor (former Commissioner U.S. Nuclear Regulatory Commission) (May 13, 2004). "Israel's Bomb". The New York Review of Books. Retrieved December 8, 2007.

[74] Burnham, David (January 27, 1978). "CIA said in 1974 Israel had A-bombs" (JPEG). *The New York Times* (image). Iran affairs. p. A5. Retrieved December 8, 2007.

[75] *Prospects for Further Proliferation of Nuclear Weapons* (PDF). Special National Intelligence Estimate. CIA. August 23, 1974. SNIE 4-1-74. Retrieved January 20, 2008.

[76] *The 22 September 1979 Event* (PDF). Interagency Intelligence Memorandum. National Security Archive. December 1979. pp. 5, 9 (paragraphs 4, 26). MORI DocID: 1108245. Retrieved November 1, 2006.

[77] "Mordechai Vanunu: The Sunday Times articles". London: The Times. April 21, 2004. Archived from the original on May 13, 2006. Retrieved July 2, 2006.

[78] "Israel's Nuclear Weapons Program". Nuclear Weapon Archive. December 10, 1997. Retrieved October 7, 2007.

[79] "Israel's Nuclear Shopping List", *The Risk Report*, Wisconsin project on nuclear arms control, **2** (4), July–August 1996, retrieved March 29, 2012

[80] Leibovitz-Dar, Sara (May 21, 2010), "This secret is fiction", *Maariv-Amusaf Le'Shabat*, NRG, pp. 10–3.

[81] "Did Israel play a role in 1979 South Africa nuclear test?". *Haaretz.com*. August 2, 2009.

[82] "The Nth Power", *Time* magazine, December 19, 1960. Retrieved July 2, 2007.

[83] Hersh 1991, p. 72.

[84] Cohen 1998a, pp. 88–9.

[85] Cohen 1991, p. 91.

[86] "One on One: Existential espionage", *Jerusalem Post*, retrieved June 4, 2011.

[87] Gallagher, Michael (July 30, 2005). *Israel and Palestine*. Black Rabbit Books. pp. 26–. ISBN 978-1-58340-605-2. Retrieved June 4, 2011.

[88] Cohen 1998a, p. 327.

[89] Cohen 1998a, p. 338.

[90] http://www.wisconsinproject.org/countries/israel/israel-aims.html

[91] "Israel has 80 nuclear warheads, report says". *The Times of Israel*.

[92] "Report: Israel halted nuclear warheads production in 2004". *ynet*.

[93] Carter says Israel has stockpile of over 300 nuclear bombs by Yoni Hirsch and Israel Hayom Staff, Israel Hayom, April 14, 2014

[94] McGreal, Chris (May 24, 2010). "Revealed: how Israel offered to sell South Africa nuclear weapons". *The Guardian*. London. Retrieved May 24, 2010.

[95] McGreal, Chris (May 24, 2010). "The memos and minutes that confirm Israel's nuclear stockpile". *The Guardian*. London. Retrieved May 24, 2010.

[96] Von Wielligh, N. & von Wielligh-Steyn, L. (2015). The Bomb – South Africa's Nuclear Weapons Programme. Pretoria: Litera.

[97] "S. African official doubts nuclear arms sale offer", *Ynet news*, May 24, 2010, retrieved June 4, 2011.

[98] "Avner Cohen: Yitzhak Rabin would have opposed sale of nuclear weapons". *The Independent*. London. May 25, 2010.

[99] The Relevance of Mordechai Vanunu's Disclosures to Israel's National Security

[100] When Ben-Gurion said no to JFK – *The Jerusalem Post*

[101] "Israel's Secret Uranium Buy". *Foreign Policy*.

[102] Israel's Nuclear Arsenal Vexed Nixon – *The New York Times*, November 29, 2007

[103] Burrows & Windrem 1994, p. 280.

[104] Cohen 1998a, pp. 273–4.

[105] *Time*, April 12, 1976, quoted in Weissman and Krosney, op. cit., 107.

[106] Tahtinen, Dale R., *The Arab-Israel Military Balance Today* (Washington, DC: American Enterprise Institute for Public Policy Research, 1973), 34.

[107] Burrows & Windrem 1994, p. 302.

[108] Kaku, op. cit., 66.

[109] Hersh 1991, p. 216.

[110] Valéry, op. cit., 807–9.

[111] Weissman & Krosney 1981, p. 109.

[112] Israel Profile. Nti.org. Retrieved on June 4, 2011.

[113] Ottenberg, Michael, "Estimating Israel's Nuclear Capabilities", *Command*, 30 (October 1994), 6–8.

[114] Pry, op. cit., 75.

[115] Pry, op. cit., 111.

[116] Data from NBC Nightly News, quoted in Milhollin, op. cit., 104.

[117] Burrows & Windrem 1994, p. 308.

[118] Data from Vanunu quoted in Milhollin, op. cit., 104.

[119] Harkavy, Robert E. "After the Gulf War: The Future of the Israeli Nuclear Strategy", *The Washington Quarterly* (Summer 1991), 164.

[120] Albright, David, Berkhout, Frans and Walker, William, *Plutonium and Highly Enriched Uranium* 1996. World Inventories, Capabilities, and Policies (New York: Stockholm International Peace Research Institute and Oxford University Press, 1997), 262–63.

[121] Hough, Harold, "Israel's Nuclear Infrastructure", *Jane's Intelligence Review* 6, no. 11 (November 1994),508.

[122] *Critical Mass: the Dangerous Race for Super-weapons in a Fragmenting World*, New York, 1994, p.308

[123] Spector, McDonough & Medeiros 1995, p. 135.

[124] Burrows & Windrem 1994, pp. 283–4.

[125] Cordesman 1996, p. 234.

[126] Norris, Robert S., William Arkin, Hans M. Kristensen, and Joshua Handler. Bulletin of the Atomic Scientists 58:5 (September/October 2002): 73–75. Israeli nuclear forces

[127] Scarborough, Rowan. *Rumsfeld's War: The Untold Story of America's Anti-Terrorist Commander*

[128] BBC News. "Israel 'has 150 nuclear weapons'", May 26, 2008. Statement by former United States president Jimmy Carter.

[129] Stockholm International Peace Research Institute (2008). *SIPRI Yearbook 2008: Armaments, Disarmament, and International Security*. United States: Oxford University Press. p. 397. ISBN 978-0-19-954895-8.

[130] Toukan, Abdullah (March 14, 2009), *Study on a Possible Israeli Strike on Iran's Nuclear Development Facilities* (PDF). CSIS.

[131] "Analysts: Israel viewed as world's 6th nuclear power". Ynet. Retrieved May 26, 2010.

[132] Kristensen, Hans M.; Norris, Robert S. (2014). "Israeli nuclear weapons, 2014" (PDF). *Bulletin of the Atomic Scientists*. **70** (6): 97–115. doi:10.1177/0096340214555409.

[133] Frantz, Douglas (October 12, 2003), "Israel Adds Fuel to Nuclear Dispute, Officials confirm that the nation can now launch atomic weapons from land, sea and air", *The Los Angeles Times*.

[134] Lennox, Duncan, ed. (January 2007), *Strategic Weapons Systems*, **46**, Jane's, pp. 82–3.

[135] Azoulay, Yuval (January 18, 2008). "Missile test 'will improve deterrence'". *Haaretz*. Retrieved June 4, 2011.

[136] Plushnick-Masti, Ramit (August 25, 2006). "Israel Buys 2 Nuclear-Capable Submarines". *The Washington Post*. Retrieved May 20, 2010.

[137] "Israel – Country Profiles – NTI". *NTI: Nuclear Threat Initiative*.

[138] Hersh 1991, p. 104.

[139] Hersh 1991, pp. 120, 173–4.

[140] Steinberg, Gerald, "Missiles", *Arms*, IL: BIU, retrieved June 4, 2011.

[141] *Commission to Assess the Ballistic Missile Threat to the United States*, FAS, retrieved June 4, 2011.

[142] *Missile Proliferation and Defences: Problems and Prospects* (PDF), MIIS, retrieved June 4, 2011.

[143] Brown, Irene (January 9, 2003). "Space Programs Thriving in Israel". *Jewish Journal*. Retrieved June 4, 2011.

[144] Hodge, Nathan (April 2, 2009). "Inside Israel's (Possible) Strike on Iran". Wired. Retrieved June 4, 2011.

[145] Butcher, Tim (January 18, 2008). "Israel test-launches nuclear-capable missile". *The Daily Telegraph*. London. Retrieved May 20, 2010.

[146] Israel carries out two-stage ballistic missile launch, Richardson, D (March 2008), *Missiles & Rockets*, Jane, **12** (3) Missing or empty |title= (help).

[147] Toukan, Abdullah (March 14, 2009), *Study on a Possible Israeli Strike on Iran's Nuclear Development Facilities* (PDF), Center for Strategic and International Studies.

[148] "Israel test-fires ballistic missile after Iran warning", *AFP*, Google, January 17, 2008, retrieved June 4, 2011.

[149] *Israel says carries out missile launching test*, Reuters, January 17, 2008.

[150] Katz, Y., "Israel test-fires long-range ballistic missile", *The Jerusalem Post*, January 17, 2008.

[151] "Israel tests new long-range missile", *USA Today*, January 17, 2008, retrieved June 4, 2011.

[152] "Report: Syria, North Korea hold high-level talks in Pyongyang", *Haaretz*, September 21, 2007, archived from the original on October 13, 2008.

[153] *Report No. 2000/09: Ballistic Missile Proliferation*, Canada: CSIS-SCRS, February 25, 2011, retrieved June 4, 2011.

[154] "Zachariah – Israel – Special Weapons Facilities", *Weapons of mass destruction*, Global security, retrieved June 4, 2011.

[155] "F-16 Falcon", *IS nukes*, CDI, retrieved June 4, 2011.

[156] "Israel's F-16 Warplanes Likely to Carry Nuclear Weapons: Report", *People daily*, CN, August 20, 2002, retrieved June 4, 2011.

[157] *Proliferation of Weapons of Mass Destruction: Assessing the Risks* (PDF). U.S. Congress Office of Technology Assessment. August 1993. OTA-ISC-559. Retrieved December 9, 2008.

[158] Norris, Robert S; Kristensen, Hans M (November–December 2004). "U.S. Nuclear Weapons in Europe, 1954–2004". *Nuclear Notebook*. Bulletin of the Atomic Scientists. Retrieved December 22, 2013. External link in |work= (help)

[159] "Israel Takes Delivery of 2 German-Built U212 Submarines", *Defense Technology News*, Google Blogger, September 30, 2009, retrieved June 4, 2011.

[160] "Dolphin Class Submarines". IL: Uri Dotan-Bochner. Archived from the original on June 28, 2006. Retrieved July 2, 2006.

[161] "USAF Counterproliferation Center: Emerging Biocruise Threat". *af.mil*.

[162] "Popeye Turbo", *Israel Special Weapons*, Global security, retrieved June 4, 2011

[163] "Nuclear Weapons Inventories of the Eight Known Nuclear Powers" (PDF). PLRC. Retrieved November 2, 2007.

[164] Ben-David 2009.

[165] Blanche, Ed (August 1, 2000), "Israel denies sub-launched missile tests", *Jane's Missile and Rockets*.

[166] "Gabriel", *Jane's Strategic Weapon Systems*, August 28, 2003.

[167] Mahnaimi, Uzi; Campbell, Matthew (June 18, 2000), "Israel Makes Nuclear Waves with Submarine Missile Test", *The Sunday Times*, London, England.

[168] *Israel Moves — Quickly — To Beef Up Its Submarine Force*, Stratfor, October 26, 2000.

[169] "Israeli nuclear submarine sailed Suez Canal to the Red Sea", *World bulletin*, July 3, 2009, retrieved June 4, 2011.

[170] "Israeli sub sails Suez, signaling reach to Iran". *Reuters*. July 3, 2009.

[171] *Electrical Equipment for Naval Vessels and Submarines* (PDF) (brochure), Siemens, retrieved June 4, 2011.

[172] "Israel Takes Delivery of 2 German-Built U212 Subs", *Defense news*, retrieved June 4, 2011.

[173] Plushnick-Masti, Ramit. "Israel Buys 2 Nuclear-Capable Submarines", *The Washington Post*, August 25, 2006. Retrieved July 4, 2007.

[174] Bergman, Ronen; Follath, Erich; Keinan, Einat; Nassauer, Otfried; Schmitt, Jorg; Stark, Holger; Weigold, Thomas; Wiegrefe, Klaus (June 4, 2012). "Israel's Deployment of Nuclear Missiles on Subs from Germany". *Der Spiegel*. Retrieved June 4, 2012.

[175] "Report: Israel fitting nuclear missiles on German-built subs". CBS News. June 4, 2012. Retrieved June 7, 2012.

[176] David Gordon Smith (June 5, 2012). "Helping Israel Defend Itself Is Germany's Duty'". *Der Spiegel*. Retrieved June 4, 2012.

[177] Hersh 1991, p. 220.

[178] Isenberg, David (January 3, 2008), "Sneak peek at a desert Armageddon", *Asia Times* (online ed.), HK, retrieved June 4, 2011.

[179] "US institute: Israel could survive nuclear war", *The Jerusalem Post*, retrieved June 4, 2011.

[180] *Electromagnetic Pulse Threats To US Military And Civilian Infrastructure*, United States: House, retrieved June 4, 2011.

[181] *Threat Posed by Electromagnetic Pulse (EMP) to U.S. Military Systems and Civil*, Global security, retrieved June 4, 2011.

[182] Cordesman, Anthony H (August 30, 2006), *Iran's Nuclear and Missile Programs: A Strategic Assessment* (PDF), Arleigh A. Burke Chair in Strategy Center for Strategic and International Studies, retrieved June 4, 2011.

[183] "Strategic Doctrine". Global Security. April 28, 2005.

[184] "Doctrine", *Israel*, FAS.

[185] Cohen 1998a, p. 343.

[186] Cohen 1998a, pp. 233–4.

[187] Cohen, Avner; Burr, William (April 30, 2006), "The Untold Story of Israel's Bomb", *The Washington Post*, p. B01.

[188] Kissinger, Henry A (July 16, 1969), *Memorandum for the President: Israeli Nuclear Program* (PDF), The White House, retrieved July 26, 2009

[189] Cohen 1998a, pp. 277, 291.

[190] "Peres admits to Israeli nuclear capability". Federation of American Scientists. July 14, 1998. Retrieved July 2, 2006.

[191] "Israel and the Bomb: Principal players". National Security Archive. Retrieved July 2, 2006.

[192] "Olmert: Iran wants nuclear weapons like Israel". *Y net news*. Retrieved December 11, 2006.

[193] "Olmert Says Israel Among Nuclear Nations". Archived from the original on December 15, 2006. Retrieved December 11, 2006.

[194] Cohen 1998a, p. 237.

[195] Hersh 1991, p. 225.

[196] Hersh 1991, pp. 227, 230.

[197] Hersh 1991, pp. 17, 216, 220, 286, 291–96.

[198] Beres, Louis René (2003), *Israel's Bomb in the Basement: Reconsidering a Vital Element of Israeli Nuclear Deterrence*, IL: ACR.

[199] Beres, Louis René (Spring 2007), "Israel's Uncertain Strategic Future", *Parameters*, United States: Army: 37–54.

[200] Hersh 1998a, p. 318.

[201] Dunn, Ross (November 3, 2002), "Sharon eyes 'Samson option' against Iraq", *The Scotsman*.

[202] Schiff, Ze'ev (May 30, 2006), *Israel Urges US Diplomacy on Iran*, Carnegie Endowment.

[203] Reiter, D (2005). "Preventive attacks against nuclear programs and the "success" at Osiraq" (PDF). *Nonproliferation Review*. **12** (2): 355–71. doi:10.1080/10736700500379008.

[204] *The Israeli Intelligence Services: Deception and Covert Action Operations*, History of War.

[205] Kessler, Glenn (September 13, 2007), "N. Korea, Syria May Be at Work on Nuclear Facility", *The Washington Post*, p. A12.

[206] *Syria Complains to UN*, JTA.

[207] Doyle, Leonard (September 18, 2007), "Syria says U.S. nuclear claims are 'false,' biased toward Israel", *Ha'aretz*, Associated Press.

[208] Brannan, Paul. "Analysis of IAEA Report on Syria: IAEA Concludes Syria "Very Likely" Built a Reactor" (PDF). Institute for Science and International Security. Retrieved May 24, 2011.

[209] Hersh, Seymour (February 11, 2008). "A Strike in the Dark". *The New Yorker*. Retrieved February 23, 2008.

[210] Mahnaimi, Uzi; Baxter, Sarah (January 7, 2007), "Revealed: Israel plans nuclear strike on Iran", *The Sunday Times*, retrieved July 3, 2007.

[211] "Israel denies planning Iran nuclear attack, UK newspaper reports Israel intends to strike up to three targets in Iran", *MSNBC*, The Associated Press, January 7, 2007.

[212] "Israel Takes Issue With Iran Weapons", *Yahoo!*, The Associated press, September 29, 2004, archived from the original on August 8, 2011.

[213] "Israeli Covert Operations in Iran", *Geopolitical Diary*, Stratfor, May 31, 2011, retrieved June 4, 2011.

[214] *Foreign Minister urges tougher UN sanctions against Iran*, Associated press, September 13, 2007.

[215] Scarborough, Rowan (February 21, 2005), "Israel pushes US on Iran nuke solution", *The Washington Times*.

[216] Coughlin, Con (February 24, 2007), "Israel seeks all clear for Iran air strike", *The Daily Telegraph*, Tel Aviv, archived from the original on November 5, 2007.

[217] Heller, Jeffrey; Trevelyan, Mark (July 7, 2009). "Analysis: Wary of naked force, Israelis eye cyberwar on Iran". *Reuters*. Retrieved November 19, 2010.

[218] "A worm in the centrifuge: An unusually sophisticated cyber-weapon is mysterious but important". *The Economist*. September 30, 2010. Retrieved March 29, 2012.

[219] "A.Q. Khan". globalsecurity.org. Retrieved April 3, 2013.

[220] Broad, William J; Markoff, John; Sanger, David E (January 15, 2011). "Israel Tests on Worm Called Crucial in Iran Nuclear Delay". *The New York Times*. Retrieved January 16, 2011.

[221] Sanger, David (September 25, 2010). "Iran Fights Malware Attacking Computers". *The New York Times*. Retrieved September 28, 2010.

[222] Cohen 1998a, pp. 300–1.

[223] Cohen 1998a, p. 315.

[224] Cohen 1998a, pp. 318–9.

[225] Jahn, George (September 25, 2007), *Israel Seeks Exemption From Atomic Rules*, The Associated Press

[226] United Nations General Assembly Session 51 *Resolution 41.Establishment of a nuclear-weapon-free zone in the region of the Middle East* **A/RES/51/41** December 10, 1996. Retrieved August 23, 2008.

[227] *United Nations General Assembly Resolution 51/41*, Jewish virtual library, December 10, 1996, retrieved June 4, 2011.

[228] Pincus, Walter (March 6, 2005), "Push for Nuclear-Free Middle East Resurfaces; Arab Nations Seek Answers About Israel", *The Washington Post*, p. A24.

[229] "Israel-Arab spat at nuclear talks", *News*, United Kingdom: BBC, September 28, 2005.

[230] *IAEA conference urges efforts for nuclear-free Mideast*, Xin Hua net, September 21, 2007.

[231] "Arab League vows to drop out of NPT if Israel admits it has nuclear weapons". *Haaretz*. March 5, 2008. Retrieved March 10, 2008.

[232] "Stronger Nuclear Nonproliferation Treaty Needed, Obama Says", *Global Security Newswire*, May 6, 2009.

[233] Lake, Eli (May 6, 2009), "Secret US-Israel Nuclear Accord In Jeopardy", *The Washington Times*, p. 1.

[234] Cohen, Avner (May 6, 2009), "Nuclear ban benefits for Israel", *The Washington Times*.

[235] Jones, General James L (May 28, 2010), *National Security Advisor on the Non-Proliferation Treaty Review Conference* (statement), White House, Office of the Press Secretary.

36.9 Bibliography

Adam Raz, The Struggle for the Bomb, Carmel Publishing House, Jerusalem 2015 [Heb.]

- Brower, Kenneth S (February 1997), "A Propensity for Conflict: Potential Scenarios and Outcomes of War in the Middle East", *Jane's Intelligence Review* (special report) (14): 14–5

- Burrows, William E; Windrem, Robert (1994), *Critical mass: the dangerous race for superweapons in a fragmenting world*, Simon and Schuster, ISBN 978-0-671-74895-1

- Cohen, Avner (1998a), *Israel and the Bomb* (hardcover), New York: Columbia University Press, ISBN 0-231-10482-0

- ——— (1998b), *Israel and the Bomb* (paperback), New York: Columbia University Press, ISBN 0-231-10483-9

- Cordesman, Anthony H (1996), *Perilous prospects: The peace process and the Arab-Israeli military balance*, Boulder, CO: Westview Press, ISBN 0-8133-2939-6

- Ben-David, Alon (October 1, 2009), "Israel seeks sixth Dolphin in light of Iranian 'threat'", *Jane's Defence Weekly*, retrieved June 4, 2011

- Farr, Warner D (September 1999), *The Third Temple's holy of holies: Israel's nuclear weapons*, The Counterproliferation Papers, Future Warfare Series, **2**, USAF Counterproliferation Center, Air War College, Air University, Maxwell Air Force Base, retrieved July 2, 2006

- Hersh, Seymour M (1991), *The Samson Option*, New York: Random House, ISBN 0-394-57006-5

- Rhodes, Richard (August 24, 2010), *The Twilight of the Bombs: Recent Challenges, New Dangers, and the Prospects for a World Without Nuclear Weapons*, Knopf Doubleday, ISBN 978-0-307-26754-2, retrieved June 4, 2011

- Spector, Leonard S; McDonough, Mark G; Medeiros, Evan S (1995), *Tracking nuclear proliferation: A guide in maps and charts*, Washington, DC: Carnegie Endowment for International Peace, ISBN 0-87003-061-2

- Weissman, Stephen 'Steve'; Krosney, Herbert (1981), *The Islamic bomb: the nuclear threat to Israel and the Middle East*, New York, NY: Times Books, p. 275, ISBN 978-0-8129-0978-4

36.10 External links

- Avner Cohen Collection, A collection of primary sources and interviews at the Nuclear Proliferation International History Project

- Beaumont, Peter; Urquhart, Conal (October 12, 2003), "Israel deploys nuclear arms in submarines", *The Observer*, United Kingdom: The Guardian

- Bisharat, George (December 2005), "Should Israel give up its nukes?", *The LA Times*, Pentagon study about nuclear nonproliferation in Middle East.

- Chirkin, Dmitry (January 15, 2003), "Marcus Klingberg, last KGB Spy to be Released in Israel", *Pravda*, RU

- Cohen, Avner, "Official documents", *Israel and the Bomb*

- ——— (Fall–Winter 2001), "Israel and Chemical/Biological Weapons: History, Deterrence, and Arms Control" (PDF), *The Nonproliferation Review*

- "Israel crosses the threshold — Israel, the bomb and the NPT in the Nixon era", *The Bulletin*, based on ——— (April 28, 2006), *documents*

- Gorman, Mark, *Bibliography of Israeli Nuclear Science Publications* (PDF), Federation of American Scientists

- Weitz, Yehiam (January 14, 2005), "History of a hot potato", *Haaretz*, IL: Tau

- "Israel", *Overview – WMD*, Nuclear Threat Initiative, May 2014.

- *The Israel Institute for Biological Research*

- "Israel", *Nuclear Files*, Nuclear Age Peace Foundation

- Nuclear Stockpiles Current information on nuclear stockpiles in Israel at Nuclear Files.org, Nuclear Age Peace Foundation

- *JIC Israel Nuclear file 1960–61* (PDF), Part 1, UK: Cabinet office

- *JIC Israel Nuclear file 1960–61* (PDF), Part 2, UK: Cabinet office

- *Time to Open the Nuclear Gates — Israel's "nuclear ambivalence" strategy*, O media

- *Annotated bibliography for the Israeli nuclear weapons program*, Alsos Digital Library on Nuclear Issues

36.11   Text and image sources, contributors, and licenses

36.11.1   Text

- **Nuclear weapon** *Source:* https://en.wikipedia.org/wiki/Nuclear_weapon?oldid=735929884 *Contributors:* AxelBoldt, Magnus Manske, TwoOneTwo, Trelvis, The Epopt, Dreamyshade, Sodium, ClaudeMuncey, Ansible, Eloquence, Mav, Wesley, Bryan Derksen, Robert Merkel, The Anome, Tarquin, AstroNomer, Taw, Manning Bartlett, Ed Poor, Alex.tan, AdamW, Andre Engels, Ted Longstaffe, Youssefsan, Arvindn, Rmhermen, Toby Bartels, SJK, Little guru, Roadrunner, Ray Van De Walker, SimonP, Maury Markowitz, Zoe, Graft, FlorianMarquardt, Hephaestos, Soulpatch, Tedernst, Olivier, Patrick, RTC, Infrogmation, JohnOwens, Michael Hardy, GABaker, Modster, Cprompt, DopefishJustin, Dante Alighieri, Norm, Dominus, Ixfd64, Bcrowell, Frank Shearar, Cameron Dewe, TakuyaMurata, GTBacchus, Dori, Eric119, Minesweeper, Alfio, Ronabop, Mkweise, Ellywa, Ahoerstemeier, Anders Feder, Snoyes, Angela, Kingturtle, Erzengel, BigFatBuddha, Aarchiba, Ugen64, Glenn, Djmutex, Vzbs34, Susurrus, Jiang, Oliezekat, Alex756, [212], Mxn, Ilyanep, Lommer, Conti, Pizza Puzzle, Rami Neudorfer, Trevor Lawson, Ehn, Vroman, Jengod, Malbi, Ec5618, Jonadab-enwiki, Timwi, David Newton, Dino, Jefelex, Daniel Quinlan, Jfeckstein, Fuzheado, Andrewman327, WhisperToMe, Zoicon5, Jessel, DJ Clayworth, Haukurth, Tpbradbury, Maximus Rex, E23-enwiki, Pacific1982, Saltine, Kaal, Nv8200pa, Tempshill, Zero0000, Omegatron, Babbler, Thue, Bevo, Xevi-enwiki, Shizhao, Topbanana, Toreau, Vaceituno, Stormie, Raul654, Pstudier, Bcorr, Jusjih, Johnleemk, Finlay McWalter, Skybunny, Owen, Stargoat, Jni, Riddley, Robbot, Ke4roh, Sander123, Astronautics-enwiki, ChrisO-enwiki, Nabeel, Fredrik, Kizor, PBS, Chris 73, Chocolateboy, Kadin2048, Romanm, Arkuat, Securiger, Lowellian, Ukuk-enwiki, Merovingian, Sverdrup, Rfc1394, Academic Challenger, SchmuckyTheCat, Texture, Meelar, Yacht, Rhombus, Bkell, Mervyn, Hadal, Victor, LX, TPK, Tsavage, Seth Ilys, Diberri, Dina, David Gerard, SimonMayer, Ancheta Wis, Alexwcovington, Benji Franklyn, DocWatson42, Christopher Parham, Oberiko, Mat-C, Sj, Kim Bruning, Inter, Tom harrison, Lupin, Ferkelparade, Fastfission, Dersen, Zigger, Karn, Peruvianllama, Everyking, No Guru, Jacob1207, Anville, Perl, Curps, Electric goat, Bensaccount, Cantus, Mike40033, Guanaco, Tom-, Jherico, Zhen Lin, Gracefool, MRubenzahl, Steven jones, Matt Crypto, Chrissmith, Bobblewik, Deus Ex, Golbez, Kandar, DontMessWithThis, Christopherlin, Hob, Stevietheman, Barneyboo, Gadfium, Knutux, LiDaobing, Sonjaaa, Quadell, Fangz, Antandrus, Tom the Goober, Beland, Estel-enwiki, Apox-enwiki, PDH, Armaced, Jossi, CaribDigita, Rdsmith4, OwenBlacker, Mitaphane, Woofles, Tothebarricades.tk, Daniel11, Mysidia, Bk0, Tyler McHenry, Anirvan, Creideiki, Neutrality, Imjustmatthew, Karl Dickman, Deglr6328, Mtnerd, Barnaby dawson, Trevor MacInnis, Zaf, Mormegil, Rfl, Freakofnurture, N328KF, Venu62, Nimbulan, DanielCD, Discospinster, Rich Farmbrough, Rhobite, FiP, Jpk, Silence, Chowells, Prateep, Dsadinoff, Xezbeth, Ponder, Ioliver, Mani1, Pavel Vozenilek, Aardark, Paul August, Stereotek, SpookyMulder, Night Gyr, Bender235, ESkog, Kaisershatner, Danny B-), Hapsiainen, Brian0918, El C, Chairboy, Aude, Shanes, Sietse Snel, RoyBoy, Triona, Bookofjude, Deanos, DarkArctic, Jburt1, Balok, Adambro, Bobo192, Smalljim, BrokenSegue, Duk, Viriditas, Serialized, Vortexrealm, Kormoran, Jag123, Scott Ritchie, Jojit fb, Kjkolb, BM, Townmouse, Bawolff, PeterisP, Naturenet, Daf, WikiLeon, Rje, Pschemp, MPerel, Sam Korn, Haham hanuka, Ral315, Ylwsub68, Nsaa, Jakew, HasharBot-enwiki, OGoncho, Jumbuck, Stephen G. Brown, Alansohn, Gary, Tablizer, Uncle.bungle, SnowFire, Mo0, 119, Atlant, Rd232, Mr Adequate, Keenan Pepper, Trainik, Joshbaumgartner, Rwoodsco, Andrew Gray, Lord Pistachio, Lectonar, MarkGallagher, Zippanova, SlimVirgin, Lightdarkness, Sligocki, Garfield226, InShaneee, Dark Shikari, Hgrenbor, Hu, Malo, Idont Havaname, Bart133, GregLindahl, Hohum, Snowolf, Melaen, ClockworkSoul, Super-Magician, Evil Monkey, Ramius, Cal 1234, RainbowOfLight, Randy Johnston, Sciurinæ, Mikeo, Pethr, Vuo, Ianblair23, DV8 2XL, LordAmeth, Stepheno, Gene Nygaard, Redvers, Admiral Valdemar, HenryLi, Dan100, Tr00st, GreatGatsby, Crosbiesmith, Feezo, Itinerant, MickWest, Sylvain Mielot, Boothy443, Reinoutr, OwenX, Woohookitty, Jannex, GrouchyDan, JarlaxleArtemis, Superstring, Jersyko, Guy M, PatGallacher, James Kemp, Nvinen, TomThe-Hand, Nameneko, JeremyA, MONGO, Nakos2208-enwiki, Tabletop, Cabhan, Firien, Bluemoose, GregorB, M412k, Petwil, Atomicarchive, Wayward, Volkz, Smartech-enwiki, Christopher Thomas, Dysepsion, GSlicer, Johndoe85839, Graham87, JiMidnite, Deltabeignet, Magister Mathematicae, Kalmia, MC MasterChef, Ligar-enwiki, Kbdank71, FreplySpang, Josh Parris, Gorrister, Rjwilmsi, Joefu, George Burgess, Phileas, Panoptical, Mystalic, Bill37212, Hiberniantears, Linuxbeak, JHMM13, Tawker, Mred64, Oblivious, Ligulem, CQJ, Frenchman113, The wub, DoubleBlue, ATLBeer, Sango123, Yamamoto Ichiro, Lcolson, Titoxd, FlaBot, Mirror Vax, RobertG, Ground Zero, A scientist, Nihiltres, Josh-enwiki, TheMelenchukSmell, Crazycomputers, Survivor, JIMBO WALES, Subterfuge-enwiki, RexNL, Gurch, Ayla, Jimbo D. Wales, RobyWayne, SweBrainz, KFP, OrbitOne, Cause of death, Jfiling, Butros, King of Hearts, Scimitar, Chobot, Hatch68, Theo Pardilla, Gangof One, Bgwhite, Digitalme, Dj Capricorn, Simesa, NSR, Gwernol, Peter Grey, Loco830, UkPaolo, The Rambling Man, YurikBot, Wavelength, TexasAndroid, Dimimimon4, Extraordinary Machine, Sceptre, Blightsoot, Hairy Dude, Jimp, RussBot, Arado, Red Slash, John Quincy Adding Machine, Majin Gojira, Anonymous editor, Splash, Alavena, Stalmannen, Anders.Warga, Anomaly1, 0nizuka the Great, Cmk5b, Akamad, CambridgeBayWeather, Shaddack, Bisqwit, Sweetwillams, GeeJo, Bullzeye, Finbarr Saunders, David R. Ingham, PaulGarner, Shanel, Nawlin-Wiki, SEWilcoBot, Wiki alf, Ceremony1968, Harrisale, WAS, Jaxl, Milo99, Robchurch, JDoorjam, Nick, Ragesoss, Anetode, Dmoss, PhilipO, Misza13, Grafikm fr, Lomn, LarryMac, Aaron Schulz, Karl Meier, DeadEyeArrow, Psy guy, Kander, Superiority, Essexmutant, Mtu, Mgnbar, Saric, FF2010, Newagelink, Georgewilliamherbert, Vonfraginoff, Enormousdude, Ali K, Lt-wiki-bot, Gtdp, Ageekgal, Theda, Jwissick, Fang Aili, Adilch, Nemu, Dspradau, Rhallanger, GraemeL, JoanneB, TBadger, CWenger, JLaTondre, Garion96, AGToth, Gorgan almighty, David Biddulph, Jack Upland, Junglecat, RG2, Mikedogg, GrinBot-enwiki, Airconswitch, Dkasak, Vreddy92, Nick-D, Sam Weber, BiH, Jade Knight, DVD R W, Kf4bdy, Marquez-enwiki, NetRoller 3D, Luk, Sycthos, Bigcheesegs, Sardanaphalus, Joshbuddy, A bit iffy, SmackBot, Looper5920, YellowMonkey, Mattarata, Tarret, Prodego, KnowledgeOfSelf, Royalguard11, Hydrogen Iodide, Melchoir, Jhartshorn, Unyoyega, Pgk, C.Fred, Bomac, Neptunius, Davewild, Wikedpedia-enwiki, CMD Beaker, Alksub, Delldot, Desk003, Sam8, Ajm81, AnOddName, Vilerage, Aivazovsky, Nscheffey, Wittylama, Alex earlier account, Gaff, Xaosflux, Yamaguchi▨▨, Zvonsully, Aksi great, Gilliam, The Gnome, Ppntori, Andy M. Wang, Psiphiorg, Afa86, The monkeyhate, Saros136, Chris the speller, Master Jay, Payam81, SlimJim, Persian Poet Gal, Lordkazan, Thumperward, Emt147, Silly rabbit, Hibernian, Imaginaryoctopus, Croquant, Sbharris, Darth Panda, A. B., Mikker, Cigale, Gsp8181, Dinnyy, Royboycrashfan, PeRshGo, Zsinj, Rogermw, Dethme0w, Can't sleep, clown will eat me, Jahiegel, Милан Јелисавчић, Jorvik, HoodedMan, Markkasan, Skidude9950, Nixeagle, OOODDD, Korinkami, Prmacn, Addshore, Interfector, Joema, Mrdempsey, Khoikhoi, WhereAmI, Digitize, Jumping cheese, Iapetus, Khukri, Makemi, Engwar, Nakon, Savidan, Loannes, Kevlar67, Shadow1, Dreadstar, Mini-Geek, Lcarscad, Polonium, TCorp, Dlamini, Daniel4004, Kotje, Whiplashxe, Edgeris, Daniel.Cardenas, Pilotguy, Prasi90, ▨▨▨, Ohconfucius, Kuzaar, Nmnogueira, WikiWitch, Rory096, Harryboyles, AAA765, Zahid Abdassabur, Dbtfz, Kreb Dragonrider, Kuru, John, HellecticMojo, Buchanan-Hermit, J 1982, Xu3w3nan, Kipala, Amenzix, Lazylaces, Sir Nicholas de Mimsy-Porpington, JorisvS, Minna Sora no Shita, CaptainVindaloo, Zarniwoot, Gevalt, JohnWittle, Scetoaux, Jaywubba1887, Syra987, Ckatz, Kkken, Slakr, Werdan7, Stwalkerster, Shangrilaista, Tasc, Mr Stephen, Nitro-X, InedibleHulk, Waggers, SandyGeorgia, Mets501, Java 109, Spook`, Unnamed01, Ryulong, Serlin, Ryanjunk, Gary Jacobsen, Balderdash707,

Aktalo, Kenny&becca, Iridescent, K, Hydra Rider, Aspuar, CapitalR, Esurnir, Aeons, Dublan, Aaron DT, Civil Engineer III, Thebigone45, Morgan Wick, Tawkerbot2, Brian53199, Dlohcierekim, Chetvorno, Zaphody3k, Benfranklinlover, Penguincornguy, ERAGON, Vikram.raja, JForget, Pigstinky, Ale jrb, Sir Vicious, TheHerbalGerbil, Pools200, Scohoust, Iced Kola, JohnCD, Grimgor79, Randalllin, 5-HT8, GHe, Dgw, Toropop, Adrienhocky16, Evilhairyhamster, Avillia, Hipdog11, Qwertyman4444, Borislav Dopudja, TJDay, PC supergeek, Abeg92, Ryan, Tkloumo, Jeffdb123, UncleBubba, Gogo Dodo, HPaul, Travelbird, Deathmak, Llort, A Softer Answer, Give Peace A Chance, Pascal.Tesson, Nate74, Noohgodno, Tawkerbot4, Roberta F., Optimist on the run, Kingthwomp, Kansas Sam, Omicronpersei8, Bascombe2, UberScienceNerd, EvocativeIntrigue, FrancoGG, Mathpianist93, Epbr123, Daa89563, PolaroidKiss, Forsaken88, Tairen125, Corsair18, Jedibob5, Sagaciousuk, Mansoorhabib, Drift~enwiki, Mojo Hand, Afitillidie13, Louis Waweru, West Brom 4ever, John254, Bobblehead, Pavel from Russia, CST, Cj67, Geostar1024, Pcbene, RamanVirk, Nick Number, Mm11, SebastianSalceek, Thedarkestshadow, Dawnseeker2000, Escarbot, Bilbobjoe, Dagingsta, Supran, KrakatoaKatie, Ialsoagree, AntiVandalBot, Yonatan, Gioto, Luna Santin, Settersr, Seaphoto, Sobaka, Opelio, Chairman Meow, Quintote, Doc Tropics, Paste, Pokemeharder, Robzz, Sweart1, Dr who1975, Arclem, Jj137, Postlewaight, Dylan Lake, LibLord, Farosdaughter, EP111, MrBill, VonV, Aliwalla, MishMich, Lorethal, Canadian-Bacon, Bigjimr, JAnDbot, Najeb, Husond, Fidelfair, FidelFair, MER-C, Skomorokh, Mark Grant, Nthep, Instinct, Arnegrim, Seddon, Rearete, Hello32020, Ribonucleic, Tengfred, Andonic, Roleplayer, TAnthony, Fluffy the Cotton Fish, LittleOldMe, Yahel Guhan, Brandox1, Magioladitis, Pedro, Bongwarrior, VoABot II, Edwardmking, Nyq, Weser, Trnj2000, Carom, J mcandrews, Ben515, Redaktor, Bfiene, Akmoilan, Avicennasis, Gblay, Bubba hotep, Animum, Adrian J. Hunter, Allstarecho, A3nm, Cpl Syx, Tokino, Spellmaster, Vssun, Glen, DerHexer, JaGa, Fulvius~enwiki, Khalid Mahmood, Hans Moravec, Saganaki-, Philbj, Tuviya, Stevepaget, FisherQueen, Leaderofearth, Hdt83, MartinBot, Mornock, Racepacket, Ninestrokes, Arjun01, One of them, Rettetast, Mschel, Nono64, PrestonH, Kentucho, Headmaster2008, MnM2324, Napalmdeth~enwiki, Zephyr21, RockMFR, Zarathura, Limongi, J.delanoy, Pharaoh of the Wizards, Loongyh, Bogey97, Hacbarton, Leaflet, Maurice Carbonaro, Fleiger, Mike.lifeguard, Menew22, Yucki8aby, Octevious, Thaurisil, Rahzvel, Sfgamfan, Tdadamemd, John11479, Aym710, Daedalus CA, Icseaturtles, Bot-Schafter, Nosfartu, Ncmvocalist, McSly, Ninjadeath, L'Aquatique, Grumpyapp, Sheahae, Reichner1000, AllanDeGroot, Pyrospirit, Detah, RenniePet, NewEnglandYankee, ChineseGoldFarmer, DadaNeem, Skrelk, Malerin, Halfvamp, Nikobro, Hrishie, Xecog, Bloodvayne, MetsFan76, Billyx1337x, Asdfasdf321, Antepenultimate, Morimura, Nukeitup2, Vanished user 39948282, Gemini1980, Natl1, Waterfox1, Cs302b, Useight, MissAtomicbomb, StoptheDatabaseState, Awesomeman42, Hmsbeagle, Phr0gor, Pistonhonda4, Idioma-bot, Wikieditor06, ACSE, Dansen3008, Happy guy of happyness, Eleron123, Mattybobo, X!, Cdmajava, JoshBuck123456789, Spartan 2.0, Nucwikigirl, VolkovBot, CN111111111, TreasuryTag, CWii, Johnfos, ABF, Jeff G., Indubitably, Lbunker, Stopping Power, HJ32, Soliloquial, Vulgarkid, Barneca, Flintsparkler, Jomorepinch, Oshwah, XavierGreen, GimmeBot, Solracm 021, SeanNovack, Maximillion Pegasus, Baldusi, Dj stone, TommyKiwi, Dchall1, Karmos, Something915, Qxz, DavidSaff, Noob wikipedian, Billy1223billy, Bloigen, Don4of4, Destroyer 2943, LeaveSleaves, Heidit, Seb az86556, UnitedStatesian, Arigato1, Cremepuff222, Hooduphodlum, Mazarin07, Tybluesum, Buffs, Sparkyrob, Feudonym, Cantiorix, Falcon8765, Tompkins818, Turgan, -ross616-, Burntsauce, Babilingbaboon, Dustybunny, Afonsecajames, NPguy, AlleborgoBot, Funeral, 682635q, Kampking13, MattW93, Worship cindy, LOTRrules, SieBot, Dusti, ShiftFn, Sonicology, Tiddly Tom, Scarian, SheepNotGoats, Jacotto, Awesome Truck Ramp, RJaguar3, Triwbe, Mcygan123, Calabraxthis, Lexicog, Bootha, Aprudhomme, Arda Xi, Dattebayo321, Keilana, Tiptoety, Oda Mari, Arbor to SJ, PeaveyStrat5, Chridd, Lagrange613, Games14pmw, Lanzarotemaps, Oxymoron83, Antonio Lopez, Byrialbot, Faradayplank, AngelOfSadness, Steven Crossin, Lightmouse, RW Marloe, Tombomp, Megansmith18, Harry the Dirty Dog, Jackal242, RouterIncident, Anchor Link Bot, Mygerardromance, Trashbird1240, Vanished User 8902317830, Nn123645, Khilon, SallyForth123, Mr. Granger, Smashville, Twinsday, Utergar~enwiki, Beeblebrox, MBK004, ClueBot, LAX, GorillaWarfare, Xilften, Snigbrook, Foxj, The Thing That Should Not Be, Ub3r n00ber, Wxyz334, Kafka Liz, Rjd0060, Dioneces, RomeijnLand II, Mickwaca, Chessy999, Rise Above the Vile, Arakunem, Magnoliasltd, Inventors, VQuakr, Cube lurker, NiD.29, Ventusa, Sasuke9031, Niceguyedc, Piledhigheranddeeper, Switchcraft, Auntof6, Ludoman, Monobi, PixelBot, Wilsone9, Zaharous, Esbboston, Ice Cold Beer, L.tak, PhySusie, Xpolygraphrightnowx, Kaiba, Dekisugi, Warrior4321, Thehelpfulone, Shimerdron, La Pianista, AbJ32, Light show, Aprock, Thingg, 9Nak, Aitias, Zilliput, Kyle2131, DJ Sturm, Party, Mahmoud-Megahid, XLinkBot, Javieranfispatria, Curby4, Gorillawataru, Bfgoobla, Lopper304, Altair1453, Bahman15, Duncan, Enigma 3, CDOG13, Osingh, Nepenthes, Lolkok, Ay nako, Atombombfootball, Dbev69, Batman278, Cstorm462, JinJian, ZooFari, Vqors, JCDenton2052, Dahobsta123, Shoemaker's Holiday, Twoolf1, Bobfran, EEng, HexaChord, Wierd al 101, Pamejudd, F-22G10, Spico1, Mbodnar101, Addbot, Cxz111, Manuel Trujillo Berges, Uruk2008, Dante4, Betterusername, CycloneGU, Irroy, Ronhjones, Mww113, Leszek Jańczuk, Black sheep997, Download, Bassbonerocks, Szooper99, Punkrockpiper, Favonian, AtheWeatherman, LinkFA-Bot, Jaydec, Joomple, Immortal Horrors or Everlasting Splendors, Amjsjc, 84user, VASANTH S.N., Erutuon, Lightbot, Krano, Apteva, Meisam, Luckas-bot, TheSuave, Yobot, 2D, Therimjob, Tohd8BohaithuGh1, TaBOT-zerem, II MusLiM HyBRiD II, Amirobot, Paepaok, DJ LoPaTa, THEN WHO WAS PHONE?, Santryl, Sorruno, WisdomFromIntrospect, Ayrton Prost, IW.HG, KNLR, Eric-Wester, AnomieBOT, Floquenbeam, VX, Jim1138, IRP, Galoubet, JackieBot, Piano non troppo, Jhjh112, Gc9580, Poiu18894012, Ulric1313, Sousapaloosa, Masterj89, Omg a llama, Flewis, Geord0, Materialscientist, Senio eilliw, Archaeopteryx, Citation bot, Optimusprimechucknorris, Neurolysis, Unh20050, Tylermweeks, Taikah, Andyconda, Chrisxmas, Dude4747, Andrewiol546, Wufei05, Ezietsman, Bobandbulider1, Ahloahlo, Parthian Scribe, Xqbot, LukeJasko5, W27138, Funnyediting, Guythatedits, Xcaliber14, Chair7, Capricorn42, Weeeeeman, Austin+mariah, Mackbot1234, Mastarhon, Kschultz15, Liquidnitrogen5000, Ooooooooyuuuu, Po.box1595, Dilholio, Ferman2727, Peaser2009, TheGunn, Kdmoss, Livrocaneca, Ricoswavez, Cockopops, Chongo713, Jordoboy123, Jhickey04, Berkeley0000, Toothdelay666, Jack conway2, ProtectionTaggingBot, ChillyMD, Kroack, Chainmaster, Surv1v4l1st, Tobby72, Trinity54, Kyteto, MathFacts, HJ Mitchell, L1ttleTr33, Operation Fiscal Jackhammer, PasswordUsername, Bambuway, Louperibot, Citation bot 1, Mimzy1990, Jonesey95, Shiva Khanal, Cos-fr, Yutsi, Fui in terra aliena, Redbeanpaste, Rotblats09, Lwiki222, Cnwilliams, Pit0001, Lightlowemon, FoxBot, TobeBot, Mercy11, Trappist the monk, Comet Tuttle, Himypiedie, Grantbonn, Mr.98, Vera.tetrix, DARTH SIDIOUS 2, RjwilmsiBot, TjBot, Jackehammond, Bhawani Gautam, Beyond My Ken, Mandolinface, DASHBot, EmausBot, John of Reading, Nima1024, Manga28, Docman500, Mk5384, Boundarylayer, Tisane, Challisrussia, Neifdude, Dcirovic, Somerwind, The Blade of the Northern Lights, JSquish, ERRORHUNT, John Cline, Quasihuman, Josve05a, Fintelia, Nicolas Eynaud, AvicAWB, Zloyvolsheb, Wikfr, Confession0791, Brandmeister, Hudson Stern, L Kensington, MonoAV, Keithgnield, Hidenori watanave, Kc0wir, Thewolfchild, Cn7abc, Tussna, Poolatino12, Hookemhornsgannon, ClueBot NG, AlbertBickford, Catlemur, Georgepauljohnringo, Spikesjb, Buklaodord, Snotbot, XHawkz, Comonline, Popcornduff, Tholme, Bibcode Bot, The lost library, Neptune's Trident, MangoWong, Usefulchanges, Lazord00d, Cadiomals, Altair, Trevayne08, Eesoov, DPL bot, Polmandc, 220 of Borg, TheGoodBadWorst, BattyBot, R3venans, Netherzone, FiveFourTwo, Cyberbot II, ChrisGualtieri, Jray310, SD5bot, John M. DiNucci, Ryay32, JYBot, Dexbot, XXzoonamiXX, Evildoer187, NagOc 945, Reatlas, Joeinwiki, Ninjaboos, Hermes 1900s, A Certain Lack of Grandeur, Godofwar1016050, Atotalstranger, Linnalid, Barjimoa, Rotaryphone111, Bennett Graff, Monkbot, Trackteur, Gronk Oz, Mhhossein, Davearthurs, Breedentials, TheGuyWhoHasAUsername, Atvica, Joshlabroski, Kymako, KenTancwell, KasparBot, Hulk576, Rightkeatsboom, Overtime.Editor, Ihsanturk, GoldCar, Wikkileaker, GreenC bot, AnimosityAnimalEdits and Anonymous: 1775

- **History of nuclear weapons** *Source:* https://en.wikipedia.org/wiki/History_of_nuclear_weapons?oldid=734038674 *Contributors:* TwoOneTwo, Trelvis, Vicki Rosenzweig, Mav, Robert Merkel, Alex.tan, Rmhermen, Roadrunner, SimonP, Comte0, Tedernst, Edward, Patrick, Crenner, Dan Koehl, 172, Kingturtle, Nikai, Jiang, Lee M, Mulad, Markb, Dino, Reddi, Pwd~enwiki, Wik, Zoicon5, Timc, DJ Clayworth, Maximus Rex, Nv8200pa, Fibonacci, Raul654, Dpbsmith, Adam Carr, David.Monniaux, Twang, Fredrik, Chris 73, Postdlf, Hemanshu, Bertie, Hadal, Diberri, Robartin, Dmn, Ancheta Wis, DocWatson42, Nikodemos, Lethe, Fastfission, Wyss, Alison, Jared R. Buckley, DO'Neil, Tom-, Get-back-world-respect, Bobblewik, Ruy Lopez, Beland, Piotrus, Andrew Kanaber, Mzajac, Eranb, Karl Dickman, Kisama, Terinthanas, N328KF, Michael Zimmermann, SpookyMulder, Bender235, Eric Forste, CanisRufus, Dgorsline, El C, Sietse Snel, Orlady, Harley peters, Wisdom89, Anthony Appleyard, 119, Eric Kvaalen, Rd232, Mr Adequate, Rwendland, DV8 2XL, SteinbDJ, Ashujo, Gmaxwell, Angr, Batintherain, Woohookitty, PoccilScript, Oliphaunt, TomTheHand, Derrida derider, JeremyA, Wikiklrsc, John Hill, Sam Coutu-Oughton, Wayward, Drbogdan, Rjwilmsi, Jake Wartenberg, Joffan, Vary, Jmcc150, Vegaswikian, Ian Dunster, Lcolson, Wikiliki, Ground Zero, Doc glasgow, Nihiltres, Gurch, Compound, King of Hearts, Manscher, Mushin, RussBot, Crazytales, Epolk, Bergsten, Groogle, Kirill Lokshin, Hydrargyrum, GeeJo, Jelay14, DarthBinky, Dureo, Mass147, Vic sinclair, Falcon9x5, Gadget850, MrBark, Kal-El, Jkelly, PTSE, CharlesHBennett, GraemeL, Junglecat, Carlosguitar, Some guy, Sacxpert, SmackBot, Kellen, Jagged 85, Gunnar.Kaestle, ScaldingHotSoup, Delldot, Cool3, Hmains, Ghosts&empties, GwydionM, Chris the speller, Hibernian, Solidusspriggan, OrphanBot, Prmacn, Zvar, Metsfanmax, Soarhead77, Qwerty0, John, Sosodank, Slinga, RomanSpa, RandomCritic, Rkmlai, Mr Stephen, Cyanidesandwich, MAG1, Jrt989, Cls14, Pjbflynn, FairuseBot, Tawkerbot2, Chetvorno, Elekas, JForget, Arathyn, Wafulz, Aherunar, Caesar Rodney, Richard Keatinge, Balloonman, MikeWren, Cydebot, Gjones0316, Chasingsol, Quibik, Christian75, DonFB, Epbr123, Andyjsmith, Headbomb, WVhybrid, Bobblehead, Tellyaddict, Nick Number, Uruiamme, AntiVandalBot, Nisselua, Yonatan, Opelio, DarkAudit, Sweart1, Modernist, C. M. Harris, SkoreKeep, Altamel, VonV, MER-C, Ericoides, VoABot II, Bfiene, Hechnal, Kevinmon, Cgingold, Chuckwatson, NatureA16, Read-write-services, STBot, R'n'B, CommonsDelinker, Brothejr, Jascal, J.delanoy, Uncle Dick, A Nobody, Thaurisil, Ownage2214, Tdadamemd, AnarchMonarch, NewEnglandYankee, Shoessss, BigHairRef, KylieTastic, Straw Cat, Hugo999, VolkovBot, Johnfos, Jim.Callahan,Orlando, Philip Trueman, TXiKiBoT, Tricky Wiki44, LeaveSleaves, Robert1947, Wyseman101, Eskovan, Spinningspark, Pentium1000, EJF, Sonicology, Smsarmad, Lightmouse, Hobartimus, Abrichte, Hockey 173, Iknowyourider, Cyfal, Blake, Randy Kryn, Martarius, MBK004, ClueBot, Antarctic-adventurer, Binksternet, Sorensen ru, Fyyer, Socrates2008, Morel, SchreiberBike, Thingg, Thehighestpower, Northwesterner1, NellieBly, Patbreen, Addbot, Fyrael, Fieldday-sunday, Smirnov, Protonk, LaaknorBot, Morning277, Chamal N, Debresser, Favonian, 5 albert square, Immortal Horrors or Everlasting Splendors, Tide rolls, Zman2134, Julia W, AnomieBOT, IRP, Tomfromwestbrom, 3BroomsticksInnkeeper, Materialscientist, Quebec99, Whatsoevernever, Dolce Vita Evita, Brutaldeluxe, Atomicgurl00, Dead Mary, FrescoBot, Ironboy11, Yghwtrrl, Pinethicket, Full-date unlinking bot, Rotblats09, Poliphile, Keri, Trappist the monk, Vrenator, Mr.98, Fastilysock, YouWillBeAssimilated, RjwilmsiBot, Hummingbird, Ajraddatz, Dewritech, Lucas hamster, Carbo1200, Gagarine, K6ka, ZéroBot, H3llBot, Donner60, Forever Dusk, ClueBot NG, Kayz911, Cdp50, Millermk, Physics is all gnomes, BG19bot, Vikasdodda, Khazar2, MadGuy7023, Crazy84007, Mogism, XXzoonamiXX, TwoTwoHello, HarveyHenkelmann, Epicgenius, Epicmanofawesome, Eating cake, Soffredo, Ol'fussandfeathers, TonyWMontana, Orduin, Badboyyy, Antisquark01, Steynrenier and Anonymous: 374

- **Timeline of nuclear weapons development** *Source:* https://en.wikipedia.org/wiki/Timeline_of_nuclear_weapons_development?oldid= 722463136 *Contributors:* Jimp, Serendipodous, Yanders, Vertium, Keallu, DH85868993, Johnfos, Go-in, Sanya3, Binksternet, Good Olfactory, Moosehadley, LilHelpa, Alvin Seville, Erik9bot, Jeffrd10, John of Reading, ClueBot NG, AlbertBickford, Snow Blizzard, BattyBot, Maxdelae-spriella and Anonymous: 32

- **Nuclear weapon design** *Source:* https://en.wikipedia.org/wiki/Nuclear_weapon_design?oldid=735006842 *Contributors:* Trelvis, Robert Merkel, The Anome, Maury Markowitz, Tedernst, Edward, Patrick, RTC, Michael Hardy, Ixfd64, TakuyaMurata, Karada, Ahoerstemeier, Darrell Greenwood, Aarchiba, Jll, Nikai, Lommer, Mulad, Charles Matthews, Warmfuzzygrrl, Zoicon5, Timc, Omegatron, Pstudier, Finlay McWalter, Mjmcb1, Jni, Riddley, Robbot, ChrisO~enwiki, Korath, Donreed, Arkuat, Securiger, Academic Challenger, Litefantastic, Bkell, Vfrickey, Gospodin david, DocWatson42, Msiebuhr, Fastfission, Karn, Leonard G., Nomad~enwiki, DO'Neil, Mboverload, Gracefool, Prosfilaes, Siroxo, Bobblewik, ConradPino, WhiteDragon, MisfitToys, Blazotron, Sean Heron, Mzajac, Bk0, Mouser, Sam Hocevar, Sonett72, Deglr6328, DMG413, Hax0rw4ng, Grunt, Mike Rosoft, Discospinster, Brianhe, Rich Farmbrough, Alsadius, Avriette, Cfailde, Bender235, FirstPrinciples, El C, Meggar, Kjkolb, Alansohn, Free Bear, Atlant, Rcbarnes, Axl, Echuck215, Wdfarmer, Wtmitchell, Rebroad, Cal 1234, DV8 2XL, Gene Ny-gaard, Alai, Richwales, Forteblast, BerserkerBen, Crosbiesmith, Optimusnauta, Joriki, Richard Arthur Norton (1958-), Vashti, Woohookitty, Nvinen, Before My Ken, Magabund, Tabletop, GregorB, MiG, SDC, GraemeLeggett, Christopher Thomas, Wikipedian231, Teemu Leisti, Rjwilmsi, Koavf, Erebus555, Joffan, Jivecat, Midwestarrival, Krash, Ian Dunster, Lcolson, Ground Zero, Dirkbike, RexNL, Gurch, BjKa, Enon, CuseMM, CStyle, GangofOne, Banaticus, JWB, Wolfmankurd, Midgley, Cliffb, Tmayes1999, Hellbus, Thoreaulylazy, Gaius Cornelius, Shaddack, NawlinWiki, Daemon8666, Welsh, LiamE, Dogcow, Gerard.Newman, Saberwyn, Lockesdonkey, Deeday-UK, Georgewilliamher-bert, Sandstein, Enormousdude, Super Rad!, Petri Krohn, CWenger, Jsplegge, Mais oui!, Allens, Carlosguitar, Yvwv, Criticality, SmackBot, Roger Hui, Gigs, Yuyudevil, KocjoBot~enwiki, Francisco de Almeida, Fulldecent, Eskimbot, ASarnat, Man with two legs, Betacommand, Chris the speller, Qwasty, Iversonali3, Fuzzform, Sbharris, Rcbutcher, Tmayes1965, Rizzardi, Rrburke, Zirconscot, Krich, E. Sn0 =31337=, Zer-well, Fedch, The PIPE, DMacks, Soarhead77, A5b, Roger.lee, Will Beback, Qwerty0, John, General Ization, Sosodank, Jrvz, Shadowlynk, Mgiganteus1, IronGargoyle, BillFlis, Dingopup, Ryulong, Rickington, Pqrstuv, Kurtycobain, Shoeofdeath, RGrimmig, Captainj, Tawkerbot2, RookZERO, Orangutan, Harold f, DJGB, Sarf, CmdrObot, Pmyteh, Scirocco6, ShelfSkewed, James5, Chrisahn, Location, MrFish, Pce3@ij.net, UncleBubba, Gogo Dodo, Chasingsol, Give Peace A Chance, Tawkerbot4, Quibik, Nabokov, Viridae, Cancun771, Epbr123, David from Dow-nunder, Uruiamme, Escarbot, AntiVandalBot, K7aay, Opelio, Barneyg, LibLord, SkoreKeep, Tangurena, JAnDbot, MER-C, The Transhu-manist, ZZninepluralZalpha, Smulthaup, Igodard, Roleplayer, Wasell, Tanada, Magioladitis, Exerda, VoABot II, Chrisempson, Nikevich, The Anomebot2, Hamiltonstone, Talon Artaine, Wdflake, Eagre, LochVoil, Wikianon, Oroso, Read-write-services, MartinBot, Jim.henderson, Red-line84, Flo422, Nono64, Gah4, J.delanoy, SaxicolousOne, Theeurocrat, Hair Commodore, DarkFalls, Coppertwig, ARTE, Olegwiki, Potatoswat-ter, OliverHarris, Entropy, King Toadsworth, George.Hutchinson, Squids and Chips, Steel1943, Signalhead, Hugo999, Johnfos, ABF, Barneca, Philip Trueman, Nono le petit robot~enwiki, JayC, Qxz, Piperh, Mzmadmike, Wingedsubmariner, Chrisj1948, SwordSmurf, Andy Dingley, Fal-con8765, Scruffy brit, NPguy, Kyle112, Gus, Camsopaint, Skipweasel, HowardMorland, SieBot, Grndrush, Permacultura, Mewikiman, Pac72, Steven Crossin, Magic9mushroom, BrightRoundCircle, WacoJacko, Fratrep, Afernand74, Rkarlsba, Anchor Link Bot, Hamiltondaniel, Anyev-erybody, DaddyWarlock, Muhends, ImageRemovalBot, MBK004, ClueBot, Binksternet, The Thing That Should Not Be, Franamax, VQuakr, Mataap, Conical Johnson, Pinkpedaller, Vegasmas, Vendeka, Razorflame, JasonAQuest, Thingg, Belchfire, Johnuniq, DumZiBoT, Noeatingal-lowed, Toolssmile34, DaL.33T, Avoided, Northwesterner1, WikHead, Cabayi, Roentgenium111, Uruk2008, Ronhjones, CanadianLinuxUser,

Aymankamelwiki, Cerabot~enwiki, Lugia2453, I am One of Many, Abishai 300, Debouch, Eyesnore, Wistchars, Zenibus, AddWittyNameHere, Slgonzalez, Royalcourtier, JaconaFrere, StarHOG, TheGuyWhoIsOnTheStreet, Monkbot, Volker Siegel, Vozul, BethNaught, Jamesnottingham, Wikiornah, WC Jay, Jaffacakemonster53, Ericwilloughby, RealFAKER, I know everything man!, Oluwa2Chainz, Deepanshu1707, Ploopkazoo, Endaine, Helmut von Moltke, Joethegreenbean, YouWantTheJoJ, OwlOtis, Natedog0011, RickyRolling, 23cartel, Dr. Thomas Boot Reincarnated, Saycheesepleasedear, Yogishan and Anonymous: 489

- **Boosted fission weapon** *Source:* https://en.wikipedia.org/wiki/Boosted_fission_weapon?oldid=735941821 *Contributors:* Maury Markowitz, Tedernst, Edward, Donreed, Fastfission, Nomad~enwiki, Urhixidur, Bender235, Rwendland, Gene Nygaard, Quale, Ewlyahoocom, JWB, Hellbus, Shaddack, Ospalh, Petri Krohn, Mais oui!, SmackBot, Elminster Aumar, Man with two legs, Jprg1966, A5b, John, Robofish, Rob1bureau, CmdrObot, Nick Number, Lklundin, JaGa, TheEgyptian, Mikek999, Tourbillon, JCavilia, Andy Dingley, Truthanado, Mirboj, MystBot, Addbot, Tsange, Luckas-bot, AnomieBOT, Saintonge235, Cyberbot II, Wikkileaker and Anonymous: 21

- **Neutron bomb** *Source:* https://en.wikipedia.org/wiki/Neutron_bomb?oldid=732684687 *Contributors:* Trelvis, Mav, Bryan Derksen, Rmhermen, Ray Van De Walker, Maury Markowitz, Stevertigo, Patrick, Michael Hardy, Cyde, Theanthrope, Delirium, Minesweeper, Tregoweth, Ellywa, William M. Connolley, Mark Foskey, Julesd, Nikai, Evercat, Lommer, Rami Neudorfer, Arteitle, Vroman, Hashar, Guaka, Adam Bishop, Malcohol, Zoicon5, Shizhao, Pakaran, Denelson83, Riddley, Lowellian, Diderot, LGagnon, Wikibot, Victor, DocWatson42, Nunh-huh, Tom harrison, Everyking, Mboverload, Jorge1000xl, Gadfium, Jamougha, Utcursch, VoX, LiDaobing, LucasVB, Beland, Eroica, Karol Langner, Oneiros, Clarknova, Æ, MementoVivere, Flex, SYSS Mouse, N328KF, Supercoop, Rupertslander, Bender235, Nharmon, Sunborn, Ben Webber, El C, RoyBoy, PatrikR, R. S. Shaw, Davidsmind, Schnolle, Still, Njaard, Joshbaumgartner, Axl, Rwendland, Hohum, Velella, Isaac, Max rspct, Pauli133, DV8 2XL, Axeman89, Crosbiesmith, Richard Arthur Norton (1958-), Woohookitty, GrouchyDan, Madmardigan53, Cyrille-Dunant, BillC, Jacobolus, Bbatsell, Chris Buckey, Kralizec!, Tmrobertson, RckmRobot, Ashmoo, Tovias, MC MasterChef, KaiMartin, Yuletide, Gordon Stangler, Amhaun01, Tajgenie, Gold Stur, Ttsalo, Ewlyahoocom, Gurch, Kolbasz, Diza, Ourboldhero, Chobot, Bgwhite, Cactus.man, Gwernol, YurikBot, Hairy Dude, Ineedbettername, Xihr, Fuzzy901, Raquel Baranow, Stephenb, C777, Gaius Cornelius, Daveswagon, Nawlin-Wiki, Slarson, Flup, MakeChooChooGoNow, MrBark, Eyal0, Daniel C, Ninly, Arthur Rubin, Aurax, JoanneB, ThunderBird, Mikkow, Allens, Drewright, SmackBot, Od Mishehu, Blue520, Clpo13, Jrockley, WayneConrad, Man with two legs, Gaff, Bluebot, Weeniemann, Persian Poet Gal, Thumperward, Hibernian, Mje, DevSolar, Gimeral, D97rolph, PointyOintment, EVula, Derek R Bullamore, Luís Felipe Braga, Giancarlo Rossi, Parrot of Doom, Will Beback, Jonnty, Esrever, Aaron Lawrence, Kuru, Alexcollins, CenozoicEra, Colin002, Fig wright, Mark Lungo, Tls, CompIsMyRx, Muadd, Therealhazel, Flaphead, QuilaBird, Hu12, Iridescent, Lord Anubis, ScwB, Jookypipe, The Letter J, Chetvorno, SkyWalker, Wpmccray, CRGreathouse, CmdrObot, Carpenoctem, Kylu, Airport 1975, Abeg92, Besieged, UncleBubba, Gogo Dodo, Hanfuzzy, Otto4711, Give Peace A Chance, Photocopier, Tkynerd, Nabokov, Optimist on the run, Bob Stein - VisiBone, Omicronpersei8, Cancun771, Smiteri, Thijs!bot, Kubanczyk, LooseArrow, Keraunos, Moonshadow Rogue, Headbomb, Kinglink, Catsmoke, Sean William, Navdar, Slive~enwiki, Jj137, Zachwoo, ARTEST4ECHO, Mccollou, Ingolfson, JAnDbot, Fil-mex91, Avaya1, Magioladitis, Bg007, SHCarter, CodeCat, Gundato, Vreemdst, LittleOldMe old, Tonbo0422, Maurice Carbonaro, Drewwiki, Thaurisil, Slow Riot, Andareed, Some Sort Of Anarchist Nutter, Matisia, Adamdaley, Cadwaladr, Петър Петров, Blood Oath Bot, Jomomm, A.Ou, Sjosa2, Emeraldcrown, Fences and windows, Philip Trueman, JayEsJay, TXiKiBoT, Chris-marsh-usa, Sintaku, IllaZilla, Tricky Wiki44, Foshowmo, LeaveSleaves, NKEISK, Doug, Wasted Sapience, NPguy, Tresiden, Scarian, WereSpielChequers, Phe-bot, Jerryobject, Jon joy 1999, Digwuren, Mandsford, Anyeverybody, PlantTrees, Sfan00 IMG, ClueBot, Qsaw, Polyamorph, Niceguyedc, Justtryn2help, DragonBot, Nymf, Crywalt, Vivio Testarossa, Sun Creator, Casi233, Levent, John Paul Parks, DumZiBoT, Faulcon DeLacy, PL290, Wgwells, Thatguyflint, Scabbed Angel, Addbot, DavidNotDave, TutterMouse, Vanished user oerjio4kdm3, Shrogen, Fluffernutter, Innocent Byproduct, Yobot, AnomieBOT, Mintrick, T34CH, ArthurBot, .45Colt, KrisBogdanov, GrouchoBot, 燃燃燃, Mattg82, Stratocracy, Cekli829, FrescoBot, HAHAHAHATHATS4HAS, Siddharth 1999, Slastic, IVAN3MAN, Neurotip, Diannaa, Reach Out to the Truth, RjwilmsiBot, EmausBot, John of Reading, Boundarylayer, Tinss, Peaceray, Wikipelli, K6ka, Illegitimate Barrister, A2soup, Canine virtuoso, SporkBot, Sailsbystars, ClueBot NG, Mesoderm, Zakblade2000, Rezabot, Oddbodz, Helpful Pixie Bot, Mbarland, BG19bot, The Mark of the Beast, Robert the Devil, TROPtastic, Nuke1st, Zedshort, Casimirck, Fixing the lie, Cutoffyourjib, Webclient101, Al Bestose, Jdc843, Jakec, Tototo30, Pietro13, ColRad85, Ajaythomas0007, Biblioworm, WC Jay, Guy Cox, Hairykrishna, KasparBot, GoldenPCMaster and Anonymous: 393

- **Radiological weapon** *Source:* https://en.wikipedia.org/wiki/Radiological_weapon?oldid=735257657 *Contributors:* TwoOneTwo, ClaudeMuncey, The Anome, Miguel~enwiki, William Avery, Roadrunner, Maury Markowitz, Alan_d, Rsabbatini, Patrick, Eric119, Bon d'une cythare, Aarchiba, Andres, Mulad, RodC, Pstudier, Sanders muc, Securiger, Jleedev, DocWatson42, Fastfission, Mboverload, Pne, Mzajac, Neutrality, Ponder, Bender235, Alansohn, Joshbaumgartner, BRW, DV8 2XL, Kazvorpal, WadeSimMiser, Urbane Legend, Gurch, YurikBot, RussBot, Crazytales, Stephenb, Palpalpalpal, Light current, NHSavage, Dr U, HereToHelp, Sardanaphalus, SmackBot, Arniebuteft, AndySayler, Nrcprm2026, A5b, Pilotguy, IgWannA, Chue03, Timetracker, Iridescent, Cnstewart, Brandizzi, Tawkerbot2, Cryptic C62, Give Peace A Chance, Dancter, Fleshwater, Honeplus, Mentifisto, K7aay, Tjmayerinsf, Sluzzelin, Deadbeef, Inks.LWC, BenB4, JimCubb, Dragonnas, Magioladitis, VoABot II, Midgrid, Cgingold, Hodja Nasreddin, Cromdog, Trumpet marietta 45750, Malinaccier, UnitedStatesian, Lamro, Falcon8765, Caltas, JabbaTheBot, RW Marloe, Francvs, Twinsday, ClueBot, Paulcmnt, Ahmed91981, TonyBallioni, Rhotel1, SkyLined, Addbot, Download, Lightbot, Legobot, Guy1890, Pganas, AnomieBOT, DemocraticLuntz, Metalhead94, Sugoi47, Lithenium, FrescoBot, Adam9389, RoyGoldsmith, BenzolBot, MastiBot, Callanecc, Dinamik-bot, Vrenator, EmausBot, AvicAWB, Arbnos, ChuispastonBot, ClueBot NG, Astatine211, Helpful Pixie Bot, Curb Chain, CitationCleanerBot, Tkbx, Cyberbot II, Jionpedia, Jray310, Limnalid, Jerryzhu2004 and Anonymous: 98

- **Nuclear electromagnetic pulse** *Source:* https://en.wikipedia.org/wiki/Nuclear_electromagnetic_pulse?oldid=733632843 *Contributors:* Magnus Manske, The Anome, Maury Markowitz, DonDaMon, KF, Twilsonb, Michael Hardy, Gabbe, Cherkash, Emperorbma, E23~enwiki, VeryVerily, Omegatron, Denelson83, Zandperl, Schusch, Greudin, Gidonb, Rhombus, Jsonitsac, Centrx, Giftlite, DocWatson42, Neuro, Junkyardprince, Antandrus, Beland, SYSS Mouse, N328KF, Rich Farmbrough, Smyth, Notinasnaid, Abelson, Antaeus Feldspar, Bender235, Rubicon, Closeapple, Kwamikagami, Aaronbrick, Harley peters, Bert Hickman, Slambo, Llywelyn, QuantumEleven, Alansohn, Luisloureiro, Daniellean, Filladdar, Rwendland, Radical Mallard, Darkskyz, BRW, Max Naylor, Drat, DV8 2XL, Abanima, Bacteria, Joriki, Alvis, Rnt20, Graham87, Drbogdan, Rjwilmsi, Hitssquad, Erebus555, Isaac Rabinovitch, Oblivious, FlaBot, Margoshot~enwiki, Nihiltres, TheDJ, Alphachimp, Johhny-turbo, RussBot, WritersCramp, Anshul, Dimblethum, Hellbus, Ioda006, Hydrargyrum, Stephenb, Gaius Cornelius, Shaddack, Bovineone, Bisqwit, Lao Wai, Brianbraddock, Janarius, Howcheng, Dureo, Adamrush, Jpbowen, Ergbert, Pyg, Antoshi, Slavik81, Dddstone, Arthur Rubin, Pádraic MacUidhir, AGToth, Ilmari Karonen, Allens, TLSuda, JDspeeder1, Patiwat, Mardus, That Guy, From That Show!, Hardscarf, SmackBot, Impaciente, Eskimbot, Fractal3, BiT, Alexisrael, Skizzik, Teemu Ruskeepää, Chris the speller, Bluebot, Crashmatrix, Jprg1966, Oli Filth, The Rogue Penguin,

Nbarth, Dual Freq, ApolloCreed, Frap, Benjamin Mako Hill, Cybercobra, Zhinker, Dreadstar, Ged UK, MusicMaker5376, Krashlandon, Mwboyer, Hypnosifl, Pseudoanonymous, Midusunknown, FaWzY, Asyndeton, Rouslan~enwiki, Dp462090, Courcelles, MrBoo, Simokon, Harold f, Eastlaw, Fvasconcellos, CndrObot, Ale jrb, Roman Dawydkin, Linus M., Requestion, Avillia, St Fan~enwiki, Oden, Kaldosh, ANTIcarrot, Red1530, Photocopier, Doug Weller, DumbBOT, Spectrun~enwiki, Cancun771, Brian1078, Thijs!bot, Begs, Headbomb, Hcobb, Nick Number, Lorcav, Escarbot, Gioto, DragonBlazer57, Lfstevens, Tjg50311, Steelpillow, JAnDbot, Emax0, Andonic, .anacondabot, Magioladitis, Withoutsin, Pixel ;-), Catgut, Cgingold, FrederikVds, Ahbushnell, CommonsDelinker, Hairchrm, Pharaoh of the Wizards, HMRothman, Astro-Hurricane001, Maurice Carbonaro, A Nobody, LordAnubisBOT, Gypsydoctor, Albinoyoda, Phuntism, Charlesblack, Spellcast, Bovineboy2008, Showjumpersam, Vyssotsky, Lexington50, Avatars10, Euphonic, Soldrinero, Martin451, George Steinmetz, Falcon8765, Logan, Deconstructhis, SieBot, Iceshark7, Meltonkt, Turbodin, Paulbrock, Falcofire, Stanleycr1, Kanonkas, Jimulacrum, Muhends, ImageRemovalBot, ClueBot, NickCT, Binksternet, Ideal gas equation, Excirial, Arjayay, Psinu, Kemosabi3000, Guitarheroguylol, DumZiBoT, XLinkBot, JediSaint, Addbot, Don'tKnowItAtAll, Eric Drexler, Ronhjones, Airsoftaddict666, Download, LaaknorBot, MetalGearJoseph, Favonian, Numsnumsnums, Tide rolls, MuZemike, Legobot, Blah28948, Luckas-bot, Yobot, Fraggle81, AnomieBOT, Rubinbot, Charlesmartin82, Aaagmnr, Lenardross, Jtmorgan, Steeplechase23, Citation bot, Xqbot, DerekFrankJames, 4twenty42o, The Evil IP address, GrouchoBot, Kstueve, Scets424, Shirik, Kyng, X5dna, Ab17aron34, Ricmetalster, GT5162, FrescoBot, 27 Juni, Stickmanx25, Galorr, Benermerut, OgreBot, Pinethicket, RedBot, Joshuachohan, Ezhuttukari, IVAN3MAN, Fama Clamosa, Mcfor, Diannaa, Bernd.Brincken, VernoWhitney, Markos Strofyllas, Nightblues13, EmausBot, John of Reading, Samstam, Bosik GN, Issac7009, Mmeijeri, Bravo Foxtrot, AvicBot, Chinaballs, Ridoking, Nik1027, Wayne Slam, Wingman4l7, NearEMPTiness, L Kensington, The rev av, Puffin, Samhoffman, DASHBotAV, Axelode, Rudolfensis, ClueBot NG, Gareth Griffith-Jones, Joefromrandb, Hon-3s-T, Blev2222, Widr, Helpful Pixie Bot, Arnavchaudhary, PTJoshua, Belac Athanasius, Altaylor5311, Ataylor21, Altaïr, Karma842w, JunoBeach, 220 of Borg, BillBucket, EdwardH, MeanMotherJr, Justincheng12345-bot, S1D3winder016, Cyberbot II, Physicsandshiz, TechProtect, ظلله, Malang49, TwoTwoHello, Mitrabarun, Jamesmcmahon0, LieutenantLatvia, YiFeiBot, Thriftylol, FtLauderGuy, Apostrostomper, WillBo and Anonymous: 363

- **Nuclear weapons delivery** *Source:* https://en.wikipedia.org/wiki/Nuclear_weapons_delivery?oldid=724336107 *Contributors:* Patrick, DocWatson42, Christopher Parham, Fastfission, Iceberg3k, Wmahan, Oneiros, Mormegil, Bender235, Bsadowski1, DV8 2XL, Woohookitty, TotoBaggins, Petwil, Lcolson, Bgwhite, Hillman, RussBot, Arado, Stalmannen, Los688, Welsh, Kvn8907, Ragesoss, Deville, Cjwright79, Mais oui!, SmackBot, John Lunney, Ohnoitsjamie, Skizzik, BigBoyRubio, Chris the speller, Init~enwiki, Xiner, Jumping cheese, General Ization, AstroChemist, JorisvS, Kevin W., Andrewrpalmer, Shoeofdeath, CMarshall, Aldis90, Thijs!bot, Kubanczyk, Nick Number, AntiVandalBot, Ingolfson, HolyT, Afaz, Bongwarrior, VoABot II, Rich257, WolfyB, STBot, CommonsDelinker, RWyn, Bogey97, McSly, NewEnglandYankee, YankeeFan90, Wiae, Wasted Sapience, SmileToday, Logan, PeterFV, Andes09, Sfan00 IMG, MBK004, ClueBot, Avenged Eightfold, Fasettle, Leonard^Bloom, Xic667, Thingg, Addbot, Lightbot, Luckas Blade, Yobot, AnomieBOT, LilHelpa, Coolguyhunksmart, JayJay, 58Extraten, FrescoBot, I dream of horses, Lamna, Skywolf24, Vrenator, DexDor, Look2See1, Boundarylayer, S trinitrotoluene, ClueBot NG, Satellizer, Nickholbrook, Hahamesswithkiernan, Sambeeni, Annajohnston, WebTV3, BattyBot, Fraulein451, Irondome, Numbermaniac, Monkbot, WC Jay, Narky Blert, Crystallizedcarbon, Editing656, Hulk576, MicahHerr, Dutral and Anonymous: 88

- **Nuclear triad** *Source:* https://en.wikipedia.org/wiki/Nuclear_triad?oldid=735578636 *Contributors:* The Anome, Enchanter, Maury Markowitz, Patrick, Julesd, Ciphergoth, Palmpilot900, Ike9898, Tpbradbury, Finlay McWalter, Joshuapaquin, D6, Nimbulan, Rich Farmbrough, Bender235, Ylee, Giraffedata, Dalillama, Vedant, TenOfAllTrades, TheAznSensation, Dr Gangrene, Kralizec!, GraemeLeggett, Nightscream, Korg, Arado, Sesquiannual, DragonHawk, Ytcracker, Cffrost, SmackBot, Nicolas Barbier, Unschool, Imzadi1979, Yamaguchi先生, Hmains, Chris the speller, Bluebot, Papa November, Hibernian, Kungming2, Colonies Chris, Ocatecir, JHunterJ, Dicklyon, Dl2000, Iridescent, CapitalR, Zaphody3k, CmdrObot, Malamockq, DumbBOT, Epbr123, Kubanczyk, Steelpillow, JAnDbot, Nyttend, Simonxag, KTo288, Maurice Carbonaro, Ash sul, Num1dgen, Dbenford, Onore Baka Sama, PaladinWhite, WereSpielChequers, LinkShadow, Herbu, Compellingelegance, Auntof6, Muro Bot, Jellyfish dave, Harman malhotra, XLinkBot, Addbot, Download, Debresser, Steed Asprey - 171, Enthusiast10, Andreasvc, Luckas-bot, Yobot, Nirvana888, Angel ivanov angelov, Dilutexyz2020, AnomieBOT, Samar60, Quebec99, LilHelpa, Johnxxx9, Frosted14, Thehelpfulbot, Lithenium, Sporath, FrescoBot, Surv1v4l1st, Vanished user aqpoi4u3tijsrfi, Bambuway, I dream of horses, Jonesey95, Trappist the monk, Samp84n, RjwilmsiBot, DexDor, Vishakh12345, John of Reading, Cos(90), AssiPunjabi, Dewritech, Winner 42, Ross2e, Jasonanaggie, Dekker451, Cogiati, Illegitimate Barrister, Anir1uph, Simpsora, Unionin, Quite vivid blur, HandsomeFella, AktiNo, Loginnigol, Snotbot, AgniKalpa, Helpful Pixie Bot, Karthiknair1991, Jjoy3646, Strike Eagle, Waynealan, Michael Barera, Pak ISPR, Mayank23yadav, RichardMills65, Standardengineer, Sghatak22, Khazar2, Irondome, Antiochus the Great, Keith 741, Gullupat, Nicky mathew, Julietdeltalima, ScrapIronIV, Chrfwow, Randhwasingh, AusLondonder, 21lima, NautilusPrime, KnightWarrior25, Deepanshu1707, MBlaze Lightning, Knab1976ed and Anonymous: 184

- **Strategic bomber** *Source:* https://en.wikipedia.org/wiki/Strategic_bomber?oldid=736032407 *Contributors:* The Anome, Fubar Obfusco, Maury Markowitz, MartinHarper, Arpingstone, HarryHenryGebel, David.Monniaux, Stargoat, Riddley, Gentgeen, Lowellian, Meelar, Jpo, Oberiko, Everyking, DO'Neil, Iceberg3k, Gdr, Rangi, Nickptar, Maikel, Discospinster, JohnRDaily, Bender235, Verbalcontract, Gershwinrb, Obradovic Goran, Octoferret, Alai, Ceyockey, Nuno Tavares, Woohookitty, GraemeLeggett, Rjwilmsi, Brendano, ViriiK, Straker, RussBot, Arado, TDogg310, Petri Krohn, Anclation~enwiki, Sailboatd2, SmackBot, 0x6adb015, Mscuthbert, Athaler, Chris the speller, Neo-Jay, Can't sleep, clown will eat me, Prmacn, TGC55, The PIPE, LUCPOL, Buckboard, HennessyC, CmdrObot, Cydebot, Fnlayson, Danrok, GSTQ21C, Aldis90, JamesAM, Honeplus, G. C. Hood, S Marshall, Bobblehead, Thadius856, AlekseyFy, Darrenhusted, Petronas, OhanaUnited, Rich257, Stephenchou0722, CommonsDelinker, Gfox88, TheTrojanHought, Ash sul, Dzerod~enwiki, TXiKiBoT, Randall uob~enwiki, Usergreatpower, TangoTheory, MikeRumex, LordAntagonist, MBK004, Binksternet, Clivemacd, Lastdingo, Voiceofbreeze, Ramif 47, Alexbot, Searcher 1990, American Imperialist, Addbot, Oldmountains, Tassedethe, Numbo3-bot, Lightbot, Greyhood, Pleclown, Enthusiast10, Luckas-bot, Mo7amedsalim, Oli26, Xqbot, Johnxxx9, Anotherclown, Jonathon A H, Midgetman433, Vanished user aqpoi4u3tijsrfi, Grand-Duc, Bambuway, BenzolBot, Joydeep ghosh, FoxBot, DexDor, FC Toronto, Joelwilliamson, Rail88, Quite vivid blur, Mikechou2, Bped1985, Heaney555z, Helpful Pixie Bot, Jay8g, Extrapolaris, Valio subaru, Farmer Brown, XXzoonamiXX, Acetotyce, Ilmarmyldmae, Wjculp, Samantha Ireland, Rhdzxjtsr and Anonymous: 111

- **Intercontinental ballistic missile** *Source:* https://en.wikipedia.org/wiki/Intercontinental_ballistic_missile?oldid=735998318 *Contributors:* TwoOneTwo, Trelvis, Eloquence, Mav, Bryan Derksen, The Anome, Guppie, Rmhermen, Roadrunner, Maury Markowitz, Daniel C. Boyer, Bobdobbs1723, Patrick, D, JohnOwens, Michael Hardy, Ixfd64, Alfio, Stan Shebs, Muriel Gottrop~enwiki, Rlandmann, DropDeadGorgias, GCarty, Mulad, Emperorbma, RickK, Audin, E23~enwiki, Tempshill, Morven, Pstudier, Robbot, Ke4roh, Astronautics~enwiki, RedWolf, Altenmann, Jondel, Bkell, Eliashedberg, HaeB, Navraj ghataura, Mattflaschen, GreatWhiteNortherner, DocWatson42, Jhf, ShaunMacPherson, Greyengine5, Fastfission, Marcika, Filceolaire, Slowking Man, Antandrus, Tim Pritlove, Melloss, Alex Cohn, Balcer, Husnock, Neutrality,

Imjustmatthew, Maikel, Trevor MacInnis, Mike Rosoft, N328KF, Jiy, Discospinster, Rich Farmbrough, Pmsyyz, Rama, Vsmith, Rupertslander, Bender235, ZeroOne, JJJJust, Kwamikagami, PhilHibbs, RoyBoy, JRM, TomStar81, Atomique~enwiki, .:Ajvol:., Apyule, Bijal d g, Dbchip, Kirk2, MARQUIS111, Towel401, Obradovic Goran, Wayfarer, Drangon, Maxim K, JohnDelano, Anthony Appleyard, Hektor, Coma28, Jeltz, Joshbaumgartner, KaranJ, Alinor, Rwendland, Cal 1234, Pauli133, BDD, Gene Nygaard, Drbreznjev, Dan100, Dennis Bratland, Falcorian, CranialNerves, Bobrayner, Simetrical, LOL, PoccilScript, Rianamit, Jaavaaguru, Kosher Fan, Ortcutt, Tabletop, Uris, GregorB, BlaiseFEgan, Kralizec!, GraemeLeggett, Palica, Digitalsuresh, Graham87, BD2412, Li-sung, Amorrow, Ketiltrout, Sjö, Rjwilmsi, Koavf, Rogerd, Soakologist, Brighterorange, Toby Douglass, FlaBot, Ground Zero, Chinfo, Ours, Vilcxjo, Russavia, Nicholasink, Chobot, Bgwhite, WriterHound, Gwernol, YurikBot, Wavelength, Abanamat, Arado, Briaboru, Hede2000, GregLoutsenko, DanMS, Subsurd, NawlinWiki, Dijxtra, Edboprima, Dudtz, Pstakem, Asherett, EEMIV, Mieciu K, Bota47, Kal-El, Cmskog, Dan Harkless, Georgewilliamherbert, Enormousdude, Petri Krohn, Allens, Benandorsqueaks, Mejor Los Indios, Isoxyl, MaeseLeon, SG, Sardanaphalus, SmackBot, KocjoBot~enwiki, Eskimbot, By78, Lamjus, Shoorick, Chris the speller, Bluebot, Jprg1966, Thumperward, Hibernian, The Rogue Penguin, Effer, Colonies Chris, WDGraham, Can't sleep, clown will eat me, Egg plant, Txinviolet, Chlewbot, OrphanBot, Joema, Drono, HeteroZellous, Flyguy649, Srjwalker, NorseOdin, Rjb uk, TTE, Ohconfucius, Will Beback, Ali 786, Hollowman512, Rockvee, Everdawn, Esw01407, John, Scetoaux, NongBot~enwiki, Illythr, Muadd, Avs5221, Rwboa22, KurtRaschke, Hoover889, Burto88, Iridescent, Kencf0618, Craigboy, IworkforNASA, Joseph Solis in Australia, MagisterLudi, Tawkerbot2, Dlohcierekim, AceKingQueenJack, CmdrObot, Ale jrb, Wafulz, Dgw, N2e, Neelix, Marqueed, PepijnvdG, Chuggsymalone, Tec15, ST47, Tkynerd, DumbBOT, Nabokov, Monksealpup, Epbr123, Kubanczyk, Sergeyy, Mpawluk, Wxstorm, N5iln, RevolverOcelotX, Bobblehead, Voltzz, Matthew Proctor, Mm11, AntiVandalBot, Arka Voltchek, Soren121, Zedla, MUSICAL, Ioeth, Youknowthatoneguy, JAnDbot, WikipedianProlific, Chanakyathegreat, TAnthony, Meeowow, Adzze, Chagai, .anacondabot, Magioladitis, Amadoraa123, VoABot II, ToMega666, Flayer, Catgut, BilCat, Damuna, Stanistani, Mark Lincoln, S3000, Walle83, Raza0007, Hdt83, MartinBot, Kiore, Alro, AlexiusHoratius, J.delanoy, Tom Paine, Aleksandr Grigoryev, Uncle Dick, Thaurisil, Tdadamemd, PaoloMarcenaro, Joman726, Thomas Larsen, Vineetmenon, Notreallydavid, Ash sul, Jackaranga, Eionm, Nat682, RVJ, Abcx123, HxG, Apocalyptic Destroyer, Vedran8080, Porckmaster, Tourbillon, LeilaniLad, Sdsds, Pwnage8, Tangerineduel, Oxfordwang, Arcyqwerty, Ferengi, Auspiv, Supertask, Raryel, Cosmos416, Gilisa, Jan Gadimzadeh~enwiki, Andy Dingley, Usergreatpower, Signsolid, SieBot, PanagosTheOther, Gerakibot, Falcofire, Lachrie, Bentogoa, Flyer22 Reborn, Fagarasan, Jasgrider, NATO UNIT 3, Steven Crossin, Lightmouse, Hobartimus, FLengyel, OKBot, Iknowyourider, Dabomb87, Segregator236, Slaporte, ImageRemovalBot, Deavenger, MBK004, ClueBot, Gits (Neo), Fyyer, The Thing That Should Not Be, Xav71176, Rjd0060, Leeveraction, Unbuttered Parsnip, Drmies, Anatidae, Easphi, Great.constantine1, Passargea, Ordinaterr, LeoFrank, Robert Skyhawk, Excirial, Jusdafax, Darkhelmet322, THE CAKE IS ALIVE, CKC01, Cunikm, Skipperboil1, Thingg, Jellyfish dave, Ovesen, Versus22, Deathmare, DumZiBoT, Harman malhotra, RajatKansal, Vivekrattan, BRPXQZME, Pwrproretaf07, Mrziggy5000, Ploughshares, Osarius, Addbot, Beamathan, Sephirothis666, Olli Niemitalo, Ronhjones, Simpsonbot, Fluffernutter, Glane23, AndersBot, Favonian, SpBot, Lightbot, OlEnglish, Gail, Greyhood, ScienceApe, Yobot, Assie Travis, Librsh, Crispmuncher, Nirvana888, Nallimbot, AnomieBOT, Rubinbot, 1exec1, Cumulus, Tavrian, Aditya, Kingpin13, Rohit spas12, Materialscientist, Druid.raul, Mr Fu.ck You, ArthurBot, LovesMacs, TinucherianBot II, Geomartin, GrouchoBot, Trongphu, Mark Schierbecker, Nx, Iceman444k, Dherky, Midgetman433, Muneer2908, Atomicgurl00, FrescoBot, Harsha363, Bharatgopal, D'ohBot, Videogamer22, Bambuway, Xaveq, Aogouguo, Gesalbte, Missile expert, Yin61289, Pinethicket, Vistaindia, Thatpage1798, King Zebu, Hamtechperson, MastiBot, Serols, Rotblats09, Poliphile, Bcs09, Jksdguiwaetrwue, Alice Muller, Trappist the monk, Mys 721tx, Neerajgoel07, Overjive, रोहिन रावन, Clarkcj12, Qazmlp1029, Draceius, Vipin3000, Yappy2bhere, Haxxor09, Tbhotch, Bamac30, Jiffles1, RjwilmsiBot, NameIsRon, Beyond My Ken, Tpanov, Chessofnerd, EmausBot, John of Reading, WikitanvirBot, Helium4, AssiPunjabi, Look2See1, Tonsitem, GoingBatty, RA0808, Pakzind, Sp33dyphil, Otthgr, Ao333, Kaakg, Chopduel, Illegitimate Barrister, F69aeb1e491308765e872de70f58e18b, Nickjf22, Anir1uph, The Nut, Victor-ny, S trinitrotoluene, Thine Antique Pen, LWG, Victory in Germany, Noodleki, Rangoon11, Mythbuster2010, Redefinecool, Generalking, ClueBot NG, Gilderien, Jenova20, Muon, O.Koslowski, Widr, Somatrix, B21O303V3941W42371, Strike Eagle, Kinaro, BG19bot, Wasbeer, Manishuvits, Peerzada111, Rafay15, Debastein, Dr. Whooves, Pritishp333, DistributorScientiae, Asd36f, HP83, Go ahead punk, BattyBot, Antares Geminorum, RichardMills65, KidA424, Director Emi, Mrt3366, Cyberbot II, Sidkoode, Tandrum, Ankit21694, Khazar2, SimranjeetSingh2507, Rakopa, Furqan123456789, Karanrawat21, Watch7, Mogism, Amu beeb, BDE1982, Lgfcd, Natureboy71, Jamesmcmahon0, Abishai 300, Poopnuke, Priyranjan87, Theemann, Mixrunya, CouvGeek, Skookum98621, Pvpoodle, Alurujaya, Headhunter FRANK, Tamlinwah, Jerryntcjc, Traitortanmay, Mfb, Ronrosano, Monkbot, Patient Zero, SupperSteve22, Shellgas, Granhil, Pototo1, Nicky mathew, TD712, Haider9019, Julietdeltalima, JLHockeyKnight, Mustafa Ispahani, Nøkkenbuer, Mooskimoomoo, Deepanshu1707, MBlaze Lightning, Jsmith7342, Harrisonrc, Majesticseaflapflap32, GreenC bot, Beretta84Fish, Dupre299 and Anonymous: 646

- **Submarine-launched ballistic missile** *Source:* https://en.wikipedia.org/wiki/Submarine-launched_ballistic_missile?oldid=731404760 *Contributors:* Mav, The Anome, Guppie, Edward, Lir, Patrick, Isomorphic, Dante Alighieri, Paul A, Alfio, Rlandmann, GCarty, [212], JidGom, Mulad, Emperorbma, David Newton, Tpbradbury, Morven, Topbanana, Romanm, Greyengine5, Fastfission, Pascal666, Comatose51, H1523702, Yardcock, Balcer, KNewman, Roberdin, Rich Farmbrough, Guanabot, Pmsyyz, Bender235, TomStar81, Longhair, Espoo, Joshbaumgartner, Alinor, Rwendland, Lofor, Kralizec!, GraemeLeggett, Jno, Graham87, NixonB, Ketiltrout, Rjwilmsi, Zambani, FlaBot, Mark83, Kolbasz, Gwernol, Borgx, Deeptrivia, Arado, Xihr, Virek, Howcheng, Charles Randles, Rwalker, Ormondroyd, PTSE, E Wing, SmackBot, Unyoyega, By78, Ottawakismet, Effer, Redline, Vladislav, Jumping cheese, The PIPE, ALR, Ohconfucius, Dammit, DI2000, Adam Keller, Thijs!bot, 1d447, Bobblehead, Miller17CU94, Arka Voltchek, Corella, Jwwil, WikipedianProlific, Chanakyathegreat, Magioladitis, Adam keller, Denorios, Theroadislong, BilCat, LorenzoB, Pikaco, Maurice Carbonaro, Cop 663, EanS 1, Buhuzu, Jackaranga, Xyl 54, Vipinhari, Thefrood, Pmarshal, Lightmouse, Moletrouser, Patrick Rogel, Russl5445, Jellyfish dave, Mayanks 098, Otherjoke~enwiki, Addbot, Pearll's sun, Nohomers48, Fireaxe888, Greyhood, Enthusiast10, Luckas-bot, Yobot, Rohitkumarindia, Materialscientist, Quebec99, Xqbot, Armbrust, Peter.thejackos, Parabellum101, FrescoBot, Bharatgopal, Whilecover-2009, Bambuway, Yin61289, Moryak, Hornet24, Lsr1980, Poliphile, Bcs09, Фёдоров, DexDor, Alph Bot, Never give in, RA0808, Sp33dyphil, Sanjeevsofteng, ZéroBot, Illegitimate Barrister, Space25689, Mdmday, Quite vivid blur, Mythbuster2010, ClueBot NG, Touchtheskywithglory, Suresh 5, Helpful Pixie Bot, Strike Eagle, Wiki13, Ninney, Harizotoh9, Pak ISPR, Ganchaoniuhe, 23haveblue, Fraulein451, Cyberbot II, ChrisGualtieri, XXzoonamiXX, Finnusertop, RobDuch, UY Scuti, Arderich, Nicky mathew, VandeMataram, Aminman, Zafer14ur8, CarolEvanoff, Nani1992, GreenC bot, Beretta84Fish and Anonymous: 95

- **Multiple independently targetable reentry vehicle** *Source:* https://en.wikipedia.org/wiki/Multiple_independently_targetable_reentry_vehicle?oldid=733710317 *Contributors:* Trelvis, The Epopt, Mav, Alex.tan, Frecklefoot, Patrick, Voidvector, Gabbe, Iluvcapra, Theresa knott, David Shay, Krellmachine, ChrisO~enwiki, Yosri, Wikibot, Elde, GreatWhiteNortherner, DocWatson42, Fastfission, Zinnmann, Golbez, Mzajac, Neutrality, Guanabot, Rama, Night Gyr, Bender235, MattTM, Maurreen, Sasquatch, Ferrierd, Voxadam, Isnow, Meneth, Graham87,

Yuriybrisk, Syndicate, Georgelazenby, Syced, FlaBot, Quicksilvre, YurikBot, Arado, PotatoSamurai, Knotnic, MrBook, Anclation~enwiki, TheDeathCard, SmackBot, KocjoBot~enwiki, Chris the speller, MalafayaBot, Hibernian, Egg plant, Glloq, Joema, A.R., Wybot, Acdx, TKarrde, Fedallah, AdultSwim, ICEBreaker, Iridescent, Alexander Iwaschkin, Primus Sheck, Nabokov, Alaibot, Aldis90, Thijs!bot, Kubanczyk, SkonesMickLoud, RevolverOcelotX, Nyq, Matt B., Kiore, Duch, JacksonVance, Greg zimmerman, Signalhead, Rockstar915, Imasleepviking, GiZiBoNG, Usergreatpower, Eskovan, Justinchudgar, Elhawarey, RobinHood70, Dpmuk, Niceguyedc, Lebrulant, Jellyfish dave, DumZiBoT, Gavinb15, SilvonenBot, Addbot, Download, SamatBot, Zorrobot, Yobot, Ptbotgourou, AnomieBOT, Capricorn42, .45Colt, Simonjon, Abcjake, Vendeae, Surv1v4l1st, MastermindPrime, Beyond My Ken, Wintonian, EmausBot, Boundarylayer, Kaakg, Yiosie2356, EWikist, JoeSperrazza, Neutrozixide, Storyofworkers, Ebehn, Whoop whoop pull up, ClueBot NG, Chrisminter, Suresh 5, Estevezj, 220 of Borg, ChrisGualtieri, TwoTwoHello, Lostromantic, JakeWi, Nicky mathew, Mustafa Ispahani, MehrdadFR, MBlaze Lightning, Got99caloriesbutbishaintone, Cantab1985 and Anonymous: 94

- **Nuclear strategy** *Source:* https://en.wikipedia.org/wiki/Nuclear_strategy?oldid=735640161 *Contributors:* Trelvis, Mav, The Anome, Koyaanis Qatsi, Guppie, Ortolan88, Edward, Patrick, JohnOwens, Fastfission, Azzurro~enwiki, Nimbulan, Bender235, SpencerWilson, Bobo192, Scott Ritchie, Perceval, Gary, Habap, Eakaplan, Nihiltres, Arado, Tfine80, Sardanaphalus, SmackBot, Reedy, Bluebot, OAS~enwiki, Tazmaniacs, Andrew Davidson, JForget, Isles, Carolmooredc, MegX, Satyabrataraic, Amaraiel, Johnfos, Strategik, Ktr101, Addbot, Yobot, AnomieBOT, ArthurBot, Lithenium, Trinity54, DrilBot, Pinethicket, Rabbabodrool, DBG Heuser, TYelliot, ClueBot NG, Iqra97, RobDuch, Abattoir666, Cganuelas, GoldCar and Anonymous: 33

- **Nuclear warfare** *Source:* https://en.wikipedia.org/wiki/Nuclear_warfare?oldid=735578787 *Contributors:* Damian Yerrick, AxelBoldt, Trelvis, Derek Ross, Mav, Bryan Derksen, Robert Merkel, Zundark, The Anome, Stephen Gilbert, Guppie, Eclecticology, Rmhermen, Enchanter, SimonP, Maury Markowitz, Graft, B4hand, Mintguy, Zippy, Ewen, Nevilley, Edward, Patrick, Smelialichu, Liftarn, Gabbe, Eurleif, Ralmin, Karada, Goatasaur, Ellywa, Ahoerstemeier, Muriel Gottrop~enwiki, Angela, Kingturtle, Cyan, Vzbs34, Jiang, Evercat, Astudent, [212], Vroman, Mulad, PaulinSaudi, Dino, JasonM~enwiki, Fuzheado, Wik, Zoicon5, Tempshill, Populus, Tophanana, JonathanDP81, Jusjih, Jni, DavidA, Mazin07, Moriori, Fredrik, Kizor, Chris 73, Moondyne, Romanm, Securiger, Dubidubno, Chris Roy, Drago9034, Yosri, Meelar, Bkell, Hadal, Robinh, Defragged, Michael Snow, Mushroom, PBP, Davidcannon, Alan Liefting, Christopher Parham, Massysett, Ferkelparade, Malcontent, Fastfission, Peruvianllama, Everyking, Jacob1207, NeoJustin, Abqwildcat, Crusty Ass, Zoney, Tweenk, Matt Crypto, Bobblewik, Neilc, Stevietheman, PeterC, Slowking Man, Antandrus, Beland, Blazotron, Mark5677, Andrew Kanaber, Rdsmith4, RetiredUser2, Sam Hocevar, JHCC, RickScott, Joyous!, Ukexpat, Karl Dickman, Trevor MacInnis, Lacrimosus, Mike Rosoft, Kingal86, Freakofnurture, N328KF, Discospinster, Twinxor, Rich Farmbrough, Guanabot, Cacycle, Rama, MeltBanana, VT hawkeye, LeoDV, Bender235, ZeroOne, Janderk, Jarsyl, Jensbn, Maclean25, Mr. Billion, El C, Kwamikagami, Shanes, Spearhead, Circeus, Apollo2011, Smalljim, WardHayesWilson, Cohesion, Maureen, Scottishneil, Homerjay, Darwinek, Physicistjedi, Towel401, DCEdwards1966, Obradovic Goran, Idleguy, MPerel, Dillee1, Jumbuck, Mrzaius, Alansohn, Gary, Borisblue, Mr Adequate, Paleorthid, Riana, Viridian, Seans Potato Business, Mailer diablo, Wdfarmer, Snowolf, Velella, Middenface, Cmapm, Meertn, H2g2bob, DV8 2XL, Dziban303, Netkinetic, Kazvorpal, Killing Vector, Tariqabjotu, CranialNerves, Ron Ritzman, Crosbiesmith, Megan1967, Kelly Martin, Jeffrey O. Gustafson, Jason Palpatine, Oliphaunt, TomTheHand, Macker, ^demon, WadeSimMiser, Hdante, Eleassar777, Uris, Wikiklrsc, Plrk, JohnC, Wayward, Rusty2005, Mandarax, MassGalactusUniversum, RichardWeiss, Graham87, Magister Mathematicae, Descendall, Jetekus, Yurik, Zzedar, Mendaliv, Calicocat, Sjö, Rjwilmsi, Eyu100, Wert0086, Lordkinbote, Toby Douglass, Nigosh, Lcolson, FlaBot, Ysangkok, Survivor, Mark83, RexNL, Gurch, Kolbasz, Jordan Elder, Srleffler, Lightamplification, JohnRThomson, Scimitar, Chobot, GangofOne, VolatileChemical, Bgwhite, Simesa, WriterHound, Banaticus, YurikBot, Blackworm, RussBot, Anonymous editor, Zafiroblue05, Kirill Lokshin, Stephenb, Ksyrie, Guslto, Wimt, NawlinWiki, Robertvan1, Tfine80, RazorICE, Nightmare X, Irishguy, Ragesoss, Anetode, Raven4x4x, Catharticflux, Saberwyn, Dbfirs, Dissolve, Barnabypage, Tzustrategy, Manjithkaini, Enormousdude, 21655, Gregzeng, Robotico, Mike Selinker, Cubic Hour, Sean Whitton, Petri Krohn, CWenger, HereToHelp, Bob Hu, Junglecat, NeilN, DVD R W, Sardanaphalus, MacsBug, Yakudza, SmackBot, XYaAsehShalomX, Moeron, Hux, PeoplesWar, Jim62sch, Aborlan, KocjoBot~enwiki, Delldot, Eskimbot, Timeshifter, Apartmento, ComaDivine, Jkp1187, Gilliam, Blather, Ohnoitsjamie, Hmains, Betacommand, Skizzik, Kevinalewis, Andy M. Wang, Maxgrin, Chris the speller, Unint, Hibernian, Iamakhilesh, Apple2gs, Mithaca, Ned Scott, Colonies Chris, A. B., Chendy, Salmar, Llwch, KaiserbBot, Brainhell, Aldaron, Krich, BostonMA, Engwar, Celardore, Fitzhugh, Soarhead77, Where, Risker, Tangsyde, Kukini, Coconuteire, NeilFraser, SirIsaacBrock, AThing, Harryboyles, Kuru, Calvados~enwiki, Writtenonsand, Drew First, Sosodank, J 1982, Ajd1992, Mygocarp, Joelmills, Shadowlynk, CaptainVindaloo, Goodnightmush, Gnevin, Majorclanger, AdAdAdAd, Ckatz, JHunterJ, MarkSutton, Special-T, Lipatden, Hypnosifl, Waggers, CharlesMartel, Fluppy, Citicat, Spacebar, Hu12, Iridescent, Sunoco, Dekaels~enwiki, Joseph Solis in Australia, ToastyMallows, DavidOaks, Happy-melon, Courcelles, Tawkerbot2, Neil Evans, Bootleg42, Ale jrb, Dycedarg, Steve.janke, Joncnunn, CWY2190, Ericanderson, Chmee2, Phædrus, Delong71487, Mapletip, Cydebot, Hydraton31, Wikiwriter706, Trasel, Mammajo, Vanished user vjhsduheuiui4t5hjri, UncleBubba, Jayen466, Give Peace A Chance, Daniel J. Leivick, SymlynX, Tawkerbot4, Quibik, HitroMilanese, Joe 1987, Omicronpersei8, Daniel Olsen, MarkJablonski, The machine512, BetacommandBot, JamesAM, Thijs!bot, Keraunos, RevolverOcelotX, Marek69, Bobblehead, J-Party, Nick Number, Matthew Proctor, Transhumanist, Dpenguinman, AntiVandalBot, Ggbroad, Luna Santin, Guy Macon, Dbrodbeck, Betdud, Carolmooredc, Prolog, Rehnn83, ASDFGHJKL, Dylan Lake, LibLord, Corella, Gdo01, Golf Bravo, Kami888, Leuqarte, Golgofrinchian, Res2216firestar, MER-C, Instinct, Lan Di, Hut 8.5, Adzze, Mifa68, Siobhan-Hansa, Acroterion, Lighthope, Magioladitis, Gekedo, Grepnork, Unused0029, Bongwarrior, VoABot II, PotentialEnergy, Adam keller, JamesBWatson, Mouchoir le Souris, Virginia Dutch, Robotman1974, Eataubm, Styrofoam1994, DerHexer, Swordsman04, Mrathel, Oroso, Gjd001, MartinBot, Arjun01, NAHID, Genghiskhanviet, R'n'B, MapleTree, Pekaje, WelshMatt, Smokizzy, J.delanoy, AstroHurricane001, Ali, UBeR, Peter Chastain, Maurice Carbonaro, Strouty13, Hodja Nasreddin, Tdadamemd, Theeurocrat, SharkD, Dontrustme, Dfoofnik, Notreallydavid, Rob322, Jarnassy, NewEnglandYankee, Matthardingu, Basketdove, Tvbrichmond, Alchemynut2, Inter16, Flag amr 5, BernardZ, CardinalDan, Wikieditor06, ACSE, Posita, Malik Shabazz, VolkovBot, Human step, Johnfos, ABF, Lt. Col. Cole, Thisisborin9, Jeff G., Disasterplanner, Barneca, Philip Trueman, XavierGreen, Alan Rockefeller, Pwnage8, Dictouray, Crohnie, SJSedlacek, Nukemason4, Oxfordwang, Clarince63, Martin451, AllGloryToTheHypnotoad, ^demonBot2, Ilyushka88, Madhero88, Larklight, MartinPackerIBM, Eskovan, Mustangtyota, Falcon8765, Vector Potential, Chodie, Mashi121994, Ronmore, Ceranthor, Unixwzrd, Spoofer25, EmxBot, HereIsNoWhy, Cryonic07, Overlord11001001, ShiftFn, OberRanks, Hanzo05, Work permit, WereSpielChequers, Polanddude, Yintan, Flyer22 Reborn, Tiptoety, Oda Mari, Le Pied-bot~enwiki, Lanzarotemaps, Lightmouse, RW Marloe, BenPhenicie, Jruderman, Onopearls, Belligero, StaticGull, Hamiltondaniel, CCHIPSS, WikiLaurent, Joelster, Denisarona, Velvetron, Explicit, Muhends, ImageRemovalBot, Loren.wilton, Martarius, Tanvir Ahmmed, MBK004, ClueBot, Emberstone878, Binksternet, Fasettle, The Thing That Should Not Be, Kavik Kang, Chessy999, VQuakr, Polyamorph, Meisme12, Indyost, Philo-sofa, Excirial, Jusdafax, Diplodoc, Erebus Morgaine, FOARP, Foundation316, NuclearWarfare, Sbfw, WillA108,

Redthoreau, SchreiberBike, O.Duke, Muro Bot, Kakofonous, Bald Zebra, Thingg, Versus22, Joedaddy09, Qwfp, Vanished User 1004, Eik Corell, Avoided, Mrrodger, ItachiSanta117, Noctibus, ZooFari, JCDenton2052, Snapperman2, Shoemaker's Holiday, Ejosse1, Femajesus, Gggh, Addbot, ERK, Stephen Fulcher, Wiki Mateo, Some jerk on the Internet, Melab-1, Hda3ku, Fi11222, Montgomery '39, Earth is over, IchWeigereMich, Chamal N, Longfetched2, Acpls, LinkFA-Bot, Arunrama, Tide rolls, Taketa, Gail, Luckas-bot, Reargun, Les boys, Hohenloh, Legobot II, Alexboi14, Newportm, Grochim, Jnivekk, Eric-Wester, AnomieBOT, Metalhead94, Noq, Nuclear-Age, Jim1138, Hadrian89, Kingpin13, CompositionII, Voodoopriest, Xqbot, Intelati, Capricorn42, BritishWatcher, Mlpearc, Cyphoidbomb, AV3000, Jane McCann, GrouchoBot, Hrdill, Brandon5485, RibotBOT, Shayanshaukat, Citationeditor, Natural Cut, WikipedianumberoneExpert, Endothermic, Operdyne7, Sesu Prime, SD5, Nenya17, BoomerAB, Gresh786, Csa.certified, Lithenium, Surv1v4l1st, Tobby72, Sotenburger, TheDarkCurrent, DivineAlpha, Xaveq, Pinethicket, I dream of horses, Omni314, Tom.Reding, Coekon, My very best wishes, Rotblats09, White Shadows, Cnwilliams, Enemenemu, Greshoops, CathySc, Comet Tuttle, Arkelweis, Zonglowe, ThunderbirdJP, Mr.98, Yankeesrule7, Tbhotch, Andrea105, Quantacoupler, Alexhobbit, John of Reading, Boboboum, ScottyBerg, Rail88, AUSTRALIIAN, Boundarylayer, Dewritech, GoingBatty, JuanAfro, Solarra, Emac15, Tommy2010, Uploadvirus, Blahamadali, Dcirovic, Jasonanaggie, Illegitimate Barrister, Josve05a, Booss6, Bollyjeff, Piano4444, IamYoshi, Lyonthunder, Idontkn0w68, Wyatthas21 3, Unreal7, Calligarus, JoeSperrazza, CSvensgaard, Niljem, Quite vivid blur, Peter Karlsen, Grampion76, Ebehn, Whoop whoop pull up, ClueBot NG, Iiii I I I, SilentScope001, Joefromrandb, Sleddog116, Vacation9, Hazhk, Bobbyb373, CaroleHenson, Names are hard to think of, Spyro019, Helpful Pixie Bot, Macelliboi2, Jjoy3646, BG19bot, Ajconst, Viggoodin, Phd8511, Flutte, Lawrencect, King Vic 2, Toccata quarta, Supaninjazombie123, Nuclearwariswrong, Conelrad79, Chuckmccaw, 1xdd0ufhgnlsoprfgd, Loriendrew, Jambombastic, WhiteRulez, WebTV3, OneCatch, BattyBot, Tcohn32, Cyberbot II, ChrisGualtieri, Mediran, Khazar2, Buxtongooner, Dexbot, Hmainsbot1, Mysterious Whisper, Syed Aamer Shah, XXzoonamiXX, Zjohn4, AldezD, Jamesx12345, Thomasedoyle2, Scottyslc, Soffredo, Alcoholic Red Squirrel, Rolf h nelson, Chinamattsunflower, DavidLeighEllis, Haminoon, Wyred, Joeletaylor, Limnalid, Jackmcbarn, CaptainMunchiesxX, Kingslove2013, Coercive Diplomacy, Fixuture, Shapirothepottedplant, Abattoir666, Ramdomedits, Britannia1234, Monkbot, Susan Macafee, Lucasjohansson, Fezzerof, WC Jay, Crystallizedcarbon, Mr. Magoo and McBarker, Patricia frisch, Guyyy23, Atvica, Rdoop, Vegitø, Whalestate, KasparBot, Aaron is ebest, Hulk576, Jodijoe, Antrangelos, Wikkileaker, Joegrainge10, Luaza1313, Jon deluxe, GreenC bot, Opacitatic, Wikipedia6969669, The Bringer of Doom and Anonymous: 873

- **On Thermonuclear War** *Source:* https://en.wikipedia.org/wiki/On_Thermonuclear_War?oldid=723736078 *Contributors:* Reddi, Bender235, Wikiklrsc, GregorB, Arado, A bit iffy, Hmains, Kevinalewis, MaxEnt, Neil916, Magioladitis, Gwern, Thaurisil, GrahamHardy, The deathmonkey, Fadesga, On Thermonuclear War, Yobot, Themfromspace, AnomieBOT, FrescoBot, Moonraker, Thinking of England, GoingBatty, Ό οΐστρος, JamKaftan, FriarTuck1981 and Anonymous: 6

- **Treaty on the Non-Proliferation of Nuclear Weapons** *Source:* https://en.wikipedia.org/wiki/Treaty_on_the_Non-Proliferation_of_Nuclear_Weapons?oldid=735983360 *Contributors:* Trelvis, Mav, Bryan Derksen, Jeronimo, SJK, Hephaestos, Boud, MartinHarper, Gabbe, Delirium, Mac, Snoyes, Aarchiba, Jiang, Kaihsu, Schneelocke, Ideyal, Vanished user 5zariu3jisj0j4irj, Timwi, VeryVerily, Shizhao, Wojeter, Cdupree, Sjorford, EdwinHJ, Xiaopo, Psychonaut, Lowellian, Yosri, Academic Challenger, CIngre, Humus sapiens, Hadal, Rho~enwiki, Mushroom, Robartin, Mattflaschen, Davidcannon, Decumanus, MaGioZal, Fastfission, Everyking, Lestatdelc, Jgritz, Get-back-world-respect, Allstar86, Luigi30, Michael Birk, Bobblewik, Blankfaze, Antandrus, Vina, Rdsmith4, Mzajac, DragonflySixtyseven, Comics, Ary29, Lacrimosus, Nimbulan, Discospinster, Twinxor, Rich Farmbrough, Avriette, Guanabot, KittySaturn, FrickFrack, Rsanchezsaez, Freestylefrappe, Smyth, Berkut, Darren Olivier, Jowr, Martpol, Wulfson~enwiki, Bender235, *drew, Kwamikagami, PhilHibbs, Shanes, Susvolans, IFaqeer, Robotje, Bawolff, Bobbis, Slipperyweasel, Lokifer, Sam Korn, Jonathunder, Leifern, Perceval, Kingsindian, Ranveig, Frodet, Uncle.bungle, V2Blast, Rd232, Geo Swan, Joshbaumgartner, ATG, ABCD, Alinor, Rwendland, Ronark, Pappa, Count Iblis, Cmapm, Jsw663, Geraldshields11, Sfacets, DV8 2XL, Joriki, Wheresthebrain, Mpatel, Firien, GregorB, Male1979, Hgd4th, Dysepsion, Magister Mathematicae, BD2412, Monk, Yurik, Kissekatt, Josh Parris, Canderson7, Rjwilmsi, Collins.mc, Grant-o, Kthejoker, Titoxd, Sky Harbor, Mariocki, WWC, RexNL, Gurch, Vneiomazza, I Am Not Willy On Wheels, Valentinian, HaZaRd, Ggb667, Chobot, Simesa, YurikBot, Wavelength, BuddyJesus, JWB, Hairy Dude, Deeptrivia, JOSEPHANTONYJ, Pburka, Loom91, Conscious, Chaser, Thoreaulylazy, Gaius Cornelius, Pseudomonas, Member, Mass147, Danlaycock, Biopresto, EEMIV, Rwalker, Scope creep, Deepak~enwiki, Kyle Cronan, Tigger69, UW, PTSE, Theda, GraemeL, Scoutersig, Katieh5584, CIreland, Mussnoon, SpLoT, Remiel, SmackBot, Hux, Xlenka~enwiki, Arthur5005, Tonyr68uk, C.Fred, Eaglizard, Michaelll, Kintetsubuffalo, Aixroot, Ohnoitsjamie, The paccagnellan, Jabbi, Ottawakismet, Xpi6, Stubblyhead, Chuyelchulo~enwiki, AntelopeInSearchOfTruth, MalafayaBot, Effer, Darth Panda, Jahiegel, Frap, Atinoda, Ajdz, Earth as one, Calbaer, Jmlk17, YankeeDoodle14, Wiz9999, Makemi, Captain Zyrain, Ala.foum, Skinnyweed, Ohconfucius, Will Beback, John, C.jeynes, Felix-felix, Joffeloff, Goodnightmush, Arborescence, IronGargoyle, Ckatz, Avs5221, Hu12, Jfix71, Peanutbutter685, Andrew Davidson, Woodgreener, Cyclopiano, Tawkerbot2, Event Horizon, CmdrObot, Sir Vicious, Blue-Haired Lawyer, Edward Vielmetti, Dub8lad1, Outriggr (2006-2009), Pgr94, Cracker017, Lixy, This1trik4u, Cydebot, Vanished user 2340rujowierfj08234irjwfw4, Master son, Tawkerbot4, Doug Weller, Thijs!bot, Wikid77, Figgie123, Faigl.ladislav, Rkessler06, RevolverOcelotX, Merbabu, Grahamdubya, Oosh, Dawnseeker2000, CTZMSC3, Metroid690, Mentifisto, AntiVandalBot, Golf Bravo, Salgueiro~enwiki, SkoreKeep, Tangurena, JAnDbot, TheBoyWonder248, MarritzN, Chanakyathegreat, Vnaynb, Andonic, Yill577, Chintanshah123, Ecki~enwiki, Magioladitis, Irishpyro, VoABot II, Alphaman, Dr B2, Nick Cooper, Jacobmal, Mike Payne, JoergenB, Stephenchou0722, Trixt, Thereen, R'n'B, Skepticus, Ssolbergj, Filll, Lizrael, Aleksandr Grigoryev, NightFalcon90909, Maurice Carbonaro, Mike.lifeguard, Thaurisil, Mamyles, ABVS1936, David r from meth productions, Abhijitsathe, Natobxl, McSly, NewEnglandYankee, Primovalentino, Corriebertus, Vanished user 39948282, VolkovBot, That-Vela-Fella, Johnfos, Jeff G., Wolfnix, Nubin wiki, TXiKiBoT, Caster23, Dchall1, Qxz, Ccwelt, Mt strasbourg, JhsBot, Mazarin07, ConservativeOpinion, Andy Dingley, ShortMunky, Albert45, Falcon8765, Bungo77, NPguy, Fudgegrad, SieBot, Bdentremont, Scarian, BotMultichill, Toddst1, Digital funeral, KoshVorlon, Lightmouse, RW Marloe, Abrichte, OKBot, Jayhawksean, Shoombooly, Toprohan, Atif.t2, Desione, Moorehaus, ClueBot, Knightrunner, Mariordo, Noone has this name yet, Bobbyque2004, The Thing That Should Not Be, Enthusiast01, Kofiannansrevenge, Mindcry, Alkamid, Leonard^Bloom, Datastat, L.tak, Nhcatsteve, Ivantheterrible1234, Joe N, Aitias, Hazmunkey, Editor2020, DumZiBoT, Afpre, Parekhharshj, Mifter, Benboy00, Good Olfactory, Milstuffxyz, Addbot, Polemarchus, Lihaas, Democracy to information, Tide rolls, Lightbot, Zaphodia, Legobot, Luckas-bot, AzureFury, Kristofferjay, QueenCake, Wonderfl, AnomieBOT, Rubinbot, Enigma Blues, Jim1138, Zoomzoom316, Kingpin13, Jacob2718, JohnnyB256, ArthurBot, Xqbot, Prevod&priredba, Khajidha, Mubhi, GrouchoBot, MerlLinkBot, Eugene-elgato, Spellage, George2001hi, FrescoBot, VS6507, Bambuway, M2545, Ahmer Jamil Khan, Slastic, Citation bot 1, Mwild96, DrilBot, Canistabbats, Isilwell, RedBot, Comancheros, Wejer, Poliphile, FoxBot, TobeBot, DixonDBot, 안용복, Vrenator, Routlee, Saurabh1984, RjwilmsiBot, Johannesf, -wrathofconn, Becritical, Pitufis, EmausBot, Nima1024, WikitanvirBot, Dewritech, Lhoaxt, Easter 8unny, L235, Wikipelli, Kboot55555, AvicBot, A2soup, BT95BT, Kris1912, Battoe19, H3llBot, PLAYBOY ROCKS!, Araliist, SporkBot, Tolly4bolly, Menikure, Jeromemoreno, Alicia wakanza, Aze0098, K. the Surveyor, Ace of Raves, Indian1985, Ganita2010, ClueBot NG,

Vjiced, Millermk, ColdWarCharlie, Mademoiselle Bunbury, Widr, Antiqueight, Prabhat1729, Watchman1806, Awasthy.aniket, Helpful Pixie Bot, Sohan516, Green786, BG19bot, Dainomite, BattyBot, KhabarNegar, Triggerhippie4, Mrt3366, Cyberbot II, Acer567345, ChrisGualtieri, Shivakumar.nm, Arcandam, Jethro B, JYBot, P3Y229, Smszmn72, Mysterious Whisper, Mogism, Esidedis, Pranavx, UNAVL, Ikesham, Timothysandole, Tanja Nikolic, Anatoly p, Builtiger, Frosty, Shandy89, Wiki.Gunjan, Soffredo, Khushank94, ArmbrustBot, Arthur goes shopping, Ugog Nizdast, XXthisguyXX, Coercive Diplomacy, Munchkin2013, Lette Sgo, Shill42, Monkbot, Susan Macafee, Leeshyukwf, Gladamas, Evissergorp, Tuneix, Leblan.evan18, Hockey8888888, KasparBot, CAPTAIN RAJU, Harish.pentapalli, Sanket Edits Wiki, Atinderbal, Baking Soda, GreenC bot, Dr. Franz sparkles, Xxxxx365, Grko3, Nakul agarwal, Siamak51 and Anonymous: 653

- **Intermediate-Range Nuclear Forces Treaty** *Source:* https://en.wikipedia.org/wiki/Intermediate-Range_Nuclear_Forces_Treaty?oldid= 735485349 *Contributors:* Bryan Derksen, Rmhermen, Patrick, Katana0182, Maximus Rex, Twang, Altenmann, Drago9034, Rsduhamel, Bkonrad, PFHLai, Kuralyov, TonyW, Neutrality, Willhsmit, Mschlindwein, N328KF, Alistair1978, XerKibard, Robotje, Nk, Lokifer, Kazuaki Shimazaki, Joshbaumgartner, Sade, ABraidotti, Cmapm, Pethr, Valip, JdforresterBot, Chobot, RobotE, Corevette, Zwobot, Paaskynen, TheKoG, Arthur Rubin, Petri Krohn, SmackBot, Unschool, Nickel Chromo~enwiki, Prodego, InverseHypercube, Gilliam, Wlmg, Hmains, Betacommand, Chris the speller, Jprg1966, Hibernian, Radchenk, Chlewbot, Ala.foum, Mathiasrex, Eivind F Øyangen, Rasmus81~enwiki, Iridescent, Outriggr (2006-2009), Cydebot, Tec15, Thijs!bot, Epbr123, The Proffesor, Hcobb, OuroborosCobra, Katiegoesmew, Sherbrooke, Seaphoto, Pemilligan, GearedBull, RP88, CommonsDelinker, Emeraldcrown, Abtinb, Rourin bushi, Mks004, Wolit, Happyme22, AlleborgoBot, Portia327, SieBot, Randy Kryn, ClueBot, Burbon2, TarzanASG, Alexbot, Mklobas, ShipFan, WikHead, MystBot, Good Olfactory, Johnkatz1972, Addbot, Polemarchus, Tide rolls, OlEnglish, 1j1z2, Luckas-bot, AnomieBOT, Rubinbot, IRP, Materialscientist, High Contrast, Obersachsebot, Locobot, Magnus242, Sporath, Jadensdad, RedBot, Full-date unlinking bot, Dinamik-bot, John of Reading, Ace of Raves, ClueBot NG, Jphill19, Garsd, BattyBot, Awooda, Comp.arch, Melcous, Ceannlann gorm, 21lima, Grko3 and Anonymous: 78

- **Strategic Offensive Reductions Treaty** *Source:* https://en.wikipedia.org/wiki/Strategic_Offensive_Reductions_Treaty?oldid=694961389 *Contributors:* Gabbe, Morwen, Sabbut, PBS, Bkonrad, Zinnmann, Vadmium, Kuralyov, Vivacissamamente, Narsil, Mr. Billion, RussBlau, Perceval, A2Kafir, Rwendland, Joriki, LOL, James Kemp, Btyner, Ketiltrout, Rillian, Jw21, YurikBot, Crazytales, Howcheng, SmackBot, Haymaker, Hmains, Jprg1966, Longboard~enwiki, Neverindoubt, Ala.foum, John, Ckatz, Joseph Solis in Australia, HelmutP, Cydebot, Thijs!bot, Kubanczyk, Headbomb, Superzohar, Yill577, Buckshot06, JaGa, Subspace1250, Dennis Myts, LogicDictates, Dzerod~enwiki, Esbruno2, Ashtray, Ipankonin, Randy Kryn, ClueBot, Binksternet, Mild Bill Hiccup, Paulcmnt, Dspark76, Good Olfactory, Addbot, Jafeluv, Polemarchus, Yobot, Rubinbot, Fc12, Zappa26, Almabot, GrouchoBot, Captain Fortran, FrescoBot, MastiBot, Full-date unlinking bot, Miracle Pen, Kris1912 and Anonymous: 31

- **Comprehensive Nuclear-Test-Ban Treaty** *Source:* https://en.wikipedia.org/wiki/Comprehensive_Nuclear-Test-Ban_Treaty?oldid= 734648548 *Contributors:* Trelvis, Mav, Bryan Derksen, Rmhermen, Gabbe, Shimmin, Kaihsu, GCarty, Schneelocke, Big Bob the Finder, Robbot, RedWolf, Tea2min, Fastfission, Poupoune5, Allstar86, Oneiros, Lacrimosus, N328KF, Rich Farmbrough, Bender235, RoyBoy, Brim, AshtonBenson, Joshbaumgartner, Ricky81682, Patrick Lucas, Mysdaao, Rwendland, Melaen, Evil Monkey, Vedant, Staeiou, Fdedio, Nightstallion, Ceyockey, Joriki, Lofor, SeventyThree, GoldRingChip, Koavf, Yamamoto Ichiro, The ARK, SonicSynergy, Subversive, Chobot, Antilived, Ahpook, YurikBot, Borgx, RussBot, Thoreaulylazy, Gaius Cornelius, CambridgeBayWeather, Danlaycock, UW, TBadger, Allens, Teryx, Victor falk, SmackBot, Reedy, Minbo, Joshfriel, QEDquid, Hibernian, Bazonka, Modest Genius, Moonsword, Engwar, Ala.foum, Kransky, Tasc, Tawkerbot2, Viennese Waltz, Boborok, Outriggr (2006-2009), Cydebot, PepijnvdG, Dancter, Mombas, Thijs!bot, Headbomb, Z10x, AntiVandalBot, Seaphoto, Chill doubt, Caper13, Yill577, Shurjil, Bencherlite, JaGa, MartinBot, Maurice Carbonaro, Thaurisil, GumboCat, JHeinonen, That-Vela-Fella, Martin451, NPguy, GenuineGoodness, SieBot, Abrichte, TheodoreB, MBK004, FelixRodriguez, ClueBot, Dpmuk, Master1228, SkeletorUK, Kofiannansrevenge, Socrates2008, Jusdafax, NuclearWarfare, Ltak, AlanM1, Il Sc0rpi0ne, Annika thunborg, Good Olfactory, Deineka, Jojhutton, Seantobin5, Scientus, Vishnava, Polemarchus, Download, LaaknorBot, Numbo3-bot, Lightbot, Legobot, Luckas-bot, Yobot, Librsh, Nallimbot, Pganas, AnomieBOT, 1exec1, WFPOhio, Flewis, Quebec99, GrouchoBot, FrescoBot, Upinews, Geheimnis128, RedBot, Yutsi, Full-date unlinking bot, Poliphile, Daspostloch, 777sms, Guatemala247, Chronulator, RjwilmsiBot, Cmlmcmillan, John of Reading, Wikipelli, Dolovis, DanKonigsbach, ClueBot NG, VAS-SAV, Rajesh1623, Green786, Doyna Yar, Geniusvinay, Keowned, Kunalkanodia77k, Davidiad, Lieutenant of Melkor, Madam Fatal, Liudvikam, STKS91, SD5bot, TestBanNow, Runlevel1, ZaidRock11, UNAVL, JigPungle, GabeIglesia, JakeWi, Khushank94, Munchkin2013, WC Jay, Jason2400, LoveToLondon, KasparBot, Vansockslayer, Grko3 and Anonymous: 156

- **List of states with nuclear weapons** *Source:* https://en.wikipedia.org/wiki/List_of_states_with_nuclear_weapons?oldid=736005911 *Contributors:* TwoOneTwo, Trelvis, Bryan Derksen, Robert Merkel, Ed Poor, Rmhermen, SimonP, Shii, Imran, Edward, Patrick, Boud, Zocky, Dante Alighieri, Wwwwolf, Ixfd64, Chinju, Ahoerstemeier, Bueller 007, Ugen64, Bogdangiusca, Vzbs34, Netsnipe, Jcolbyk, BOARshevik, Jiang, Kaihsu, Dwo, Vroman, RodC, Charles Matthews, Guaka, Daniel Quinlan, Savantpol, Snickerdo, Wik, Steinsky, DJ Clayworth, Tpbradbury, Imc, Saltine, VeryVerily, SEWilco, Zero0000, MiLo28, Joy, Pakaran, Hajor, Owen, Carlossuarez46, Kulkuri, Denelson83, PuzzletChung, Robbot, Dale Arnett, Korath, Donreed, Moondyne, Naddy, Modulatum, Securiger, Lowellian, Bkell, Xanzzibar, Dmn, Katarzyna, Peter L, Dave6, Centrx, DocWatson42, MaGioZal, Akadruid, Elconde, Nichalp, Fudoreaper, Aratuk, Fastfission, Hokanomono, Alterego, Everyking, Lestatdelc, Niteowlneils, Sukh, Duncharris, Xinoph, Tom-, Sundar, Get-back-world-respect, Gzornenplatz, AdamJacobMuller, Deus Ex, Auximines, Telso, Geni, Slowking Man, GeneralPatton, Antandrus, ClockworkLunch, Vina, RetiredUser2, PFHLai, Cornischong, Eranb, Sam Hocevar, Rangi, Neutrality, Shadypalm88, Irpen, Demiurge, Genesis, Canterbury Tail, Mike Rosoft, Blorg, Mormegil, Miborovsky, Nimbulan, Spiko-carpediem~enwiki, Discospinster, Rhobite, Cfailde, Smyth, Berkut, Darren Olivier, SpookyMulder, Edgarde, Bender235, Swid, Kelvinc, Plugwash, Brian0918, Mr. Billion, *drew, FirstPrinciples, El C, DamianFinol, AreJay, RoyBoy, Zaidiwaqas, Jpgordon, Thuresson, Ce garcon, Bobo192, Surcouf, Circeus, Viriditas, Cmdrjameson, Dpaajones, Scottishneil, Kjkolb, Jman, Ardric47, Idleguy, AppleJuggler, Supersexyspacemonkey, Perceval, Caleb Erikson, Kvaks, Petdance, Ranveig, Bob rulz, Mrzaius, Alansohn, MatthewWilcox, Ricardo monteiro, Patthek, Rd232, Andrew Gray, Bz2, Rfgdxm, Ynhockey, Redfarmer, Mysdaao, Malbrecht, Rwendland, Marianocecowski, Wtmitchell, Velella, Frankman, Vedant, Mikeo, Capitalistpiglet, TheAznSensation, Sfacets, Mmsarfraz, DV8 2XL, Axeman89, Instantnood, Nightstallion, LukeSurl, Xtopher, Ultramarine, Zntrip, Thryduulf, Kelly Martin, ScottDavis, TomTheHand, Benbest, Roscommon, Zonath, Prashanthns, Jacj, Cedrus-Libani, Silophant, Lovro, Raguks, Monk, Zzedar, Jclemens, Pmj, Ciroa, Sjö, Rjwilmsi, Carwil, Coemgenus, Nightscream, Mbahrami, Koavf, Miros~enwiki, Mystalic, Moosh88, PinchasC, Amire80, MZMcBride, Salleman, Gudeldar, Oblivious, DonSiano, ElKevbo, Gadha, Czalex, MarnetteD, Lcolson, Bobstay, Mirror Vax, SchuminWeb, Ground Zero, Survivor, Gurch, TheDJ, Leonardwee, Tu160m, Zotel, Taichi, Slow Graffiti, Chobot, Topstar, Benlisquare, DVdm, Bgwhite, Will Lakeman, Bomb319, Adonsick, Wjfox2005, Sophitus, Jernejl, YurikBot,

JWB, RussBot, Arado, Sillybilly, TheDoober, Xihr, Anders.Warga, DanMS, Thoreaulylazy, Mythsearcher, Million Little Gods, Lord Voldemort, Okedem, Gaius Cornelius, Rsrikanth05, Ergzay, Gcapp1959, Anomalocaris, Madcoverboy, Liastnir, Johann Wolfgang, Welsh, Dudtz, MSully4321, Lexicon, Retired username, Elektrocrow, Aaron Schulz, Barjcasi, Gadget850, MrBark, Deepak–enwiki, Emesik, Slicing, Dan Harkless, Tigger69, Georgewilliamherbert, Bobbytomatoseed, StuRat, Grintoul, Sanmarcos, Smoggyrob, Ziggur, E Wing, JQF, Petri Krohn, Chrishmt0423, Jor70, Smurrayinchester, HereToHelp, Andjam, Mais oui!, AGToth, Katieh5584, Jeffreymcmanus, NeilN, Carlosguitar, John Broughton, Asterion, Victor falk, Sardanaphalus, Jesus geek, SmackBot, Dweller, KnowledgeOfSelf, Olorin28, WilyD, Davewild, Cla68, Flying Canuck, Kintetsubuffalo, Mauls, Toonmon2005, Wikikris, Gilliam, Brianski, Jushi, Skizzik, Larryincinci, The monkeyhate, Valley2city, Chris the speller, Emperorj, Bluebot, Schwitzer, Thom2002, Master of Puppets, SchfiftyThree, Hibernian, Xx236, Effer, Kungming2, DHNbot–enwiki, A. B., Thief12, Zsinj, Can't sleep, clown will eat me, RandyKaelber, Gatherton, Lantrix, JonHarder, OneEuropeanHeart, SundarBot, Stevenmitchell, Jmlk17, MrRadioGuy, Cybercobra, Sirgregmac, Engwar, Volksgeist, Asiir, Legaleagle86, Dreadstar, Allansteel, Geoffr, Zerwell, TCorp, Megalophias, Tim merm jones, TopRank, Bhludzin, Raabbasi, Sckchui, WikiWitch, Rory096, Unre4L, Dunkerya, Kuru, John, Adavidw, Phatmatt12188, DmitriyR–enwiki, Amenzix, CenozoicEra, Stattouk, JorisvS, Accurizer, Joffeloff, Scetoaux, Jochietoch, The Man in Question, Slakr, Hvn0413, Noah Salzman, Publicus, New Guy The 17th, Interlingua, Quarty–enwiki, Ambuj.Saxena, Ryulong, Fluppy, Odin of Valhalla, Andrwsc, Naumz, Esoltas, Fjbex, JoeBot, Shoeofdeath, Digitalsurgeon, Bockspur, Twas Now, Corsair Armada, CapitalR, Jamesandra, Captainj, Maelor, Adambiswanger1, Woodshed, Mountainhawk, FairuseBot, Jvol, PGSable, Plasma Twa 2, Aristotle1990, FatalError, KshitijS, CmdrObot, Such–enwiki, Tanthalas39, Scaremonger, W guice, Im.a.lumberjack, KnightLago, Goatchurch, WeggeBot, Logical2u, DanNavon, Lixy, Karenjc, Fnlayson, Hydraton31, Samuell, Reywas92, ST47, Give Peace A Chance, Tawkerbot4, Roberta F., DumbBOT, Blindman shady, Daven200520, Kippered, Mtpaley, Zalgo, Grandi, Cubfanpgh, Aldis90, TheMann, Epbr123, Wikid77, Duke333, Craggyisland, Keraunos, CynicalMe, Mojo Hand, Purple Paint, Woody, Randomfrenchie, Evanescenceboy–enwiki, Turkeyphant, Mnemeson, RamanVirk, Matthew Proctor, Aleemhaider, Dr. Zaret, The Person Who Is Strange, Lexino, AntiVandalBot, MiNeOuT, Vamsae, PastryTarget, Gon4z, Quintman, Yerkschmerk, Postlewaight, Rdavi404, Coviecarbine, Segiterrus, Spearman, THEunique, Tillman, SkoreKeep, L1X, Ingolfson, MikeLynch, DOSGuy, JAnDbot, Chaitanya.lala, Husond, Ndyguy, MER-C, Avaya1, Nthep, Jonemerson, Chanakyathegreat, Nthre06, Gerash77, Hello32020, Plm209, Colotfox, Dscotese, Sarah777, Aekhal, Anoop anooprs, Exo81, Magioladitis, VoABot II, Red aRRow, Izwzyzx, Dentren, Camhusmj38, MastCell, JamesBWatson, FrankSui, Ling.Nut, Otterfan, Cgingold, WeeWillieWiki, Shame On You, Cliché Online, MojoTas, Adrian J. Hunter, Me...™, Spellmaster, Raoul11, Nielswik, JaGa, Pluto.2006, D.h, NatureA16, Raza0007, MartinBot, Tearsofmyguitar, A suyash, Ashwin.hariharan, Kateshortforbob, CommonsDelinker, PrestonH, RockMFR, J.delanoy, Superman1888, DarKnEs5 WaRri0r, RatSkrew, Hans Dunkelberg, Bazkit, Maurice Carbonaro, Mohsin Ahmed, Nmbtho, Alexinc9, Katalaveno, Nosfartu, McSly, Andywebby, Nukemason2, Mikael Häggström, Hennessey, Patrick, Newtman, SuperN, Potatoswatter, KylieTastic, Joshua Issac, Wcd480, DH85868993, TopGun, Tahirakram, Tim Riches, Joanee Woolee, WLRoss, MalikCarr, Wikieditor06, Goalie1998, RaulCovita, VolkovBot, Tourbillon, That-Vela-Fella, The Duke of Waltham, TheQuandry, AlnoktaBOT, Teledildonix314, Ryan032, Barneca, Technopat, Tomsega, Rei-bot, Guillaume2303, Ceorl, Z.E.R.O., Fantasi, Lradrama, DennyColt, Martin451, Triang–enwiki, Leafyplant, Evening Times, Pdimond, Ussr 1991, Vereinigen, Lerdthenerd, Usergreatpower, Wasted Sapience, RuleBrittania, Hanskarlperez, Kermanshahi, Master of the Oríchalcos, NPguy, Bluedenim, Arjun024, SieBot, WalkedTheLine, VK35, Sonicology, AC1, Tresiden, Ashoup, Thatlot!!, Professor Schmidly, Flyer22 Reborn, Emilfarb, Oxymoron83, KathrynLybarger, Hobartimus, Astrale01, Jruderman, Star-of-David92, Allenlau1994, Jacob.jose, Anyeverybody, Paulinho28, Altzinn, WikiLaurent, Andrei Sakharov, Denisarona, Segregator236, Rlest, ImageRemovalBot, SallyForth123, Mender, ClueBot, Rumping, GorillaWarfare, Noorkhanuk85, Gits (Neo), Neile smith, Cencini, GordonBrownforPresident, The Thing That Should Not Be, Rjd0060, General Epitaph, Rdeman, Enthusiast01, Frmorrison, Hbanerjee3000, Boing! said Zebedee, LD1989, Niceguyedc, Slashtom, Akashsoham, Harland1, LizardJr8, Edknol, Auntof6, NuclearVacuum, DragonBot, Ktr101, Excirial, Aussie matu, Alexbot, Uraza, Chitresh verma, MatttK, Vin Kaleu, TheGreenEditor, The Founders Intent, Rhododendrites, Jotterbot, Promethean, Micronie, Thehelpfulone, Jdkkp, Geo0910, Jai Dixit, Aitias, Ranjithsutari, Mjrules09, Scorpionfangs, Thomas91091, ALEXF971, Sustain4people, DumZiBoT, Getitright13, Kiensvay, Mandeep 619, Helixweb, XLinkBot, Pakman290, Cooke star, Skarebo, Satyenpandey, Dwilso, Fheidener, SelfQ, Sargcullen1, Milstuffxyz, Addbot, Manaspunhani, L33tb0b, Roentgenium111, Tsunami0603, Captain Ref Desk, Nohomers48, Scientus, Conningcris, Lihaas, Fireaxe888, Tide rolls, Zorrobot, Bobby no hair, Enthusiast10, Luckas-bot, Yobot, AzureFury, Fraggle81, WatcherZero, Evans1982, Fenrir-of-the-Shadows, Nirvana888, EnochBethany, AastraPhone, Reenem, Jalal0, Thelonerex, Harsha850, Da Vinci Nanjing, AnomieBOT, SaaHc2B, Cullen123, Nuclear-Age, RanEagle, Jim1138, Galoubet, Zoomzoom316, Jiangyu911, Sz-iwbot, Materialscientist, Deviljin60, Citation bot, Samar60, E2eamon, Dewan357, ArthurBot, LilHelpa, Enok, Xqbot, Supersaiyan474, Nasnema, DSisyphBot, Sazarin–enwiki, Johnxxx9, Kajowi, Howyesno, Walter J. Rotelmayer, Mianhammad59, Ruodyssey, Parabellum101, SassoBot, Kieryh, UplinkAnsh, MerlLinkBot, Jaconway88, AsiBakshish, Flarkins, Atomicgurl00, FrescoBot, Meepmoo, GiW, User F203, Mistakefinder, D'ohBot, K.Khokhar, Bambuway, Ahmer Jamil Khan, Slastic, Citation bot 1, Aogouguo, OreL.D, Sopher99, LilShaddie, Redrose64, Pinethicket, I dream of horses, Andynct, ImageTagBot, Beao, Turian, Poliphile, Mercy11, كاشف عقيل, Callanecc, HWAskiBRAVO, Edinwiki, Mr.98, Jeffrd10, Diannaa, Alan261296, Breein1007, RjwilmsiBot, VernoWhitney, EmausBot, Orphan Wiki, Ghostofnemo, Rail88, High Tinker, Pmilind, DaiNipponTeikoku93, Slightsmile, Wikipelli, K6ka, Lifecommand1, Qmoz, Werieth, Vicboy, Prayerfortheworld, Sundostund, Josve05a, Shuipzv3, Kanghu, ObscureReality, Celestialroad, Stemoc, Raggot, Mitchelljam, FortEuropa, Gz33, Sparatokos, Wayne Slam, Donner60, Quite vivid blur, Golfcourseairhorn, TurtleMelody, Chris857, ClamDip, Russiaiskingyo, ClueBot NG, Michaelmas1957, Redemptionless, CocuBot, Unused000704, Ehtuchison2012, Telemachus.forward, Spartan7W, Coco bean112, Aight 2009, Bobbyb373, Widr, CostaDax, ClaireFanch, Mtking, Syedwarsi1, Helpful Pixie Bot, Somatrix, Jjoy3646, Titodutta, Hexnutdriver, Doyna Yar, Soufle, BG19bot, Narayan89, Mouloud47, Froginvestor, Tjl1128, Deathhard3, Phd8511, FiveColourMap, Iamiam2011, Jeancey, Pregabalin Nights, Writ Keeper, Snow Blizzard, Account.ka.naam, Dhruvgupta97, Klilidiplomus, Achowat, Orangewhitegreen, Jakebarrington, BattyBot, Iloverussia, Findsivro, Standardengineer, Talonvor, Mrt3366, Cyberbot II, ChrisGualtieri, Nick.mon, Min2winit, SD5bot, Kmj1980, Futurist110, Dexbot, FoCuSandLeArN, XXzoonamiXX, TwoTwoHello, Encyclopedia Of The Universe, Lugia2453, CaSJer, Frosty, Irbananaking, Nomian, Tablespite27, Wywin, NightShadow23, Obaid Raza, Gotitsaz, NelxonT, Moony22, Vanamonde93, Fouzia Mushtaq, TheTallSomething, Eyesnore, Supersaiyen312, DavidLeighEllis, Antiochus the Great, Comp.arch, CptAnthony, Ugog Nizdast, Nissan Silvia, Ginsuloft, AddWittyNameHere, Arkhan21, Stamptrader, Wikiguy3000, Glc72, JaconaFrere, Imhungry1234, Ol'fussandfeathers, 1990'sguy, Argovian, WikiWinters, Zachverb, Unician, WC Jay, Money Man 327, ChamithN, Krishnachaitan, EMefectivo, Mighaduh, Cynulliad, NatoArmy1960, Tuneix, Verdantfields, Nikhil ballia, Leblan.evan18, Asliammy, Randhwasingh, 1776004789█████, Juneymb, Brunov07, Braintan, Hulk576, Lokisis, Ninjasquirrell12, Gadarboy2015, BD2412bot, Piano Concerto in F Minor, Andrewh3153, Foreach n everyday, Firebrace, Seeker4621, Debanjands4, JacobSun, Abhiwind, LuigiPortaro29, Tpdwkouaa, Baking Soda, Nuwikipedia, Mmfaizanfaiz, Nitspark, Nitzter, Spartacus!, Sudheerbolla, Tammykim, Steynrenier, ███, Comaman679 and Anonymous: 1384

- **Nuclear Tipping Point** *Source:* https://en.wikipedia.org/wiki/Nuclear_Tipping_Point?oldid=723049332 *Contributors:* Webhat, Evolauxia,

Gary, Rjwilmsi, SmackBot, Gobonobo, HolyT, Shawn in Montreal, Johnfos, Bovineboy2008, Randy Kryn, Good Olfactory, AnomieBOT, Fortdj33, Webzenmaster, Jonkerz, Target for Today and Skr15081997

- **Nuclear disarmament** *Source:* https://en.wikipedia.org/wiki/Nuclear_disarmament?oldid=732857482 *Contributors:* AxelBoldt, Trelvis, The Epopt, Ed Poor, Enchanter, William Avery, SimonP, Edward, Patrick, RTC, Boud, Ellywa, GCarty, Kbk, Katana0182, Imc, Itai, Raul654, Hadal, DocWatson42, Fennec, Fastfission, Jackol, Pne, Bobblewik, Quadell, Kuralyov, Sam Hocevar, Qubex, ErikvDijk, Kooo, Bender235, Livajo, Kwamikagami, Alpheus, Alansohn, Sciqui fox, Guy Harris, Arthena, Rwendland, DV8 2XL, Markaci, Lapsed Pacifist, Bluemoose, BD2412, Reisio, Daniel Collins, Bhadani, Yamamoto Ichiro, Gumbagumba, Ground Zero, Mister Matt, Gurch, John Maynard Friedman, Ggb667, Chobot, RussBot, Anders.Warga, Kirill Lokshin, Nirvana2013, Kietotheworld, Milt, Wikivek, Exodio, JoanneB, Kungfuadam, Victor falk, SmackBot, Marktreut, Gilliam, Bluebot, Ottawakismet, Fshoutofdawater, Ala.foum, Enr-v, Schnarr, Skinnyweed, John, Lapaz, Btg2290, Minna Sora no Shita, Joffeloff, Ckatz, MTSbot~enwiki, Linkspamremover, Tar7arus, ERAGON, DeLarge, Outriggr (2006-2009), Cydebot, Dancter, Jonrudder, Prof75, Mombas, Añoranza, Ante Aikio, Escarbot, Golf Bravo, Ndyguy, Timwright, Kaobear, MER-C, NE2, VoABot II, Weser, Yandman, Nyttend, KConWiki, Player 03, Nbauman, Maurice Carbonaro, Thaurisil, Tdadamemd, NewEnglandYankee, Davidswayn, Histruth, TreasuryTag, Johnfos, Jack Rabitt, Pelarmian, Irish Pearl, NPguy, Malcolmxl5, Flyer22 Reborn, Megansmith18, Abrichte, Encykeiz, Randy Kryn, Epthorn, ClueBot, Drmies, Dupreem, Klrichar, Dspark76, Arjayay, ParaGreen13, Achoffman, XLinkBot, Johnkatz1972, That-guyflint, Stephen Fulcher, Imeriki al-Shimoni, TwilightSamus, MrOllie, Download, Redheylin, Lightbot, Legobot, Yobot, Pganas, AnomieBOT, Dwayne, Ulric1313, 469, LilHelpa, Xqbot, Johnxxx9, Grim23, Bodinagamin, Atomicgurl00, FrescoBot, LucienBOT, LittleWink, Liam83119, Rotblats09, LilyKitty, Mr.98, Melnyk24, Boundarylayer, Lhoaxt, Sujaybagi, Ilaydarocks, DavidMCEddy, Arrala, ClueBot NG, Vkurtz, Widr, ClaireFanch, BG19bot, Jonreidhunt, ISTB351, HIDECCHI001, BattyBot, Cyberbot II, ChrisGualtieri, Mysterious Whisper, ShinhyeongYou, XXzoonamiXX, Hermes 1900s, Eyesnore, Restlesstruth, Gravuritas, Coercive Diplomacy, Susan Macafee, Grand'mere Eugene, Crystallized-carbon, Weegeerunner, PeterTheFourth, CAPTAIN RAJU, Tuhermanavendepupusas, GreenC bot, Toasymaster69 and Anonymous: 150

- **History of the anti-nuclear movement** *Source:* https://en.wikipedia.org/wiki/History_of_the_anti-nuclear_movement?oldid=732019074 *Contributors:* Rich Farmbrough, Bender235, Woohookitty, Drbogdan, Simesa, Limulus, John1, Mombas, CommonsDelinker, Johnfos, Malcolmxl5, Mild Bill Hiccup, Ulric1313, Materialscientist, Widr, Helpful Pixie Bot, Seergenius, Dexbot, XXzoonamiXX, Hermes 1900s and Anonymous: 6

- **Effects of nuclear explosions on human health** *Source:* https://en.wikipedia.org/wiki/Effects_of_nuclear_explosions_on_human_health?oldid=733895774 *Contributors:* Wikster E, Gscshoyru, Bender235, Nsaa, Woohookitty, Sburke, stevenfruitsmaak, Bedford, DVdm, Wavelength, Ytrottier, Shaddack, Wknight94, Theda, ASmartKid, SmackBot, MARussellPESE, Timeshifter, Hmains, Chris the speller, Hibernian, Kokot.kokotisko, MJBoa, JamesAM, Epbr123, Qwyrxian, DPdH, Weinzierl, EagleFan, Beagel, CommonsDelinker, Nono64, Kavanagh21, Belegur, Johnfos, Philip Trueman, AllGloryToTheHypnotoad, Mdeweydiii, Tomwhite56, Oda Mari, Beeblebrox, ClueBot, Addbot, Queenmomcat, Rmorrisons, Yobot, Byu123, AnomieBOT, Jim1138, Aaagmnr, Haleyga, LilHelpa, Grey ghost, Locobot, Oliviervdst, Trinity54, Dendereon, Kingky, Ripchip Bot, Ajraddatz, Boundarylayer, Bailite, Donner60, ClueBot NG, Gilderien, Wbm1058, BattyBot, Biosthmors, Mdann52, Kehkou, Hillbillyholiday, Epicgenius, DavidLeighEllis, Qstem, Quenhitran, Manul, Davearthurs, Lourdes and Anonymous: 82

- **Peaceful nuclear explosion** *Source:* https://en.wikipedia.org/wiki/Peaceful_nuclear_explosion?oldid=735077357 *Contributors:* Patrick, Michael Hardy, Katana0182, Altenmann, Auric, Victor, Fastfission, Paulscrawl, Reflex Reaction, Stocksy, Bender235, Max rspct, Curtis Autery, DV8 2XL, GregorB, Emerson7, Sjö, The wub, Roboto de Ajvol, JarrahTree, Limulus, Los688, Alex Bakharev, TDogg310, Wikipeditor, Jonathan.s.kt, David Straub, SmackBot, Ohnoitsjamie, Fuzzform, CSWarren, Vladislav, Thomas Connor, DinosaursLoveExistence, Dcamp314, Hrimfaxi, A5b, John, Gobonobo, KuharJ2, Cydebot, Matthew Proctor, Chubbles, Albany NY, Emax0, Hut 8.5, Dricherby, Nyttend, Commons-Delinker, Leaflet, DadaNeem, BernardZ, Johnfos, MCTales, NPguy, MattW93, Afernand74, PixelBot, SchreiberBike, Addbot, Diptanshu.D, 84user, Lightbot, Yobot, Pequeño Aceite, AnomieBOT, ThaddeusB, Gilo1969, Boundarylayer, Schroep, Whodarep08, Bulwersator, Vceinc, Smm201'0, Jphill19, BG19bot, Davidiad, BattyBot, MassimoScipio, STKS91, TestBanNow, GabeIglesia, Lgfcd, WC Jay and Anonymous: 45

- **Nuclear weapons and the United States** *Source:* https://en.wikipedia.org/wiki/Nuclear_weapons_and_the_United_States?oldid=734027843 *Contributors:* Rmhermen, AdamRetchless, Jdlh, Edward, Patrick, Kosebamse, Jebba, Andrewa, PaulinSaudi, Katana0182, Tpbradbury, Fibonacci, Vfrickey, DocWatson42, Christopher Parham, Jhf, Fastfission, Marcika, Gadfium, Xtreambar, Vincom2, Brianhe, Rich Farmbrough, Bender235, Ylee, Kwamikagami, Evand, EmilJ, Jwlee, Prainog, Brim, Kevinh456, BillyTFried, Nsaa, Alansohn, Arvedui, Batmanand, Stack, RainbowOfLight, Cmapm, W7KyzmJt, DV8 2XL, Dziban303, Johntex, Crosbiesmith, Wikiklrsc, I64s, Maartenvdbent, Tutmosis, Wayward, Graham87, Kbdank71, Rjwilmsi, Coemgenus, Koavf, Jivecat, Ian Dunster, Yamamoto Ichiro, GünniX, RexNL, DVdm, VolatileChemical, Cornellrockey, Noclador, RussBot, Arado, Chris Capoccia, Anders.Warga, Shaddack, Rsrikanth05, Scfitch, NawlinWiki, Worldruler20, Ragesoss, Moe Epsilon, Brat32, Spheroide, Ageekgal, Curpsbot-unicodify, Johnpseudo, Nekura, Amberrock, Luk, SmackBot, Gilliam, Blather, Ohnoitsjamie, Hmains, Akanksh, Chris the speller, Jibbajabba, Rakela, Jprg1966, Audigex, CSWarren, Rlevse, Chendy, Can't sleep, clown will eat me, Metallurgist, Lantrix, Jonathanleblang, Prmacn, AlexJP, Engwar, Dreadstar, The PIPE, DMacks, Bhoy Wonder, Qwerty0, Gailim, ABurness, John, EDUCA33E, Elon~enwiki, Mgiganteus1, Rkmlai, Bless sins, Slakr, Publicus, Caiaffa, Pjbflynn, Tawkerbot2, WildWeathel, Dgw, ShelfSkewed, HonztheBusDriver, Cygni13, MikeWren, A876, DaMang111, Mtpt, Kornmonkie, Aldis90, Mombas, Thijs!bot, Biruitorul, Kubanczyk, Savasci, WVhybrid, Woody, Nick Number, Dawnseeker2000, Sherbrooke, Hires an editor, AntiVandalBot, Carolmooredc, Dark-Audit, The doctor23, SkoreKeep, The Fifth Horseman, Sonicsuns, AlmostReadytoFly, VoABot II, Inmate20, Steven Walling, Rich257, Twsx, Brusegadi, Cgingold, Ascraeus~enwiki, CommonsDelinker, Nono64, J.delanoy, Dispenser, Apostle12, Trumpet marietta 45750, Pikaverive, Ndunruh, Tatrgel, Andy Marchbanks, WRE451, VolkovBot, Johnfos, LeilaniLad, Xenophrenic, Wassermann~enwiki, Robert1947, Anarchangel, Zachjeli, RuleBrittania, Bahamut0013, K. Aainsqatsi, Jasonquick, Flyingkotaekwondoman, Sonicology, France3470, Oxymoron83, Lightmouse, Yyyooo, Abrichte, WacoJacko, Fratrep, Onopearls, Cyfal, Anchor Link Bot, Slaporte, Mkunichi, MBK004, Fasettle, The Thing That Should Not Be, Drmies, Strategik, Ktr101, Excirial, Socrates2008, Jusdafax, PixelBot, Coinmanj, NuclearWarfare, Redthoreau, ChrisHodgesUK, Thingg, Rossen4, DumZiBoT, Zombie Hunter Smurf, Astrofreak92, Northwesterner1, Addbot, DOI bot, Landon1980, Fluffernutter, Marrsman, LaaknorBot, Htews, Legobot, Luckas-bot, AzureFury, Ptbotgourou, Senator Palpatine, Fraggle81, Politicsnme, Vortico, AnomieBOT, DemocraticLuntz, Jim1138, BlackBerryHill, Nemesis63, Materialscientist, Spw0766, Mr. Military, Riotrocket8676, Anotherclown, Mark Schierbecker, Frost111, Erujui12, Snkntr01, MerlLinkBot, Shadowjams, Gnomsovet, RightCowLeftCoast, FrescoBot, Army46Q, Michael93555, Airborne84, Citation bot 1, GaussianCopula, Pinethicket, Beao, Tresiqusiiix, Rotblats09, Cnwilliams, Ninja Auditor, Trappist the monk, A p3rson, Mysticquill, Adi4094, Lalnose, RjwilmsiBot, Acather96, Look2See1, Tommy2010, Dcirovic, Mz7, Evanh2008, Lesswealth, Josve05a, MUCHERS22, Omar-Toons, H3llBot, Rdhulljr, Δ, Brandmeister, Donutcity, Magicvan, Woolfy123, ClueBot NG, Joefromrandb, 1chewy2,

Arkie77, Spartan7W, Widr, North Atlanticist Usonian, Shrug-shrug, SoreJordan004, Calabe1992, BG19bot, Smarky111, Frze, Extrapolaris, Piguy101, BattyBot, Iloverussia, Cyberbot II, Khazar2, XXzoonamiXX, Happyseeu, Redalert2fan, RotlinkBot, Copyright Troll, Staceydolxx, Jamesmcmahon0, Tentinator, Kharkiv07, Limnalid, Jpmickyd, JaconaFrere, Abattoir666, Antideregister, Robertvincentswain, Susan Macafee, Roberto9999999999, Dr.aazamparvez, SmartHunter1, UglowT, Anonyprem, Kellpet11, CaptainCarlosdeCorona, Peter SamFan, Jehoiachin12 and Anonymous: 307

- **Nuclear weapons and the United Kingdom** *Source:* https://en.wikipedia.org/wiki/Nuclear_weapons_and_the_United_Kingdom?oldid= 734676238 *Contributors:* William Avery, GABaker, Darkwind, Jll, Tpbradbury, Andrew Yong, PBS, Timrollpickering, Bkell, DocWatson42, Fastfission, MSGJ, Grant65, Bobblewik, Geni, DragonflySixtyseven, Icairns, Hammersfan, Cynical, Lacrimosus, Mike Rosoft, N328KF, Dead-lock, Rich Farmbrough, YUL89YYZ, Cromis, Bender235, Ylee, Clue, Dpaajones, AKGhetto, Rd232, Rwendland, Radical Mallard, Max rspct, Benson85, Saga City, LukeSurl, Crosbiesmith, FrancisTyers, Woohookitty, PoccilScript, James Kemp, Z303, TreveX, GraemeLeggett, Graham87, BD2412, Rjwilmsi, Jmcc150, Bomble, Ligulem, Palpatine, Ian Dunster, Ground Zero, Mark83, Bgwhite, Jimp, RussBot, Arado, Anders.Warga, Alex Bakharev, Hawkeye7, Grafen, Rjensen, Howcheng, Patrick Neylan, Djm1279, Pyrotec, Roche-Kerr, Georgewilliamher-bert, Lynbarn, Mais oui!, Otto ter Haar, Victor falk, SmackBot, Britannicus, Royalguard11, David.Mestel, AndyZ, Mauls, Chris the speller, Thom2002, Achmelvic, Hibernian, Colonies Chris, Chendy, Pickle UK, Engwar, Parrot of Doom, Ohconfucius, Khazar, John, Jamestown, Elon-enwiki, MilborneOne, IronGargoyle, Mr Stephen, AxG, Burto88, Iridescent, JoannaSerah, FairuseBot, Newsnightmeirion, CmdrObot, Mattbr, Makeemlighter, Banedon, Wbd, Goatchurch, Markleci, Richard Keatinge, Nabokov, Gnfnrf, Brian.Burnell, Gonzo fan2007, Mtpa-ley, Kubanczyk, Betdud, SkoreKeep, Shlgww, Doctorhawkes, Unused0029, Vernon39, Buckshot06, Steven Walling, MCG, Sm8900, Mau-rice Carbonaro, Notreallydavid, AdamBMorgan, Youngjim, Vanished user 39948282, Sir rupert orangepeel, George.Hutchinson, Iliff, Signal-head, Hugo999, Sam Blacketer, VolkovBot, Johnfos, Guthrum, A man2, From-cary, Peter K Burian, Billinghurst, Usergreatpower, Bungo77, StAnselm, Lucasbfrbot, Freeman501, Lightmouse, Abrichte, Fratrep, Duffy2032, Jw2034, Tanvir Ahmmed, Fasettle, Awg1010, Enthusiast01, Ahtclide, Mild Bill Hiccup, Niceguyedc, Arunsingh16, Aitias, ShipFan, DumZiBoT, Addbot, The Bushranger, Luckas-bot, Yobot, AnomieBOT, Zangar, Materialscientist, Citation bot, Gilo1969, WotWeiller, Anotherclown, Gnomsovet, Dougofborg, Heroicrelics, RightCowLeftCoast, FrescoBot, Grand-Duc, Bambuway, Citation bot 1, Pinethicket, Qwertyuiop1994, RedBot, Rotblats09, Vometia, 10987sa, Izzwizzwoo, Jami-etw, RjwilmsiBot, Jakerin, John of Reading, Orphan Wiki, NorthernKnightNo1, Yattum, ZéroBot, Brianm358, H3llBot, HammerFilmFan, BNSF1995, Brigade Piron, Horresco, Cooper 25, ClueBot NG, Armouredduck, Calisthenis, FightingMac, Helpful Pixie Bot, BG19bot, Phd8511, InExcelsisDeo, Silvrous, Pregabalin Nights, BattyBot, JoshuSasori, Cyberbot II, ChrisGualtieri, Khazar2, Dexbot, Mogism, Jamesx12345, Rob984, PhantomTech, Antiochus the Great, Kharkiv07, Monkbot, Naelalozma, Julietdeltalima, BenjaminBluesilk, ThaBigCheese99, Fire-brace, Cantab1985, LoveEverybodyUnconditionally, P3G4SuSuOuT, Navalnewsup and Anonymous: 174

- **Russia and weapons of mass destruction** *Source:*   https://en.wikipedia.org/wiki/Russia_and_weapons_of_mass_destruction?oldid= 733439382 *Contributors:* Rmhermen, Dante Alighieri, Gabbe, Tannin, IZAK, Ronabop, Bogdangiusca, Topbanana, Finlay McWalter, Owen, DocWatson42, Oberiko, Fastfission, Everyking, Ezhiki, Get-back-world-respect, Rjyanco, Deus Ex, Beland, Thorwald, Discospinster, Rich Farmbrough, TomPreuss, Bender235, Jag123, Larryv, Anittas, Rwendland, J Heath, Firsfron, Slazenger (usurped), BD2412, Ketiltrout, Rjwilmsi, Idaltu, King of Hearts, WriterHound, Splash, Dudtz, Wknight94, 21655, Abune, Mais oui!, Sacxpert, Sardanaphalus, SmackBot, Ohnoitsjamie, Hmains, This username is just a name nobody's using or will want 657657, Chris the speller, Ottawakismet, Koliokolio, Hibernian, Muzi, Pi-lotguy, John, Iliev, Marco polo, Tasc, Dicklyon, Optakeover, EdC-enwiki, Odin of Valhalla, Hu12, Majora4, SkyWalker, Cydebot, Epbr123, James086, JustAGal, MarkV, Superzohar, MER-C, Fetchcomms, Sarah777, Shadiac, Hellerick, R'n'B, J.delanoy, Uncle Dick, Thaurisil, Hodja Nasreddin, Garret Beaumain, Bernard S. Jansen, Jevansen, TheNewPhobia, Homologeo, X!, Ilya1166, Oshwah, SteveStrummer, Someguy1221, Natg 19, Mouse is back, Turgan, EJF, Flyer22 Reborn, CarlosPn, Onopearls, ClueBot, LAX, Strategik, Mild Bill Hiccup, Nanobear-enwiki, Mcfly2008, Excirial, L.tak, Pyrofork, Trulystand700, DumZiBoT, Addbot, Krawndawg, Tcncv, Nohomers48, LatitudeBot, Jarble, Greyhood, Luckas-bot, Yobot, AzureFury, Ptbotgourou, Angel ivanov angelov, THEN WHO WAS PHONE?, AnomieBOT, FeelSunny, Jim1138, Piano non troppo, Leosls, Racastremus, Fusioned Capacity, Mynameinc, GrouchoBot, Amaury, Doulos Christos, Zemant, Gnomsovet, Spongefrog, LucienBOT, Trust Is All You Need, Eightofnine, James Cusens, Pinethicket, RedBot, HowardJWilk, Serols, Rotblats09, Mercy11, Lufen1987, Reaper Eternal, TheGrimReaper NS, MaseJr8990, Lenin1055, EmausBot, Orphan Wiki, Challisrussia, H3llBot, EWikist, Brandmeister, Ab-hishekitmbm, KZfan, ClueBot NG, Catlemur, Adair2324, ORCZORR, BlitZx SiN, Dm68k, Helpful Pixie Bot, Bgarner123456789, Wasbeer, Chrisvik12, Extrapolaris, Yowanvista, Qwerty951753, Ichek, Occidentaloccidental, DenzilUK, Iloverussia, Cyberbot II, RCFrank, EagerTod-dler39, PepeEscobar, Hwr007, TwoTwoHello, Lucassandershapiro, Eckshotgunz, PussBroad, Kharkiv07, Fnordson, Mandruss, Sam Sailor, Lim-nalid, Ol'fussandfeathers, Abattoir666, Vladimir.putina, SmartHunter1, WC Jay, Inthefastlane, Kionay, Hulk576, Yankeesgiantsrangersknicks and Anonymous: 206

- **China and weapons of mass destruction** *Source:* https://en.wikipedia.org/wiki/China_and_weapons_of_mass_destruction?oldid=735068894 *Contributors:* Alex.tan, Rmhermen, Roadrunner, Dante Alighieri, Gabbe, Jiang, Kaihsu, Haukurth, Furrykef, Topbanana, Korath, Oberiko, Fast-fission, TDC, Everyking, Get-back-world-respect, Gzornenplatz, Rjyanco, Deus Ex, Huaiwei, Cynix, Neutrality, Bhugh, Atchom, Alistair1978, Bender235, Ylee, Mr. Billion, Surcouf, Giraffedata, Sasquatch, Holdek, Sherurcij, Rwendland, Hohum, Hypo, Crosbiesmith, Robertl234, Ipcellon, John Hill, BD2412, Rjwilmsi, Koavf, Hitssquad, FayssalF, Kallemax, Ground Zero, Kolbasz, Chobot, Benlisquare, JWB, RussBot, Dreammaker182, Arado, John Smith's, Anders.Warga, Stephenb, Grafen, Megapixie, Danlaycock, Efreeti, Dspradau, SkerHawx, Sardanaphalus, SmackBot, YellowMonkey, Hux, Squiddy, Chris the speller, Bluebot, Ottawakismet, Ddrfreak103, Hibernian, Thewho-enwiki, Muzi, Bolivian Unicyclist, Nakon, Esw01407, JorisvS, Publicus, EdC-enwiki, Hu12, Quaeler, Joseph Solis in Australia, RekishiEJ, FairuseBot, Logical2u, Zyxi, AndrewHowse, Cydebot, Spylab, Aldis90, RevolverOcelotX, Hcobb, Nick Number, Luna Santin, Amberina, SkoreKeep, Aliwalla, Ingolfson, Albany NY, VoABot II, IkonicDeath, Walle83, R'n'B, KTo288, Thedeadlypython, Skier Dude, Bernard S. Jansen, DadaNeem, Veritek83, John-fos, Kyle the bot, Turgan, Dncdncdnc, Lowbart, Interchange88, Happysailor, CarlosPn, PianoKeys, Gomeying, Oneforlogic, ClueBot, Fasettle, Wikievil666, Arunsingh16, Puchiko, Alexbot, Socrates2008, Coinmanj, Hadooooookin, Dekisugi, Zappa711, Rosywounds, DumZiBoT, Crazy Boris with a red beard, Otherjoke-enwiki, Addbot, Quercus solaris, Lightbot, Jarble, ScienceApe, Luckas-bot, Yobot, AnomieBOT, Leosls, Blitzoace, Capricorn42, Chen Guangming, Mynameinc, RadiX, GrouchoBot, Parabellum101, Omar77, Gnomsovet, Atomicgurl00, RedBot, Full-date unlinking bot, Poliphile, Mr.98, RjwilmsiBot, EmausBot, John of Reading, TomahawkHunter, IronChloride, Dewritech, Wikipelli, H3llBot, L1A1 FAL, ClueBot NG, Redemptionless, JesseW900, Gcorral, Oddbodz, Helpful Pixie Bot, BG19bot, Phd8511, BattyBot, Hsasar, Choy4311, Cyberbot II, EuroCarGT, Dexbot, Frosty, Marvell1x1, Jamesmcmahon0, FreeWorldAdvocate, EvergreenFir, Kharkiv07, Limnalid, KOT-TOK, EMP90, Abattoir666, ICPSGWU, JW19335762743, YeOldeGentleman, Queso.robusto, Warrior Covert, Dr. Franz sparkles and Anonymous: 129

- **India and weapons of mass destruction** *Source:* https://en.wikipedia.org/wiki/India_and_weapons_of_mass_destruction?oldid=734843039 *Contributors:* Rmhermen, Edward, Patrick, Gabbe, Karada, Docu, Andrewa, Aarchiba, PaulinSaudi, Selket, Schutz, Oberiko, MSGJ, Obli, Mboverload, Bobblewik, Edcolins, Golbez, Utcursch, Oneiros, Supadawg, Efriedman, Grunt, Shahab, Moverton, Rich Farmbrough, Tirthajyoti, Ashwatham, Stereotek, Bender235, Mr. Billion, Livajo, Dhoom, AreJay, Maureen, Bijal d g, Chirag, A2Kafir, Alansohn, Mac Davis, Rwendland, Wtmitchell, Vedant, DV8 2XL, Kelly Martin, Kosher Fan, Marudubshinki, Deltabeignet, Toba1, Rjwilmsi, Koavf, Drench, Dar-Ape, Saksham, Ian Pitchford, Tu160m, Bgwhite, Deeptrivia, RussBot, Hornplease, Limulus, Anders.Warga, Thoreaulylazy, Gaius Cornelius, Rsrikanth05, Kimchi.sg, Asherett, Thiseye, Brandon, Snkutty, Varun dt, Danlaycock, DeadEyeArrow, Cyberwizmj, Tigger69, LarryLACa, American2, Itake, PTSE, Josh3580, Anakinskywalker, Allens, Abhishekmathur, Sardanaphalus, SmackBot, Monkeyblue, Mysterius, Chackojoseph, Carl.bunderson, Chris the speller, Bluebot, Ottawakismet, Sujithk, Hibernian, Effer, ACupOfCoffee, Rama's Arrow, Sgt Pinback, NYKevin, TKB, Easwarno1, Jmlk17, MrRadioGuy, Arun Philip, Savidan, Valenciano, Legaleagle86, Andrew c, Natebjones, Ohconfucius, Archit Patel, Pizzadeliveryboy, Sambot, Esw01407, John, Soumyasch, MilborneOne, Sir Nicholas de Mimsy-Porpington, Evenios, Rajesh Rao, Shyamsunder, Smashingpumpkins, Rcowlagi, AdultSwim, EdC~enwiki, Skapur, Hxnagara, Twas Now, Eluchil404, FairuseBot, Who-Said?, SkyWalker, CmdrObot, Yourdeadin, Goatchurch, DumbBOT, Editor at Large, Cancun771, JamesAM, Thijs!bot, Qwyrxian, Revolve-rOcelotX, Rosarinagazo, Nick Number, Asen y2k, AntiVandalBot, Indivisible, QuiteUnusual, Indotan, Chaitanya.lala, Saddysan, Chanakyathegreat, KuwarOnline, Andonic, Ryan4314, Raanoo, Magioladitis, Askari Mark, Flayer, Aye Carumba Fajita Pizza, Avicennasis, Adrian J. Hunter, Cranium1, Michael.fernando, S3000, Raza0007, MartinBot, EyeSerene, Mr.Falcon, Arjun01, R'n'B, CommonsDelinker, Tgeairn, J.delanoy, Abecedare, Maurice Carbonaro, Natobxl, Apurv1980, Ash sul, AntiSpamBot, Master shepherd, Tatrgel, Olegwiki, Shoessss, Joshua Issac, Bilalspike, S (usurped also), Funandtrvl, Wikieditor06, Nigel Ish, RaulCovita, Hersfold, Ktalon, Nubin wiki, Oshwah, Shrao, Davehi1, Vipinhari, Vishwas008, Detroit4, Nukemason4, Sniperz11, Doug, Vladsinger, Nirmarun, Falcon8765, NPguy, Chrisphmb, Arjun024, SieBot, Sonicology, Miremare, M.thoriyan, Abhishikt, Vmrgrsergr, Aspects, Fratrep, Maelgwnbot, Anchor Link Bot, Punitpankaj, Joel Rennie, ImageRemovalBot, Dlrohrer2003, MBK004, ClueBot, Fasettle, Lamoonia, Hornet35, Mild Bill Hiccup, Great.constantine1, Shovon76, Otolemur crassicaudatus, Cirt, Auntof6, Vidhyardhi, Lartoven, Sun Creator, Another Believer, DumZiBoT, Crazyrobin4u, XLinkBot, Jovianeye, Richard-of-Earth, Voltigeur, Good Olfactory, UnknownForEver, MatthewVanitas, GDibyendu, Addbot, Manaspunhani, Lakshmim 84, CL, Lihaas, Debresser, CarTick, Jossejonathan, Steed Asprey - 171, Lightbot, Ias2008, Jarble, The Bushranger, Ben Ben, Enthusiast10, Luckasbot, Yobot, Reenem, Dilutexyz2020, AnomieBOT, Amityadav8, Jim1138, Zoomzoom316, Homoatrox, Vindastra, Julnap, Leosls, Blitzoace, Materialscientist, Deviljin60, Druid.raul, Samar60, FreeRangeFrog, Apoorv020, Johnxxx9, Mynameinc, MerlLinkBot, Mittal.fdk, Gnomsovet, Pkhagah, Operdyne7, Thehelpfulbot, Atomicgurl00, FrescoBot, Harsha363, Bharatgopal, Teckgeek, Nosedown, HJ Mitchell, Manoij, Bambuway, Lordharrypotter, I dream of horses, Elockid, ImageTagBot, Jschnur, Serols, Σ, Pinochet (3), Full-date unlinking bot, Joydeep ghosh, Deskitemssnake, Rotblats09, SkyMachine, Mainmahan, Trappist the monk, AMuraliKumar, SeoMac, Goodboy2009, Mr.98, Reaper Eternal, Diannaa, Hari7478, RjwilmsiBot, Sethemanuel, EmausBot, John of Reading, Pradhankk, Courcelles is travelling, GoingBatty, Sujaybagi, Sanjeevsofteng, Tommy2010, Kkm010, Rijuroy, Anir1uph, Abdul.125, Labnoor, Power125, PA.TheOne, Aniqrandhawa, Donner60, Puffin, Pak125, Orange Suede Sofa, K. the Surveyor, Gshashank18, Indian1985, Sven Manguard, Chiragmarwaha, Whoop whoop pull up, ClueBot NG, Pradyumnas741, Catlemur, Waleed genius1, Rajeshjha 103, Buklaodord, Crimemaster007, Twillisjr, AgniKalpa, Widr, Truth1Please, Helpful Pixie Bot, Jjoy3646, Strike Eagle, Honorprevails123, DBigXray, IridiumIs77, Lowercase sigmabot, Sudev123, Robinson James, Hknair, Hypd09, Raghav10089, Saumya.nar.14, Ninney, FiveColourMap, Ravigodse87, Ddineshk, Beetelaces, Rahul 06, Pak ISPR, DistributorScientiae, Klilidiplomus, Ssaxe01s, Orangewhitegreen, BattyBot, Pranavswarup76, Dec22, Mrt3366, FiveFourTwo, Cyberbot II, ChrisGualtieri, 2011kdp, Rkb76in, Mediran, Nandakishore221, BrightStarSky, Anurag2k12, BigJolly9, Bharat9090, Sharkkiller2050, Shantanu Chakraborty, TwoTwoHello, Vipulkmr99, Donnowin1, Neelkamala, Cawhee, Rockthecassbach, Capitals00, Raviteja338, Jodosma, Rld, Soffredo, Evano1van, Giant4627, Sandeep dhiran, Tusharmod, Antiochus the Great, Hankspanks, Alurujaya, Limnalid, Rekhil Jose, Luxure, Harshkumar989, Inphynite, Xs9xxc49, Dr.aazamparvez, UglowT, Nicky mathew, Josan420, SUBhAA CooL, Hungarian historian, Randhwasingh, Tiger7253, AusLondonder, Firebrace, InternetArchiveBot, Thepolyeditor 11, Sudheerbolla, Mohitraj mhr and Anonymous: 614

- **Pakistan and weapons of mass destruction** *Source:* https://en.wikipedia.org/wiki/Pakistan_and_weapons_of_mass_destruction?oldid=735163259 *Contributors:* Rmhermen, Edward, Patrick, RTC, Gabbe, Paul A, Ahoerstemeier, Morven, Owen, Sjorford, RedWolf, Donreed, Pretzelpaws, Kawai~enwiki, Edcolins, Golbez, Mr impossible, Sam Hocevar, Faraz, Imjustmatthew, Mike Rosoft, Jayjg, Vsmith, Modargo, Bender235, ESkog, Cmdrjameson, Giraffedata, Andrewbadr, BillyTFried, Idleguy, Sam Korn, Perceval, Rwendland, Oneliner, Atomicthumbs, Vedant, Mmsarfraz, DV8 2XL, Gene Nygaard, Woohookitty, Milen~enwiki, BlaiseFEgan, Gimboid13, Mandarax, Dpv, Ketiltrout, Toba1, Rjwilmsi, Applepie~enwiki, Ground Zero, Kolbasz, UkPaolo, RussBot, Koffieyahoo, Siddiqui, Welsh, Retired username, Dmoss, Danlaycock, MySchizoBuddy, Szhaider, Smaines, Mail2amitabha, Dan Harkless, BlueZenith, Mercenary2k, Taptee, David Biddulph, Allens, Tiger888, Street Scholar, Sardanaphalus, SmackBot, Innocentmind, YellowMonkey, K-UNIT, Sam8, HeartofaDog, Gilliam, Hmains, Chris the speller, Bluebot, Persian Poet Gal, Master of Puppets, ACupOfCoffee, Redline, Minister of Darkness, MTBradley, AKMask, Gracenotes, Salin, InnocentMind, Wes!, Fuhghettaboutit, AndyBQ, جلب, Ohconfucius, Ali 786, Yahya01, Terminator50, Euchiasmus, Fast track, Zmustafa, Green Giant, JHunterJ, Beetstra, Interlingua, EdC~enwiki, Pedrora, DI2000, Simon12, Iridescent, Clarityfiend, FairuseBot, Tere naam, CmdrObot, JohnCD, Leujohn, The Enslaver, MikeWren, Fnlayson, DumbBOT, Cancun771, NadirAli, Aldis90, Daa89563, Barticus88, David from Downunder, N5iln, Gralo, Sandbreak, Hcobb, Nick Number, AnAj, Corella, Trakesht, Pknightru, Ndyguy, Samar, Scythian1, Hodgetts, Mkashif, Kidal, Ryan4314, Dildar Hussain, Russianmissile, Prateek sterling, Sushant gupta, Hullaballoo Wolfowitz, JamesBWatson, Puddhe, Sodabottle, Adrian J. Hunter, Custodiet ipsos custodes, Adil zia, S3000, Kgeis, Raza0007, MartinBot, CommonsDelinker, Razor mc, Mfarooqumer, Hans Dunkelberg, Zuhair siddiqui, Mianhassan, Katalaveno, Plasticup, Usman, Muhammad, MatthewBurton, Rumpelstiltskin223, Jamesontai, TopGun, Signalhead, VolkovBot, Jeff G., Webkami, Aslamt, Uch, Mrghumman, Razzsic, Mazarin07, Gilisa, Meters, Lamro, Falcon8765, Drutt, C45207, Master of the Oríchalcos, Jrevill79, NPguy, Arjun024, StAnselm, RaneGuruprasad, Smsarmad, LibStar, Adityagupta101, TSCL, Vmrgrsergr, Benea, Lightmouse, Anchor Link Bot, Mr. Stradivarius, JL-Bot, Joel Rennie, Mr. Granger, Zear+shauna, ClueBot, Noorkhanuk85, The Thing That Should Not Be, Lamoonia, Swapnils2106, Silence Will Speak, DarthRad, CasualObserver'48, Drmies, Mild Bill Hiccup, Shovon76, Hasantheman123, Jusdafax, Wonder scorpio2005, SyedNaqvi90, Versus22, DumZiBoT, SilvonenBot, Good Olfactory, UnknownForEver, MatthewVanitas, Addbot, Jawadqamar, Download, Debresser, Tassedethe, Jarble, Ben Ben, Yobot, Worldbruce, Fmrauch, Umairtunio, Jalal0, Bbb23, Qaatil, AnomieBOT, Gpakistan, Ipatrol, Nadir Usman Ishaq, Materialscientist, Deviljin60, Zahab, Quebec99, LilHelpa, Drilnoth, Mynameinc, AbdulIshaq, GrouchoBot, Hj108, Wikireader41, UplinkAnsh, AzanGun, Gnomsovet, Jonathon A H, Mughalnz, Plot Spoiler, Midgetman433, FrescoBot, Ironboy11, Barsamin, OgreBot, Citation bot 1, Zuhayer171288, I dream of horses, Elockid, Calmer Waters, Just a guy from the KP, TRBP, Deskitemssnake, Rotblats09, Ahsaninam, Super Saiyan 3, Trappist the monk, Joakimm, LogAntiLog, Gajus, Scorched Earth76, Dabamizan48, Anikkuttan, Zink Dawg, Hari7478, RjwilmsiBot, Gould363, EmausBot, John of Reading, AlphaGamma1991,

Dewritech, NorthernKnightNo1, Dcirovic, NadirAwan, Mar4d, Anir1uph, H3llBot, Abdul.125, Power125, Δ, Zuggernaut, Haemetite, K. the Surveyor, FeatherPluma, Petrb, ClueBot NG, Catlemur, Tricolor Truce, Waleed genius1, Abby1fu, Lord Roem, The Master of Mayhem, Snot-bot, Frietjes, 149AFK, Helpful Pixie Bot, Emperitor, Strike Eagle, KLBot2, BG19bot, Sahara4u, Darkness Shines, Bilaljshahid, Sageam1, Pak ISPR, FALCON-786-, 220 of Borg, Aisteco, Bcary, BattyBot, Zhaofeng Li, FiveFourTwo, Cyberbot II, Khazar2, Versova, BrightStarSky, Mo-gism, PeerBaba, Timothysandole, The Anonymouse, Palestine194, Faizan, Maxx786, Surfer43, Evano1van, Kharkiv07, Oculedextra, Limnalid, Keeperedsad, OccultZone, Stamptrader, Rtaimoorkhan, I.Bhardwaj, Aamir Leghari, Ethically Yours, Abattoir666, Xs9xxc49, Monkbot, Sameer malik1, Vieque, Charles Bill Bob, The Prince Sky legend, Owais Khursheed, Je.est.un.autre, IRNBarren, Sizzling hnny, Randhwasingh, Qw-ertyw123, Timothyjosephwood, Srednuas Lenoroc, Bob chasm, The Quixotic Potato, SHAMSbALOCH321, Qzd, Wolframemanager, GreenC bot and Anonymous: 492

- **Nuclear weapons and Israel** *Source:* https://en.wikipedia.org/wiki/Nuclear_weapons_and_Israel?oldid=735856810 *Contributors:* Ijon, Ineuw, Ehn, ChicXulub, Robert Brockway, Oneiros, Ttyre, Now3d, Rich Farmbrough, Alistair1978, Bender235, Swid, Mind the gap, Ylee, El C, Ertly, Uncle.bungle, Rwendland, Wtmitchell, Deacon of Pndapetzim, Jg325, Henrik, Pol098, Lapsed Pacifist, GregorB, BD2412, Rjwilmsi, Carwil, Koavf, Wikiliki, 0kmck4gmja, Atrix20, Zotel, Chobot, GangofOne, Bgwhite, Wavelength, JWB, Arado, Hornplease, Anders.Warga, Anoma-locaris, Joshdboz, Emesik, SamuelRiv, Avraham, Bondegezou, Nick-D, Borisbaran, SmackBot, Cla68, IronDuke, Gilliam, Hmains, Chris the speller, Thom2002, Snori, Hibernian, RayAYang, Tewfik, Shuki, Calbaer, Cybercobra, Salamurai, Phinn, John, JoshuaZ, NYCJosh, Ckatz, Makyen, Publicus, Martian.knight, EdC~enwiki, Noleander, FairuseBot, Newsnightmeirion, CmdrObot, Goatchurch, Necessary Evil, Quibik, Nabokov, Nishidani, NorwegianBlue, Hcobb, Afabbro, Chacor, Carolmooredc, Jj137, TimVickers, Bjenks, CombatWombat42, Avaya1, Al-bany NY, Flayer, Valerius Tygart, Veriss1, DadaNeem, BernardZ, Izno, Eclipsemullet, Ryan032, Dchall1, Steven J. Anderson, Seb az86556, Gilisa, Kermanshahi, NPguy, Swliv, Stephendcole, WRK, Gilinsky, Sohelpme, Lightmouse, Joegelman, Joel Rennie, Fâtimâh bint Fulâni, ImageRemovalBot, Sfan00 IMG, Knightrunner, NickCT, Deedub1983, Ignorance is strength, CasualObserver'48, BlackLukes, Masterbloore-gard, Muhandes, 7&6=thirteen, Bothtones7, Ra2007, WikiDao, Addbot, Heavenlyblue, Jdelanoy, Xoloki, Zellfaze, Nohomers48, Ronhjones, CanadianLinuxUser, OliverTwisted, Wikifan12345, רודג. Legobot, EnochBethany, Reenem, AnomieBOT, DemocraticLuntz, Wikieditorofto-day, Ulric1313, Materialscientist, Citation bot, Kasaalan, LilHelpa, Obersachsebot, Trahelliven, Mynameinc, MerlLinkBot, Gnomsovet, Plot Spoiler, Dailycare, FrescoBot, Eyal3400, Cs32en, Citation bot 1, Elockid, Poliocretes, Jonesey95, Monkeymanman, Full-date unlinking bot, Sh33pl0re, Rotblats09, Enemenemu, Trappist the monk, SeoMac, Prom000, Mr.98, Diannaa, WillNess, Breein1007, RjwilmsiBot, 8digits, ZéroBot, A2soup, Oncenawhile, Medeis, H3llBot, Edith Smitters, Guyziv, Metawizard, Donner60, Titeuf06, Whoop whoop pull up, SochenZar, ClueBot NG, Eynbein, Vjiced, Widr, ClaireFanch, Helpful Pixie Bot, Lowercase sigmabot, Oleg-ch, Mrt3366, Cyberbot II, IPWAI, Khazar2, Ekren, Unique Ubiquitous, Futurist110, Niv062, YasBot, TwoTwoHello, Neverdozin, Palli3000, Shlomi s, Lgfcd, Chicago Style (without pants), Jamesmcmahon0, Babramow, Pssymoneyweed247, Napy65, Kharkiv07, Finnusertop, EMP90, Ol'fussandfeathers, Abattoir666, Monkbot, WC Jay, GreenC bot, Steynrenier, Grko3, Siamak51 and Anonymous: 160

36.11.2   Images

- **File:1957-06-03_British_H-Bomb_(claimed).ogv** *Source:*   https://upload.wikimedia.org/wikipedia/commons/2/28/1957-06-03_British_ H-Bomb_%28claimed%29.ogv *License:* Public domain *Contributors:* http://archive.org/details/1957-06-03_British_H-Bomb (first 48s of news-reel) *Original artist:* Universal International Newsreel

- **File:4_Babur_Cruise_Missiles_on_a_Truck_at_IDEAS_2008.jpg** *Source:*   https://upload.wikimedia.org/wikipedia/commons/e/ef/4_ Babur_Cruise_Missiles_on_a_Truck_at_IDEAS_2008.jpg *License:* CC BY-SA 3.0 *Contributors:* Own work *Original artist:* SyedNaqvi90 (talk) (Uploads)

- **File:ANTIAKW.jpg** *Source:* https://upload.wikimedia.org/wikipedia/commons/4/40/ANTIAKW.jpg *License:* CC BY-SA 2.0 de *Contributors:* Own work *Original artist:* Hans Weingartz (Leonce49 at de.wikipedia)

- **File:Able_crossroads.jpg** *Source:* https://upload.wikimedia.org/wikipedia/commons/f/f4/Able_crossroads.jpg *License:* Public domain *Con-tributors:* ? *Original artist:* ?

- **File:Agni-II_missile_(Republic_Day_Parade_2004).jpeg** *Source:*   https://upload.wikimedia.org/wikipedia/commons/9/96/Agni-II_ missile_%28Republic_Day_Parade_2004%29.jpeg *License:* CC BY 3.0 br *Contributors:* http://img.radiobras.gov.br/Aberto/index.php/ Imagens.Principal.120.0.2004-01-31
  Original artist: Antônio Milena (ABr)

- **File:Agni_Missile_Range_comparison.svg** *Source:*   https://upload.wikimedia.org/wikipedia/commons/a/a7/Agni_Missile_Range_ comparison.svg *License:* CC BY-SA 3.0 *Contributors:* another Wikimedia image *Original artist:* Michael. SVG conversion by Srikar Kashyap

- **File:All_proposed_routes.PNG** *Source:* https://upload.wikimedia.org/wikipedia/commons/a/a7/All_proposed_routes.PNG *License:* CC BY-SA 3.0 *Contributors:* Own work *Original artist:* AlwaysUnite

- **File:Alpha_1_racetrack,_Uranium_235_electromagnetic_separation_plant,_Manhattan_Project,_Y-12_Oak_Ridge.jpg**
  Source: https://upload.wikimedia.org/wikipedia/commons/c/ce/Alpha_1_racetrack%2C_Uranium_235_electromagnetic_separation_ plant%2C_Manhattan_Project%2C_Y-12_Oak_Ridge.jpg *License:*   Public domain *Contributors:*   Retrieved September 27, 2014 from   <a data-x-rel='nofollow' class='external text' href='https://ia902303.us.archive.org/26/items/ManhattanDistrictHistory/ MDH-B5V03-Electromagnetic-Design.pdf'>Leslie R. Groves, Ed.   (~1948) *Manhattan District History*, Manhattan Project, U.S. Army Corps of Engineers, Book V: Electromagnetic Project, Vol. 3: Design, Appendix C: Photograph No. 6: Alpha 1 Racetrack</a>, declassified version, on Internet Archive *Original artist:* Leslie R. Groves

- **File:Ambox_current_red.svg** *Source:* https://upload.wikimedia.org/wikipedia/commons/9/98/Ambox_current_red.svg *License:* CC0 *Contrib-utors:* self-made, inspired by Gnome globe current event.svg, using Information icon3.svg and Earth clip art.svg *Original artist:* Vipersnake151, penubag, Tkgd2007 (clock)

- **File:Ambox_globe_content.svg** *Source:* https://upload.wikimedia.org/wikipedia/commons/b/bd/Ambox_globe_content.svg *License:* Public domain *Contributors:* Own work, using File:Information icon3.svg and File:Earth clip art.svg *Original artist:* penubag

- **File:Bundesarchiv_Bild_101I-668-7161-31A,_Flugzeug_Heinkel_He_177.jpg** *Source:* https://upload.wikimedia.org/wikipedia/commons/b/b8/Bundesarchiv_Bild_101I-668-7161-31A%2C_Flugzeug_Heinkel_He_177.jpg *License:* CC BY-SA 3.0 de *Contributors:* This image was provided to Wikimedia Commons by the German Federal Archive (Deutsches Bundesarchiv) as part of a cooperation project. The German Federal Archive guarantees an authentic representation only using the originals (negative and/or positive), resp. the digitalization of the originals as provided by the Digital Image Archive. *Original artist:* Linden

- **File:Bush_and_Putin_signing_SORT.jpg** *Source:* https://upload.wikimedia.org/wikipedia/commons/4/4b/Bush_and_Putin_signing_SORT.jpg *License:* Public domain *Contributors:* whitehouse.gov, President Bush, Russian President Putin Sign Nuclear Arms Treaty *Original artist:* White House photo

- **File:CTBT_Participation.svg** *Source:* https://upload.wikimedia.org/wikipedia/commons/b/be/CTBT_Participation.svg *License:* CC BY-SA 3.0 *Contributors:*

- BlankMap-World6._compact.svg *Original artist:*

- derivative work: Allstar86 (<a href='//commons.wikimedia.org/wiki/User_talk:Allstar86' title='User talk:Allstar86'>talk</a>)

- **File:Castle_Bravo_Shrimp_composite.png** *Source:* https://upload.wikimedia.org/wikipedia/commons/3/39/Castle_Bravo_Shrimp_composite.png *License:* Public domain *Contributors:* U.S. Government photos, published in Chuck Hansen, *Swords of Armageddon*, 1995, Volume IV, Figures IV-21, 22, & 24. *Original artist:* U.S. Government

- **File:Chicxulub_impact_-_artist_impression.jpg** *Source:* https://upload.wikimedia.org/wikipedia/commons/8/8c/Chicxulub_impact_-_artist_impression.jpg *License:* Public domain *Contributors:* http://www.jpl.nasa.gov/releases/98/yucatan.html *Original artist:* Donald E. Davis

- **File:ChinaABomb_2.jpg** *Source:* https://upload.wikimedia.org/wikipedia/commons/e/e9/ChinaABomb_2.jpg *License:* Public domain *Contributors:* http://nuclearweaponarchive.org/China/ChinaTesting.html *Original artist:* An US intelligence satellite

- **File:China_Emblem_PLA.svg** *Source:* https://upload.wikimedia.org/wikipedia/commons/3/34/China_Emblem_PLA.svg *License:* Public domain *Contributors:* Originally from zh.wikipedia; description page is/was China Emblem PLA.svg. *Original artist:* Original uploader was Nicolau at zh.wikipedia

- **File:Chinese_nuclear_bomb_-_A2923.jpg** *Source:* https://upload.wikimedia.org/wikipedia/commons/f/fe/Chinese_nuclear_bomb_-_A2923.jpg *License:* Public domain *Contributors:* Own work (Own photo) *Original artist:* Megapixie - Max Smith

- **File:Commons-logo.svg** *Source:* https://upload.wikimedia.org/wikipedia/en/4/4a/Commons-logo.svg *License:* CC-BY-SA-3.0 *Contributors:* ? *Original artist:* ?

- **File:ConstellationGPS.gif** *Source:* https://upload.wikimedia.org/wikipedia/commons/9/9c/ConstellationGPS.gif *License:* Public domain *Contributors:* Transferred from en.wikipedia *Original artist:* Original uploader was El pak at en.wikipedia

- **File:Convair_B-36_Peacemaker.jpg** *Source:* https://upload.wikimedia.org/wikipedia/commons/5/5a/Convair_B-36_Peacemaker.jpg *License:* Public domain *Contributors:* http://www.af.mil/News/Photos.aspx?igphoto=2000594578 *Original artist:* U.S. Air Force photo

- **File:Crossroads_baker_explosion.jpg** *Source:* https://upload.wikimedia.org/wikipedia/commons/2/21/Crossroads_baker_explosion.jpg *License:* Public domain *Contributors:* http://www.dtra.mil/press_resources/photo_library/CS/CS-1.cfm *Original artist:* U.S. Army Photographic Signal Corps

- **File:Cuban_missiles.jpg** *Source:* https://upload.wikimedia.org/wikipedia/commons/5/57/Cuban_missiles.jpg *License:* Public domain *Contributors:* National Archives.gov *Original artist:* see above

- **File:DASO_Trident_missile_test_firing_on-board_HMS_Vigilant_MOD_45159461.jpg** *Source:* https://upload.wikimedia.org/wikipedia/commons/f/fa/DASO_Trident_missile_test_firing_on-board_HMS_Vigilant_MOD_45159461.jpg *License:* OGL *Contributors:*

- Photo http://www.defenceimagery.mod.uk/fotoweb/fwbin/download.dll/45153802.jpg *Original artist:* PO(Phot) Simmo Simpson

- **File:DavyCrockettBomb.jpg** *Source:* https://upload.wikimedia.org/wikipedia/commons/4/43/DavyCrockettBomb.jpg *License:* Public domain *Contributors:* Chuck Hansen, *The Swords of Armageddon: U.S. Nuclear Weapons Development Since 1945* (Sunnyvale, CA: Chukelea Publications, 1995).[1] *Original artist:* US government DOD and/or DOE photograph

- **File:DemonstrationRaketenStationierung1982.jpg** *Source:* https://upload.wikimedia.org/wikipedia/commons/8/8b/DemonstrationRaketenStationierung1982.jpg *License:* Public domain *Contributors:* http://www.defenseimagery.mil/imagery.html#guid=984f5767b25da01f5c79c0bcce1aea28d0529096 *Original artist:* MSGT DON SUTHERLAND

- **File:Deuterium-tritium_fusion.svg** *Source:* https://upload.wikimedia.org/wikipedia/commons/3/3b/Deuterium-tritium_fusion.svg *License:* Public domain *Contributors:* Own work, based on w:File:D-t-fusion.png *Original artist:* Wykis

- **File:Dnepr_rocket_lift-off_1.jpg** *Source:* https://upload.wikimedia.org/wikipedia/commons/4/47/Dnepr_rocket_lift-off_1.jpg *License:* CC BY 2.5 *Contributors:* ? *Original artist:* ?

- **File:EMP_mechanism.png** *Source:* https://upload.wikimedia.org/wikipedia/commons/2/24/EMP_mechanism.png *License:* Public domain *Contributors:* Wikipedia in english, page "High-altitude nuclear explosion" *Original artist:* User:Photocopier

- **File:Ed_White_with_Space_Gun_maneuvering_unit.jpg** *Source:* https://upload.wikimedia.org/wikipedia/commons/0/01/Ed_White_with_Space_Gun_maneuvering_unit.jpg *License:* Public domain *Contributors:* NASA Gemini 4 Page *Original artist:* James McDivitt

- **File:Edit-clear.svg** *Source:* https://upload.wikimedia.org/wikipedia/en/f/f2/Edit-clear.svg *License:* Public domain *Contributors:* The *Tango! Desktop Project*. *Original artist:*

  The people from the Tango! project. And according to the meta-data in the file, specifically: "Andreas Nilsson, and Jakub Steiner (although minimally)."

- **File:EdwardTeller1958_fewer_smudges.jpg** *Source:* https://upload.wikimedia.org/wikipedia/commons/e/ef/EdwardTeller1958_fewer_smudges.jpg *License:* CC BY-SA 3.0 *Contributors:*

- EdwardTeller1958.jpg *Original artist:* w:User:Greg L, Papa Lima Whiskey

- **File:Edward_Teller_&_Stanislaw_Ulam_1951_On_Heterocatalytic_Detonations_-_Secret_of_hydrogen_bomb_-_p_1.png** *Source:* https://upload.wikimedia.org/wikipedia/commons/4/40/Edward_Teller_%26_Stanislaw_Ulam_1951_On_Heterocatalytic_Detonations_-_Secret_of_hydrogen_bomb_-_p_1.png *License:* Public domain *Contributors:* Retrieved October 6, 2014 from <a data-x-rel='nofollow' class='external text' href='http://www.nuclearnonproliferation.org/LAMS1225.pdf'>Edward Teller, Stanislaw Ulam, *On Heterocatalytic Detonations I: Hydrodynamic Lenses and Radiation Mirrors*, Report LAMS-1225, Los Alamos Scientific Laboratory, March 9, 1951, declassified version, p. 1</a> on Nuclear Nonproliferation Institute website *Original artist:* Edward Teller and Stanislaw M. Ulam

- **File:Edward_Teller_&_Stanislaw_Ulam_1951_On_Heterocatalytic_Detonations_-_Secret_of_hydrogen_bomb_-_p_3.png** *Source:* https://upload.wikimedia.org/wikipedia/commons/3/30/Edward_Teller_%26_Stanislaw_Ulam_1951_On_Heterocatalytic_Detonations_-_Secret_of_hydrogen_bomb_-_p_3.png *License:* Public domain *Contributors:* Retrieved October 6, 2014 from <a data-x-rel='nofollow' class='external text' href='http://www.nuclearnonproliferation.org/LAMS1225.pdf'>Edward Teller, Stanislaw Ulam, *On Heterocatalytic Detonations I: Hydrodynamic Lenses and Radiation Mirrors*, Report LAMS-1225, Los Alamos Scientific Laboratory, March 9, 1951, declassified version, p. 3</a> on Nuclear Nonproliferation Institute website *Original artist:* Edward Teller and Stanislaw M. Ulam

- **File:Edward_Teller_(1958)-LLNL.jpg** *Source:* https://upload.wikimedia.org/wikipedia/commons/c/cf/Edward_Teller_%281958%29-LLNL.jpg *License:* Public domain *Contributors:* ? *Original artist:* ?

- **File:Einsteinium.jpg** *Source:* https://upload.wikimedia.org/wikipedia/commons/5/55/Einsteinium.jpg *License:* Public domain *Contributors:* [1], Haire, Richard G. (2006). "Einsteinium". In Morss; Edelstein, Norman M.; Fuger, Jean. The Chemistry of the Actinide and Transactinide Elements (3rd ed.). Dordrecht, The Netherlands: Springer Science+Business Media. ISBN 1-4020-3555-1. p. 1580 *Original artist:* Haire, R. G., US Department of Energy. Touched up by Materialscientist at en.wikipedia.

- **File:Embracing_the_base,_Greenham_Common_December_1982_-_geograph.org.uk_-_759090.jpg** *Source:* https://upload.wikimedia.org/wikipedia/commons/b/b2/Embracing_the_base%2C_Greenham_Common_December_1982_-_geograph.org.uk_-_759090.jpg *License:* CC BY-SA 2.0 *Contributors:* From geograph.org.uk; transferred by User:Skinsmoke using geograph_org2commons. *Original artist:* ceridwen

- **File:Ensign_of_the_Indian_Air_Force.svg** *Source:* https://upload.wikimedia.org/wikipedia/commons/c/c2/Ensign_of_the_Indian_Air_Force.svg *License:* Public domain *Contributors:* ? *Original artist:* ?

- **File:Essais_nucleaires_manif.jpg** *Source:* https://upload.wikimedia.org/wikipedia/commons/9/93/Essais_nucleaires_manif.jpg *License:* CC-BY-SA-3.0 *Contributors:* Community of the Ark of Lanza del Vasto. *Original artist:* Community of the Ark of Lanza del Vasto.

- **File:Europe-UK.svg** *Source:* https://upload.wikimedia.org/wikipedia/commons/5/5b/Europe-UK.svg *License:* CC BY-SA 4.0 *Contributors:*

- File:Location European nation states.svg *Original artist:* Rob984

- **File:Exercise_Desert_Rock_I_(Buster-Jangle_Dog)_002.jpg** *Source:* https://upload.wikimedia.org/wikipedia/commons/5/50/Exercise_Desert_Rock_I_%28Buster-Jangle_Dog%29_002.jpg *License:* Public domain *Contributors:* http://www.dtra.mil/press_resources/photo_library/CS/CS-3.cfm *Original artist:* Federal Government of the United States

- **File:Fallout_shelter.jpg** *Source:* https://upload.wikimedia.org/wikipedia/commons/e/e6/Fallout_shelter.jpg *License:* CC-BY-SA-3.0 *Contributors:* Transferred from en.wikipedia to Commons. *Original artist:* The original uploader was Ex11e at English Wikipedia

- **File:Fat_man.jpg** *Source:* https://upload.wikimedia.org/wikipedia/commons/c/c2/Fat_man.jpg *License:* Public domain *Contributors:* U.S. Department of Defense *Original artist:* U.S. Department of Defense

- **File:Fission_bomb_assembly_methods.svg** *Source:* https://upload.wikimedia.org/wikipedia/commons/c/cb/Fission_bomb_assembly_methods.svg *License:* Public domain *Contributors:* Own work *Original artist:* Fastfission

- **File:Flag_of_Belgium_(civil).svg** *Source:* https://upload.wikimedia.org/wikipedia/commons/9/92/Flag_of_Belgium_%28civil%29.svg *License:* Public domain *Contributors:* ? *Original artist:* ?

- **File:Flag_of_China.svg** *Source:* https://upload.wikimedia.org/wikipedia/commons/f/fa/Flag_of_the_People%27s_Republic_of_China.svg *License:* Public domain *Contributors:* Own work, http://www.protocol.gov.hk/flags/eng/n_flag/design.html *Original artist:* Drawn by User:SKopp, redrawn by User:Denelson83 and User:Zscout370

- **File:Flag_of_France.svg** *Source:* https://upload.wikimedia.org/wikipedia/en/c/c3/Flag_of_France.svg *License:* PD *Contributors:* ? *Original artist:* ?

- **File:Flag_of_German_Reich_(1935–1945).svg** *Source:* https://upload.wikimedia.org/wikipedia/commons/9/99/Flag_of_German_Reich_%281935%E2%80%931945%29.svg *License:* Public domain *Contributors:* Own work *Original artist:* Fornax

- **File:Flag_of_Germany.svg** *Source:* https://upload.wikimedia.org/wikipedia/en/b/ba/Flag_of_Germany.svg *License:* PD *Contributors:* ? *Original artist:* ?

- **File:Flag_of_IAEA.svg** *Source:* https://upload.wikimedia.org/wikipedia/commons/5/54/Flag_of_IAEA.svg *License:* Public domain *Contributors:* Flag code: [1] *Original artist:* IAEA

- **File:Flag_of_India.svg** *Source:* https://upload.wikimedia.org/wikipedia/en/4/41/Flag_of_India.svg *License:* Public domain *Contributors:* ? *Original artist:* ?

- **File:Flag_of_Indian_Army.svg** *Source:* https://upload.wikimedia.org/wikipedia/commons/e/ea/Flag_of_Indian_Army.svg *License:* CC-BY-SA-3.0 *Contributors:* extracted from File:Flag_of_Indian_Army.svg *Original artist:* Fred the Oyster

- **File:Flag_of_Israel.svg** *Source:* https://upload.wikimedia.org/wikipedia/commons/d/d4/Flag_of_Israel.svg *License:* Public domain *Contributors:* http://www.mfa.gov.il/MFA/History/Modern%20History/Israel%20at%2050/The%20Flag%20and%20the%20Emblem *Original artist:* "The Provisional Council of State Proclamation of the Flag of the State of Israel" of 25 Tishrei 5709 (28 October 1948) provides the official specification for the design of the Israeli flag.

- **File:Location_Russia.svg** *Source:* https://upload.wikimedia.org/wikipedia/commons/3/37/Location_Russia.svg *License:* Public domain *Contributors:* Own work, based on Image:BlankMap-World6.svg *Original artist:* User:Kelvinc

- **File:M110_Column.JPEG** *Source:* https://upload.wikimedia.org/wikipedia/commons/9/9a/M110_Column.JPEG *License:* Public domain *Contributors:* http://www.defenselink.mil/multimedia/ *Original artist:* BRAM DE JONG / Dirk Van Laer

- **File:MGR-1_Honest_John_rocket.jpg** *Source:* https://upload.wikimedia.org/wikipedia/commons/b/b3/MGR-1_Honest_John_rocket.jpg *License:* Public domain *Contributors:* http://www.redstone.army.mil/history/archives/missiles/missiles0007.html *Original artist:* US military

- **File:MK6_TITAN_II.jpg** *Source:* https://upload.wikimedia.org/wikipedia/commons/3/36/MK6_TITAN_II.jpg *License:* Public domain *Contributors:* ? *Original artist:* ?

- **File:MMIII_C5_airdrop(Oct_1974).jpg** *Source:* https://upload.wikimedia.org/wikipedia/en/2/27/MMIII_C5_airdrop%28Oct_1974%29.jpg *License:* PD *Contributors:*
USAF Historical archives
Original artist:
USAF Minuteman Program Office (crossrich (talk) 15:30, 10 June 2011 (UTC))

- **File:Malabar_2012_INS_Satpura_(F-48).jpg** *Source:* https://upload.wikimedia.org/wikipedia/commons/4/4f/Malabar_2012_INS_Satpura_%28F-48%29.jpg *License:* Public domain *Contributors:* http://www.navy.mil/view_image.asp?id=121565 *Original artist:* U.S. Navy photo by Mass Communication Specialist 3rd Class Christopher Farrington

- **File:Medium_emblem_of_the_Armed_Forces_of_the_Russian_Federation_(27.01.1997-present).svg** *Source:* https://upload.wikimedia.org/wikipedia/commons/0/09/Medium_emblem_of_the_Armed_Forces_of_the_Russian_Federation_%2827.01.1997-present%29.svg *License:* Public domain *Contributors:*

- www.mil.ru - Эмблема Вооруженных Сил Российской Федерации *Original artist:* F l a n k e r

- **File:Merge-arrows.svg** *Source:* https://upload.wikimedia.org/wikipedia/commons/5/52/Merge-arrows.svg *License:* Public domain *Contributors:* ? *Original artist:* ?

- **File:Military_truck_carrying_IRBMs_of_Pakistani_Army.jpg** *Source:* https://upload.wikimedia.org/wikipedia/commons/0/07/Military_truck_carrying_IRBMs_of_Pakistani_Army.jpg *License:* CC BY-SA 3.0 *Contributors:* Own work by the original uploader *Original artist:* User:SyedNaqvi90

- **File:Minuteman3launch.jpg** *Source:* https://upload.wikimedia.org/wikipedia/commons/6/60/Minuteman3launch.jpg *License:* Public domain *Contributors:* http://www.af.mil/News/Photos.aspx?igphoto=2000597905 *Original artist:* U.S. Air Force photo

- **File:Minuteman_III_MIRV_path.svg** *Source:* https://upload.wikimedia.org/wikipedia/commons/f/ff/Minuteman_III_MIRV_path.svg *License:* Public domain *Contributors:* http://www.nukestrat.com/us/afn/Minuteman.pdf. Completely re-drawn and re-worked from scratch by Fastfission in Inkscape. *Original artist:* Fastfission

- **File:Miss_launch_veh.jpg** *Source:* https://upload.wikimedia.org/wikipedia/commons/e/ed/Miss_launch_veh.jpg *License:* Public domain *Contributors:* http://www.defenseimagery.mil/imagery.html#guid=84296161aac277c7db856ad81937d0b75a790e96 (From Soviet Military Power 1985)
Original artist: Edward L. Cooper

- **File:Mk17_bomb.jpg** *Source:* https://upload.wikimedia.org/wikipedia/commons/7/7d/Mk17_bomb.jpg *License:* Public domain *Contributors:* ? *Original artist:* ?

- **File:NASA-project-orion-artist.jpg** *Source:* https://upload.wikimedia.org/wikipedia/commons/6/61/NASA-project-orion-artist.jpg *License:* Public domain *Contributors:* http://mix.msfc.nasa.gov/abstracts.php?p=704 *Original artist:* NASA

- **File:NNSA-NSO-787.jpg** *Source:* https://upload.wikimedia.org/wikipedia/commons/e/eb/NNSA-NSO-787.jpg *License:* Public domain *Contributors:* This image is available from the National Nuclear Security Administration Nevada Site Office Photo Library under ID 787. *Original artist:* National Nuclear Security Administration / Nevada Site Office

- **File:NPT_Effective.svg** *Source:* https://upload.wikimedia.org/wikipedia/commons/d/d4/NPT_Effective.svg *License:* CC BY-SA 3.0 *Contributors:* Own work *Original artist:* JWB

- **File:NPT_parties.svg** *Source:* https://upload.wikimedia.org/wikipedia/commons/b/ba/NPT_parties.svg *License:* CC BY-SA 3.0 *Contributors:* This file was derived from: NPT Participation.svg
Original artist: File:NPT Participation.svg: Allstar86, L.tak, Danlaycock

- **File:Nagasakibomb.jpg** *Source:* https://upload.wikimedia.org/wikipedia/commons/e/e0/Nagasakibomb.jpg *License:* Public domain *Contributors:* http://www.archives.gov/research/military/ww2/photos/images/ww2-163.jpg National Archives image (208-N-43888) *Original artist:* Charles Levy from one of the B-29 Superfortresses used in the attack.

- **File:Naval_Ensign_of_India.svg** *Source:* https://upload.wikimedia.org/wikipedia/commons/3/35/Naval_Ensign_of_India.svg *License:* Public domain *Contributors:* [1] *Original artist:* Original upload by Denelson83, most recent version by Fry1989.

- **File:Naval_Jack_of_Pakistan.svg** *Source:* https://upload.wikimedia.org/wikipedia/commons/0/07/Naval_Jack_of_Pakistan.svg *License:* Public domain *Contributors:* ? *Original artist:* ?

- **File:Neutron_radiation_weighting_factor_as_a_function_of_kinetic_energy.gif** *Source:* https://upload.wikimedia.org/wikipedia/commons/5/5f/Neutron_radiation_weighting_factor_as_a_function_of_kinetic_energy.gif *License:* CC BY-SA 3.0 *Contributors:* Made this graph using OpenOffice based on data from reference documents *Original artist:* Ytrottier

- **File:Neutroncrosssectionboron.png** *Source:* https://upload.wikimedia.org/wikipedia/commons/5/5c/Neutroncrosssectionboron.png *License:* Public domain *Contributors:* w:Image:Neutroncrosssectionboron.jpg *Original artist:* wikipedia:en:user:Cadmium

- **File:SLBM_Comparison.jpg** *Source:* https://upload.wikimedia.org/wikipedia/commons/6/6a/SLBM_Comparison.jpg *License:* Public domain *Contributors:* http://www.mda.mil/mdalink/bcmt/slbm_1.htm *Original artist:* MDA-file

- **File:SS-24-DIA.jpg** *Source:* https://upload.wikimedia.org/wikipedia/commons/0/05/SS-24-DIA.jpg *License:* Public domain *Contributors:* http://www.dia.mil/history/art/series_one.html *Original artist:* Edward L. Cooper

- **File:SS-24_silo_destruction.jpg** *Source:* https://upload.wikimedia.org/wikipedia/commons/9/9c/SS-24_silo_destruction.jpg *License:* Public domain *Contributors:* ? *Original artist:* ?

- **File:Sedan_Plowshare_Crater.jpg** *Source:* https://upload.wikimedia.org/wikipedia/commons/b/b6/Sedan_Plowshare_Crater.jpg *License:* Public domain *Contributors:* This image is available from the National Nuclear Security Administration Nevada Site Office Photo Library under number NF-12187. *Original artist:* Federal Government of the United States

- **File:ShaktiBomb.jpg** *Source:* https://upload.wikimedia.org/wikipedia/en/2/2a/ShaktiBomb.jpg *License:* Fair use *Contributors:*
Government of India (note higher resolution versions on nuclearweaponarchive.org) *Original artist:* ?

- **File:Shippingport_Reactor.jpg** *Source:* https://upload.wikimedia.org/wikipedia/commons/0/0c/Shippingport_Reactor.jpg *License:* Public domain *Contributors:* http://www.mbe.doe.gov/me70/history/photos.htm (originally; now 404. Mirrored on [1]) *Original artist:* ?

- **File:Sir_Mark_Oliphant.jpg** *Source:* https://upload.wikimedia.org/wikipedia/commons/3/34/Sir_Mark_Oliphant.jpg *License:* Public domain *Contributors:* http://www.portrait.gov.au/static/coll_741Sir+Mark+Oliphant.php *Original artist:* Bassano Ltd

- **File:Smiling_Sun_English_Language.svg** *Source:* https://upload.wikimedia.org/wikipedia/commons/9/9e/Smiling_Sun_English_Language.svg *License:* GFDL 1.2 *Contributors:* https://commons.wikimedia.org/wiki/File:Smiling_Sun_-_English.jpg *Original artist:* Anne Lund

- **File:Snle-snle-ng-svg.svg** *Source:* https://upload.wikimedia.org/wikipedia/commons/6/6f/Snle-snle-ng-svg.svg *License:* CC-BY-SA-3.0 *Contributors:* Own work *Original artist:* Vector version by Dake. Based on a drawing by Rama.

- **File:Sound-icon.svg** *Source:* https://upload.wikimedia.org/wikipedia/commons/4/47/Sound-icon.svg *License:* LGPL *Contributors:* Derivative work from Silsor's versio *Original artist:* Crystal SVG icon set

- **File:South_African_nuclear_bomb_casings.jpg** *Source:* https://upload.wikimedia.org/wikipedia/en/1/10/South_African_nuclear_bomb_casings.jpg *License:* Fair use *Contributors:*
[1], original source photojournalist Mungo Poore *Original artist:* ?

- **File:Steel_balls_png.png** *Source:* https://upload.wikimedia.org/wikipedia/commons/f/fc/Steel_balls_png.png *License:* Public domain *Contributors:* Original in http://en.wikipedia.org/wiki/File:Steel_balls_png.png *Original artist:* Illustration. Author/artist B.Burnell. Copyright B.Burnell.

- **File:Swedish_Atomic_Bomb.png** *Source:* https://upload.wikimedia.org/wikipedia/commons/1/17/Swedish_Atomic_Bomb.png *License:* Public domain *Contributors:* From T. Magnusson, "Design and Effects of Atomic Weapons," Kosmos, Fysika Uppsatser, 34 180 (1956) Sweden. Translated and printed as JPRS:8295, Office of Technical Services, U.S. Dept of Commerce. This diagram was reprinted as Figure 8 in Robert W. Selden "An Introduction to Fission Explosives," UCID-15554, Lawrence Radiation Laboratory, Livermore, California, July 1969. Line drawing scanned from "History of the Swedish Atom Bomb," Ny Teknik No. 17, 25 April 1985, cleaned up, re-labeled, and colored by Howard Morland, 2007. *Original artist:* HowardMorland

- **File:Symbol_book_class2.svg** *Source:* https://upload.wikimedia.org/wikipedia/commons/8/89/Symbol_book_class2.svg *License:* CC BY-SA 2.5 *Contributors:* Mad by Lokal_Profil by combining: *Original artist:* Lokal_Profil

- **File:Símbolo_radiación.png** *Source:* https://upload.wikimedia.org/wikipedia/commons/8/81/S%C3%ADmbolo_radiaci%C3%B3n.png *License:* CC BY 3.0 *Contributors:* Own work *Original artist:* Mr. Tamagotchi

- **File:Teller-Ulam_device.png** *Source:* https://upload.wikimedia.org/wikipedia/commons/8/8c/Teller-Ulam_device.png *License:* Public domain *Contributors:* ? *Original artist:* ?

- **File:Teller-Ulam_device_3D.svg** *Source:* https://upload.wikimedia.org/wikipedia/commons/c/c1/Teller-Ulam_device_3D.svg *License:* Public domain *Contributors:* ? *Original artist:* ?

- **File:TellerUlamAblation.png** *Source:* https://upload.wikimedia.org/wikipedia/commons/4/43/TellerUlamAblation.png *License:* CC BY-SA 2.5 *Contributors:* ? *Original artist:* ?

- **File:Text_document_with_red_question_mark.svg** *Source:* https://upload.wikimedia.org/wikipedia/commons/a/a4/Text_document_with_red_question_mark.svg *License:* Public domain *Contributors:* Created by bdesham with Inkscape; based upon Text-x-generic.svg from the Tango project. *Original artist:* Benjamin D. Esham (bdesham)

- **File:The_patient'{}s_skin_is_burned_in_a_pattern_corresponding_to_the_dark_portions_of_a_kimono_-_NARA_-_519686.jpg** *Source:* https://upload.wikimedia.org/wikipedia/commons/e/e9/The_patient%27s_skin_is_burned_in_a_pattern_corresponding_to_the_dark_portions_of_a_kimono_-_NARA_-_519686.jpg *License:* Public domain *Contributors:* U.S. National Archives and Records Administration *Original artist:* Unknown<a href='//www.wikidata.org/wiki/Q4233718' title='wikidata:Q4233718'><img alt='wikidata:Q4233718' src='https://upload.wikimedia.org/wikipedia/commons/thumb/f/ff/Wikidata-logo.svg/20px-Wikidata-logo.svg.png' width='20' height='11' srcset='https://upload.wikimedia.org/wikipedia/commons/thumb/f/ff/Wikidata-logo.svg/30px-Wikidata-logo.svg.png 1.5x, https://upload.wikimedia.org/wikipedia/commons/thumb/f/ff/Wikidata-logo.svg/40px-Wikidata-logo.svg.png 2x' data-file-width='1050' data-file-height='590' /></a> or not provided

- **File:Titan_II_launch.jpg** *Source:* https://upload.wikimedia.org/wikipedia/commons/9/93/Titan_II_launch.jpg *License:* Public domain *Contributors:* U.S. DefenseImagery photo VIRIN: DF-ST-84-06932; National Museum of the U.S. Air Force photo 140124-F-DW547-006 *Original artist:* U.S. DoD

- **File:Tomahawk_Block_IV_cruise_missile.jpg** *Source:* https://upload.wikimedia.org/wikipedia/commons/5/55/Tomahawk_Block_IV_cruise_missile.jpg *License:* Public domain *Contributors:* ? *Original artist:* ?

- **File:YellowSunBomb1.JPG** *Source:* https://upload.wikimedia.org/wikipedia/commons/5/56/YellowSunBomb1.JPG *License:* CC BY-SA 3.0 *Contributors:* Own work (Original text: *I (Nabokov (talk)) created this work entirely by myself.*) *Original artist:* Nabokov (talk). Required citation is "Photo by Tom Oates, 2011".

- **File:Самолет_"Илья_Муромец".jpg** *Source:* https://upload.wikimedia.org/wikipedia/commons/f/fb/%D0%A1%D0%B0% D0%BC%D0%BE%D0%BB%D0%B5%D1%82_%22%D0%98%D0%BB%D1%8C%D1%8F_%D0%9C%D1%83%D1%80% D0%BE%D0%BC%D0%B5%D1%86%22.jpg *License:* Public domain *Contributors:* http://fandavion.free.fr/sikorsky-images.htm *Original artist:* Unknown<a href='//www.wikidata.org/wiki/Q4233718' title='wikidata:Q4233718'><img alt='wikidata:Q4233718' src='https://upload.wikimedia.org/wikipedia/commons/thumb/f/ff/Wikidata-logo.svg/20px-Wikidata-logo.svg.png' width='20' height='11' srcset='https://upload.wikimedia.org/wikipedia/commons/thumb/f/ff/Wikidata-logo.svg/30px-Wikidata-logo.svg.png 1.5x, https://upload.wikimedia.org/wikipedia/commons/thumb/f/ff/Wikidata-logo.svg/40px-Wikidata-logo.svg.png 2x' data-file-width='1050' data-file-height='590' /></a>

- **File:Тягач_МЗКТ−79221_(комплекс_Тополь-М).jpg** *Source:* https://upload.wikimedia.org/wikipedia/commons/d/dd/%D0%A2% D1%8F%D0%B3%D0%B0%D1%87_%D0%9C%D0%97%D0%9A%D0%A2-79221_%28%D0%BA%D0%BE%D0%BC%D0%BF% D0%BB%D0%B5%D0%BA%D1%81_%D0%A2%D0%BE%D0%BF%D0%BE%D0%BB%D1%8C-%D0%9C%29.jpg *License:* CC BY-SA 3.0 *Contributors:* Own work *Original artist:* ru:Участник:Goodvint

36.11.3   Content license